JEFFREY K. GURNEY

AUTOMATED VEHICLE LAW

LEGAL LIABILITY, REGULATION, AND DATA SECURITY

Cover design by Tahiti Spears/ABA Design

Printed in the United States of America.
24 23 22 21 20 5 4 3 2 1

A catalog record for this book is available from the Library of Congress.

Discounts are available for books ordered in bulk. Special consideration is given to state bars, CLE programs, and other bar-related organizations. Inquire at Book Publishing, ABA Publishing, American Bar Association, 321 N. Clark Street, Chicago, Illinois 60654-7598.

www.ShopABA.org

Contents

Acknowledgments v
Preface vii

1. Part 1: The Automated Vehicle Industry 1

Chapter 1 The Industry 3

Chapter 2 Levels of Automation 11

Chapter 3 Automation Hardware and Software 21

Chapter 4 Effects of Automated Vehicles 31

2. Part 2: The Regulation of Automated Vehicles 43

Chapter 5 The Federal Regulatory Apparatus 45

Chapter 6 State Regulation of Motor Vehicles and Driving 89

Chapter 7 Automated Vehicle Guidance and Laws 101

3. Part 3: Civil Liability 119

Chapter 8 Liability for Automated Vehicle Crashes 123

Chapter 9 Level 2 Driving Automation Systems 167

Chapter 10 Level 3 Automated Driving Systems 175

Chapter 11 Level 4 and Level 5 Automated Driving Systems 183

Chapter 12 Immunity for Level 4 and Level 5 Crashes 201

Chapter 13 Automated Rental Vehicles 209

Chapter 14 Automated Ridesharing Vehicles 217

Chapter 15 State and Local Government Liability Risks 225

4. Part 4: Data Privacy and Cybersecurity 233

Chapter 16 Federal Regulation of Data Privacy and Security 235

Chapter 17 State Data Privacy and Security Laws 267

Chapter 18 Liability for Cybersecurity Breaches and Cyberattacks 291

5. Part 5: Criminal Law 303

Chapter 19 Automobile Crimes 305

Chapter 20 Physical and Digital Interference with Automated Vehicles 339

Chapter 21 The Right to Seize and Search Highly Automated Vehicles and Their Occupants 361

Chapter 22 Searches of Automated Vehicle Data 381

Table of Authorities 397

About the Author 429

Acknowledgments

It was approximately eight years ago that my law review provided a list of law article topics for members to write a paper on during their 2L year. Some unknown member of the law review had included the Google self-driving car as a topic in that list. I do not know who had included that topic, but after my then-girlfriend, now wife, perused that list and suggested the Google self-driving car topic, a chain of events began that has culminated with this book. I am thankful to that unknown person.

The fire that was lit by that article remained burning due to the many people whom I have met with about automated vehicles. From invitations to write new articles or give talks, to many conversations with the wonderful industry members, groups flying from around the world to meet with me to discuss some legal aspect of automated vehicles, and clients, each new conversation or project led to the organic growth of this book. I am so thankful for all of those opportunities and conversations, many of which have impacted the book in one way or another.

I would like to thank the many people who provided feedback on this book and help through the book process, including Victoria Voisin, Nicholas Gurney, Andrew Gurney, David Owen, Steve Lending, Patrick Cleary, Colton Driver, Sam Kherat, Jon Riehl, Sam Voisin, Miriam Williamson, Samer Roshdy, Sheria Clarke, James Perl, Alex Sarch, and Ryan Abbott. I am thankful for the SciTech division of the American Bar Association and, in particular, Sarah (Sam) Orwig and Sarah McMillan, as well as Courtney Coffman, the project manager for the book, and the copyeditors at Lachina.

A book of this sort could not have been possible without the guidance, help, and insight of my family and friends. Two family members have gone well above their call during my immersion into automated vehicles: my older brother, Nick, and my wife, Tory. They have been sounding boards on automated vehicles, traveled with me to conferences, and read far more of my drafts than they ever should have over the past decade. I am especially thankful to Tory for her support and love throughout the process of this book and throughout all my projects into automated vehicles. Through the long hours, you were and remain my guiding light. My dogs, Barney and Ozzy, also provided incredible companionship throughout this process.

Jeffrey K. Gurney
Greenville, South Carolina

Preface

About eight years ago, I first began thinking and writing about legal issues confronting automated vehicles. Automated vehicles were nascent and, to many, seemed like a pipe dream. My fascination with automated vehicles started because of the opportunities they present to revolutionize society and improve our well-being. The legal field is not immune to the revolutionary effects of automated vehicles. Fundamental assumptions underlying our jurisprudence will be overturned.

Over the years, I have been fortunate that so many people from around the world have asked me diverse, and sometimes obscure, legal questions about automated vehicles. Much of my focus has been concerned with the effects that automated vehicles will have on the law, and the effects the law will have on automated vehicles. Eventually, ideas and solutions began to coalesce into a jurisprudence that I call automated vehicle law.

Automated vehicle law is a growing and rapidly changing area of law. It will be one of the legal community's first significant encounters with artificial intelligence and its effects on existing doctrines. This book is framed around five areas of automated vehicle law: (1) background on automated vehicles, (2) the regulation of automated vehicles, (3) civil liability for automated vehicle crashes, (4) data security and privacy, and (5) criminal law.

Before delving into the substance of the book, I want to identify a few stylistic items. In this book, I use the term "automated vehicles" to describe these vehicles. Popular names include "autonomous vehicles," "self-driving cars," "robocars," and "driverless vehicles." These names are generally not used in this book for different reasons. Scholars have incorrectly called certain vehicles equipped with automation features "self-driving" or "driverless." Though there may eventually be self-driving or driverless vehicles on the roadway, consumer technology is generally not currently at that level. The other word—"autonomous"—has been used in the context of automated driving, with many, including myself in the past, referring to automated vehicles as "autonomous vehicles" or "autonomous cars" or "autonomous driving." The usage of "autonomous vehicles" is common in everyday conversation about vehicle automation, so it may be that the phrase has long passed its tipping point. However, "autonomous" derives from "autonomy," which means "1: the quality or state of being self-governing . . . 2. Self-driving freedom and especially moral independence . . . 3: a self-governing state."[1] Therefore, from a definitional standpoint, we would expect an "autonomous vehicle" to have a mind of its own, including the ability to disobey its users, which none of us expect automated vehicles to have (or do we?). For the sake of clarification and being technically correct in describing automated vehicles, I do not use the words "autonomous vehicles" and instead use "automated vehicles." At one point in the future, we may begin discussing "autonomous vehicles," but that day is not today.

I refer to automated vehicles as levels of automation, such as a "level 4 automated vehicle." This is not technically correct. A vehicle is not a level of automation, and an automated vehicle may have various automation or human driving capabilities. The correct way to refer to such a vehicle would be to say "a vehicle equipped with a level 4 automated driving system." When the level 4 system is engaged, the correct reference would be to say "a vehicle operated by a level 4 automated driving system" or to continually say "the vehicle with a level 4 automated driving system engaged." These are more precise than saying "a level 4 automated vehicle"; however, they are wordier. I have chosen the conciseness of referring to vehicles as levels of automation. Unless the context states otherwise, my use of a specified level of automated vehicle means a vehicle with that specific level of driving automation system engaged.

Although I use the levels of automation, I also use shorthand forms for the levels of automation. "Highly automated vehicles" is a term I use to refer principally to level 4 or level 5 automated vehicles. "Lower levels of automation" refers to level 1, level 2, and level 3 driving automation systems. Sometimes the context of highly automated vehicles will include level 3 automated driving systems. As we will see, level 3 teeters between level 2 and level 4. For some legal analyses, categorizing level 3 with the level 4 and level 5 automated driving systems makes more sense; for other analyses, categorizing level 3 with the level 1 and level 2 driving automation systems makes more sense.

This book is intended as a reference or a guide for automated vehicle law. Any legal analysis of complex topics like civil liability for crashes, criminal law, and regulation requires a review of the underlying facts. Facts are the most crucial part of any legal analysis. This book is intended to provide you with legal authority and context to apply the law to the new factual situations created by automated vehicles.

---

1. *Autonomy*, Merriam-Webster, https://www.merriam-webster.com/dictionary/autonomy (last visited Mar. 30, 2020).

Part 1

The Automated Vehicle Industry

INTRODUCTION

The automated vehicle industry has limited current commercial activity, but many people expect the industry to grow and be vibrant. Commercialization of automated vehicles could create a significant industry and result in cost savings by reducing crashes, decreasing fuel consumption, and increasing productivity. Cruise, an automated vehicle company that General Motors acquired, has estimated that the global automated vehicle industry can be an $8 trillion market.[1] Most revenue is expected to come from providing automated ridesharing and freight services. Other organizations have given significant predictions for the annual revenue of the automated vehicle industry.[2] Automated vehicles will also create significant savings. Morgan Stanley has estimated that

1. Michael Wayland, *GM's Cruise Values Autonomous Vehicle Industry at $8 Trillion*, CNBC (Feb. 5, 2020), https://www.cnbc.com/2020/02/05/gms-cruise-values-autonomous-vehicle-industry-at-8-trillion.html.

2. *See, e.g.*, Sam Unsted, *Robo-Taxi Industry Could Be Worth $2 Trillion by 2030, UBS Says*, Bloomberg (May 23, 2019), https://www.bloomberg.com/news/articles/2019-05-23/robo-taxi-industry-could-be-worth-2-trillion-by-2030-ubs-says (UBS Group AG predicting that the global automated ridesharing industry could be worth more than $2 trillion a year); David Z. Morris, *Driverless Cars Will Be Part of a $7 Trillion Market by 2050*, Fortune (June 3, 2017), https://fortune.com/2017/06/03/autonomous-vehicles-market/ ($7 trillion market by 2050).

annual savings could be $1.3 trillion per year once automated vehicles reach market saturation.[3]

As the automated vehicle industry shifts from research and development to commercialization, attorneys will have a significant role in shaping policy, design, and deployment; suing or defending crashes of automated vehicles; protecting privacy rights and data collection; and prosecuting or defending persons who use these vehicles. Part 1 is a primer on the automated vehicle industry, automated vehicle terminology, the automated vehicle technology stack, and the effects of automation. Chapter 1 explores the history of the automated vehicle industry, the current industry, and the expected means of deploying automated vehicles. Chapter 2 introduces the levels of automation, the basic terminology used in the automated vehicle industry. Chapter 3 explores the technology stack and artificial intelligence that comprise automated vehicles and the testing of these vehicles. Chapter 4 outlines the potential benefits and negative side effects of automated vehicles.

3. Morgan Stanley Research Global, *Autonomous Cars: Self-Driving the New Auto Industry Paradigm*, Morgan Stanley (Nov. 6, 2013), https://orfe.princeton.edu/~alaink/SmartDrivingCars/PDFs/Nov2013MORGAN-STANLEY-BLUE-PAPER-AUTONOMOUS-CARS%EF%BC%9A-SELF-DRIVING-THE-NEW-AUTO-INDUSTRY-PARADIGM.pdf.

Chapter 1

The Industry

Automated vehicles have been a futuristic idea imagined at least as far back as Leonardo Da Vinci. Da Vinci designed a self-propelled cart that used springs to move.[4] In 1939, General Motors sponsored Norman Bel Geddes's Futurama exhibit at the World's Fair in New York that depicted a world of driverless vehicles.[5] For decades, the focus on developing a vehicle that drove itself was based on electronic or smart highways.[6] The costs and difficulties of implementing a nationwide electronic highway ended this idea. These shortcomings led people to focus on developing automated vehicles that do not require smart infrastructure.

Automated vehicles are an actual reality today. From automated ridesharing services in Phoenix and Las Vegas to automated shuttles on university campuses and in cities,[7] companies are deploying automated vehicles for public use. This chapter begins by providing a background on the Defense Advanced Research Projects Agency (DARPA) Grand Challenges and the current state of the industry. The chapter focuses primarily on the methods of deploying automated vehicles.

---

4. Wired Brand Lab, *A Brief History of Autonomous Vehicle Technology*, Wired, https://www.wired.com/brandlab/2016/03/a-brief-history-of-autonomous-vehicle-technology/ (last visited Mar. 22, 2020).
5. Hod Lipson & Melba Kurman, Driverless: Intelligent Cars and the Road Ahead 108–12 (2016).
6. *Id.* at 116–22.
7. *See* Alexandria Sage, *Waymo Unveils Self-Driving Taxi Service in Arizona for Paying Customers*, Reuters (Dec. 5, 2018), https://www.reuters.com/article/us-waymo-selfdriving-focus/waymo-unveils-self-driving-taxi-service-in-arizona-for-paying-customers-idUSKBN1O41M2.

I. THE DARPA GRAND CHALLENGES

The modern automated vehicle is typically traced to the DARPA Grand Challenges that began in 2004.[8] In the National Defense Authorization Act for Fiscal Year 2001, Congress set a goal that a third of military vehicles of the United States Armed Forces be driverless by 2015.[9] In the National Defense Authorization Act for Fiscal Year 2003, Congress enabled military departments to provide prizes "to promote science, mathematics, engineering, or technology education in support of the missions of the U.S. Department of Defense."[10] DARPA took the cues from Congress and instituted the first Grand Challenge to create a competitive atmosphere for designing driverless technologies.

DARPA hosted the first Grand Challenge on March 13, 2004. DARPA offered a $1 million prize to the team that could navigate a 142-mile track in the Mojave Desert in the shortest time. To prevent preprogramming to a set map, DARPA did not provide the track map until the morning of the race. Fifteen self-driving vehicles competed. None of the self-driving cars made it further than 7.32 miles.

DARPA hosted another Grand Challenge the following year in the Mojave Desert on October 8, 2005, and increased the top prize to $2 million. One hundred ninety-five groups applied to enter the race, 40 teams were selected for the qualification, and 23 teams competed in the actual race. Five teams completed the course. The 2005 competition involved two frontrunners—Stanford's team led by Sebastian Thurn and Michael Montermelo and Carnegie Mellon's team led by Red Whittaker and Chris Urmson. Stanford entered one vehicle, Stanley, and Carnegie Mellon entered Sandstorm and H1ghlander. Stanley won, followed by Sandstorm, H1ghlander, Team Gray by the Gray Insurance Company, and Team Terramax by the Oshkosh Truck Corporation.

The success of the 2005 Grand Challenge led DARPA to host an Urban Challenge on November 3, 2007. The Urban Challenge was a 60-mile course hosted on the George Air Force Base in California. The Urban Challenge featured a $2 million first-place prize, a $1 million second-place prize, and a $500,000 third-place prize. DARPA required teams to comply with traffic laws and be able to merge into and negotiate with traffic. DARPA also provided funding for some teams. After narrowing down the field significantly, DARPA allowed 11 teams to compete. Stanford and Carnegie Mellon were again the frontrunners of this competition. Six teams finished the race. Carnegie Mellon won,

---

8. For an excellent discussion of the DARPA Grand Challenges, see Lawrence D. Burns & Christopher Shulgan, Autonomy: The Quest to Build the Driverless Car—And How It Will Reshape Our World (2018).

9. National Defense Authorization Act for Fiscal Year 2001, Pub. L. No. 106-398, § 217 (2000).

10. National Defense Authorization Act for Fiscal Year 2003, Pub. L. No. 107-314, § 2374b (2002).

followed by Stanford, Virginia Tech, MIT, the University of Pennsylvania, and Cornell University.

Following the success of the Urban Challenge, DARPA ended the Grand Challenges. The automated vehicle industry might have returned to academic research institutions and the fire created by the Grand Challenges could have begun to ember had it not been for Sergey Brin and Larry Page, the cofounders of Google. Brin and Page recruited some of the top participants from the DARPA Grand Challenges to build an automated vehicle for Google. This team included Sebastian Thurn, Chris Urmson, Anthony Levandowski, Dimitri Dolgov, and Michael Montemerlo. The project, officially known as Project Chauffeur but popularly called the Google self-driving car project, began in 2009. Project Chauffeur was the preeminent and first significant team developing automated vehicles for commercial purposes. The team built a cute vehicle known as Firefly that was a low-speed, fully self-driving vehicle. Project Chauffeur eventually became Waymo LLC, a leader in the automated vehicle industry.

II. THE AUTOMATED VEHICLE INDUSTRY

The automated vehicle industry includes companies that are developing driving automation systems, components, or vehicles, addressing potential problems like cybersecurity, and creating eventual services for the deployment of automated vehicles. In 2017, the Brookings Institute found that $80 billion had already been invested in the automated vehicle industry.[11]

Companies developing automated vehicles are generally segmented based on the sophistication of the driving automation system. The automated vehicle industry classifies driving automation systems into six levels, which are described in Chapter 2. The lower levels of automation, levels 1, 2, and 3, all rely on human drivers for at least one component of driving. Level 4 and level 5 automated vehicles remove humans from the loop.

With lower levels of automation, the automakers and Tier 1 suppliers are the principal developers. These lower levels of automation include adaptive cruise control, a level 1 driving automation system, or level 2 driving automation systems like Tesla Motors's Autopilot, General Motors's Super Cruise, Volvo's Pilot Assist, or Nissan's ProPilot Assist. The automakers have been the principal developers of level 3 automated driving systems, though no company has deployed these systems.

---

11. Cameron F. Kerry & Jack Karsten, *Gauging Investment in Self-Driving Cars*, Brookings (Oct. 16, 2017), https://www.brookings.edu/research/gauging-investment-in-self-driving-cars/.

Highly automated vehicles, level 4 and level 5 automated vehicles, are not currently available for sale to consumers. These automation systems are developed principally by companies explicitly formed to create them. These companies include Waymo, Cruise, Aurora, Argo AI, Voyage, Aptiv, Nutonomy, Zoox, Easy Mile, and Nuro. Other nontraditional automotive supply chain companies are also researching and developing automated vehicles, such as Uber, Lyft, Apple, Intel, Baidu, and Amazon.[12] Many of these companies, like Waymo and Aurora, are focused primarily on building a "driver"—the software and algorithms necessary to drive without creating an actual vehicle. Companies like Nuro, Cruise, and Zoox are building specially designed vehicles for their automated driving systems.

Except for Tesla Motors, most of the automakers were late to the highly automated vehicle industry. The automakers have taken one or a combination of three approaches to highly automated vehicles. Some automakers have made specific investments in or acquired some of the companies dedicated to developing highly automated vehicles. General Motors acquired Cruise for $1 billion, and Cruise has since received additional investments from Honda. Argo AI has received a $1 billion investment from Ford Motor Company and a $2.6 billion investment from Volkswagen. Some automakers have developed strategic partnerships, either among themselves and suppliers or with automated vehicle companies. Fiat Chrysler Automobiles and Jaguar have both agreed to supply vehicles to Waymo. Additionally, Waymo has entered into an agreement with Volvo to be its exclusive supplier of level 4 automated driving systems.[13] BMW Group, Intel, Mobileye (now a division of Intel), and Fiat Chrysler Automobiles have entered into a memorandum of understanding to develop an automated driving system.[14] Finally, many automakers, such as Mercedes Benz, Nissan, and Tesla Motors, are developing automated driving systems for their vehicles.

III. MODES OF DEPLOYING AUTOMATED VEHICLES

Companies are pursuing automation for different means and goals. Some companies are developing automated driving systems but are agnostic to their

---

12. Amazon has recently agreed to purchase Zoox, a leading automated vehicle company dedicated to designing an automated ridesharing vehicle. The sale allows Amazon to walk if Zoox fails to retain a sufficient number of employees. Stephen Nellis & Jane Lanhee Lee, *Exclusive: Amazon Plans at Least $100 Million to Keep Zoox Talent after $1.3 Billion Deal*, REUTERS (July 9, 2020), https://www.reuters.com/article/us-amazon-com-zoox-exclusive/exclusive-amazon-plans-at-least-100-million-to-keep-zoox-talent-after-1-3-billion-deal-idUSKBN24A3C8. If this sale closes, Amazon will have made a significant splash into the automated vehicle industry and, more generally, automated ridesharing.

13. Waymo, *Partnering with Volvo Car Group to Scale the Waymo Driver*, WAYMO BLOG (June 25, 2020), https://blog.waymo.com/2020/06/partnering-with-volvo-car-group-to.html.

14. *Fiat Chrysler Automobiles to Join BMW Group, Intel and Mobileye in Developing Autonomous Driving Platform*, FIAT CHRYSLER AUTOS., https://www.fcagroup.com/en-US/media_center/fca_press_release/2017/august/Pages/Fiat_Chrysler_Automobiles_to_Join_BMW_Group,_Intel_and_Mobileye_in_Developing_Autonomous_Driving_Platform.aspx (last visited Mar. 22, 2020).

ultimate deployment. Other companies are developing automated driving systems for a specific purpose, such as automated ridesharing, driverless shuttles, freight, or delivery bots. The method of deploying automated vehicles will impact legal analysis. Different regulatory regimes apply to some deployments but not others. Liability for crashes involving automated vehicles depends upon their ownership, use, and control. Responsibility for traffic infractions and Fourth Amendment rights may be affected by deployment and use. Data ownership and privacy rights can also turn on deployment. This section describes some ways that companies may deploy automated vehicles.

A. Private Ownership and Quasi-Ownership of Automated Vehicles

Automakers are currently selling vehicles equipped with level 1 and level 2 driving automation systems, and they will eventually consider offering highly automated vehicles for private sale to consumers. These privately owned highly automated vehicles may have different basic capabilities for human driving. Some highly automated vehicles could lack steering wheels, brake pedals, and accelerator pedals, effectively not allowing for human driving in any manner. In contrast, others may have human-driving components. The first privately owned highly automated vehicles will likely have human-driving components because they will be limited by operational design domains—the operating conditions in which the system can safely function.

One of the most significant risks for vehicle manufacturers in selling highly automated vehicles to consumers is about the obligations and duties that these manufacturers may incur for an indefinite time. For example, when does a manufacturer's responsibility to update and maintain software end? What if some people own their automated vehicles for 30 or 40 years? Automakers need to consider these questions before selling automated vehicles to avoid consumer suits for retroactively changing their products postsale.

The other problem that these manufacturers may encounter concerns their liability risk. It will remain to be seen whether the manufacturers of level 4 and level 5 automated vehicles may bear a more significant share of liability for crashes caused by their automated driving systems, a topic described in Part 3. If manufacturers bear responsibility, they will want greater control over maintenance, repairs, and alterations of these vehicles. Hardware and computing systems improve over time, and manufacturers may want to cycle out old hardware.

Manufacturers could retain control over their vehicles by leasing, licensing, or using a subscription model. Subscription models have become popular in recent years, with automakers offering subscriptions for specific models of their vehicles. Automated vehicles may help further the development and use of these subscription models.

B. Automated Ridesharing, Robo-Taxis, On-Demand Transportation Networks

Many companies intend to offer automated vehicles in an on-demand transportation network, robo-taxi, or automated ridesharing service, similar to services currently offered by transportation network companies (TNCs), like Uber and Lyft. Waymo currently provides this service with its Waymo One service in the Phoenix metropolitan area, and Aptiv, in coordination with Lyft, offers automated ridesharing in Las Vegas.

There are two broad ways of deploying automated ridesharing vehicles that could necessitate different legal analyses. Companies could develop or purchase large fleets of automated vehicles to use within their automated ridesharing service. The most significant advantage of this approach is consistency. The automated ridesharing company retains complete control over its services and vehicles. The most significant drawbacks are the capital required to own a fleet of automated vehicles and the different regulatory hurdles for a company-owned ridesharing fleet. Another model, like the one Tesla Motors is pursuing,[15] involves using privately owned automated vehicles to operate a ridesharing business. Under this model, the ridesharing company could be the automaker itself (or an affiliate thereof) or an independent third-party company, like Uber or Lyft, that connects privately owned automated vehicles with passengers.

C. Automated Rental Vehicles and Car Sharing

Car rental companies will own and offer automated vehicles for rental. With highly automated vehicles, these car rental companies may resemble ridesharing and vice versa. The only distinction between a traditional car rental company and ridesharing will (likely) be the length of the rental—a distinction that may not matter legally. Car sharing will also resemble the automated ridesharing companies.

D. Shuttle Services and Microtransit

Companies are already deploying automated shuttles on our roads. These shuttles typically follow a premapped short route that transports passengers along a corridor. Automated shuttles are currently being developed by Navya, Easymile, Local Motors, and Auro Robotics. Shuttles could be used on college campuses or by cities and towns to replace human-driven buses and other types of public transportation.

Automated vehicles could also solve the last-mile transit problem for existing public transportation routes or provide more dynamic routes than traditional

---

15. *See* Kirsten Korosec, *Tesla Plans to Launch a Robotaxi Network in 2020*, TECHCRUNCH (Apr. 22, 2019), https://techcrunch.com/2019/04/22/tesla-plans-to-launch-a-robotaxi-network-in-2020/.

public transportation. Automated vehicles could fill the gap between traditional public transit endpoints and the person's destination. Some passengers, such as the elderly and those suffering from disabilities, may not practically be able to use public transportation that has an endpoint away from their homes and destinations.

E. Freight Trucking

Freight trucking is an obvious early choice for automated vehicles. The trucking labor market is experiencing labor shortages and failing to attract younger drivers[16]—while, at the same time, wages and benefits for truck drivers are high costs of trucking.[17] Human drivers are also incapable of driving around the clock (and the Federal Motor Carrier Safety Regulations impose hour restrictions).

There are multiple models for how automated trucking could be deployed. Peloton Trucking is currently developing platooning to deploy driverless trucks behind a lead human driver. Tu Simple is developing level 4 automated trucks that do not need human drivers. Automated trucking could also use a teleoperator (a remote driver) or human drivers to operate automated trucks outside of their operational design domains.

F. Last-Mile Delivery Services

Automated technologies will help develop vehicles or bots that can deliver goods on a "last-mile" delivery service, such as pizza or e-commerce delivery. These bots may operate on roads or sidewalks. The National Highway Traffic Safety Administration granted its first exemption to a vehicle design for the highly automated vehicle industry to Nuro for its driverless delivery vehicle, R2X.[18] Other companies, like Postmates, FedEx, and Amazon, are also developing last-mile delivery bots.

---

16. Am. Transp. Research Inst., *An Analysis of the Operational Costs of Trucking: 2018 Update* 22 (Oct. 2018), http://truckingresearch.org/wp-content/uploads/2018/10/ATRI-Operational-Costs-of-Trucking-2018.pdf ("[T]he American Trucking Associations (ATA) estimated that the driver shortage would reach 50,000 drivers by the end of 2017, and if current trends were to continue, the shortage could grow to 174,000 drivers by 2026."); *but see* Stephen V. Burks & Kristen Monaco, *Is the U.S. Labor Market for Truck Drivers Broken?*, Monthly Labor Rev., U.S. Bureau of Labor Statistics (Mar. 2019), https://doi.org/10.21916/mlr.2019.5.

17. According to a report prepared by the American Transportation Research Institute, driver wages and benefits constitute an average cost of 72.9 cents per mile. Am. Transp. Research Inst., *supra* note 16.

18. See Chapter 5 for a discussion of this exemption.

Chapter 2

Levels of Automation

In its Taxonomy and Definitions for Terms Related to Driving Automation Systems for On-Road Motor Vehicles, published in January 2014, the Society of Automotive Engineers International (SAE) released its now-famous six levels for describing vehicle automation.[1] The United States Department of Transportation has adopted and uses the SAE levels of automation.[2] In September 2016, SAE issued a revised Taxonomy and Definitions for Terms Related to Driving Automation Systems for On-Road Motor Vehicles,[3] which was again revised and reissued on June 15, 2018.[4] One of the underlying purposes of the six levels of automation is to classify automated technology such that it can be readily discussed among professionals from different industries.[5]

This book uses the June 2018 revised version of the Taxonomy and Definitions for Terms Related to Driving Automation Systems for On-Road Motor Vehicles (the Taxonomy) to describe the SAE levels. The Taxonomy "describes

1. *Taxonomy and Definitions for Terms Related to On-Road Motor Vehicle Automated Driving Systems*, Document J3016, SAE INT'L (Jan. 16, 2014).

2. *See* NAT'L HIGHWAY TRAFFIC SAFETY ADMIN., U.S. DEP'T OF TRANSP., FEDERAL AUTOMATED VEHICLES POLICY: ACCELERATING THE NEXT REVOLUTION IN ROADWAY SAFETY 9–10 (2016).

3. *Taxonomy and Definitions for Terms Related to On-Road Motor Vehicle Automated Driving Systems*, Document J3016, SAE INT'L (Sept. 2016).

4. *Taxonomy and Definitions for Terms Related to On-Road Motor Vehicle Automated Driving Systems*, Document J3016, SAE INT'L (June 15, 2018) [hereinafter *Taxonomy*].

5. *See id.* at 1 ("This document has been developed according to the following guiding principles, namely, it should: . . . [b]e useful across disciplines, including engineering, law, media, public discourse. . . . Standardizing levels of driving automation and supporting terms serves several purposes, including . . . [p]roviding clarity and stability in communications on the topic of driving automation, as well as a useful short-hand that saves considerable time and effort.").

motor vehicle driving automation systems that perform part or all of the dynamic driving task on a sustained basis . . . [to] provide[] a taxonomy with detailed definitions for six levels of driving automation, ranging from no driving automation (level 0) to full driving automation (level 5)."[6]

The levels of automation are not without controversy.[7] There are at least two problems with the levels of automation. First, by identifying certain technologies as being a specific level of automation, people may erroneously be lulled into a false sense of trust by their "automated vehicle." People should not consider a vehicle equipped with adaptive cruise control an automated vehicle, even though such feature is a level 1 automated feature according to SAE or, perhaps, a vehicle equipped with a level 2 driving automation system. Second, by using levels of automation (0 to 5), SAE treats automation—or causes people to perceive automation—as a continuum in which a level 2 driving automation system is just two steps from being a level 4 automated driving system. What this means is that, in its attempt to provide clarity on the language used in the industry, SAE may have created confusion and increased risks for nonindustry participants and the public.

Notwithstanding the shortcomings of the SAE levels, an attorney should not begin to engage in the automated vehicle industry without a mastery of these levels. Almost every sophisticated conversation with attorneys, engineers, or regulators about automated vehicles assumes that participants will know what a speaker is discussing when he or she mentions a level of automation. An attorney who does not understand the levels of automation will be at a disadvantage. Because of the importance of the industry terminology, this chapter dives deeply into the definitions included in the Taxonomy.

In reviewing the levels of automation, it is helpful to remember that technical requirements and legal requirements are not synonymous. There may be a technical requirement under the Taxonomy but not a corresponding legal requirement. This disconnect can arise for numerous reasons, such as due to product marketing or manufacturer representations. Another reason for the disconnect between legal and technical requirements can be that courts or legislatures decline to make the technical requirements the law. Simply stated, there is no requirement that courts consult, review, or consider the Taxonomy when rendering decisions (though they should).

---

6. *Id.* at 2.

7. *See, e.g.*, Alex Roy, *How the Language of Self-Driving Is Killing Us*, THE DRIVE (May 1, 2018), https://www.thedrive.com/opinion/20495/how-the-language-of-self-driving-is-killing-us.

I. A DEFINITIONAL BACKGROUND

The Taxonomy is, as one would expect, definition driven. This section introduces some definitions to help provide a foundation for the SAE levels. The Taxonomy uses numerous acronyms, but this book does not use those acronyms because some of the acronyms invite confusion. For example, the acronym for an automated driving system (ADS), the system used in levels 3, 4, and 5 automated vehicles, is too close to the acronym for advanced driver assistance systems (ADASs). ADASs are the current systems used in vehicles like automated lane keeping and automatic emergency braking.

To establish a level of driving, the Taxonomy examines the expected[8] roles played by three primary actors in the dynamic driving task and the dynamic driving task fallback: the human driver, the driving automation system, and other vehicle components. The dynamic driving task is "[a]ll of the real-time operational and tactical functions required to operate a vehicle in on-road traffic, excluding the strategic functions such as trip scheduling and selection of destinations and waypoints."[9] The Taxonomy lists six nonexclusive examples of operational and tactical subtasks:

1. Lateral vehicle motion control via steering (operational);
2. Longitudinal vehicle motion control via acceleration and deceleration (operational);
3. Monitoring the driving environment via object and event detection, recognition, classification, and response preparation (operational and tactical);
4. Object and event response execution (operational and tactical);
5. Maneuver planning (tactical); and
6. Enhancing conspicuity via lighting, signaling, gesturing, etc. (tactical).[10]

The Taxonomy further categorizes the dynamic driving task into two principal categories: (1) the sustained lateral and longitudinal vehicle motion control (subtasks 1 and 2 above), and (2) the object and event detection and response (subtasks 3 and 4 above).

The Taxonomy also distinguishes levels of automation, in part, based on whether the human user or the driving automation system performs the dynamic driving task fallback. The dynamic driving task fallback is defined as follows:

---

8. The Taxonomy is clear that it does not apply to the actual role—an acknowledgment that one actor (likely the human) may not live up to his or her expected role. *Taxonomy, supra* note 4, at 2.

9. *Id.* at 6.

10. *Id.*

> The response by the user to either perform the [dynamic driving task] or achieve a minimal risk condition after occurrence of a [dynamic driving task] performance-relevant system failure(s) or upon [operational design domain] exit, or the response by an [automated driving system] to achieve minimal risk condition, given the same circumstances.[11]

The Taxonomy also distinguishes among levels of automation if a driving automation system is constrained to an operational design domain, which is the "[o]perating conditions under which a given driving automation system or feature thereof is designed to function."[12] An operational design domain "includes, but is not limited to, environmental, geographical, and time-of-day restrictions, and/or the requisite presence or absence of certain traffic or roadway characteristics."[13] Unlimited operational design domain is the ability for a driving automation system to "perform the entire [dynamic driving task] and [dynamic driving task] fallback . . . in all driver-manageable on-road driving situations."[14]

Before delving into the specific levels, the concept of "monitor" is important to introduce at this point. Monitor is "[a] general term referencing a range of functions involving real-time human or machine sensing and processing of data used to operate a vehicle, or to support its operation."[15] The Taxonomy does not consider the "condition of being receptive to alerts or other indicators" as monitoring.[16] Receptivity means "[a]n aspect of consciousness characterized by a person's ability to reliably and appropriately focus his/her attention in response to a stimulus."[17] The Taxonomy delineates four types of monitoring: "(1) monitor the user; (2) monitor the driving environment; (3) monitor vehicle performance; and (4) monitor driving automation system performance."[18]

Monitor the user means "[t]he activities and/or automated routines designed to assess whether and to what degree the user is performing the role specified for him/her."[19] The Taxonomy notes that "[u]ser monitoring in the context of driving automation is most likely to be deployed as a countermeasure for misuse or abuse (including over-reliance due to complacency) of a driving automation system but may also serve some other purposes."[20] The Taxonomy further

---

11. *Id.* at 7.
12. *Id.* at 14.
13. *Id.*
14. *Id.* at 18.
15. *Id.* at 12.
16. *Id.* at 17. The Taxonomy illustrates the distinction between monitoring and being receptive with the following example: "A person who becomes aware of a fire alarm or a telephone ringing may not necessarily have been monitoring the fire alarm or the telephone." *Id.*
17. *Id.* at 14.
18. *Id.* at 12.
19. *Id.* at 13.
20. *Id.*

notes that "user monitoring is primarily useful for levels 2 and 3, as evidence from the field on the use of level 1 features has not identified significant incidence of misuse or abuse of driving automation technology, and above these levels the [automated driving system] is by definition capable of achieving a minimal risk condition automatically."[21] Misuse or abuse of level 2 and level 3 technology is described in Chapters 9 and 10, respectively.

Monitor the driving environment means "[t]he activities and/or automated routines that accomplish real-time roadway environmental object and event detection, recognition, classification, and response preparation (excluding actual response), as needed to operate a vehicle."[22]

Monitor vehicle performance means "[t]he activities and/or automated routines that accomplish real-time evaluation of the vehicle performance, and response preparation, as needed to operate a vehicle."[23] At levels 4 and 5, the automated driving system is responsible for monitoring the vehicle performance, while the human driver is responsible at levels 1 and 2. At level 3, the human must be "receptive to vehicle conditions that adversely affect performance of the [dynamic driving task]."[24]

Monitor driving automation system performance refers to "[t]he activities and/or automated routines for evaluating whether the driving automation system is performing part or all of the [dynamic driving task] appropriately."[25] "At levels 1 and 2, the driver monitors the driving automation system's performance," while "[a]t higher levels of driving automation (levels 3-5), the [automated driving system] monitors its own performance of the complete [dynamic driving task]."[26]

II. THE SAE LEVELS

The Taxonomy categorizes driving automation systems based on five factors:

1. Whether the driving automation system performs either the longitudinal or the lateral vehicle motion control subtask of the [dynamic driving task].
2. Whether the driving automation system performs both the longitudinal and the lateral vehicle motion control subtasks of the [dynamic driving task] simultaneously.

---

21. *Id.*
22. *Id.*
23. *Id.*
24. *Id.*
25. *Id.*
26. *Id.*

3. Whether the driving automation system also performs the object and event detection and response subtask of the [dynamic driving task].
4. Whether the driving automation system also performs the [dynamic driving task] fallback.
5. Whether the driving automation system is limited by an operational design domain.[27]

Before examining the specific SAE levels, it is essential to keep in mind three points. A vehicle is not an SAE level—an automated feature (or lack thereof) is an SAE level. For simplicity, this book will, at times, describe an automated vehicle by delineating the specific level, even though this is not technically correct. A vehicle may have various automated features of differing SAE levels.[28] Additionally, there is no requirement that a higher SAE level have all of the lower levels.

A. SAE Level 0

At level 0, the automation feature (or lack thereof) is one in which a driving automation system "[d]oes not perform any part of the [dynamic driving task] on a sustained basis."[29] The traditional driver is responsible for the entire dynamic driving task and the dynamic driving task fallback. Almost every vehicle throughout history lacked automation features. Some automation or technological features in vehicles today, such as automatic transmissions or power steering, are considered level 0 features. So long as the feature does not perform any part of the dynamic driving task on a *sustained*[30] basis, then those automation features are not considered driving automation systems for purposes of the Taxonomy.[31] Also, a vehicle may be equipped with higher levels of automation and have capabilities for human driving.

B. SAE Level 1

A level 1 driving automation system performs either the longitudinal vehicle motion control subtask or the lateral vehicle motion control subtask. It can only perform, at most, a limited aspect of the object and event detection and response. During the use of a level 1 driving automation system, the human driver has the following responsibilities: (1) performance of the remaining

27. *Id.* at 18.

28. *See id.* at 5 ("A given driving automation system may have multiple features, each associated with a particular level of driving automation and [operational design domain].").

29. *Id.* at 21.

30. Sustained means "[p]erformance of part or all of the [dynamic driving task] both between and across external events, including responding to external events and continuing performance of part or all of the [dynamic driving task] in the absence of external events." *Id.* at 15.

31. *Id.*

dynamic driving task; (2) supervision of the driving automation system and the responsibility to "intervene[] as necessary to maintain safe operation of the vehicle"; (3) the responsibility to determine the appropriate time to engage or disengage the driving automation system; and (4) to be available any time to "immediately perform[] the entire [dynamic driving task] whenever required or desired."[32] The Taxonomy considers a level 1 driving automation system a "driver support feature." An example of a level 1 feature is adaptive cruise control.

C. SAE Level 2

A level 2 driving automation system performs both the longitudinal vehicle motion control subtask and the lateral vehicle motion control subtask. Similar to a level 1 driving automation system, a level 2 driving automation system can only perform a limited aspect of the object and event detection and response. The driver using a level 2 driving automation system has the same responsibilities as a driver using a level 1 driving automation system. The Taxonomy refers to a level 2 driving automation system as a "driver support feature." Tesla's Autopilot and General Motors's Super Cruise are examples of level 2 driving automation systems.

D. SAE Level 3

Level 3, unlike levels 1 and 2, fundamentally changes the role of the underlying driver. The Taxonomy introduces two new terms at level 3: automated driving system and the dynamic driving task fallback-ready user. The automated driving system is the name of the driving automation system for levels 3 through 5. The dynamic driving task fallback-ready user is the human located in the driver's seat. At level 3, while engaged, the automated driving system

> (1) performs the entire [dynamic driving task]; (2) determines whether [the operational design domain] limits are about to be exceeded and, if so, issues a timely request to intervene to the [dynamic driving task] fallback-ready user; (3) determines whether there is a [dynamic driving task] performance-relevant system failure of the [automated driving system] and, if so, issues a timely request to intervene to the [dynamic driving task] fallback-ready user; (4) disengages an appropriate time after issuing a request to intervene; and (5) disengages immediately upon driver request.[33]

32. *Id.* at 21.
33. *Id.* at 22.

The dynamic driving task fallback-ready user also has responsibilities while using a level 3 automated driving system. The dynamic driving task fallback-ready user must

> (1) [be] receptive to a request to intervene and respond[] by performing [the dynamic driving task] fallback in a timely manner; (2) [be] receptive to [dynamic driving task] system failures in vehicle systems and, upon occurrence, perform[] [the dynamic driving task] fallback in a timely manner; (3) determine[] whether and how to achieve a minimal risk condition; and (4) become[] the driver upon requesting disengagement of the [automated driving system].

Although the defining feature of level 3 is that the automated driving system cannot perform the dynamic driving task fallback in all situations, this does not mean that the automated driving system is incapable of performing the dynamic driving task fallback at all.

As of the date of this book, no manufacturer deploys a level 3 automated driving system. Audi teased a level 3 automated driving system with its traffic jam assist but never ended up offering the feature. Many automated vehicle companies and experts view the handoff from the automated driving system to the dynamic driving task fallback-ready user as too difficult of a problem to solve, at least in comparison to higher levels of automation.[34] This problem is referred to as the Handoff Problem in the industry and is described in Chapter 10. Simply stated, the Handoff Problem is the problem of transitioning the level 3 automated driving system to the human driver in a safe and timely manner.

E. SAE Level 4

Level 4 represents high driving automation and is the first level where humans are removed from the loop. The level 4 automated driving system

> performs the entire [dynamic driving task]; may issue a timely request to intervene; performs [the dynamic driving task] fallback and transitions automatically to a minimal risk condition when: (a) a [dynamic driving task] performance-relevant system failure occurs or (b) a user does not respond to a request to intervene or (c) a user requests that it achieve a minimal risk condition; disengages, if appropriate, only after: (y) it achieves a minimal risk condition or (z) a driver is performing the [dynamic driving task].[35]

---

34. See Chapter 10 for a discussion of these beliefs.
35. *Taxonomy*, *supra* note 4, at 22.

The human user of a level 4 automated driving system is referred to by one of two terms: passenger or driverless operation dispatcher.[36] A passenger is "[a] user in a vehicle who has no role in the operation of [the] vehicle" while the automated driving system is engaged. A driverless operation dispatcher is "[a] user(s) who dispatches an [automated driving system] equipped vehicle(s) in driverless operation."[37] Thus, level 4 is the first automation level where the vehicle could be driverless. While a level 4 automated driving system is engaged, the passenger or driverless operation dispatcher

> [n]eed not perform [the dynamic driving task] or [the dynamic driving task] fallback; need not determine whether and how to achieve a minimal risk condition; may perform [the dynamic driving task] fallback following a request to intervene; may request that the automated driving system disengage and may achieve a minimal risk condition after it is disengaged; may become the driver after a requested disengagement.[38]

The role of the user will, of course, depend upon whether the user is present (i.e., a passenger) or not present (i.e., a driverless operation dispatcher) in the car. It is important to note that, from a technical standpoint, the user is not required to perform the dynamic driving task fallback, even if the user is requested to intervene. The user is also not required "to supervise a level 4 [automated driving system] or be receptive to intervene while the [automated driving system] is engaged."[39]

The major distinguishing feature between level 4 and level 5 is that a level 4 automated driving system has a limited operational design domain.[40] Thus, although a level 4 automated driving system can perform all of the necessary driving tasks in its operational design domain, such operational design domain is limited in one respect or another. Waymo currently deploys a level 4 automated driving system in an automated ridesharing capacity within a limited operational design domain in the Phoenix suburbs. Initial highly automated vehicles will almost certainly be level 4 systems because they will have an operational design domain.

F. SAE Level 5

SAE level 5 represents full driving automation. While engaged, the roles of the automated driving system and the passenger or driverless operation dispatcher

36. *Id.*
37. *Id.* at 17.
38. *Id.* at 22.
39. *Id.* at 25.
40. *Id.* at 19.

remain the same as they were at level 4. The defining feature of level 5 is that the automated driving system has an unlimited operational design domain. Thus, "there are no design-based weather, time-of-day, or geographical restrictions[41] on where and when the [automated driving system] can operate the vehicle."[42] Level 5 is an aspirational goal for the automated vehicle industry.

---

41. A level 5 automated driving system can have geographic limitations and retain its level 5 status if those limitations are due to business or legal reasons. *Id.* at 33.

42. *Id.* at 25. This does not mean that the automated driving system can operate in any and all environments. Like a human driver, there are some situations where a level 5 automated driving system could not operate, such as during a "white-out snow storm, flooded roads, [and] glare ice." *Id.*

Chapter 3

Automation Hardware and Software

Attorneys who practice automated vehicle law need to understand how automated vehicles perceive the world, make decisions, and control those decisions. Attorneys who do not understand the technology and testing of automated vehicles may miss claims or lose arguments. This chapter is a primer on automated vehicle functionality, hardware, machine learning, and testing.

Automated vehicles operate similarly to human drivers. Although people may not realize that they follow a set routine while driving, almost every drive follows a common theme. Humans generally develop a plan for the drives. Typically, that decision is to arrive at a specific location. Once goals have been set, drivers are then tasked with perceiving the world and determining how they will act to accomplish their goals. Drivers use situational awareness to see the world and make decisions that help them achieve their ultimate goals. Humans perceive the world through our senses and technology, such as global positioning systems or communication applications like Waze. Drivers detect obstacles on their drives that create miniature goals. A slow vehicle in front of a driver may create an impediment to the ultimate goal for which the driver develops the miniature goal of passing the slow driver. After perceiving the world and developing their plans, drivers control their vehicles and accomplish their goals.

Automated vehicles operate similarly. Automated vehicles will be assigned a goal by their users—typically, automated vehicles need to go to specific locations. Once automated vehicles have their primary goals for their drives, they will then plan how they will accomplish them. The planning stage requires automated vehicles to know their locations and their targets' locations—achieved through a combination of maps and geolocation devices. Automated vehicles then need to perceive the world to gain situational awareness of their surroundings. Automated vehicles perceive the world using technologies, such as cameras, radars, light detection and ranging (LiDAR), maps, and artificial intelligence. Automated vehicles then use the information perceived through their technology stack to plan how to effectuate their goals. Finally, driving automation systems use their mechanical components to accomplish their drives and goals.

This chapter is broken down into three sections. The first section examines the automated vehicle stack. The second section introduces how artificial intelligence is used in automated vehicles. Finally, this chapter concludes by describing the testing of automated vehicles.

I. AUTOMATED VEHICLE STACK

Not all automated vehicles utilize the same hardware or software suites. Differences abound throughout the levels of automation. For level 2 driving automation systems, companies disagree on the types of driver monitoring systems and the need for geofencing. At higher levels of automation, a significant difference is that Tesla Motors, through Elon Musk, has been outspoken about its belief that LiDAR is an unnecessary (potentially redundant) component of a highly automated vehicle. Tesla Motors's view is one that most significant players in the automated vehicle industry disagree with, and most believe that LiDAR currently is a necessary component for a safe automated driving system.

These disagreements are ripe for lawyers to second-guess design decisions when these vehicles crash, and, given the difference of beliefs, there will always be expert witnesses on both sides who can attest to why a given system is defective or not defective. Tesla Motors may very well be right that LiDAR is unnecessary for the safe operation of highly automated vehicles. Still, it may not be hard for a plaintiff's attorney to show that LiDAR could have helped prevent a crash, and widespread industry practice may be quite persuasive to juries.

Redundancy is a word used by people describing automated driving stacks. Sometimes this word is used to describe things that are not redundant. Redundancy in design means that there are fail-safes that perform the same or similar functions. If one system fails, the other system's performance means that one failure does not create gaps. A power generator at a hospital would be a redundant technology to the main power source. Redundancy is not the use of different technologies to address technological holes or shortcomings of a given technology. For example, 3D maps and GPSs perform complementary functions and are not redundant systems.

This section does not describe the placement of technology on automated vehicles. Technology placement is an essential issue in designing automated vehicles and in litigation. Attorneys will want their expert witnesses to explore whether the placement of technology impacted the safety of the automated vehicle.

A. High-Definition 3D Maps

High-definition 3D maps serve numerous functions for automated vehicles. These preloaded maps help with localization—the automated vehicle's ability to know precisely where it is in the world—and establish a baseline for sensor readings. Automated vehicle companies typically develop their maps with cameras and LiDARs and then have humans annotate them. Given the difficulty in obtaining highly detailed, up-to-date 3D maps of every roadway, companies like Tesla Motors and Mobileye are obtaining data from cameras installed on privately owned vehicles.

B. Global Navigation Satellite Systems

GPSs are used by automated vehicles for localization and navigation. A GPS allows an automated vehicle to determine its location by receiving signals from satellites located in medium Earth orbit. Differential GPSs are systems that provide positional corrections to GPSs to increase accuracy; differential GPSs use a fixed reference point on Earth to calculate positional errors. GPSs that are used in phones or as navigation systems in conventional vehicles are not accurate enough for automated vehicles—mobile phone GPSs are accurate within about 4.9 meters.[1] Automated vehicle GPSs need to be accurate within centimeters.

C. Inertial Measurement Unit

Inertial Measurement Units (IMUs) are important safety technologies for automated vehicles. Unlike other sensors, such as GPSs, digital cameras, radars, and LiDARs, that interact with the external environment of automated vehicles, IMUs do not rely on the external environment. They are self-contained systems that monitor the vehicle's system and location. IMUs help enable precise localization for an automated vehicle and can provide localization when a GPS fails to obtain a signal. IMUs generally consist of accelerometers, which measure acceleration, and gyroscopes, which measure and maintain velocity and orientation.

1. *GPS Accuracy*, GPS.Gov, https://www.gps.gov/systems/gps/performance/accuracy/ (last visited Feb. 8, 2020).

D. Digital Cameras

Digital cameras provide a driving automation system with a clear view of the world by using computer vision to mimic human vision. Vision is probably the most important sense for driving because a large portion of the driving task is visual.[2] Digital cameras detect color, signs, signals, lane markings, people, vehicles, and objects. Companies use cameras that measure different distances and widths. For example, Tesla Motors's Autopilot uses eight cameras to provide images around the entire vehicle: a narrow-forward camera that measures up to 250 meters, a main-forward camera with a max distance of 150 meters, a wide-forward camera, forward-looking side cameras, backward-looking side cameras, and a rearview camera.[3]

Digital cameras require highly sophisticated computer vision to identify the objects within the images. Computer vision, a type of artificial intelligence, converts data provided by digital cameras into a constantly updating and moving 3D map of the vehicle's surroundings. Some companies, like Tesla Motors, believe that digital cameras and computer vision are sufficient for a highly automated vehicle to navigate the roads safely without needing LiDAR.

E. Radar

Radio detection and ranging (radar) is a component of automated vehicle design. Radar detects physical objects by emitting electromagnetic waves and measuring the "echo" created when those waves bounce off objects. Radar transmits signals using millimeter radio waves of 24, 77, or 79 GHz bandwidth. Radars are currently used in most modern vehicles that have adaptive cruise control or advanced driver-assistance systems. Automobiles generally use long-range and short-range radars. Long-range radar is used to detect objects and measure speed, distance, and velocity of objects and vehicles. Short-range radar is used for close-vehicle proximity, like blind-spot detection. Radars also do not need to be only forward-facing. Rear-facing radar helps detect vehicles behind an automated vehicle.

F. LiDAR

LiDAR is the spinning gadget on top of automated vehicles. LiDAR transmits lasers and measures objects using light. LiDAR paints a very detailed 3D map of the vehicle's surroundings. LiDAR's original purpose was 3D mapping—LiDAR has also become effective at detecting objects and people and their

---

2. Michael Sivak, *The Information That Drivers Use: Is It Indeed 90% Visual?*, 25 Perception 1081 (1996) (analyzing observations about whether vision equates to 90 percent of the driving task).

3. *Future of Driving*, Tesla, https://www.tesla.com/autopilot (last visited Mar. 22, 2020).

actions and movements. LiDARs detect changes in the base map by identifying objects that were not located in the base map.

LiDAR can capture a vehicle's entire surroundings, as does the spinning "coffee can" LiDAR gadget on top of many automated vehicles, or capture a single direction. LiDAR is an expensive new technology for vehicles, currently costing thousands of dollars per unit. Unless and until the cost of LiDAR decreases significantly, it may be cost prohibitive for privately owned automated vehicles. LiDAR struggles in certain weather conditions like snow, rain, or fog, as well as with certain surfaces. LiDAR has difficulty with close proximity to a vehicle.

LiDAR has a significant advantage over cameras in measuring distances. LiDARs can provide the true distance between an object and the LiDAR. In contrast, a camera measures distance like humans: through perspective. Cameras must use significant computing power to analyze images to process distances.

Some companies are dedicated solely to developing LiDAR for automated vehicles. These companies include Velodyne Lidar, Ouster, and Luminar Technologies. Automated driving companies, such as Waymo, Cruise, and Argo AI, have acquired LiDAR companies or are otherwise developing their LiDARs.

G. Other Sensors

Automated vehicles could use other sensors, such as sonar or thermal cameras. Sonar, sound navigation and ranging, measures distance using sound waves. Sonar is slower than radar and becomes weaker over long distances. Sonar is more precise than radar at close distances and is often used for parking and other close-range activities for vehicles. Thermal cameras could be an effective technology for automated vehicles in detecting and classifying people, animals, and objects. A thermal camera measures objects and people based on their heat emissions.

H. Connected Automated Vehicles

Automated vehicles do not have to be connected vehicles. A connected vehicle is one that can communicate with infrastructure or other vehicles. Connectivity is expected to make vehicles—whether human-driven or automated—safer and help with congestion and roadway throughput. Connectivity can improve functionality for technologies and increase situational awareness for automated vehicles. Connectivity could provide important information to automated vehicles about other automated vehicles or human-driver behaviors for connected conventional vehicles.

Discussions about connectivity typically concern the types and the means of communication. Vehicles can use these types of communications:

vehicle-to-vehicle (V2V), vehicle-to-infrastructure (V2I), vehicle-to-pedestrian (V2P), or vehicle-to-everything (V2X). For example, in a V2I communication, a traffic signal could share its current signal state with vehicles. The means of these communications could be satellite, cellular, 5G, or dedicated short-range communication (DSRC). DSRC is the most prominent communication technology discussed in the automotive industry. DSRC "is a two-way wireless radio service that provides a high speed, reliable, short-range communication media for use among moving vehicles" and infrastructure.[4] DSRC sends a basic safety message (BSM) that includes vehicle data, such as its position, speed, heading, steering angle, and brake status.[5] A DSRC-equipped vehicle broadcasts this BSM every 100 milliseconds on the 5.9 GHz spectrum band.[6]

I. Safety Drivers

An essential component of an automated driving system's testing stack is the human safety driver. For initial testing, the safety driver may be the most critical component. The most infamous safety driver failing was the tragic Uber crash that killed a pedestrian while the safety driver stared into her lap. Safety drivers generally provide two crucial functions to an automated vehicle program. They monitor the system and identify its irregularities so that problems can be identified and corrected. They also serve as the fallback for system mishaps, problems, and failures. In this regard, safety drivers are what makes automated vehicle testing on public roads less radical and risky. Some states require a safety driver for public testing of automated vehicles.

Automated vehicle companies have safety driver selection and training procedures. The Automated Vehicle Safety Consortium, consisting of major industry players, issued its best practices for safety drivers in November 2019 (Best Practices).[7] The Best Practices provides guidance on driver characteristics and training, such as conducting background checks on drivers, ensuring basic driving competency and a safety mindset, and providing training on the automated driving system and driving training, as well as guidance on pre-, during, and post-trip requirements and protocols to follow.

---

4. U.S. Dep't of Transp., Status of the Dedicated Short-Range Communications Technology and Applications: Report to Congress 16 (July 2015), https://www.its.dot.gov/research_archives/connected_vehicle/pdf/DSRCReportCongress_FINAL_23NOV2015.pdf.

5. *Id.* at 16–17.

6. *Id.*

7. Automated Vehicle Safety Consortium, *AVSC Best Practice for In-Vehicle Fallback Test Driver Selection, Training, and Oversight Procedures for Automated Vehicles under Test* (Nov. 2019), *available at* https://saemobilus.sae.org/content/AVSC00001201911/.

J. Human Drivers

An essential "component" of a level 1, level 2, and level 3 driving automation system is the human driver. The human driver is responsible for driving the vehicle while using a level 1 driving automation system and is responsible for supervising and being able to retake control at any time while using a level 2 driving automation system. The human driver must be able to execute the dynamic driving task fallback for level 3 automated driving systems. The human driver is an unnecessary component of a level 4 and level 5 automated driving system. Automakers need to design their driving automation systems to foster the function of this "component" of their driving automation systems.

K. Driver Monitoring Systems

Driver monitoring systems (DMSs) have become popular topics following crashes involving level 2 automated vehicles where drivers were distracted. These deaths and crashes, coupled with abundant, consistent human factor studies showing that level 2 driving automation systems create opportunities for drivers to become distracted,[8] have led many people and the National Transportation Safety Board to believe that a level 2 driving automation system should not be deployed without a DMS. Given that drivers using level 2 driving automation systems have an obligation under the Taxonomy (and the owner's manual) to supervise the drive, these drivers are seemingly misusing these systems by not supervising their drives. A DMS is the industry solution to that misuse. A DMS is a system that monitors the driver or otherwise keeps the driver engaged to fulfill his or her duty to supervise the drive.

Manufacturers have deployed different types of DMSs. General Motors's Super Cruise uses eye tracking to monitor the driver's eyes. Tesla Motors's Autopilot requires drivers to keep their hands on the steering wheel.

Although DMSs are not currently used for human driving, the use of DMSs may have significant safety benefits. As vehicles equipped with level 2 driving automation systems include DMSs, automakers will need to consider whether to have DMSs active at all times or just when the level 2 system is engaged. The inclusion of DMSs in these vehicles may create new opportunities for product liability claims by people harmed by human drivers.

L. Geofencing

Geofencing or fencing is another term that is frequently used in the automated vehicle industry. Geofencing is a virtual perimeter; it can be used for limiting operations. Geofencing can be used to program an automated vehicle to comply

8. See Chapter 9 for a discussion of this topic.

with traffic laws in different jurisdictions or to restrict an automated vehicle's use to a set geographic location or a vehicle's operation design domain. Geofencing is often used when a specific area has been premapped for automated vehicle use.

Geofencing should not be confused with an operational design domain. A driving automation system may have a specified operational design domain, but the manufacturer may not have geofenced its driving automation system to that operational design domain. Some operational design domains may be easier to geofence than others. Geographic operational design domains are static. Other operational design domains are not static, such as those dependent upon weather, and are harder to geofence.

M. Platooning

Platooning is the process of automated vehicles sharing information to coordinate movements. Platooned vehicles use V2V communications to coordinate their operations. Once in a platoon, vehicles can travel in a cluster rather than individually. Platooning increases road throughput and capacity.

II. ARTIFICIAL INTELLIGENCE: THE BRAIN BEHIND AUTOMATED VEHICLES

The inherent limitations that prevented automated vehicles from becoming a reality until recently were twofold: computers have historically been literal machines, and the computing power necessary to automate vehicles in a useable manner was not available. Computers are programmed to perform specific tasks given specific parameters. Computers would not know how to handle new parameters or new variables without further instructions. This limitation created a problem for computer drivers. Every driver knows that roadways are filled with new parameters or new variables. Although the roads we drive on every day remain the same, our experiences on those roads change every time. There may be different vehicles and different drivers, pedestrians, bicyclists, skateboarders, e-scooters, or animals on or about the road; trash and garbage may litter the road; or weather conditions may create diverse driving experiences. Because our roadways are not constants, computers historically have been ill equipped to navigate this ever-changing environment.

Artificial intelligence (AI) changes that calculus. Automated vehicles use different AI techniques, including, most prominently, deep learning. Computer vision and machine learning are two of the more prominent deep learning techniques used for automated vehicles. Computer vision is the process of computers seeing, detecting, and perceiving objects from images. Computer vision analyzes images captured by digital cameras and other sensors and determines what objects are within those images. Computer vision performs object detection and recognition for automated vehicles by detecting lanes, traffic signs and signals, bystanders, vehicles, and other objects. Machine learning is the process of software learning through detecting and extrapolating patterns. Through machine learning, automated vehicles can learn without the need for explicit programming for every situation the vehicle encounters. Automated vehicles rely upon deep learning for many tasks, including perception, localization, and path planning.

Sensor fusion is also one of the more important components of AI for automated vehicles. Sensor fusion is the process of "fusing" the information collected by different technologies to create a better understanding of the world. Sensor fusion can be used to extract information from each technology's strengths. Sensor fusion algorithms parse data collected from the automated vehicle stack to create a singular, clear vision of the world.

Deep learning algorithms are taught using supervised, unsupervised, or reinforcement learning. Supervised learning uses inputs and outputs, and the algorithm learns the function by determining how to turn the input into an output. With unsupervised learning, the algorithm is given the input data but not a corresponding output; the algorithm discovers new outputs. Unsupervised algorithms can use clustering or rule learning. Reinforcement is the process of rewarding or punishing systems (referred to as agents) for favorable or unfavorable decisions in a given environment. The agent's goal is to maximize its total reward.

Two of the more popular deep learning methodologies are convolutional neural networks (CNNs) and recurrent neural networks (RNNs). There are many other deep learning methodologies used for automated vehicles, such as YOLO (you only look once).

CNNs are used principally for computer vision and image analyses. Digital cameras capture images that are viewed by the software as a series of pixels. CNNs use convolutional layers to transform inputs to the next layer. Each convolutional layer uses filters to detect patterns and extract features of an image before passing the output (which becomes a new input) to the next layer, where the process continues.

RNNs are algorithms that detect patterns in sequences of data. RNNs, unlike regular neural networks, use memory and leverage past experiences to interpret the present and predict the future. Long short-term memory networks (LSTMs) are a type of RNN. An LSTM uses gates to control the flow of information into the algorithm by determining which data is important and is kept and which data is unimportant and discarded. With the LSTM, the RNN can leverage important data for longer.

Because deep learning relies upon the dataset provided to the algorithms, data is the gasoline of automated vehicles. Automated vehicle companies feed their systems data from numerous sources, including publicly available data sets, real-world data from operating their vehicles on public roads, simulations, and their technology stack.

III. TESTING AUTOMATED VEHICLES

Automated vehicles are tested in three different areas: test tracks, public roads, and simulation. Many developers have test tracks, and many other test tracks are available, such as those located at universities. At one time, the United States Department of Transportation had identified ten automated vehicle proving grounds. Typically, companies use test tracks to verify their system's safety before deploying it on public roads. Automated vehicles are also tested on public roadways. Jurisdictions impose different requirements before a company can publicly test its automated vehicles. Simulation is used to teach an automated driving system and enables a company to have its system "drive" significant miles in a short time and encounter corner cases. Simulation also allows a company to test subtle differences. For example, variables like a pedestrian's shirt color, clothes, height, and weight can be changed.

The use of these testing components in an automated vehicle program tells very little about the effectiveness of the miles driven in simulation or the real world for the ultimate deployment of automated vehicles. For the attorney trying an automated vehicle case or the regulator determining whether to permit automated vehicles on the roadways, much importance should be placed on the integrity of a testing regimen. Did the manufacturer utilize simulated and real-world miles? If the manufacturer used simulation, importance should be placed on the quality of that simulation. A billion simulation miles may not matter if the simulation model itself was deficient in some material way, such as through the design of the simulated world, data input into the simulated world, or the tweaks to an automated driving system based on the simulations. For real-world miles driven, the inquiry should focus on where those miles were driven. A manufacturer that only tests in Arizona may have an excellent system for Arizona or similar climates but may be ill equipped, without more testing, to deploy vehicles in locations with different climates, such as Alaska. Likewise, a manufacturer who primarily tests on highways or simple roads may be ill equipped, without more testing, to deploy vehicles in chaotic cities with more challenging roads, like San Francisco. Not all miles are created equal, and savvy litigants will pry into the testing regimens when litigating automated vehicle crashes.

Chapter 4

Effects of Automated Vehicles

Every new technology promises new benefits—otherwise there would be no reason for adopting the technology—and has adverse effects or creates new risks, intentionally or unintentionally—otherwise there would be no need to regulate the technology. Good policy depends upon a firm understanding of the beneficial and negative effects of a given technology. Good policy arguments in litigation depend upon this information as well. An attorney who desires to practice automated vehicle law needs to have a firm grasp of the effects of automation and the nuances that abound from these effects. Automated vehicles are not guaranteed to deliver on any of their perceived benefits. Understanding why they may not deliver on those benefits or the nuances of a perceived benefit, such as reducing congestion, can allow policymakers to develop better policy and those who advise the industry to serve clients' interests better.

The expected effects of a given automated feature depend upon its technological capabilities. When most scholars and the media speak of the benefits of automated vehicles, they assume level 4 or level 5 automated vehicles. However, automation is not going to only exist at levels 4 and 5—vehicles today utilize level 1 and level 2 driving automation systems.

I. REDUCED CRASHES

A primary purpose of the automobile when it was first introduced in the late 19th century was to increase safety.[1] As Frank Duryea, the coinventor of the

1. Jerry L. Mashaw & David L. Harfst, The Struggle for Auto Safety 28–29 (1990).

first gasoline-powered automobile in the United States, remarked: "The horse is a willful, unreliable brute. The ever recurring accidents due to horses which are daily set forth in the papers prove that the horse is a dangerous motor and not the docile pet of the poet. The mechanical motor is his superior in many respects, and when its superiority has become better known his inferiority will be more apparent."[2] Within decades of the introduction of the automobile, people became well accustomed to the new risks created by automobiles. The principal new risk was that human error replaced horse error. Indeed, "[w]ithin twenty-five years of the first known automobile fatality, motor vehicle mishaps became the nation's leading cause of accidental death."[3]

According to the National Highway Traffic Safety Administration (NHTSA), 36,560 people died on United States roadways in 2018—a 2.4 percent reduction from the 37,473 deaths in 2017.[4] NHTSA estimates that there were a total of 6,734,000 crashes in 2018.[5] The World Health Organization reports that there are 1.35 million road traffic fatalities worldwide each year, making traffic deaths the eighth leading cause of death for people of all ages and the leading cause of death for people aged 5–29.[6]

Human driver error is the overwhelming cause of automobile crashes. Between 2005 and 2007, NHTSA conducted a National Motor Vehicle Crash Causation Survey (Crash Survey).[7] The Crash Survey used 5,471 crashes that met specific criteria. The Crash Survey classified the critical precrash event—"the circumstances that led to [a] vehicle's first impact in the crash sequence"—into four categories: (1) driver, (2) vehicle, (3) roadway, and (4) atmospheric condition. The Crash Survey was only able to identify a critical precrash event in 5,361 of the examined crashes. The Crash Survey then estimated the total number of crashes in each category.

The Crash Survey was careful to state that the critical precrash event does not necessarily equate to the cause of the crash. The critical precrash event tells *how* but not necessarily *why* a crash occurred. For example, the critical precrash event could have been a slow reaction time, but the underlying "cause" of the crash may have been the driver's intoxication. The *why* in many of these critical precrash events, however, presupposes human error.

---

2. *Id.*
3. *Id.* at 43.
4. Nat'l Highway Traffic Safety Admin., *2018 Fatal Motor Vehicle Crashes: Overview*, U.S. Dep't of Transp. (Oct. 2019), https://crashstats.nhtsa.dot.gov/Api/Public/ViewPublication/812826.
5. Nat'l Highway Traffic Safety Admin., *Police-Reported Motor Vehicle Traffic Crashes in 2018*, U.S. Dep't of Transp. (Nov. 2019), https://crashstats.nhtsa.dot.gov/Api/Public/ViewPublication/812860.
6. World Health Org., Global Status Report on Road Safety 2018, *available at* https://apps.who.int/iris/bitstream/handle/10665/276462/9789241565684-eng.pdf?ua=1.
7. Nat'l Highway Traffic Safety Admin., *National Motor Vehicle Crash Causation Survey: Report to Congress*, U.S. Dep't of Transp. (July 2008), https://crashstats.nhtsa.dot.gov/Api/Public/ViewPublication/811059.

Drivers were the critical reason for an estimated 94 percent of crashes. NHTSA broke down the critical precrash events attributed to drivers into the following subcategories:[8]

Critical Precrash Event	Unweighted Number	Weighted Number
Recognition Error	2,094	828,308
Decision Error	1,752	695,516
Performance Error	510	210,143
Nonperformance Error (sleep, heartache, etc.)	369	145,844
Other	371	162,132
Total	5,096	2,041,943

Level 4 and level 5 automated vehicles should significantly reduce crashes because they have numerous advantages over human drivers. First, automated driving systems will not suffer from the main problems that underlie many human-driver crashes: distraction, intoxication, fatigue, or drowsiness. Second, automated driving systems are expected to have better situational awareness and reaction times than human drivers,[9] which may reduce the number and severity of crashes. Third, automated vehicle companies can update their other automated driving systems to fix similar errors in those systems.[10] Human drivers commit the same mistakes that others have committed, and some human drivers routinely repeat their own mistakes. These factors forecast a favorable impact of highly automated vehicles on the frequency and severity of automobile crashes.

But an expectation of safety benefits does not mean that society should blindly believe that *all* automated technologies in any given deployment—whether based on the manufacturer, the type of driver, or the level of automation—will provide these same safety benefits.

---

8. *Id.*

9. *See, e.g.*, David C. Vladeck, *Machines without Principals: Liability Rules and Artificial Intelligence*, 89 Wash. L. Rev. 117, 126 (2014) (noting that the Google car "'sees' everything in the vicinity; reacts at speeds humans cannot match; and constantly checks the performance of every component in the vehicle to ensure that it is functioning properly"); Damien A. Riehl, *Car Minus Driver, Part I Autonomous Vehicles Driving Regulation, Liability and Policy*, 73 J. Mo. B. 208, 210 (2017) ("Automated driving systems can also have super-human qualities: 360-degree vision; 100 percent alert time; constant communication with the road, traffic lights, and other cars; 'sight' through fog and darkness; and universal, system-wide routing for traffic flow optimization. Computers react faster: Humans' reaction time is approximately 1.5 seconds, while computers' reaction times are measured in milliseconds (and, per Moore's Law, improving exponentially).").

10. Bryant Walker Smith, *Automated Driving and Product Liability*, 2017 Mich. St. L. Rev. 1, 15–16 (2017).

Some levels of automation could possibly prove to be equally or more dangerous than human driving and are not safety features.[11] For example, level 1 and level 2 driving automation systems are generally convenience features and are not intended to provide direct safety benefits to consumers—other systems, such as advanced driver assistance systems, are the safety systems in these vehicles. It may be that certain deployment methods of level 2 automation (or at least a level 2 driving automation system that fails to address misuse) increase the frequency of crashes.[12] The Handoff Problem at level 3 may mean that a level 3 automated driving system that relies heavily on the dynamic driving task fallback-ready user may be less safe than human driving.[13]

Automated vehicles may also create new and different types of crashes, and therefore predicting a decrease in crashes based on the kinds of crashes caused today may not adequately reflect the crashes that may be caused tomorrow. Software bugs and failures may become a more significant cause of crashes. Crashes caused by malicious cyberattacks may increase in the context of automated vehicles. Different vehicle configurations could create more danger than existing vehicle configurations. Additionally, automated vehicles will likely increase vehicle miles traveled (VMT), which means that a safer automated vehicle may not positively impact roadway safety. By way of example, if automated vehicles are twice as safe as human drivers but lead to twice as many miles driven, then their safety benefits may be negligible at best.

Finally, the impact of automated vehicles on traffic fatalities may be marginal until they penetrate and have operational design domains in rural communities or low-income countries. A significant percentage of fatal crashes in the United States occur in rural communities, not in cities where people expect that highly automated vehicles will first become prominent. "In 2012, 19 percent of the US population lived in rural areas but rural road fatalities accounted for 54 percent of all fatalities."[14] Likewise, the World Health Organization reports that low-income countries have three times higher roadway fatality death rates than high-income countries.[15] Until highly automated vehicles penetrate rural markets and low-income countries, the number of vehicle-related deaths will remain significant.

---

11. *See* James M. Anderson et al., *Autonomous Vehicle Technology: A Guide for Policymakers*, RAND CORP. (2016), at 16, https://www.rand.org/pubs/research_reports/RR443-2.html (noting that "Level 2 and 3 automation . . . might increase some categories of crashes if consumers rely upon the systems too much"); HOD LIPSON & MELBA KURMAN, DRIVERLESS: INTELLIGENT CARS AND THE ROAD AHEAD 61 (2016) ("Clearly there's a tipping point at which autonomous driving technologies will actually create more danger for human drivers rather than less.").

12. See Chapter 9 for a discussion of misuse.

13. The Handoff Problem is discussed in Chapter 10.

14. Fed. Highway Admin., *Local and Rural Road Safety Program*, U.S. DEP'T OF TRANSP., https://safety.fhwa.dot.gov/local_rural/ (last visited Mar. 22, 2020).

15. WORLD HEALTH ORG., *supra* note 6.

II. HAPPINESS

Because most people do not enjoy their daily commutes (and other times when they have to drive), allowing people to engage in other activities while using highly automated vehicles will increase their happiness. Annually, drivers in the United States spend approximately 50 billion hours driving.[16] "Most of these driving hours are neither productive nor provide positive utility."[17] Such strides to increase people's happiness can only (legally and safely) begin to occur at levels 3, 4, and 5 when the human is no longer required to supervise the drive.

III. MOBILITY BENEFITS

A significant promise of automated vehicles is increased mobility for people who cannot currently drive, such as persons suffering from physical or legal disabilities.[18] Some manufacturers have teased this potential benefit by permitting disabled persons to test their automated vehicles. The promise of increased mobility is a strong reason for supporting the introduction and development of automated vehicles, but one should not falsely assume that automation guarantees this promise. First, these mobility benefits are only possible at level 4 and level 5, where humans are removed from the loop. Second, these mobility benefits may depend on private ownership of level 4 and level 5 automated vehicles. Third, even if manufacturers sell level 4 and level 5 automated vehicles, persons suffering from physical or legal disabilities may not be able to afford the technology.

IV. VEHICLE DOWNTIME

Currently, vehicles are parked approximately 95 percent of the time.[19] Our vehicles sit in parking spots throughout our workdays and while we sleep. Given how expensive automobiles are, vehicle downtime does not make much sense. Automated ridesharing vehicles could significantly reduce vehicle downtime. Some people may even forgo purchasing vehicles to avoid the substantial

16. Saptarshi Das et al., *Impacts of Autonomous Vehicles on Consumers' Time-Use Patterns*, 8 CHALLENGES 32 (2017).

17. *Id.*

18. John W. Zipp, *The Road Will Never Be the Same: A Reexamination of Tort Liability for Autonomous Vehicles*, 43 TRANSP. L.J. 137, 148 (2016).

19. *See, e.g.*, David Z. Morris, *Today's Cars Are Parked 95% of the Time*, FORTUNE (Mar. 13, 2016), https://fortune.com/2016/03/13/cars-parked-95-percent-of-time/.

downtime and use automated ridesharing vehicles. Alternatively, personally owned highly automated vehicles could be shared and rented when they otherwise would be idly parked.

V. VEHICLE MILES TRAVELED

In 2018, people traveled 3.21 trillion miles on American roadways.[20] Highly automated vehicles are expected to increase VMT due to several factors, the most significant of which is that the cost of driving will decrease for highly automated vehicles. Because users of highly automated vehicles can engage in other activities, the nonfinancial costs of "driving" will decrease. As vehicle designs shift from conventional to occupant-favorable, being in a vehicle may be no different than being in a traditional room of a home or a train car. Day trips, weekend trips, and long commutes may be more appealing with highly automated vehicles. Other factors that may cause VMT to increase include the potential lack of downtown parking as traditional parking lots and garages are repurposed and empty miles driven by automated ridesharing vehicles.

Automated ridesharing vehicles could also help minimize an increase in VMT in two ways. The first way is that the use of automated ridesharing vehicles may reflect the actual financial cost of driving.[21] Conventional vehicles have high fixed costs, such as their initial purchase price, annual registration fees, insurance, and variable costs, such as fuel costs or traditional wear and tear. Automated ridesharing vehicles will require companies to charge consumers a price per mile that reflects the actual financial cost of driving. The second way that automated ridesharing vehicles could help minimize an increase in VMT would be if people shared rides with other passengers.

VI. CONGESTION

The effects of vehicle automation on congestion are unclear, but automated vehicles will probably increase congestion. Traffic congestion can be either recurrent—"congestion that occurs in the same time and location on a daily basis"—or nonrecurrent—delays due to "isolated events or limited-duration circumstances."[22] Recurrent delays "are the result of prevailing travel patterns in which the number of vehicles trying to use a road at the same time exceeds

---

20. Alternative Fuels Data Ctr., *Annual Vehicle Miles Traveled in the United States*, U.S. Dep't of Energy, https://afdc.energy.gov/data/10315 (last visited Feb. 16, 2020).

21. Todd Litman, *Autonomous Vehicles Implementation Predictions: Implications for Transport Planning*, Victoria Transp. Pol'y Inst. 21 (Mar. 18, 2019).

22. Anderson et al., *supra* note 11, at 22–23.

the road's capacity."[23] Nonrecurrent delays "act to either reduce capacity or create a surge in demand."[24] Examples of nonrecurrent delays include traffic crashes, concerts, sporting events, weather, or construction.[25] Nonrecurrent delays account for approximately half of congestion.[26]

Four broad and somewhat conflicting effects on congestion are expected to occur due to the use of automated vehicles: (1) increased and more efficient throughput, (2) reduction in the number of crashes, (3) reduction in the delay of traffic surrounding crashes (rubbernecking), and (4) increased VMT.[27] Automated vehicles should increase throughput on roadways because they brake and accelerate more efficiently and can platoon.[28] Traffic crashes account for approximately a quarter of congestion.[29] The crash itself creates congestion, and then rubbernecking leads to further congestion. Automated vehicles should reduce crashes. As described earlier, VMT is expected to increase.

Regardless of the effects of highly automated vehicles on congestion, the costs of congestion for levels 3, 4, and 5 automated vehicles will decrease.[30] Traffic congestion involves "wasted time, excess fuel consumption, increased emission of local air pollutants and greenhouse gases, driver stress, diminished quality of life, and reduced economic efficiency."[31] Many of these costs are reduced by highly automated vehicles that allow people to engage in other activities instead of navigating the traffic jam.

---

23. *Id.*

24. *Id.* at 23.

25. *Id.*

26. Fed. Highway Admin., *Reducing Non-Recurring Congestion*, U.S. Dep't of Transp., https://ops.fhwa.dot.gov/program_areas/reduce-non-cong.htm (last visited July 7, 2018).

27. Anderson et al., *supra* note 11, at 17.

28. Raphael E. Stern et al., *Dissipation of Stop-and-Go Waves via Control of Autonomous Vehicles: Field Experiments*, 89 Transp. Research Part C: Emerging Techs. 205–21 (2018); *see* Sven A. Beiker, *Legal Aspects of Autonomous Driving*, 52 Santa Clara L. Rev. 1145 (2012) ("Autonomous driving technology can help to harmonize traffic flow by controlling individual automobiles more precisely through anticipation and inter-vehicle collaboration."); Anderson et al., *supra* note 11, at 21 (stating that automated vehicles could "support higher vehicle throughput rates on existing roads").

29. Anderson et al., *supra* note 11, at 23.

30. *Id.* at 24 ("In contrast to remaining uncertainty regarding the effects of [automated vehicles] on traffic congestion, the technology appears almost certain to offer major benefits in terms of reducing the costs associated with traffic congestion, particularly with respect to [higher levels of automation].").

31. *Id.* For those interested in traffic congestion statistics and the costs, the Texas A&M Transportation Institute annually publishes its *Urban Mobility Report*, which can be found at https://mobility.tamu.edu/.

VII. ENVIRONMENTAL BENEFITS

In 2016, vehicle transportation accounted for approximately 23.24 percent of the greenhouse gas pollution in the United States.[32] Automated vehicles may affect the environmental impact of motor vehicles and driving based on three factors: (1) fuel efficiency, (2) the power source of automated vehicles, and (3) increased VMT.[33] There should be fuel efficiency gains at all levels of automation. For low levels of automation, such as levels 1 and 2, fuel efficiency should improve by optimized driving through "smooth and gradual acceleration and deceleration."[34] At higher levels of automation, these effects will be more profound, and platooning could also result in higher fuel efficiency.[35] Also, safer driving may result in lighter vehicles, and lighter vehicles could lead to higher fuel efficiency.[36] Automated vehicles will likely use electricity, and not petroleum, as their power source.[37] A significant increase in VMT could reduce environmental benefits or result in a net negative impact, especially if automated vehicles use petroleum.[38] Depending upon the interplay of these factors, there may be significant, moderate, or minor environmental benefits. Alternatively, automated vehicles could prove to be environmentally

32. *Green Vehicle Guide: Fast Facts on Transportation Greenhouse Gas Emissions*, U.S. Envtl. Prot. Agency, https://www.epa.gov/greenvehicles/fast-facts-transportation-greenhouse-gas-emissions (last visited Apr. 30, 2020). The transportation sector was the largest producer of greenhouse gas emissions in the United States, accounting for approximately 28 percent of all such emissions. *Id.* The transportation sector "includes the movement of people and goods by cars, trucks, trains, ships, airplanes, and other vehicles." *Sources of Greenhouse Gas Emissions, Transportation Sector Emissions*, U.S. Envtl. Prot. Agency, https://www.epa.gov/ghgemissions/sources-greenhouse-gas-emissions#transportation (last visited Apr. 30, 2020). Light-duty vehicles represented 60 percent of the greenhouse gas emissions of the transportation sector, and medium- and heavy-duty trucks accounted for 23 percent of such emissions. *Green Vehicle Guide, supra.* The EPA noted that there has been an increase in greenhouse gas emissions from the transportation sector since 1990 due to a 45 percent increase in VMT.

33. Anderson et al., *supra* note 11, at 28.

34. *Id.* at 29.

35. *Id.* at 29–30.

36. *Id.* at 30–32; *see also* Lipson & Kurman, *supra* note 11, at 30 ("As accident rates drop significantly, driverless cars could be lighter and smaller, and therefore more fuel efficient.").

37. Anderson et al., *supra* note 11, at 32–36.

38. *See* Justin Worland, *Self-Driving Cars Could Help Save the Environment—or Ruin It. It Depends on Us*, Time (Sept. 8, 2016) ("The biggest potential downside of driverless cars for the environment is that automation could dramatically increase the total number of miles traveled by American vehicles."). One study commissioned by the National Laboratory of the United States Department of Energy noted that if only energy benefits are considered, automation could result in fuel savings of up to 90 percent, but if only energy increases of automation are considered, automation could result in upwards of a 250 percent increase in energy use. Austin Brown, Brittany Repac & Jeffrey Gonder, *Autonomous Vehicles Have a Wide Range of Possible Energy Impacts*, Nat'l Renewable Energy Lab. (July 16, 2013), https://www.nrel.gov/docs/fy13osti/59210.pdf.

neutral, or worse, result in minor, moderate, or significant harm to the environment compared to conventional driving.[39]

VIII. LAND USE

Automated vehicles, especially highly automated vehicles, will impact land use. Highly automated vehicles will lower the transaction costs of commuting by allowing occupants to engage in other activities during their commutes.[40] People may be more willing to live in the suburbs or rural areas, leading to urban or suburban sprawl. Driverless delivery bots and their access points may significantly change the location of stores.

In the United States, there are over one billion parking spots—these spots, spread horizontally, would cover an area bigger than the state of Connecticut.[41] Some cities dedicate one-third of their land to parking places.[42] Level 4 and level 5 automated vehicles will impact the need for parking in two broad ways. Level 4 and level 5 automated vehicles may lead to fewer vehicles, and people may abandon personal ownership of vehicles or reduce their ownership. Spouses may share one automated vehicle instead of owning two conventional vehicles. The reduction of vehicle ownership will mean that fewer parking spots are needed. The second significant effect of these highly automated vehicles is that they do not need parking located at their occupants' destinations. An automated vehicle could drop off its occupants and park at a remote location, enabling cities to reclaim precious real estate and obviating the need for centrally located parking.

IX. JOBS

Highly automated vehicles may "have enormous impacts on the workforce, some negative and positive, some subtle and surprising."[43] Highly automated

---

39. *See* Alex Davies, *So, Nobody Knows How Much Self-Driving Cars Will Pollute*, Wired (Nov. 18, 2016), https://www.wired.com/2016/11/nobody-knows-much-self-driving-cars-will-pollute/ (quoting Christopher Grundler, the head of the Office of Transportation and Air Quality at the EPA, as saying that "the technology could create 'utopia or dystopia for the environment'").

40. Anderson et al., *supra* note 11, at 25–26.

41. Eric Phillips, *The Future of Autonomous Vehicles in American Cities*, 21 N.Y.U. J. Legis. & Pub. Pol'y 287, 302 (2018).

42. Nick Bilton, *Disruptions: How Driverless Cars Could Reshape Cities*, N.Y. Times (July 7, 2013), https://bits.blogs.nytimes.com/2013/07/07/disruptions-how-driverless-cars-could-reshape-cities/.

43. Nidhi Kalra, *What Autonomous Vehicles Could Mean for American Workers*, Rand (Aug. 29, 2017), https://www.rand.org/blog/2017/08/what-autonomous-vehicles-could-mean-for-american-workers.html; Melissa L. Griffin, *Steering (or Not) through the Social and Legal Implications of Autonomous*

vehicles may significantly displace transportation workers. Excluding drivers for transportation network companies, approximately four million Americans are professional drivers, such as truckers, chauffeurs, or bus drivers.[44] Highly automated vehicles will likely displace these careers.[45] Highly automated vehicles may adversely affect the "crash economy"—insurance companies, health care, lawyers, and automobile repair shops.[46] The negative effects of job losses could be offset through retention by public or private entities or through skill development. These workforces will not disappear overnight, and people who may have entered into these industries may see the writing on the wall and resist making one of these industries their long-term career choice.[47] Although highly automated vehicles may impact these workforces, they will create new workforces.[48] New jobs may be created directly by the automated vehicle industry, such as software or engineering jobs or safety driver jobs, or indirectly by other companies that redeploy saved capital into new sectors.

X. GOVERNMENT REVENUE

Highly automated vehicles have the potential to significantly alter or eliminate sources of transportation revenues for governments. State and local governments have made budgetary decisions based on historic assumptions that people will need parking, commit traffic offenses, have driver's licenses, own vehicles, and use vehicles operated by internal combustion engines. Highly automated vehicles could undermine many of these historical assumptions. Highly automated vehicles may be less dependent upon centrally located parking, and automated ridesharing vehicles may not need any public parking. Highly automated vehicles should commit fewer traffic violations. People may not need driver's licenses to use highly automated vehicles. People may purchase fewer vehicles or stop purchasing them altogether. Automated vehicles probably will be electric vehicles.

State and local governments need to review their budgets to determine which portions may be at risk with highly automated vehicles or certain deployments of highly automated vehicles. Jurisdictions will need to consider new funding mechanisms, such as a VMT tax, a congestion tax, toll roads, or fees (such as a ridesharing fee).

Vehicles, 11 J. Bus. Entrepreneurship & L. 81, 101 (2018) ("Autonomous vehicles will both displace and create jobs; however, there is much debate about whether these numbers will balance out.").

44. Kalra, *supra* note 43.

45. *Id.*

46. *Id.*

47. *See* Griffin, *supra* note 43, at 103.

48. *Id.*

XI. PRIVACY

Automated technologies will adversely affect personal privacy of their users and owners.[49] "Driverless cars will have the technological capacity to collect a full run-down on both passengers' and pedestrians' identities and habits, including where they travel and with whom."[50] One prediction is that automated vehicles will generate approximately four terabytes of data every hour and a half of operation.[51] Outward-facing sensors will capture data on vehicle surroundings and bystanders. Inward-facing sensors will collect data on vehicle occupants. Vehicles will always know the location of their occupants.

All of these new data points will create new opportunities for commercial parties and government entities to obtain personal information and data about people and the world. Data privacy and cybersecurity are detailed in Part 4. Chapter 22 explores government access to automated vehicle data.

XII. FREEDOM

Highly automated vehicles may negatively impact freedom for many people, especially if human driving is directly (through government action) or indirectly (through insurance markets) outlawed. Driving has been an emblem of freedom for people—the freedom to go anywhere at any time.[52] If human driving is eventually banned, personal autonomy and freedom will be adversely affected. However, level 4 and level 5 automated vehicles may increase freedom for people who currently cannot drive, such as disabled persons or minors, by allowing them not to depend upon others for their transportation needs.[53]

---

49. *See, e.g.*, William J. Kohler & Alex Colbert-Taylor, *Current Law and Potential Legal Issues Pertaining to Automated, Autonomous and Connected Vehicles*, 31 Santa Clara High Tech. L.J. 99, 120 (2015) ("Far more profusely than today's vehicles, mature and market-ready autonomous vehicles will generate and broadcast personal data, the use and storage of which will implicate important privacy rights in complicated ways that will likely have to be faced well before Level 3 and Level 4 vehicles become a commercial reality."); Dorothy Glancy, *Privacy in Autonomous Vehicles*, 52 Santa Clara L. Rev. 1171, 1179–80 (2012) ("With regard to personal information privacy, [self-contained and interdependent] autonomous vehicles are likely to have highly detailed continuous data regarding vehicle location, as well as information about where the user wanted to go, did go, and what the user could have seen along the way.").

50. Lipson & Kurman, *supra* note 11, at 237.

51. Kathy Winter, *For Self-Driving Cars, There's Big Meaning behind One Big Number: 4 Terabytes*, Intel Newsroom (Apr. 14, 2017), https://newsroom.intel.com/editorials/self-driving-cars-big-meaning-behind-one-number-4-terabytes/#gs.pmoas9 (last visited Jan. 4, 2020).

52. This differs from other modes of transportation that are dependent upon people's schedules. Charles C.W. Cooke, *The War on Driving to Come*, Nat'l Rev. (Dec. 30, 2017), https://www.nationalreview.com/2017/12/autonomous-vehicles-will-spark-government-efforts-ban-driving/; *see also Human Driving Manifesto*, http://humandriving.org/manifesto (last visited Feb. 8, 2020).

53. *See, e.g.*, Glancy, *supra* note 49, at 1186 ("For some people, autonomous vehicles could enable more autonomy than they now have.").

Part 2

The Regulation of Automated Vehicles

INTRODUCTION

Motor vehicles and driving are heavily regulated. Through decades of regulation, the federal government, state governments, and local governments have defined their roles. The federal government's principal role with respect to noncommercial motor vehicles is the regulation of vehicle design through the promulgation of Federal Motor Vehicle Safety Standards. The federal government does not generally regulate noncommercial drivers. State governments, for their part, have typically regulated everything else and, in some respects, vehicle design. With the driving function moving from humans to onboard vehicle systems, the federal government's historic regulation of design will begin to cover driving capability as well.

Chapter 5 describes the current federal regulatory apparatus for the regulation of automated vehicles. As will be seen, automated vehicles are not unregulated by the federal government, even in the absence of a specific federal automated vehicle law. The National Highway Traffic Safety Administration broadly regulates the design of motor vehicles and motor vehicle equipment, and such authority extends to automated vehicles and their components. The Federal Motor Carrier Safety Administration has broad authority over commercial motor vehicles, including automated commercial motor vehicles. The

National Transportation Safety Board has significant latitude to investigate roadway crashes and has exercised that authority to investigate automated vehicle crashes.

States have more diverse laws regulating motor vehicles, driving, and roadways. Each state has, among other things, a motor vehicle code, requirements and regulations for drivers, insurance regimes, and franchise regulations. Chapter 6 canvasses state laws that could interplay with automated vehicles and that states should consider revising for highly automated vehicles.

Chapter 7 explores the automated vehicle guidance and laws. Congress has not enacted an automated vehicle law, and much of the federal guidance on automated vehicles has come in the form of policy statements issued by the United States Department of Transportation. Many states have enacted automated vehicle laws. The Uniform Law Commission has proposed a uniform state law on highly automated vehicles. Finally, Chapter 7 concludes by considering whether automated vehicles are legal in the absence of an automated vehicle law.

Chapter 5

The Federal Regulatory Apparatus

Given the role of motor vehicles in our economy and the effects that motor vehicles and our roadways have on interstate commerce, the federal government's involvement in regulating motor vehicles and our roadways is unsurprising. What is, perhaps, surprising is the limited nature of such involvement. For the most part, the federal government has taken an almost nonexistent role in regulating noncommercial driving, an isolated role in regulating noncommercial motor vehicles, and a broad, significant role in regulating commercial motor vehicles and driving. The federal government has allowed states and their subdivisions to regulate all other aspects of driving, motor vehicles, and roadways. This chapter explores the areas where the federal government has decided to regulate and reviews ancillary agencies and laws that might apply to automated vehicles. This chapter begins with the United States' international obligations under the Convention on Road Traffic, which was signed in Geneva in 1949 (the Geneva Convention).[1]

I. INTERNATIONAL OBLIGATIONS

The only treaty relating to motor vehicles and driving to which the United States is a signatory is a good place to start the exploration of the federal

1. Convention on Road Traffic, Geneva, Sept. 19, 1949, 125 U.N.T.S. 3, *available at* https://treaties.un.org/pages/ViewDetailsV.aspx?src=TREATY&mtdsg_no=XI-B-1&chapter=11&Temp=mtdsg5&clang=_en [hereinafter Geneva Convention]. Scholar Bryant Walker Smith conducted a very comprehensive analysis of the Geneva Convention and its applicability; Bryant Walker Smith, *Automated Vehicles Are Probably Legal in the United States*, 1 Tex. A&M L. Rev. 411, 424–57 (2014).

government's role in regulating automated vehicles. The Geneva Convention, as a treaty ratified by the United States, requires the United States to interpret and enforce it in good faith[2] and, domestically, has the effect of federal law, is enforceable in every state, and preempts inconsistent state laws.[3] There are 98 parties to the Geneva Convention. The United States is not a signatory to another prominent treaty on road traffic, the Convention on Road Traffic signed at Vienna in 1968 (the Vienna Convention); there are 78 parties to the Vienna Convention.[4]

The Geneva Convention was generally established to create uniform rules to further "the development and safety of international road traffic."[5] Among other things, the Geneva Convention provides basic rules for drivers and driving of motor vehicles and animal convoys, signage and signals, and rear license plates or other identifiers. Perhaps most importantly, the Geneva Convention provides that licensed drivers of one contracting state are permitted to drive within other contracting states.

The Geneva Convention gained prominence in the automated vehicle space when it was realized that Articles 8 and 10 might complicate or create potential barriers for highly automated vehicles.[6] Article 8 provides as follows:

> 1. Every vehicle or combination of vehicles proceeding as a unit shall have a driver.
>
> 2. Draught, pack or saddle animals shall have a driver, and cattle shall be accompanied, except in special areas which shall be marked at the points of entry.
>
> 3. Convoys of vehicles and animals shall have the number of drivers prescribed by domestic regulations.
>
> 4. Convoys shall, if necessary, be divided into sections of moderate length, and be sufficiently spaced out for the convenience of traffic. This provision does not apply to regions where migration of nomads occurs.
>
> 5. Drivers shall at all times be able to control their vehicles or guide their animals. When approaching other road users, they shall take such precautions as may be required for the safety of the latter.

---

2. Vienna Convention on the Law of Treatises art. 26, May 23, 1969, 1155 U.N.T.S. 331, 8 I.L.M. 679 ("Every treaty in force is binding upon the parties to it and must be performed by them in good faith.").

3. U.S. Const. art. VI.

4. Convention on Road Traffic, Vienna, Nov. 8, 1968, 1042 U.N.T.S. 17, *available at* https://treaties.un.org/Pages/ViewDetailsIII.aspx?src=TREATY&mtdsg_no=XI-B-19&chapter=11&Temp=mtdsg3&clang=_en [hereinafter Vienna Convention]. For signatories to the Vienna Convention and the Geneva Convention, the Vienna Convention superseded the Geneva Convention.

5. Geneva Convention, preamble.

6. *See* Smith, *supra* note 1.

Conceivably, Article 8 could prohibit driverless automated vehicles by requiring drivers who can control the vehicles. However, as Bryant Walker Smith identified, Article 8 uses the term "vehicle" instead of the defined term "motor vehicle."[7] According to Smith, the use of vehicles when coupled with the reference to animals and the history behind Article 8 suggests that Article 8 "was addressed to unsupervised animals (including animal-powered vehicles) rather than to unsupervised automobiles."[8] The Smith interpretation has since been called the "commonly accepted interpretation" of Article 8.[9]

Under Article 10, "[t]he driver of a vehicle shall at all times have its speed under control and shall drive in a reasonable and prudent manner." Other articles contain references to drivers. "Driver," in turn, is defined as "any person who drives a vehicle . . . or who is in actual physical control of the same."[10]

The broad definition of a driver makes it easier for the Geneva Convention to accommodate automated vehicles. As Smith acknowledged, use of the word "any" implies that "[a]n automated vehicle might therefore have multiple simultaneous drivers."[11] NHTSA and states have broadly interpreted the meaning of "driver" for highly automated vehicles to include more than just the traditional human driver—although NHTSA has since changed its position.[12]

Ideally, the contracting parties would amend the Geneva Convention to clarify its application to highly automated vehicles. Thus far, the contracting parties have been unable to garner sufficient support to amend the Geneva Convention. One difficulty is that, for an amendment to become effective, a certain percentage of the contracting parties need to approve the amendment within four months of transmission to the other contracting parties.[13] Another

---

7. Smith, *supra* note 1, at 425. "'Motor vehicle' means any self-propelled vehicle normally used for the transport of persons or goods upon a road, other than vehicles running on rails or connected to electric conductors. Any State bound by annex 1 shall exclude from this definition cycles fitted with an auxiliary engine of the type described in that annex." Geneva Convention, art. 4.

8. Smith, *supra* note 1, at 428.

9. Anders Eugennson et al., *Environmental, Safety, Legal and Societal Implications of Autonomous Driving Systems*, https://www-esv.nhtsa.dot.gov/Proceedings/23/files/23ESV-000467.PDF (last visited May 1, 2020).

10. Geneva Convention, art. 4.

11. Smith, *supra* note 1, at 434.

12. See Chapter 7 for a discussion of the state statutes. NHTSA had initially considered that the automated driving system could be a driver. Letter from Paul A. Hemmersbaugh, Chief Counsel, NHTSA, to Chris Urmson, Google (Feb. 4, 2016), *available at* https://isearch.nhtsa.gov/files/Google%20--%20compiled%20response%20to%2012%20Nov%20%2015%20interp%20request%20--%204%20Feb%2016%20final.htm. However, in a notice of proposed rulemaking on March 30, 2020, NHTSA declined that interpretation for the definition of "driver" in part 571. *See* Nat'l Highway Traffic Safety Admin., *Occupant Protection for Automated Driving Systems*, 85 Fed. Reg. 17,624 (Mar. 30, 2020).

13. Geneva Convention, art. 31.

challenge is that, for many of the contracting parties, the Geneva Convention is no longer binding because they have ratified the Vienna Convention.

The Vienna Convention also contains potentially complicating language for certain deployments of highly automated vehicles. Article 8 of the Vienna Convention includes requirements that a vehicle have a driver who "possess[es] the necessary physical and mental ability and be in a fit physical and mental condition to drive"; that a driver of a "power-driven vehicle" "possess the knowledge and skill necessary for driving the vehicle"; that a driver control the vehicle; and that distracted driving be prohibited. Article 13 contains requirements that could impede certain modes of automated driving:

> 1. Every driver of a vehicle shall in all circumstances have his vehicle under control so as to be able to exercise due and proper care and to be at all times in a position to perform all manoeuvres required of him. He shall, when adjusting the speed of his vehicle, pay constant regard to the circumstances, in particular the lie of the land, the state of the road, the condition and load of his vehicle, the weather conditions and the density of traffic, so as to be able to stop his range of forward vision and short of any foreseeable obstruction. . . .
>
> 3. The driver of a vehicle moving behind another vehicle shall keep at a sufficient distance from that other vehicle to avoid collision if the vehicle in front should suddenly slow down or stop.

The Vienna Convention with its more liberal amendment process—any amendment is effective unless more than one-third of the contracting parties reject the amendment or request a conference on the amendment[14]—has since been amended to provide that "[v]ehicle systems which influence the way vehicles are driven shall be deemed to be in conformity with paragraph 5 of [Article 8] and with paragraph 1 of Article 13, so long as the vehicles conform with United Nation vehicle regulations or a driver can override or turn off the vehicle system."[15]

II. FEDERAL REGULATION OF VEHICLES AND DRIVING

The federal government's treaty obligations concerning driving may be minimal. The federal government, for its part, has focused its regulatory attention on areas where there is a significant need for uniformity in the United States,

14. Vienna Convention, art. 49.

15. U.N. Econ. & Soc. Council, Inland Transp. Comm., Report of the Sixty-Eighth Session of the Working Party on Road Traffic Safety (Apr. 17, 2014), https://www.unece.org/fileadmin/DAM/trans/doc/2014/wp1/ECE-TRANS-WP1-145e.pdf. These amendments went into force on March 23, 2016.

such as commercial driving or motor vehicle design. This section focuses on Congress's enactments and the United States Department of Transportation's (USDOT) preeminence in motor vehicle design, highways, and commercial motor vehicles. This section concludes by discussing the National Transportation Safety Board (NTSB), an independent investigatory agency for civil transportation accidents.

In the mid-1960s, roadway fatalities and vehicle design became inextricably intertwined, and the need for federal regulations became apparent when Ralph Nader's book, *Unsafe at Any Speed*, was published. In 1965, the year of its publication, 47,089 people died on American roadways, tragically increasing to 50,894 people the following year.[16] In 1966, Congress enacted the National Traffic and Motor Vehicle Safety Act of 1966 (as amended, the Motor Vehicle Safety Act) and the Highway Safety Act of 1966. The Motor Vehicle Safety Act had a very simple purpose: the reduction of traffic crashes and injuries caused by traffic crashes. Congress initially empowered the Department of Commerce to enforce the Motor Vehicle Safety Act, but Congress changed course by creating the USDOT a month later.

The Motor Vehicle Safety Act created predecessor bureaus to the agencies responsible for regulating motor vehicle designs, highways, and commercial motor vehicles. The first of these, the National Highway Safety Bureau, soon became the National Highway Traffic Safety Administration (NHTSA). The Bureau of Public Roads eventually became the Federal Highway Administration (FHWA).[17] The last of these bureaus, the Bureau of Motor Carrier Safety, became the Federal Motor Carrier Safety Administration (FMCSA) in 1999. This section discusses these three regulatory agencies and their roles in automated vehicles.

A. NHTSA

NHTSA is the primary regulator of motor vehicle safety in the country. The Motor Vehicle Safety Act grants NHTSA specific powers to promote motor vehicle safety, including the power to (1) set Federal Motor Vehicle Safety Standards (FMVSS or Safety Standards) for motor vehicles and motor vehicle equipment; (2) investigate and recall motor vehicles and motor vehicle equipment that have safety defects or fail to comply with the FMVSS; (3) enforce fuel economy standards; and (4) manage and control the New Car Assessment Program.

---

16. Fed. Highway Admin., *Motor Vehicle Traffic Fatalities, 1900–2007*, U.S. Dep't of Transp., https://www.fhwa.dot.gov/policyinformation/statistics/2007/pdf/fi200.pdf (last visited Mar. 8, 2020).

17. FHWA is sort of a successor agency to a number of historical agencies that monitored roads and highways; these prior names were the Office of Road Inquiry, the Office of Public Roads, and then the Bureau of Public Roads.

Though NHTSA's statutory authority predates the advent of automated vehicles, NHTSA's authority is broad enough to include automated vehicles and their components.[18] NHTSA has regulatory power over "motor vehicles" and "motor vehicle equipment." "Motor vehicle" is defined broadly to mean "a vehicle driven or drawn by mechanical power and manufactured primarily for use on public streets, roads, and highways, but does not include a vehicle operated only on a rail line."[19] In comparison, all automated vehicles will be mechanically powered, and most will be operated on public roadways. An automated vehicle would be a motor vehicle for purposes of the Motor Vehicle Safety Act. Additionally, "motor vehicle equipment" is broadly defined as follows:

> "motor vehicle equipment" means—
>
> (A) any system, part, or component of a motor vehicle as originally manufactured;
>
> (B) any similar part or component manufactured or sold for replacement or improvement of a system, part, or component, or as an accessory or addition to a motor vehicle; or
>
> (C) any device or an article or apparel, including a motorcycle helmet and excluding medicine or eyeglasses prescribed by a licensed practitioner, that—
>
> (i) is not a system, part, or component of a motor vehicle; and
>
> (ii) is manufactured, sold, delivered, or offered to be sold for use on public streets, roads, and highways with the apparent purpose of safeguarding users of motor vehicles against risk of accident, injury, or death.[20]

For automated vehicles, (A) and (B) are the relevant provisions for determining the breadth of the definition. These two subsections capture three different categories of items of motor vehicle equipment: "(1) systems, parts, and components installed or included in a vehicle, (2) replacements and

---

18. Stephen P. Wood, Jesse Chang, Thomas Healy & John Wood, Symposium, *The Potential Regulatory Challenges of Increasingly Autonomous Motor Vehicles*, 52 Santa Clara L. Rev. 1423, 1438–41 (2012); *see also* Nat'l Highway Traffic Safety Admin., *NHTSA Enforcement Guidance Bulletin 2016-02: Safety-Related Defects and Automated Safety Technologies*, 81 Fed. Reg. 65,705-01 (Sept. 23, 2016) (describing NHTSA's position).

19. 49 U.S.C. § 30102(a)(7) (2016). Courts have considered whether vehicles that do not primarily use public streets, roads, and highways are considered "motor vehicles" under the statutory definition. *See* Koehring Co. v. Adams, 452 F. Supp. 635, 638 (E.D. Wis. 1978), *aff'd*, 605 F.2d 280 (7th Cir. 1979) (declining to find that mobile cranes, mobile excavators, and mobile well drills that operate on public roads between jobs were "motor vehicles" because their highway use was sporadic and incidental); Util. Equip., Inc. v. Morbark Indus., Inc., 308 Or. 209, 779 P.2d 139, 143–44 (1989) (holding that a mobile wood chipper is not a motor vehicle under the Motor Vehicle Safety Act).

20. 49 U.S.C. § 30102(a)(8).

improvements to those systems, parts, and components, and (3) accessories and additions to motor vehicles."[21] For vehicles developed initially as automated vehicles, all of the sensors used to perceive the world and the driving automation system are included in the first category of items.[22] The second category of items captures aftermarket devices, including, but not limited to, automated retrofits on existing conventional vehicles or replacement sensors or systems.[23] NHTSA also believes this category includes updates to the driving automation system because they are improvements.[24] The third category of items includes two similar but distinct types of items. Additions are items connected to the vehicle that do not merely replace a system. An accessory is "'motor vehicle equipment' if a substantial portion of its expected use were in conjunction with motor vehicles."[25]

NHTSA has expressed its belief that automated technologies are motor vehicle equipment:

> With respect to current and emerging automated motor vehicle safety technologies, NHTSA considers such technologies (including systems and equipment) to be motor vehicle equipment, whether they are offered to the public as part of a new motor vehicle (as original equipment) or as an after-market replacement(s) of or improvement(s) to original equipment. NHTSA also considers software (including, but not necessarily limited to, the programs, instructions, code, and data used to operate computers and related devices), and after-market software updates, to be motor vehicle equipment within the meaning of the Safety Act. Software that enables devices not located in or on the motor vehicle to connect to the motor vehicle or its systems could, in some circumstances, also be considered motor vehicle equipment.[26]

This section focuses on how NHTSA can exercise its statutory authority to regulate automated vehicles. NHTSA has already exercised its authority in issuing letters requesting information from companies, seeking to prevent

---

21. Wood et al., *supra* note 18, at 1441.

22. *Id.* at 1441–45. NHTSA has broadly interpreted this definition to include vehicle communication systems like General Motors's OnStar. *Id.; see* Letter from Anthony M. Cooke, Chief Counsel, NHTSA, to Ashley G. Alley, Off. of General Counsel, Gov't Accountability Off. (July 19, 2007).

23. Wood et al., *supra* note 18, at 1442–43.

24. *Id.* at 1443.

25. *Id.* at 1446.

26. Nat'l Highway Traffic Safety Admin., *NHTSA Enforcement Guidance Bulletin 2016-02: Safety-Related Defects and Automated Safety Technologies*, 81 Fed. Reg. 65,705-01, 65,707 (Sept. 23, 2016).

"unlawful" operation of automated shuttles,[27] investigating crashes,[28] and causing EasyMile to cease its automated shuttle operations in ten states following an incident.[29]

1. FMVSS

a. Overview of the FMVSS

The Motor Vehicle Safety Act requires NHTSA to set Safety Standards for new motor vehicles and motor vehicle equipment.[30] Safety Standards generally establish a performance standard and leave the design choices to attain that performance standard to the industry—though, at times, Safety Standards have mandated certain design choices.[31] Safety Standards can also require the use of new technology and the improvement of existing technology.[32] Each Safety Standard is required to meet three basic requirements: the Safety Standard must be "practicable, meet the need for motor vehicle safety, and be stated in objective terms."[33] These basic requirements are substantive (in addition to

---

27. Ryan Felton, *Feds Order Company to Stop Shuttling Florida Kids to School in Autonomous Bus*, JALOPNIK (Oct. 22, 2018), https://jalopnik.com/feds-order-company-to-stop-shuttling-florida-kids-to-sc-1829906624?rev=1540215965818&utm_source=jalopnik_twitter&utm_campaign=socialflow_jalopnik_twitter&utm_medium=socialflow.

28. *US Safety Agency Opens Probe into Fatal Tesla Model 3 Crash in Indiana*, CNBC (Jan. 8, 2020), https://www.cnbc.com/2020/01/08/us-safety-agency-opens-probe-into-fatal-tesla-crash-in-indiana.html (noting that NHTSA's special crash investigation team has reviewed 23 crashes involving vehicles using advanced driver features).

29. Aarian Marshall, *The Feds Ban a Self-Driving Shuttle Fleet from Carrying People*, WIRED (Feb. 29, 2020), https://www.wired.com/story/feds-ban-self-driving-shuttle-fleet-carrying-people/.

30. 49 U.S.C. § 30111(a).

31. *See, e.g.*, Chrysler Corp. v. Dep't of Transp., 472 F.2d 659, 664 n.4 (6th Cir. 1972) (stating that NHTSA is not precluded from developing Safety Standards that, by their implication, mandate a specific design); Washington v. Dep't of Transp., 84 F.3d 1222, 1224 (10th Cir. 1996) (noting that the line between performance and design can be difficult to apply, especially when the "specification of [performance] effectively entails, or severely constrains, the choice of [design]").

32. *See, e.g.*, *Chrysler Corp.*, 472 F.2d at 671–73 (describing the legislative history and stating that the Motor Vehicle Safety Act allows NHTSA to require new technology and the improvement of new technology so long as the substantive requirements of objectivity, practicability, and need for motor vehicle safety are satisfied); Pac. Legal Found. v. Dep't of Transp., 593 F.2d 1338, 1344 (D.C. Cir. 1979) (citing *Chrysler Corp.* for the proposition that the Motor Vehicle Safety Act "authorizes safety standards that push the automobile industry beyond present engineering capabilities. . . ."); Motor Vehicle Mfrs. Ass'n of U.S., Inc. v. State Farm Mut. Auto. Ins. Co., 463 U.S. 29, 49, 103 S. Ct. 2856, 77 L. Ed. 2d 443 (1983) (citing *Chrysler* to state that the Motor Vehicle Safety Act "intended that safety standards not depend on current technology and could be 'technology-forcing' in the sense of inducing the development of superior safety design").

33. 49 U.S.C. § 30111(a). In addition to the three core requirements, the Motor Vehicle Safety Act requires NHTSA to "(1) consider relevant available motor vehicle safety information"; (2) consult with other agencies and state or interstate authorities; (3) "consider whether a proposed standard is reasonable, practicable, and appropriate for the particular type of motor vehicle or motor vehicle equipment for which it is prescribed;" and (4) consider whether the Safety Standard furthers the purposes and policies of the Motor Vehicle Safety Act. 49 U.S.C. § 30111(b).

the procedural requirements imposed by the Administrative Procedures Act[34]) and have provided litigants with opportunities to challenge Safety Standards.[35]

Courts have established three bases by which a Safety Standard must be practicable: technologically practicable, economically practicable, or practicable based on public acceptance and use.[36] The two most prominent means by which litigants have challenged a Safety Standard based on practicability are whether the Safety Standard is technologically or economically practicable. A Safety Standard is technologically practicable if it "offer[s] the regulated party a chance to demonstrate compliance."[37] Economic impracticability is difficult to show. One court has noted that practicability may limit NHTSA's ability to "impose standards so imperative as to put a manufacturer out of business," but NHTSA can otherwise require manufacturers to divert more resources than they desire for automobile safety.[38] The D.C. Circuit Court of Appeals has stated that considering public reaction and response to a Safety Standard is part of NHTSA's obligation to determine that a Safety Standard is practicable.[39]

The second requirement is that the Safety Standard must meet the need for motor vehicle safety. "Motor vehicle safety" is "the performance of a motor vehicle or motor vehicle equipment in a way that protects the public against unreasonable risk of accidents occurring because of the design, construction, or performance of a motor vehicle, and against unreasonable risk of death or injury in an accident, and includes nonoperational safety of a motor vehicle."[40] A Safety Standard does not meet the need for motor vehicle safety if there is only a "remote relation between the requirements and motor vehicle safety."[41] Simply stated, "there must be a nexus between the safety problem and the stan-

34. Safety Standards are promulgated using the informal rulemaking procedures of the Administrative Procedures Act. *See, e.g., Motor Vehicle Mfrs. Ass'n of U.S., Inc.*, 463 U.S. at 41; Auto. Parts & Accessories Ass'n v. Boyd, 407 F.2d 330, 335–37 (D.C. Cir. 1968). Effectively, this means that the procedural review of these standards is based on whether the agency's action is "arbitrary, capricious, an abuse of discretion, or otherwise not in accordance with law." *Motor Vehicle Mfrs. Ass'n of U.S., Inc.*, 463 U.S. at 41 (quoting 5 U.S.C. § 706(2)(A)).

35. *See, e.g.*, Nat'l Truck Equip. Ass'n v. Nat'l Highway Traffic Safety Admin., 711 F.3d 662, 670–71 (6th Cir. 2013); *see also Chrysler Corp.*, 472 F.2d at 671 (stating that the applicable standard "must meet all statutorily prescribed criteria [in the Motor Vehicle Safety Act], and the Agency must have complied with all applicable procedural requirements [in the Administrative Procedures Act]").

36. Wood et al., *supra* note 18, at 1454–57; *see also* Paccar, Inc. v. Nat'l Highway Traffic Safety Admin., 573 F.2d 632, 635 n.5 (9th Cir. 1978) ("'Practicable' is defined to 'require consideration of all relevant factors, including technological ability to achieve the goal of a particular standard as well as consideration of economic factors.'" (quoting H.R. Rep. No. 1776, 89th Cong., 2d Sess., at 16 (1966))).

37. Nat'l Truck Equip. Ass'n v. Nat'l Highway Traffic Safety Admin., 919 F.2d 1148, 1153 (6th Cir. 1990); *see also Paccar, Inc.*, 573 F.2d at 644 (finding that NHTSA's testing procedure makes it such that the manufacturers cannot determine compliance with the Safety Standard).

38. *Chrysler Corp.*, 472 F.2d at 672–73.

39. *Pac. Legal Found.*, 593 F.2d at 1345–46.

40. 49 U.S.C. § 30102(a)(9).

41. Nat'l Tire Dealers & Retreaders Ass'n, Inc. v. Brinegar, 491 F.2d 31, 33 (D.C. Cir. 1974).

dard."[42] In examining this prong, courts have considered the purpose and evidence considered by NHTSA in adopting the Safety Standard.[43]

Objectivity requires objectiveness in testing and in the performance standard. The need for objectivity is significant because manufacturers are required to self-certify that their vehicles or equipment satisfies the Safety Standards, and such self-certification becomes nearly impossible when the tests are subjective.[44] Courts have generally found three components to objectivity.[45] First, the testing method in a Safety Standard "must be capable of producing identical results when test conditions are exactly duplicated."[46] Second, the identical results must "be decisively demonstrable by performing a rational test procedure."[47] Third, compliance with the Safety Standard must be "based upon the readings obtained from measuring instruments as opposed to the subjective opinions of human beings."[48]

A person seeking to challenge a Safety Standard needs to file an action in the United States Circuit Court of Appeals where its principal place of business is located within 59 days of the issuance of the order.[49] The Ninth Circuit has held that a Safety Standard is "'issued' on the date that the regulation is made available for public inspection."[50] A Safety Standard that is not challenged within that 59-day period becomes effective. NHTSA is authorized to specify the effective date of a Safety Standard.[51]

The FMVSS are currently published in the Code of Federal Regulations in 49 C.F.R. § 571 with the corresponding Safety Standard's number following. For example, Safety Standard 101 on Controls and Displays is published at 49 C.F.R. § 571.101. NHTSA categorizes the Safety Standards into three categories: (1) the 100-series are crash avoidance Safety Standards, (2) the 200-series are crashworthiness Safety Standards, and (3) the 300-series are postcrash survivability Safety Standards. Several Safety Standards are vehicle-type dependent, differentiating between, for example, passenger vehicles, multipurpose passenger

42. Wood et al., *supra* note 18, at 1451.

43. *Nat'l Truck Equip. Ass'n*, 711 F.3d at 674; Vehicle Equip. Safety Comm'n v. Nat'l Highway Traffic Safety Admin., 611 F.2d 53, 54 (4th Cir. 1979); *Nat'l Tire Dealers & Retreaders Ass'n*, 491 F.2d at 35–37; H & H Tire Co. v. U.S. Dep't of Transp., 471 F.2d 350, 354–55 (7th Cir. 1972).

44. *Chrysler Corp.*, 472 F.2d at 675.

45. *Id.* at 676; *Paccar, Inc.*, 573 F.2d at 644 (citing the three factors from *Chrysler Corp.*).

46. *Chrysler Corp.*, 472 F.2d at 676.

47. *Id.* This component of objectivity does not require perfection of the testing or measuring device. B.F. Goodrich Co. v. Dep't of Transp., 541 F.2d 1178, 1188 (6th Cir. 1976).

48. *Chrysler Corp.*, 472 F.2d at 676; *see also Paccar, Inc.*, 573 F.2d at 644.

49. 49 U.S.C. § 30161; *see also* Contini by Contini v. Hyundai Motor Co., 876 F. Supp. 540, 546 (S.D.N.Y. 1995) (discussing how a challenge to a Safety Standard was improperly raised to the district court).

50. Pub. Citizen Inc. v. Mineta, 343 F.3d 1159, 1167 (9th Cir. 2003).

51. 49 U.S.C. § 30111(d).

vehicles, low-speed vehicles, trucks, trailers, buses, school buses, and motorcycles. There are currently 73 Safety Standards in effect.

Unless a general, specific, or other exemption[52] applies, manufacturers are prohibited from selling, offering to sell, or introducing or deploying any motor vehicle or item of motor vehicle equipment that does not comply with the FMVSS.[53] A manufacturer or distributor of motor vehicle equipment is required to certify at delivery that the product complies with the FMVSS.[54] The certification must be affixed to the vehicle or the equipment (or on the package).[55] If the product does not comply with the FMVSS, then the Motor Vehicle Safety Act requires the manufacturer to recall the product.[56]

The Motor Vehicle Safety Act empowers NHTSA to grant temporary exemptions from complying with a Safety Standard.[57] A manufacturer seeking a general exemption must demonstrate to NHTSA that it satisfies one of the following criteria for a general exemption:

> (A) an exemption is consistent with the public interest and this chapter or chapter 325 of this title (as applicable); and
>
> (B)(i) compliance with the standard would cause substantial economic hardship to a manufacturer that has tried to comply with the standard in good faith;
>
> (ii) the exemption would make easier the development or field evaluation of a new motor vehicle safety feature providing a safety level at least equal to the safety level of the standard;
>
> (iii) the exemption would make the development or field evaluation of a low-emission motor vehicle easier and would not unreasonably lower the safety level of that vehicle; or
>
> (iv) compliance with the standard would prevent the manufacturer from selling a motor vehicle with an overall safety level at least equal to the overall safety level of nonexempt vehicles.[58]

For a (B)(i) exemption, the manufacturer cannot produce more than 10,000 vehicles per year, and the exemption cannot exceed three years.[59] For the other exemptions, the manufacturer cannot sell more than 2,500 vehicles per year,

---

52. The other exemptions include those listed in 49 U.S.C. § 30161(b).
53. *Id.* § 30112(a).
54. *Id.* § 30115.
55. *Id.*
56. *Id.* § 30116.
57. *Id.* § 30113(b)(1).
58. *Id.* § 30113(b)(3).
59. *Id.* § 30113(d), (e).

and the exemption cannot exceed two years.[60] The exemption can be renewed on reapplication by the manufacturer.[61]

The Motor Vehicle Safety Act also grants NHTSA authority to exempt a specific vehicle or item of motor vehicle equipment, if NHTSA determines that such exemption is "necessary for research, investigations, demonstrations, training, competitive racing events, show, or display."[62] Manufacturers that have historically manufactured and distributed compliant motor vehicles can introduce noncompliant vehicles without the need to obtain an exemption "for purposes of testing or evaluation by a manufacturer that agrees not to sell or offer for sale the motor vehicle at the conclusion of the testing or evaluation."[63] Given the extent to which automated vehicles are being used in demonstrations, research, training, and testing, these exemptions or exclusions are useful to traditional automakers.

International manufacturers that design their vehicles for their own countries may not have incorporated the FMVSS in their designs. Section 30141 empowers NHTSA to allow for the sale of noncompliant vehicles that can be made to comply after making certain findings about the vehicle after public notice and comment.

People can submit a request for interpretation to NHTSA's Chief Counsel asking for its opinion on the application of the FMVSS to a specific motor vehicle or item of motor vehicle equipment. NHTSA publishes these interpretations, which can provide helpful guidance when advising clients. NHTSA maintains its database of these interpretations at https://isearch.nhtsa.gov/.

b. Application of the FMVSS to Automated Vehicles

None of these Safety Standards explicitly references automated vehicles or driving automation systems. NHTSA commissioned the John A. Volpe National Transportation Systems Center (the Volpe Center) to review the Safety Standards to determine which Safety Standards may create barriers or complications for automated vehicles.[64] The Volpe Center found that the existing Safety Standards create ambiguities and obstacles for unique designs. NHTSA has stated the following about the Safety Standards: "it is a manufacturer's design of a motor vehicle without manual driving controls, design of a motor vehicle with novel seating configurations or orientations, or a covered party's disabling

60. *Id.*

61. *Id.* § 30113(b)(1).

62. *Id.* § 30114(a).

63. *Id.* § 30112(b)(10).

64. *See* Anita Kim, David Perlman, Dan Bogard & Ryan Harrington, *Review of Federal Motor Vehicle Safety Standards (FMVSS) for Automated Vehicles: Identifying Potential Barriers and Challenges for the Certification of Automated Vehicles Using Existing FMVSS,* U.S. Dep't of Transp. John A. Volpe Nat'l Transp. Sys. Ctr. (Preliminary Report, Mar. 2016) [hereinafter Volpe Report].

of any part of a device or element of design of a motor vehicle or motor vehicle equipment that is currently in compliance with applicable FMVSS, that could complicate the compliance of the vehicle to the existing FMVSS—not solely the inclusion of the hardware and software that make up an [automated driving system]."[65] Many of the current Safety Standards operate under a fundamental assumption: each vehicle has a human driver that is necessary for the safe operation of the vehicle and the driver "will be human, will sit in the vehicle's left front seat to drive . . . and will need certain controls to be accessible and telltales and other displays to be viewable in order to do the driving."[66]

(i) The Driver and Operator Safety Standards

The FMVSS refer to drivers in several different ways:[67] references to the visibility of the driver,[68] the operability by a driver,[69] the identification of control use by reference to a driver,[70] the driver's physical state or anatomy,[71] the driver's

---

65. Nat'l Highway Traffic Safety Admin., *Removing Regulatory Barriers for Vehicles with Automated Driving Systems*, 84 Fed. Reg. 24,433, 24,435–36 (May 28, 2019).

66. Nat'l Highway Traffic Safety Admin., *Removing Regulatory Barriers for Vehicles with Automated Driving Systems*, 83 Fed. Reg. 2607, 2608–09 (Jan. 18, 2018).

67. The list that follows and the examples used in the footnotes are taken from the Volpe Report.

68. For example, Safety Standard 101 provides in pertinent part:

> The telltales and indicators listed in Table 1 and Table 2 and their identification must be located so that, when activated, they are visible to a driver under the conditions of S5.6.1 and S5.6.2.

49 C.F.R. § 571.101 S5.1.2.

69. For example, Safety Standard 101 provides in pertinent part:

> The controls listed in Table 1 and in Table 2 must be located so they are operable by the driver under the conditions of S5.6.2.

Id. § 571.101 S5.1.1.

70. For example, Safety Standard 135 provides in pertinent part:

> The service brakes shall be activated by means of a foot control. The control of the parking brake shall be independent of the service brake control, and may be either a hand or foot control.

Id. § 571.135 S5.3.1.

71. For example, Safety Standard 101 provides in pertinent part:

> S5.6 Conditions [under which displays must be visible] S5.6.1 The driver has adapted to the ambient light roadway conditions. S5.6.2 The driver is restrained by the seat belts installed in accordance with 49 CFR 571.208 and adjusted in accordance with the vehicle manufacturer's instructions.

Id. § 571.101 S5.6.

location in the vehicle,[72] and the capabilities of a driver.[73] Other Safety Standards use the word "operator" instead of "driver."[74] According to the Volpe Report, 33 Safety Standards reference the driver.[75]

The FMVSS define a driver as "the occupant of a motor vehicle seated immediately behind the steering control system."[76] This definition, at least for certain designs of level 4 and level 5 automated vehicles, could be nearly impossible for a human occupant to satisfy. In response to a letter from Google about its Firefly design, NHTSA stated that in situations where no human occupant can drive an automated vehicle, "it is more reasonable to identify the 'driver' as whatever (as opposed to whoever) *is* doing the driving."[77] NHTSA stated that "an item of motor vehicle equipment, the [automated driving system], is actually driving the vehicle." NHTSA also confirmed that the term "operator" as used in the FMVSS refers to the automated driving system.[78] However, NHTSA did not allow that determination to mean that the automated driving system could automatically satisfy Safety Standards that reference the driver or operator.

In one example, NHTSA declined to state that the automated driving system satisfies Safety Standard 135 because the explicit language of that Safety Standard would prohibit that interpretation. Safety Standard 135 provides, in pertinent part, that "[t]he service brakes shall be activated by means of a foot control. The control of the parking brake shall be independent of the service

---

72. For example, Safety Standard 109 provides in pertinent part:

> Each vehicle, except for a trailer or incomplete vehicle, shall show the information specified in S4.3 (a) through (g), and may show, at the manufacturer's option, the information specified in S4.3 (h) and (i), on a placard permanently affixed to the driver's side B-pillar.

Id. § 571.109 S4.3.

73. For example, Safety Standard 105 provides in pertinent part:

> The control force used for the baseline check stops or snubs shall be not less than 10 pounds, nor more than 60 pounds.

Id. § 571.105 S5.1.4.3(a).

74. For example, Safety Standard 108 defines "turn signal operating unit" to mean "an operating unit that is part of a turn signal system by which the operator of a vehicle causes the signal units to function." *Id.* § 571.108.

75. Volpe Report, *supra* note 64, at 12.

76. 49 C.F.R. § 571.3.

77. Letter from Paul A. Hemmersbaugh, Chief Counsel, NHTSA, to Chris Urmson, Google (Feb. 4, 2016), *available at* https://isearch.nhtsa.gov/files/Google%20--%20compiled%20response%20to%2012%20Nov%20%2015%20interp%20request%20--%204%20Feb%2016%20final.htm ("No human occupant of the [self-driving vehicle] could meet the definition of 'driver' in Section 571.3 given Google's described motor vehicle design—even if it were possible for a human occupant to determine the location of Google's steering control system, and sit 'immediately behind' it, that human occupant would not be capable of actually driving the vehicle as described by Google.")

78. *Id.*; *see also* Volpe Report, *supra* note 64, at 8 (stating that "[t]he terms 'driver' and 'operator' are used interchangeably" in the FMVSS).

brake control, and may be either a hand or foot control."[79] NHTSA acknowledged that the automated driving system could perform these functions but stated that the plain language is clear: this Safety Standard requires hand or foot control to activate the brakes. Until NHTSA commences rulemaking to consider a new means to satisfy the Safety Standard, Google would need an exemption from this Safety Standard.

In other examples where the language was not as explicit as in Safety Standard 135, NHTSA declined to issue interpretations that an automated driving system complies with Safety Standards until NHTSA developed a standard and testing procedure through rulemaking. Thus, although NHTSA has indicated a willingness to interpret an automated driving system as a driver or operator, this interpretation did not alleviate the need to seek exemptions for automated vehicles.

NHTSA has since decided not to interpret the term "driver" to include an automated driving system.[80] In its notice of proposed rulemaking in March 2020, NHTSA interpreted the term "driver" as defined in part 571.3 to mean "the occupant of a motor vehicle seated immediately behind the steering control system." NHTSA declined to reconsider the definition of "driver" or to adopt an interpretation that included an automated driving system, but NHTSA did seek input on how to modify the definition of "driver" and how to address nonhuman drivers.

(ii) The Vehicle Design Safety Standards

Based on the human driver assumption, many Safety Standards require certain designs of motor vehicles—some that explicitly reference a driver or operator, such as Safety Standard 135 described above, and others that do not.[81] These types of Safety Standards may prohibit certain automated vehicle designs. According to the Volpe Report, 32 Safety Standards "contain performance specifications, test procedures, or equipment requirements that present potential barriers to the certification of one or more AV concepts."[82]

Certain of the Safety Standards can create problems for automation features, regardless of there being significant high levels of automation, unique interior designs, or the removal of certain human driving mechanisms. NHTSA has issued an interpretation of whether BMW's Park Assistant System complies

---

79. 49 U.S.C. § 571.135 S5.3.1.

80. *See* Nat'l Highway Traffic Safety Admin., *Occupant Protection for Automated Driving Systems*, 85 Fed. Reg. 17,624 (Mar. 30, 2020).

81. The use of "design" may not be the correct nuance but is used to simplify the analysis. In reality, the Safety Standards, with certain exceptions, assume that vehicles have certain features and prescribe performance standards and not designs. It is the assumptions about certain features and the performance standards that effectively "prohibit" a vehicle without certain features.

82. Volpe Report, *supra* note 64, at 12.

with the brake transmission shift interlock requirement in Standard 114.[83] The system is a remote-controlled parking system that is activated by a driver after exiting and turning off the vehicle. The driver activates the system by holding a button on the key fob during the entire parking maneuver. Safety Standard 114 provides, in pertinent part, the following:

> S5.3 Brake transmission shift interlock. Each motor vehicle manufactured on or after September 1, 2010 with a GVWR of 4,536 kilograms (10,000 pounds) or less with an automatic transmission that includes a "park" position shall be equipped with a system that requires the service brake to be depressed before the transmission can be shifted out of "park." This system shall function in any starting system key position in which the transmission can be shifted out of "park."

BMW sought an interpretation that brake depression by the electronic stability control satisfies this part of Safety Standard 114. NHTSA examined the underlying goal of this part—prevention of occupants, especially children, from shifting the transmission out of park—and the definition of "depress." NHTSA found that the definition of "depress" did not require a human to depress the brakes and that the electronic stability control's depression furthers the goals of the Safety Standard. NHTSA ruled favorably that the system satisfies this part of Safety Standard 114.

(iii) Testing Methods

NHTSA has identified and sought comments on certain testing methods in the Safety Standards that may be difficult to perform depending upon the configuration of an automated vehicle.[84] For example, NHTSA identified that current testing mechanisms might not be capable of being performed with automated vehicles that lack human controls or have unique seating arrangements.[85] These problems create difficulties for manufacturers in determining whether they can self-certify that their vehicles or items of motor vehicle equipment comply with the Safety Standards and for NHTSA in testing compliance with the Safety Standards.

---

83. Letter from Paul A. Hemmersbaugh, Chief Counsel to BMW of N. Am., LLC (Jan. 4, 2016), *available at* https://isearch.nhtsa.gov/files/15-005347%20BMW%20Brake%20Transmission%20Shift%20Interlock%20v5.htm.

84. Nat'l Highway Traffic Safety Admin., *Removing Regulatory Barriers for Vehicles with Automated Driving Systems*, 83 Fed. Reg. 2607-01 (Jan. 18, 2018); Nat'l Highway Traffic Safety Admin., *Removing Regulatory Barriers for Vehicles with Automated Driving Systems*, 84 Fed. Reg. 24,433-01 (May 28, 2019).

85. Nat'l Highway Traffic Safety Admin., *Removing Regulatory Barriers for Vehicles with Automated Driving Systems*, 83 Fed. Reg. 2607-01, 2609 (Jan. 18, 2018).

c. NHTSA's Actions to Revise the FMVSS

NHTSA understands that certain design decisions and deployment methods of automated vehicles will be unable to satisfy current Safety Standards.[86] NHTSA sought public comment in January 2018 on revising and amending the Safety Standards to accommodate the removal of human controls and unique configurations within automated vehicles.[87] On May 28, 2019, NHTSA issued an advance notice of proposed rulemaking on revising and amending certain of these Safety Standards to address automated vehicles without human-driving controls.[88] On March 16, 2020, NHTSA issued a notice of proposed rulemaking to modernize certain crashworthiness (200-series) Safety Standards "to remove unnecessary barriers to Automated Driving System-equipped vehicles (ADS-equipped vehicles) and the unconventional interior designs that are expected to accompany these vehicles, including the lack of driving controls."[89] If the proposed rulemaking is ultimately adopted, many barriers and unnecessary requirements for highly automated vehicles may be removed.

2. Limitations of Authority under FMVSS

Though NHTSA has broad authority to promulgate Safety Standards, there are numerous limitations on NHTSA's authority that may be relevant to automated vehicles and driving automation systems. In these circumstances, regulation, if any, will need to come from states or their subdivisions. NHTSA does not regulate driver competency or training. The Motor Vehicle Safety Act has limited applicability to repairs and alterations of vehicles by prohibiting manufacturers, distributors, dealers, rental companies, or motor vehicle repair businesses from making certain devices inoperative.[90] The Motor Vehicle Safety Act does not prevent vehicle owners or their friends[91] from physically or digitally (such as through jailbreaking) altering automated vehicles. In addition, NHTSA

86. *See generally* Nat'l Highway Traffic Safety Admin., *Occupant Protection for Automated Driving Systems*, 85 Fed. Reg. 17,624 (Mar. 30, 2020); Nat'l Highway Traffic Safety Admin., *Removing Regulatory Barriers for Vehicles with Automated Driving Systems*, 84 Fed. Reg. 24,433-01 (May 28, 2019); Nat'l Highway Traffic Safety Admin., *Removing Regulatory Barriers for Vehicles with Automated Driving Systems*, 83 Fed. Reg. 2607-01 (Jan. 18, 2018).

87. Nat'l Highway Traffic Safety Admin., *Removing Regulatory Barriers for Vehicles with Automated Driving Systems*, 83 Fed. Reg. 2607-01 (Jan. 18, 2018).

88. Nat'l Highway Traffic Safety Admin., *Removing Regulatory Barriers for Vehicles with Automated Driving Systems*, 84 Fed. Reg. 24,433 (May 28, 2019).

89. Nat'l Highway Traffic Safety Admin., *Occupant Protection for Automated Driving Systems*, 85 Fed. Reg. 17,624 (Mar. 30, 2020).

90. *See* 49 U.S.C. § 30122.

91. "Motor vehicle repair business" is defined to mean "a person holding itself out to the public to repair for compensation a motor vehicle or motor vehicle equipment." *Id.* § 30122(a).

cannot mandate retroactive equipping of new motor vehicle equipment or set inspection standards for noncommercial vehicles.[92]

3. Representing Automated Vehicle Companies

The Safety Standards may prohibit certain automated vehicle designs in the absence of regulatory or congressional action. Manufacturers of these types of automated vehicles will need to consider three regulatory relief mechanisms: interpretations, exemptions, and petitions. NHTSA published a guide on these tools and provided additional guidance in its Federal Automated Vehicle Policy (AV 1.0).[93] Alternatively, a manufacturer could design its vehicle to qualify as a low-speed vehicle. This section describes these options.

a. Interpretation Process

The interpretation process can be a helpful way for manufacturers of innovative technologies to seek guidance from NHTSA about the application of certain Safety Standards to a particular motor vehicle technology. Interpretations cannot change or affect a Safety Standard substantively; instead, interpretations provide guidance on how NHTSA "believes existing law applies to the requestor's motor vehicle or motor vehicle equipment."[94]

Before submitting an interpretation request, the requestor should search for existing interpretations to determine whether NHTSA has previously interpreted the specific issue or if there are comparable favorable or unfavorable rulings. An interpretation request should ask a clear, specific question and not be open-ended. The requestor should provide its opinion on how NHTSA should interpret the statute or rule, support its interpretation with facts and laws, and explain why the current statute or regulation is unclear. Bald, unsupported statements that a given interpretation furthers safety are unfavored, and the requestor needs to tie its specific motor vehicle or motor vehicle equipment and its proposed interpretation to the safety-related implications of both. A request should also provide a clear, well-supported legal argument about why NHTSA should issue an interpretation.

---

92. Wood et al., *supra* note 18, at 1436.

93. U.S. Dep't of Transp. & Nat'l Highway Traffic Safety Admin., *Federal Automated Vehicles Policy: Accelerating the Next Revolution in Roadway Safety* (Sept. 2016), https://www.transportation.gov/sites/dot.gov/files/docs/AV%20policy%20guidance%20PDF.pdf. Chapter 7 details the Federal Automated Vehicle Policy. NHTSA has also published a separate guide on its regulatory tools. *See* Nat'l Highway Traffic Safety Admin., *Understanding NHTSA's Regulatory Tools: Instructions, Practical Guidance, and Assistance for Entities Seeking to Employ NHTSA's Regulatory Tools* [hereinafter The Guide], https://www.nhtsa.gov/sites/nhtsa.dot.gov/files/documents/understanding_nhtsas_current_regulatory_tools-tag.pdf (last visited Apr. 5, 2020).

94. The Guide, *supra* note 93.

If NHTSA denies a request for an interpretation, the requestor can send a subsequent request providing additional information or arguments. In AV 1.0, NHTSA said it would attempt to respond to a simple interpretation request within 60 days and a complex request within 90 days.

b. Exemption Process

An alternative approach is for a manufacturer to seek a temporary general exemption from the application of certain Safety Standards to automated vehicles or motor vehicle equipment. There are four categories for which a manufacturer can seek exemption: (1) substantial economic hardship, (2) development or field evaluation of a new motor vehicle safety feature, (3) development or field evaluation of a low-emission motor vehicle, and (4) overall safety level of exempted vehicle at least equal to overall safety level of nonexempt vehicles. NHTSA also has the authority to grant a specific exemption to a vehicle used for research, demonstrations, testing, training, racing, show, or display.[95] The applicant must clearly explain in its application how it satisfies one of the above criteria.

The application process for exemption is set forth in 49 C.F.R. part 555. The application itself must meet certain formal requirements outlined in 49 C.F.R. § 555.5(b).[96] The applicant must set forth the specific criteria by which it seeks exemption and provide information required for each basis.[97] The applicant must state which Safety Standard or part thereof from which it seeks exemption and the desired time of such exemption. The applicant must set forth the reasons why the exemption is in the public interest and, as applicable, consistent with the objectives of chapters 301 or 325 of title 49 of the United States Code. An applicant can request that certain information or data be withheld from the public.

If an application fails to include sufficient information for NHTSA to assess the application, NHTSA notifies the applicant of the deficiencies and provides the applicant with an opportunity to supplement. If an application contains sufficient information such that NHTSA will not categorically deny it, the public is allowed to comment on the application. Interested parties can discuss applications for exemption with NHTSA officials, but NHTSA will not have a formal public hearing or proceeding about an application.[98]

---

95. 49 U.S.C. § 30114.

96. These requirements include being written in the English language, the need to submit three copies to the administrator of NHTSA, and a requirement for the applicant to identify itself and its place of organization, as well as the more substantive components of the application.

97. *See* 49 C.F.R. § 555.6.

98. *Id.* § 555.7(b), (c).

NHTSA prioritizes reviews of applications that advance safety. Thus, in crafting an application for exemption, it is important to emphasize the safety benefits of the desired design. NHTSA will publish its decision and its justifications in the Federal Register.

NHTSA received a petition from General Motors, LLC on January 11, 2018, and a petition from Nuro, Inc. on October 23, 2018, for exemptions from particular Safety Standards that the petitioners believed were unnecessary for their automated vehicle designs.[99] The petitions are for two different deployments of automated vehicles. Nuro's exemption is for a driverless, low-speed delivery vehicle, called an R2X, that operates on roadways without any occupants. In contrast, General Motors's petition is for human-occupied, non-low-speed vehicles called zero-emission autonomous vehicles or ZEAVs. ZEAVs would not have steering wheels or brake or accelerator pedals.

Based on these differences, it is unsurprising that NHTSA ruled on the Nuro petition first. As of April 1, 2020, NHTSA still had not decided General Motors's petition. On February 6, 2020, NHTSA granted Nuro's petition for a temporary exemption from three requirements under the Safety Standards.[100] Nuro sought an exemption from the need to have a windshield that complies with Safety Standard 205; the linger time, deactivation, and other requirements for the backup camera; and the requirement that the vehicle contain interior and exterior mirrors. Nuro relied upon the low-emission motor vehicle ground for asking for the petition, and NHTSA agreed with that basis and also thought that the petition could be granted because "compliance with the standard would prevent the manufacturer from selling a motor vehicle with an overall safety level at least equal to the overall safety level of nonexempt vehicles."

One of the important questions in the automated vehicle petitions was whether NHTSA would consider the safety of the automated driving system itself. Some commentators suggested that NHTSA should consider the functionality and capabilities of the automated driving system, while other commentators thought that such inquiries were irrelevant to the petition. NHTSA found that the automated driving system "does not factor into the comparative safety findings NHTSA must make to grant an exemption under either the [equivalent overall safety] or [low-emission vehicle] basis in this instance." NHTSA stated that its inquiry was about whether a compliant R2X would be

99. Nuro, Inc., Petition for Exemption from Certain Provisions of Federal Motor Vehicle Safety Standard, No. 500 (Oct. 19, 2018), *available at* https://www.nhtsa.gov/sites/nhtsa.dot.gov/files/documents/nuro_petition.pdf; General Motors, LLC, Petition under 49 U.S.C. § 30113 and 49 C.F.R. Part 555 to Advance Safety and Zero-Emission Vehicles through Technology That Achieves the Safety Purpose of the FMVSS (Jan. 11, 2018), *available at* https://www.nhtsa.gov/sites/nhtsa.dot.gov/files/documents/gm_petition.pdf.

100. Nat'l Highway Traffic Safety Admin., *Nuro, Inc.; Grant of Temporary Exemption for a Low-Speed Vehicle with an Automated Driving System*, 85 Fed. Reg. 7826 (Feb. 11, 2020).

safer than a noncompliant R2X. NHTSA was careful to state that it may, in other cases, consider the automated driving system before granting the exemption.

NHTSA attached some strings to its favorable ruling. NHTSA imposed data reporting requirements on Nuro to enable oversight and allow NHTSA to learn from the operations. NHTSA also required Nuro to report cybersecurity risks and cease operations if its vehicles suffer a cybersecurity risk related to safety until it is remedied. Nuro must certify that it engaged with local communities and obtained necessary state and local approvals before deploying its R2Xs in jurisdictions.

c. *Petition for Rulemaking*

An interested person can petition NHTSA to engage in rulemaking by issuing, amending, or revoking a Safety Standard.[101] Generally, NHTSA considers granting sound, well-supported petitions that "would promote safety if, in the Agency's judgment, the Agency would be able to develop and issue a sound, well-supported proposed rule, including regulatory text with performance requirements and test procedures, without conducting more than minimal additional research (e.g., to establish a sound basis for taking the recommended action or to develop and validate performance requirements, test conditions, or test procedures)."[102] NHTSA has expressed its concerns over the amount of resources that are allocated to answering and reviewing petitions and has indicated that it "must be particularly careful in deciding whether a submission qualifies as a petition and whether to grant each petition."[103]

NHTSA reviews each properly filed petition.[104] NHTSA can hold a public meeting about the petition by publishing a notice in the Federal Register.[105] Interested persons can participate and submit written or oral presentations to NHTSA.[106] NHTSA must notify the petitioner of its decision within 120 days of the receipt of the petition.[107] A grant of the petition means that a rulemaking proceeding by NHTSA is commenced.[108] A denial and the reasons are published in the Federal Register.[109]

---

101. 49 C.F.R. part 552.
102. The Guide, *supra* note 93.
103. U.S. Dep't of Transp., *Federal Automated Vehicles Policy: Accelerating the Next Revolution in Roadway Safety* 66 (Sept. 2016), https://www.transportation.gov/sites/dot.gov/files/docs/AV%20policy%20guidance%20PDF.pdf.
104. 49 C.F.R. § 552.6.
105. *Id.* § 552.7.
106. *Id.*
107. *Id.* § 552.8.
108. *Id.* § 552.9(b).
109. *Id.* § 552.10.

Petitions have been used in recent years to attempt to have NHTSA mandate automatic emergency braking (AEB) in commercial trucks and consumer vehicles.[110] NHTSA denied the consumer vehicle petition but granted the commercial truck petition. Petitions should not be filed without care because NHTSA's decision not to grant a petition can, as one court has recently held, have a preemptive effect on state product liability lawsuits that seek to mandate the identical relief sought in the petition.[111] Thus, for those petitions that seek to force technological change, groups should consider whether a petition for rulemaking, with the potential preemptive effect of a denial by NHTSA, or a state product liability action is the best course to impose the change.

d. Low-Speed Vehicle

Another potential option is to develop an automated vehicle that can be classified as a low-speed vehicle. A low-speed vehicle is a vehicle that weighs less than 3,000 pounds and cannot exceed 25 miles per hour.[112] Low-speed vehicles are not required to comply with every Safety Standard and have only the limited requirements outlined in Safety Standard 500.[113] The inherent problem with pursuing low-speed vehicle classification is the limited market for vehicles that travel no faster than 25 miles per hour. Safety Standard 500 is not a perfect fit even for low-speed automated vehicles, as was evidenced by Nuro's petition for exemption.

Google relied on Safety Standard 500 when developing its Firefly automated vehicle. The Firefly lacked a steering wheel and a dashboard. The vehicle had two seats and looked like a pod. The Firefly's max speed was 25 miles per hour, allowing it to fall within Safety Standard 500.

4. The Notice and Recall Power: Noncompliance with the FMVSS or Safety Defects

The Motor Vehicle Safety Act requires manufacturers to notify consumers and others and recall motor vehicles or items of motor vehicle equipment when they do not comply with a Safety Standard or suffer from defects related to motor vehicle safety. NHTSA has stated that it intends to use the notice and

---

110. *See* Nat'l Highway Traffic Safety Admin., *Federal Motor Vehicle Safety Standards; Automatic Emergency Braking,* 80 Fed. Reg. 62,487 (Oct. 16, 2015) (commercial trucking petition); Nat'l Highway Traffic Safety Admin., *Federal Motor Vehicle Safety Standards; Automatic Emergency Braking,* 82 Fed. Reg. 8391 (Jan. 18, 2017) (consumer vehicles).

111. Dashi v. Nissan N. Am., Inc., 247 Ariz. 56, 445 P.3d 13 (Ct. App. 2019), *review denied* (Jan. 7, 2020). *Dashi* is explored in detail in Chapter 8.

112. 49 C.F.R. § 571.3.

113. *Id.* § 571.500.

recall mechanism to regulate automated safety technologies.[114] This section explores the notice and recall remedy and the meaning of noncompliance and defects related to motor vehicle safety.

a. The Notice and Recall Remedy

Manufacturers and NHTSA are under a simultaneous obligation to identify safety defects and noncompliances with a Safety Standard. The vast majority are identified by manufacturers.[115] This section briefly outlines the respective processes and responsibilities for each.

(i) Manufacturer

The vehicle manufacturer is responsible for any safety defects or noncompliance in the vehicle or an item of original equipment.[116] The manufacturer of replacement equipment is responsible for defects and noncompliance in the replacement equipment.[117] These manufacturers are under a good-faith obligation to report safety defects or noncompliance.[118] If the manufacturer believes that the defect or noncompliance is inconsequential to motor vehicle safety, the manufacturer can apply for an exemption from the recall provisions.[119]

Once a manufacturer determines that a defect or noncompliance exists, the manufacturer must submit a report to NHTSA within five working days of the determination.[120] The report, known as a Part 573 Safety Recall Report, must provide substantial information concerning, among other things, the vehicles or equipment affected, the defect or noncompliance and its severity and discovery, and the proposed recall solution.[121] To the extent that such information is not available within five working days of the determination, the manufacturer must provide the information within five working days of confirming its accuracy.

(ii) NHTSA

NHTSA's investigatory process into safety defects and noncompliance can begin from several sources. NHTSA maintains a vehicle safety hotline that

---

114. Nat'l Highway Traffic Safety Admin., *NHTSA Enforcement Guidance Bulletin 2016-02: Safety-Related Defects and Automated Safety Technologies*, 81 Fed. Reg. 65,705-01 (Sept. 23, 2016).

115. Nat'l Highway Traffic Safety Admin., *Motor Vehicle Safety Defects and Recalls: What Every Vehicle Owner Should Know*, U.S. Dep't of Transp. 10, https://www-odi.nhtsa.dot.gov/recalls/documents/MVDefectsandRecalls.pdf (last visited Aug. 11, 2019).

116. 49 C.F.R. § 573.5(a).

117. *Id.* § 573.5(b).

118. 49 U.S.C. § 30118(c).

119. *Id.* § 30118(d).

120. 49 C.F.R. § 573.6.

121. *Id.*

allows consumers to report defects and an online complaint portal. These complaints, whether verbal or electronic, are cataloged into NHTSA's consumer complaint database according to the year, make, model, and affected item of motor vehicle equipment. NHTSA publishes these reports online. Each report is reviewed and screened by technical staff in the Office of Defects Investigation (ODI) to determine whether to begin an investigation.

An interested person can also petition NHTSA to commence a proceeding to decide whether to issue an order about a safety-related defect or noncompliance.[122] The petition process is almost identical to the one described above for rulemaking. If NHTSA grants a petition, ODI begins an investigatory proceeding.[123]

NHTSA's Office of Vehicle Safety Compliance (OVSC) investigates compliance with the Safety Standards.[124] NHTSA has broad information-gathering powers and authority to conduct inspections and investigations.[125] NHTSA can issue subpoenas, conduct information-gathering hearings, have administrative depositions, issue general or special orders, and request the production of documents and things.[126] Following a motor vehicle accident, NHTSA can request records of the manufacturer, distributor, or dealer, inspect the premises where the vehicle or equipment is manufactured, the distribution housing, or the dealership, and inspect the vehicle or motor vehicle equipment involved in the accident.[127]

NHTSA can also learn about defects or noncompliance through reporting requirements imposed on manufacturers.[128] Manufacturers must provide NHTSA's Early Warning Division with copies of any external communications about a defect (including a malfunction), regardless of whether the defect is safety-related.[129] Manufacturers are also required to provide NHTSA with a copy of external communications "relating to a customer satisfaction campaign, consumer advisory, recall, or other safety activity involving the repair or replacement of motor vehicles or equipment."[130] Manufacturers are required to report safety recalls or safety campaigns in foreign countries.[131] Manufacturers of more than a certain specified number of vehicles have quarterly reporting requirements about the production of vehicles by the manufacturer, incidents involving death or injury involving a vehicle up to nine model years before

122. 49 C.F.R. part 552.
123. *Id.* § 552.9(b).
124. *Id.* § 554.4.
125. 49 U.S.C. § 30166; 49 C.F.R. part 510.
126. 49 C.F.R. § 510.3.
127. 49 U.S.C. § 30166.
128. *See generally* 49 C.F.R. part 579.
129. *Id.* § 579.5(a).
130. *Id.* § 579.5(b).
131. *Id.* § 579.11.

the date of the report, claims relating to property damage and warranties, and consumer complaints and field reports.[132]

Based on all of the sources of information, the ODI can determine whether to investigate a specific safety defect.[133] The OVSC conducts investigations into noncompliance with the Safety Standards.[134] ODI and OVSC can open an investigation on their own or by granting a petition to open the investigation. The manufacturer of the vehicle is notified immediately upon an investigation being opened. During the investigative process, the relevant office obtains information from the manufacturer, and the manufacturer can present its views and evidence concerning noncompliance or defect. If ODI believes there is a safety-related defect or OVSC believes there is noncompliance, the relevant office will ask the manufacturer to conduct a recall during the investigative process.

Before discussion of the notification and recall mechanism, there are two important considerations about the process. First, NHTSA can settle and resolve disputes before a final determination without imposing the notification and recall mechanism.[135] Second, the attorney general's ability to compel notice and recall is not necessarily dependent upon NHTSA's final determination. In a dispute over GM's X-cars, NHTSA did not issue a final determination before referring the matter to the attorney general to bring an enforcement action. GM challenged the attorney general's ability to bring the suit before the completion of the administrative process, and the court ruled that the attorney general need not wait on a final determination.[136]

(iii) Process Following Defect or Noncompliance Determination

Upon a final determination by NHTSA of either noncompliance with a Safety Standard or a safety defect, the manufacturer that disagrees with the final determination has two options: (1) file suit, or (2) force the attorney general to enforce the final determination. A manufacturer that challenges NHTSA's final determination can be subject to civil penalties if it does not prevail in

---

132. *Id.* § 579.21.

133. *Id.* § 554.5.

134. *Id.* § 554.4.

135. *See* Ctr. for Auto Safety, Inc. v. Lewis, 685 F.2d 656, 663 (D.C. Cir. 1982) (holding that "in the absence of a final determination of defect . . . the Motor Vehicle Safety Act does not forbid the Secretary to settle a defect investigation on the basis of a remedy other than" the notification and recall remedy). This ability to settle an initial defect determination is not absolute because courts have permitted other interested parties to challenge NHTSA's settlement under the Administrative Procedures Act. *Center for Auto Safety, Inc.*, 685 F.2d at 663–64.

136. United States v. Gen. Motors Corp., 574 F. Supp. 1047, 1049 (D.D.C. 1983) ("[T]he Court is unable to ascribe to Congress an intent to require the nation's chief law enforcement officer to await the end of the administrative process before going to court on his own to redress a violation of the Act which he may perceive to present an imminent danger to the public.").

the action.[137] The amount of the civil penalties depends upon the number of violations but is otherwise presently capped at approximately $111 million, increased for inflation.[138] Of the thousands of recalls that have been instituted under the Motor Vehicle Safety Act, very few cases have ever been litigated.

Manufacturers are required to include specific information in the notice and provide the notice and the envelope to NHTSA at least five business days before its mailing.[139] Generally, the notice to consumers must be sent within 60 days after the manufacturer files its report or the final determination by NHTSA.[140] A manufacturer must also notify dealers and distributors of the safety-related defect or noncompliance.[141] A manufacturer must remedy the defect or noncompliance at no cost upon the presentment of the vehicle or the equipment for remedy.[142] A manufacturer can remedy the defect by (1) repairing the vehicle or equipment, (2) replacing the vehicle or equipment with an identical or reasonably equivalent vehicle or equipment, or (3) refunding the purchase price (but there can be an allowance for depreciation).[143]

b. Noncompliance and Safety Defect

The notice and recall obligations are imposed when a motor vehicle or an item of motor vehicle equipment either (1) fails to technically comply with the Safety Standard or (2) has a defect related to motor vehicle safety. This second recall

---

137. 49 U.S.C. § 30121(b).

138. *Id.* § 30165.

139. 49 C.F.R. § 577.5. The notification procedures are outlined in section 30119 and part 577 of the regulations. The envelope must include SAFETY RECALL NOTICE, and the notice must state IMPORTANT SAFETY RECALL at the top of the notification. The notice must indicate the reason—noncompliance with a Safety Standard or safety defect—for the recall and provide a clear description of the problem. *Id.* § 577.5(c), (e). The manufacturer must provide the consumer with an evaluation of the motor vehicle safety risk due to the noncompliance or defect. *Id.* § 577.5(f). The notice must identify the recall remedy that is available to the consumer. *Id.* § 577.5(g). In the event that the manufacturer challenges a final determination by NHTSA, the manufacturer must provide an interim notice to inform consumers of NHTSA's final determination. *Id.* § 577.6.

140. *Id.* § 577.7. NHTSA can require that the notice be sent sooner in the need of public interest. *Id.* The notice must be mailed by first-class mail to each vehicle owner based on the state records or the manufacturer's records. If the manufacturer is unable to identify an owner, the manufacturer can notify the most recent purchaser. If, based on the manufacturer's quarterly reports, NHTSA determines that an inadequate number of consumers have remedied their vehicles or items of equipment, NHTSA can require a follow-up notification. *Id.* § 577.10.

141. *Id.* § 577.13. This notice must advise the dealer and the distributor that federal law prohibits the sale or lease of new vehicles covered by the notification until the noncompliance or defect has been remedied. *Id.* § 577.13(b). The prohibitions on sales and leases are set forth in sections 573.11 and 573.12. This notice must be sent within a reasonable time from the determination of the defect or noncompliance. *Id.* § 577.7(c). When there is an "immediate and substantial threat to motor vehicle safety," the manufacturer must provide the notice within three business days if sent by mail or five business days if by electronic transmission. *Id.*

142. 49 U.S.C. § 30120(a).

143. *Id.*

reason imposes an additional obligation on manufacturers above and beyond mere compliance with the Safety Standards.[144] A motor vehicle or item of motor vehicle equipment may comply with a Safety Standard but still suffer from a safety defect, or, conversely, a motor vehicle or item of motor vehicle equipment may not have a safety defect but fail to comply with a Safety Standard. In either case, the motor vehicle or item of motor vehicle equipment is subject to recall.

Noncompliance with a Safety Standard is relatively straightforward, though Chrysler Corporation challenged an alleged noncompliance and won in the D.C. Circuit in the 1990s.[145] With more frequency (but still very rarely), manufacturers have combated whether their vehicles contain defects related to motor vehicle safety.[146] NHTSA must establish two elements to warrant a recall based on a safety defect: (1) the motor vehicle or item of motor vehicle equipment suffers from a defect and (2) such defect is related to safety.[147]

5. Meaning of "Defect"

The Motor Vehicle Safety Act defines "defects" to "include[] any defect in performance, construction, a component, or material of a motor vehicle or motor vehicle equipment."[148] Because the definition of "defect" does not require an "engineering, metallurgical, or manufacturing failure[]," courts have ruled that a defect can be "based exclusively on the performance record of the vehicle or component."[149] In one case, the D.C. Circuit established the following general rule for when a defect exists:

> [A] vehicle or component "contains a defect" if it is subject to a significant number of failures in normal operation, including failures either occurring during specified use or resulting from owner abuse (including inadequate maintenance) that is reasonably foreseeable (ordinary abuse), but

---

144. *See, e.g.*, United States v. Chrysler Corp., 158 F.3d 1350, 1351 (D.C. Cir. 1998) (stating that noncompliance with a Safety Standard and a safety defect are distinct).

145. *Chrysler Corp.*, 158 F.3d 1350.

146. *See, e.g.*, United States v. Gen. Motors Corp., 518 F.2d 420 (D.C. Cir. 1975) (Kelsey-Hayes wheels); United States v. Ford Motor Co., 453 F. Supp. 1240 (D.D.C. 1978) (windshield wipers); United States v. Gen. Motors Corp., 561 F.2d 923 (D.C. Cir. 1977) (pitman arms); United States v. Gen. Motors Corp., 565 F.2d 754 (D.C. Cir. 1977) (carburetors); United States v. Ford Motor Co., 574 F.2d 534 (D.C. Cir. 1978) (seat-pin bracket in the driver's seat); United States v. Gen. Motors Corp., 656 F. Supp. 1555 (D.D.C. 1987), *aff'd*, 841 F.2d 400 (D.C. Cir. 1988) (X-cars); United States v. Gen. Motors Corp., 841 F.2d 400 (D.C. Cir. 1988) (X-cars).

147. 49 U.S.C. § 30118.

148. *Id.* § 30102(a)(3).

149. *Gen. Motors Corp.*, 518 F.2d at 432. But in the X-cars case, the district court, the trier of fact, found, and the D.C. Circuit affirmed, that consumer complaints were insufficient, in and of themselves, in that case to prove the existence of a defect. *Gen. Motors Corp.*, 841 F.2d at 412–13.

> excluding failures attributable to normal deterioration of a component as a result of age and wear.[150]

In examining whether a defect exists, the D.C. Circuit examines three general factors: "the nature of the component involved, the circumstances in which the failures occurred, and the number of failures experienced."[151]

There must be a significant number of failures to warrant the drastic mechanism of notification and recall.[152] A vehicle is not defective if there have been only a de minimis number of failures.[153]

A product can be defective under this standard even if the defect is due to foreseeable misuse or other failures to follow instructions by the driver or vehicle owner.[154] Not all defects under the Motor Vehicle Safety Act will be viable product liability actions, and not all successful product liability actions will trigger the notification and remedy mechanism under the Motor Vehicle Safety Act.[155] "[G]ross and unforeseeable owner abuse or unforeseeable neglect of vehicle maintenance" is a defense to a defect investigation by NHTSA.[156]

6. Relation to Motor Vehicle Safety

The Motor Vehicle Safety Act defines "motor vehicle safety" as "the performance of a motor vehicle or motor vehicle equipment in a way that protects the public against unreasonable risk of accident occurring because of the design, construction, or performance of a motor vehicle, and against unreasonable risk of death or injury in an accident, and includes nonoperational safety of a motor vehicle."[157]

In the 1970s, NHTSA successfully established a per se theory of when a defect is related to motor vehicle safety:

> In our view, where a defect—a term used in the sense of an "error or mistake"—has been established in a motor vehicle, and where this defect

---

150. *Gen. Motors Corp.*, 518 F.2d at 427.

151. *Id.*

152. *Id.* at 438; *see also* United States v. Gen. Motors Corp., 561 F.2d 923, 929 (D.C. Cir. 1977) (J. Leventhal concurring and dissenting in part) (describing how "some products are bound to be 'lemons'" and that "total elimination" of all defects "would produce a purchase price so prohibitive that it cannot be taken as the contemplation of Congress").

153. *Gen. Motors Corp.*, 518 F.2d at 438 n.84.

154. *Id.* at 434.

155. *See id.* ("The Act was designed as a preventive measure 'supplementary of and in addition to the common law of negligence and products liability.' Determinations of fault or liability relevant to the award of damages at common law are not controlling on the interpretation of the scope of this prophylactic defect notification legislation." (quoting Larsen v. Gen. Motors Corp., 391 F.2d 495, 506 (8th Cir. 1968))).

156. *Gen. Motors Corp.*, 518 F.2d at 438.

157. 49 U.S.C. § 30102(a)(9).

> results in hazards as potentially dangerous as a sudden engine fire, and where there is no dispute that at least some such hazards, in this case fires, can definitely be expected to occur in the future, then the defect must be viewed as one "related to motor vehicle safety," and the Act's basic purpose of protecting the public requires that notification be provided.[158]

Under this per se theory, it does not matter if the manufacturer can establish that only a negligible amount of failures will occur in the future.[159] A manufacturer's only defense, in these instances, is "by showing that the defect occurs only in circumstances that render it non-dangerous."[160]

When the per se theory is not applicable, courts have examined three factors in determining the reasonableness of the risk to motor vehicle safety: "(1) the severity of the harm it threatens; (2) the frequency with which that harm occurs in the threatened population relative to its incidence in the general population; and (3) the economic, social, and safety consequences of reducing the risk to a so-called 'reasonable' level."[161]

7. Penalties

The Motor Vehicle Safety Act imposes significant civil penalties on manufacturers.[162] The civil penalties include penalties for failing to comply with the Safety Standards or an order of NHTSA or for having a motor vehicle safety defect. There are civil penalties for knowingly and willingly submitting materially false or misleading information to NHTSA. Additionally, if a person knowingly provides false or misleading information intending to mislead NHTSA and the defects have caused serious bodily injury or death, the person is subject to criminal penalties and imprisonment not to exceed 15 years.[163]

8. New Car Assessment Program

The New Car Assessment Program (NCAP) is another tool that NHTSA can use to help foster growth in automated technologies or to deter the use of less adequate automated technologies. The idea behind NCAP is that by providing consumers with information about the safety ratings and features of vehicles, manufacturers will be encouraged to design safer vehicles.[164] NHTSA

---

158. United States v. Gen. Motors Corp., 565 F.2d 754, 758 (D.C. Cir. 1977).

159. *Id.*

160. United States v. Gen. Motors Corp., 841 F.2d 400, 415 (D.C. Cir. 1988) (citing United States v. Gen. Motors Corp., 561 F.2d 923, 930–31 (D.C. Cir. 1977)).

161. United States v. Gen. Motors Corp., 656 F. Supp. 1555, 1578 (D.D.C. 1987).

162. 49 U.S.C. § 30165.

163. *Id.* § 30170.

164. Nat'l Highway Safety Admin., DOT HS 810 698, The New Car Assessment Program Suggested Approaches for Future Program Enhancements 3 (Jan. 2007).

currently rates the crashworthiness of vehicles by issuing a rating of up to five stars to a vehicle.[165] NHTSA gives each vehicle a vehicle safety score that is a combination of its score on three criteria: frontal crash, side crash, and rollover resistance.[166] NHTSA's testing is not limited to the testing mechanisms in the Safety Standards, and NHTSA can use the NCAP to promote automated technologies, even if NHTSA has not issued a Safety Standard on those automated technologies.[167]

B. Federal Highway Administration

FHWA is the federal agency responsible for coordinating highway transportation throughout the country.[168] FHWA's role in our nation's highway system allows it to coordinate state and local governments regarding the use of automated vehicles and connected infrastructure. Although FHWA's main purpose is the coordination of the highway system in the United States, FHWA conducts numerous other activities and helps facilitate discussions on roads, infrastructure, and traffic devices throughout the country. FHWA has developed the CARMA platform to support research and development of automated vehicles and connected infrastructure.

An important role of FHWA is to establish and maintain the Manual on Uniform Traffic Control Devices (MUTCD).[169] The MUTCD is the "national standard for all traffic control devices installed on any street, highway, bikeway, or private road open to public travel."[170] States are required to follow the MUTCD in federal aid highway contracts. Many states have adopted the MUTCD (but some supplement it) through statutory enactments or regulatory pronouncements,[171] and the MUTCD can be relevant in negligence cases.[172]

---

165. Wood et al., *supra* note 18, at 1493–94.

166. *Id.* at 1494.

167. *Id.* at 1493.

168. The FHWA derives its authority from title 23 (Highways) and title 49 (Transportation).

169. *See, e.g.*, 23 C.F.R. § 655.603(a).

170. U.S. Dep't of Transp., Manual on Uniform Traffic Control Devices for Streets and Highways 1-1 (2009 ed., rev. May 2012), *available at* https://mutcd.fhwa.dot.gov/pdfs/2009r1r2/mutcd2009r1r2edition.pdf.

171. *See generally* Fed. Highway Admin., *MUTCDs & Traffic Control Devices Information by State*, U.S. Dep't of Transp., https://mutcd.fhwa.dot.gov/resources/state_info/index.htm (last visited Aug. 14, 2019). The following states have adopted the MUTCD: Ala. Code § 32-5A-30; Ariz. Rev. Stat. Ann. § 28-641; Colo. Rev. Stat. § 42-4-104; 625 Ill. Comp. Stat. Ann. 5/11-301; Ind. Code § 9-21-2-2; Kan. Stat. Ann. § 8-2003 (West); Me. Stat. tit. 29-A, § 2075; Mont. Code Ann. § 61-8-202 (West); N.C. Gen. Stat. § 136-30; N.D. Cent. Code § 39-13-06; Okla. Stat. tit. 74, § 2285; 75 Pa. Cons. Stat. Ann. § 6121; S.C. Dep't of Transp., *Supplemental MUTCD*, https://www.scdot.org/business/pdf/permits-supplement_mutcd.pdf (last visited May 4, 2020) (noting that the SCDOT adopted the MUTCD as supplemented by the SCDOT); Utah Code Ann. § 41-6a-301 (West); Vt. Stat. Ann. tit. 23, § 1025 (West); Wyo. Stat. Ann. § 31-5-112 (West).

172. Eva Marie Mancuso & Sonja L. Deyoe, *Considerations before Filing Suit—Road Construction Contractor*, 5 Litigating Tort Cases § 57:31 (Sept. 2019 update).

FHWA is studying automated vehicle technologies to revise the MUTCD to address these technologies. FHWA will play a significant role in coordinating the use of connected infrastructure and, to the extent that automated vehicles rely less on traditional traffic devices and signage, the shift to traffic devices and signage that are more useful to automated vehicles.

C. Federal Motor Carrier Safety Administration

The FMCSA is the primary federal agency with oversight of commercial motor vehicles (CMVs) and their drivers and operations in interstate commerce. The FMCSA also works with state agencies. Among other things, the FMCSA maintains and enforces the Federal Motor Carrier Safety Regulations (the FMCSRs), the Commercial Driver's License Program, the Motor Carrier Safety Assistance Program,[173] and the Hazardous Materials Program.

Because driving automation systems will be introduced and used in commercial motor vehicles, the FMCSA will have a substantial role in the deployment of automated commercial motor vehicles. The FMCSA's role could be permissive or prohibitive to the use of automated technologies in commercial motor vehicles, though, at least as of now,[174] the FMCSA appears to desire to foster the introduction of automated vehicles. Among other things, the FMCSA has issued a request for comments on which FMCSRs are a barrier to the deployment of automated driving systems;[175] issued an advance notice of proposed rulemaking on the use of automated driving systems in commercial motor vehicles;[176] tasked the Motor Carrier Safety Advisory Committee with presenting recommendations on whether the FMCSRs are appropriate for the introduction of highly automated commercial vehicles;[177] commissioned a

173. The Motor Carrier Safety Assistance Program (MCSAP) "is a Federal grant program that provides financial assistance to States to reduce the number and severity of accidents and hazardous materials incidents involving commercial motor vehicles." 49 C.F.R. § 350.101(a).

174. At one point, the FMCSA expressed its belief that the FMCSRs require that "a trained commercial driver must be behind the wheel at all times, regardless of any automated driving technologies available on the CMV, unless a petition for a waiver or exemption has been granted." Nat'l Highway Traffic Safety Admin., *Automated Driving Systems 2.0: A Vision for Safety* 2, U.S. Dep't of Transp. (Sept. 12, 2017), https://www.nhtsa.gov/sites/nhtsa.dot.gov/files/documents/13069a-ads2.0_090617_v9a_tag.pdf.

175. Fed. Motor Carrier Safety Admin., *Request for Comments Concerning Federal Motor Carrier Safety Regulations (FMCSRs) Which May Be a Barrier to the Safe Testing and Deployment of Automated Driving Systems-Equipped Commercial Motor Vehicles on Public Roads*, 83 Fed. Reg. 12,933-01 (Mar. 26, 2018).

176. Fed. Motor Carrier Safety Admin., *Safe Integration of Automated Driving Systems-Equipped Commercial Motor Vehicles*, 84 Fed. Reg. 24,449-01 (May 28, 2019).

177. Letter from John Lannan to the Honorable Raymond P. Martinez, MCSAC Task 17-01: Highly Automated Commercial Vehicles (Oct. 16, 2018), https://www.fmcsa.dot.gov/advisory-committees/mcsac/mcsac-task-17-1-final-report.

review of the FMCSRs by the John A. Volpe National Transportation Systems Center;[178] and hosted several listening sessions with stakeholders.

1. The FMCSRs

The FMCSA administers the FMCSRs, which broadly regulate CMVs that transport property and passengers, as well as CMV owners and their drivers and employees.[179] In the USDOT's third automated vehicles policy (AV 3.0), the FMCSA expressed how it intended to exercise its oversight of CMVs equipped with automated driving systems.[180] The FMCSA would first consider whether the commercial motor vehicle meets the equipment standard and the parts and accessories standard. The FMCSA would then consider whether the motor carrier complies with the operational requirements in the FMCSRs.

The FMCSRs set a minimum standard for equipment in CMVs.[181] Equipment standards exist for many parts of a CMV, including, but not limited to, tires, speedometers, exhaust systems, brakes, fuel systems, lamps, and floors. The minimum equipment standards are just that—they set minimum standards for the equipment covered. The FMCSRs acknowledge that CMVs may be equipped with equipment and accessories that are not covered by the FMCSRs and contain a catchall provision. Under the catchall, additional equipment and accessories are permitted so long as they do not decrease the safety of operation of a CMV.[182] Once added to a CMV, these additional parts and accessories must "be in safe and proper operating condition at all times."[183] The FMCSA has indicated that, because there is not a specific standard for automated driving systems, it will analyze the deployment of these systems under the additional equipment and accessories standard and consider whether these systems decrease safety. If these systems decrease safety, the FMCSA may impose fines, and if these systems create an imminent hazard, the FMCSA can also place the vehicle out of service.[184]

---

178. Volpe Center, *Review of the Federal Motor Carrier Safety Regulations for Automated Commercial Vehicles, Preliminary Assessment of Interpretation and Enforcement Challenges, Questions, and Gaps* (Mar. 2018).

179. The FMCSRs are located in subchapter B of chapter III of title 49.

180. U.S. Dep't of Transp., *Automated Vehicles 3.0: Preparing for the Future of Transportation* (Oct. 2019) [hereinafter AV 3.0]. AV 3.0 is described in Chapter 7.

181. 49 C.F.R. part 393.

182. *Id.* § 393.3.

183. *Id.* § 396.3(a)(1).

184. AV 3.0, *supra* note 180, at 9.

The FMCSRs contain numerous operational requirements for motor carriers, including those on driver education and training,[185] qualifications of drivers,[186] commercial driver's licenses (CDLs),[187] alcohol and controlled substance use and testing,[188] safety fitness of motor carriers,[189] hours of service,[190]

---

185. Part 380 provides certain requirements concerning driver training and education, including the need for a "driver who wishes to operate a [longer combination vehicle to] first take and successfully complete an LCV driver-training program that provides the knowledge and skills necessary to operate an LCV." 49 C.F.R. § 380.107. This testing includes both a written knowledge test and a skills component. *Id.* § 380.109.

186. The FMCSRs impose certain qualifications on drivers of CMVs and requirements on employers before employing drivers. 49 C.F.R. part 391. Drivers must meet certain requirements, including being 21 years old, having the ability to speak English, having the ability to safely operate the CMV the driver desires to drive, possessing the appropriate CDL, and meeting certain physical qualifications. The physical qualifications prohibit people who suffer from certain physical or mental disorders from operating a CMV. *Id.* § 391.41.

187. The FMCSRs establish and enforce CDL requirements in part 383. Every person who "operates a commercial motor vehicle" is required to possess a CDL. Operators must meet certain requirements prior to being eligible to obtain a CDL and pass a test with a state or, as applicable, a third-party tester. The FMCSRs impose certain knowledge and skill requirements before an operator can obtain a CDL. An operator can also become disqualified for committing certain traffic violations or other violations.

188. Part 381 regulates controlled substance and alcohol use and testing for CMV drivers. Drivers are restricted from using alcohol while performing safety-sensitive functions or four hours prior to such performance, as well as prohibited from alcohol use following certain types of crashes, for which postaccident testing is required, until the earlier of such test or eight hours. *Id.* § 383.201 *et seq.* The FMCSRs require employers to test drivers in specific situations, such as before their first drive (subject to limited exceptions), following certain accidents, randomly, and when they have reasonable suspicion that a driver has violated the prohibitions.

189. Part 385 imposes a safety fitness standard on motor carriers and provides a safety rating. The safety fitness standard is based on motor carriers' safety management controls and compliance with safety requirements in the FMCSRs. A motor carrier that receives an unsatisfactory rating is prohibited from operating a CMV, and federal agencies are prohibited from using such motor carrier.

190. Drivers are also regulated on the amount of time they can spend in service in a given day and week. 49 C.F.R. part 395. Drivers of property-carrying CMVs are limited to 14-hour days, with 11 such hours being driving hours. *Id.* § 395.3. Drivers are required to take at least a 30-minute break for every eight hours of driving. *Id.* Drivers are prohibited from being on duty more than 60 hours within seven consecutive days or 70 hours during eight consecutive days. *Id.* Drivers are required to keep records to substantiate their on-duty, driving, sleeper berth, and off-duty time. *Id.* § 395.8. The FMCSA has mandated that drivers use an electronic logging device that records data that helps FMCSA verify compliance with the hour restrictions. *Id.* § 395.20 *et seq.*

driving behavior,[191] and inspection, repair, and maintenance of CMVs.[192] In AV 3.0, the FMCSA stated that it would not apply human-specific operational FMCSRs to automated driving systems that do not require a human operator. Thus, to the extent that a CMV is equipped with a level 5 automated driving system that does not ever need a human operator, the physical qualification requirements, hours-of-service requirements, CDLs, and drug testing would not apply. Level 4 automated driving systems may require a human operator if the vehicles must operate outside of their operational design domain (e.g., a level 4 automated driving system that has an operational design domain of interstates may need a human operator when the CMV is operated on roadways that are not interstates). The human operators and eventual drivers of these level 4 commercial automated vehicles are not relieved from the human-specific FMCSRs. Because level 1, level 2, and level 3 driving automation systems always require a human operator within their operational design domains, CMVs equipped with these systems would still need to comply with the FMCSRs.

2. *FMCSRs Applied to Automated CMVs*

Because the FMCSRs were, for the most part, promulgated prior to the advent of automated vehicles, the FMCSRs and their comprehensive breadth on all aspects of CMVs, from the equipment itself to the drivers, create ambiguities and potential complications when applied to driving automation systems. The FMCSA is aware of these ambiguities and complications and commissioned the John A. Volpe National Transportation Systems Center to review

---

191. Drivers of CMVs are also regulated at the federal level by the FMCSRs. 49 C.F.R. part 392. Generally, CMVs must be operated in compliance with the laws of the jurisdiction where they are operated, unless the FMCSRs impose greater requirements than such jurisdiction, in which case the FMCSRs must be followed. *Id.* § 392.2. Drivers are prohibited from operating CMVs when their abilities or alertness is impaired, whether through fatigue, illness, or otherwise, such that operation of the CMVs is unsafe. *Id.* § 392.3. Drivers are prohibited from possessing alcohol while on duty. *Id.* § 392.5. Drivers are also prohibited from driving CMVs unless they are satisfied that certain parts of the CMV—such as the brakes, tires, horn, parking brakes, and windshield wipers—are in good working condition. *Id.* § 392.7. Drivers also have certain responsibilities with respect to the cargo, including requirements for periodic inspection of the cargo during a drive. *Id.* § 392.9. Other requirements pertain to instructions at railroad crossings, hazardous conditions, and the need to use a seat belt. *Id.* §§ 392.10, 392.11, 392.14, 392.16. When a CMV is stopped on a highway, the driver must immediately activate warning signal flashers and place warning devices to alert incoming traffic. *Id.* § 392.22. Drivers are prohibited from texting or using handheld mobile telephones while driving. *Id.* §§ 392.81, 392.82.

192. Motor carriers and drivers are required to systematically inspect, repair, and maintain CMVs. 49 C.F.R. part 396. Drivers must create a daily report of the condition of the equipment and identify any defects or deficiencies and provide this report to the motor carrier. *Id.* § 396.11. Each CMV must be inspected at least annually by inspectors meeting specified qualifications. *Id.* §§ 396.18, 396.19. FMCSA agents are permitted to enter and inspect any CMV. *Id.* § 396.9.

potentially problematic FMCSRs (the Volpe FMCSR Review).[193] The FMCSA has also identified these FMCSRs and has begun a discussion with the public about potential amendments and interpretations of the FMCSRs for level 4 and level 5 automated driving systems.

The Volpe FMCSR Review examined the complications that exist under seven different modes of deploying CMVs (from monitoring lower-level systems to remote supervision of level 5 CMVs). The Volpe FMCSR Review stated that deployments "that retain direct human driver involvement in control of the vehicle, at least occasionally, will face the fewest challenges in complying with the current FMCSRs."[194] Conversely, those deployments that do not have a human operator face the most significant challenges to complying with the FMCSRs.[195] For example, simple things like the need to immediately place warning devices after a CMV stops on a highway create problems for CMVs that do not have human operators. The Volpe Center believed that many interpretive challenges depend upon the identity of the "driver" or "operator" of the automated CMV.

In its May 28, 2019, advanced notice of proposed rulemaking, the FMCSA stated that the terms "driver" and "operator" can include an automated driving system and that it "will interpret its regulations to no longer assume that the CMV driver is always a human or that a human is present onboard a commercial vehicle during its operation, provided that the vehicle is equipped with a level 4 or level 5 [automated driving system] and is operating within its [operational design domain] (in the case of Level 4)."[196] With that interpretation, the FMCSA acknowledged that it will need to consider rules for situations where the automated driving system driver or operator is incapable of performing the human obligations or where the human obligations are no longer needed, such as hours-of-service limitations. However, for the most part,[197] FMCSA expressed a belief that its existing rules should apply when there is a human operator in the vehicle.

---

193. *See* David Perlman et al., *Review of the Federal Motor Carrier Safety Regulations for Automated Commercial Vehicles: Preliminary Assessment of Interpretation and Enforcement Challenges, Questions, and Gaps*, John A. Volpe Nat'l Transp. Sys. Ctr. (Mar. 2018), https://rosap.ntl.bts.gov/view/dot/35426.

194. *Id.* at 8.

195. *Id.*

196. Fed. Motor Carrier Safety Admin., *Safe Integration of Automated Driving Systems-Equipped Commercial Motor Vehicles*, 84 Fed. Reg. 24,449-01, 24,453 (May 28, 2019).

197. A notable exception to allowing the status quo to continue is that the FMCSA believes that drivers can, when in a sleeper berth, treat their time as off-duty similar to the role as a team driver. *Id.* at 24,454. This is a significantly favorable interpretation for the use of automated CMVs, because they can operate without the constraints of the hours-of-service limitations.

3. Waiver and Exemption

For those carriers that want specific FMCSRs not to apply, the FMCSRs contain procedures for seeking waivers or exemptions from their application.[198] Waivers and exemptions differ by the length of their relief—a waiver provides three months of relief, while an exemption provides relief for up to five years and can be renewed.[199] The FMCSA has the authority to grant waivers and exemptions only for certain FMCSRs.[200] Under either option, the requestor needs to submit a written request to the FMCSA explaining the need for the waiver or exemption, the FMCSRs for which the waiver or exemption is needed, and other information including the safety effects of the waiver or exemption and the impact if the waiver or exemption is not granted.[201] Requests for exemptions, but not waivers, are published in the Federal Register so that the public can comment on the requests.[202]

4. State Review and Preemption

States are required to annually analyze their laws and the laws of their subdivisions to determine whether they are compatible with the FMCSRs.[203] States must submit their results to the FMCSA by August 1 each year.[204] Generally, the FMCSRs do not broadly preempt state law or prevent states from enacting safety regulations.[205] The FMCSA has the power to preempt state laws and regulations.[206] State laws that are determined by the FMCSA to be less stringent than an FMSCR are preempted.[207] State laws or regulations that are in addition to or more stringent than an FMCSR are preempted if the FMCSA determines that the laws or regulations have no safety benefits or are incompatible with an FMCSR, or that their enforcement unreasonably burdens interstate com-

---

198. 49 C.F.R. part 381.

199. *Id.* §§ 381.200, 381.300.

200. These parts and sections are part 382 (Controlled Substances and Alcohol Use and Testing); part 383 (Commercial Driver's License Standards; Requirements and Penalties); section 390.19 (Motor Carrier Identification Report); section 390.21 (Marking of Commercial Motor Vehicles); part 391 (Qualifications of Drivers); part 392 (Driving of Commercial Motor Vehicles); part 393 (Parts and Accessories Necessary for Safe Operation); part 395 (Hours of Service of Drivers); part 396 (Inspection, Repair, and Maintenance) (except section 396.25); and part 399 (Step, Handhold and Deck Requirements). 49 C.F.R. §§ 381.200, 381.300.

201. 49 C.F.R. §§ 381.210, 381.310.

202. *Id.* § 381.315.

203. *Id.* § 355.21.

204. *Id.* § 355.23.

205. *See* 49 C.F.R. § 390.9 ("Except as otherwise specifically indicated, subchapter B of this chapter is not intended to preclude States or subdivisions thereof from establishing or enforcing State or local laws relating to safety, the compliance with which would not prevent full compliance with these regulations by the person subject thereto.").

206. 49 U.S.C. § 31141.

207. *Id.* § 31141(c)(3).

merce.[208] To preempt a state law, the FMCSA initiates a rulemaking proceeding under part 389 of the FMCSRs.[209]

Preemption can also occur in the CMV context under the Federal Aviation Administration Authorization Act of 1994.[210] Known as FAAA preemption, the statute prohibits states or political subdivisions from "enact[ing] a law, regulation, or other provision hav[ing] the force and effect of law related to a price, route, or service of" motor carriers of property.[211] This language was modeled after the language in the Airline Deregulation Act of 1978,[212] and the United States Supreme Court utilizes its judicial interpretations of the Airline Deregulation Act of 1978 in interpreting the FAAA preemption.[213] This language is based on the belief and goal that "competitive market forces" on rates, routes, and services "stimulate efficiency, innovation, and low prices, as well as variety and quality."[214] The Supreme Court has described FAAA preemption as follows:

> (1) that "[s]tate enforcement actions having a connection with, or reference to," carrier "'rates, routes, or services' are pre-empted"; (2) that such pre-emption may occur even if a state law's effect on rates, routes, or services "is only indirect"; (3) that, in respect to pre-emption, it makes no difference whether a state law is "consistent" or "inconsistent" with federal regulation; and (4) that pre-emption occurs at least where state laws have a "significant impact" related to Congress' deregulatory and pre-emption-related objectives.[215]

State laws that only have "tenuous, remote, or peripheral" effects are not preempted.[216] The FAAA exempts state safety regulations, highway regulations based on the size or weight of the truck or the hazardous materials carried

208. *Id.* § 31141(c)(4). In weighing whether a state law or regulation unreasonably burdens interstate commerce, the FMCSA can consider the effects such law or regulation would have if implemented in all 50 states. *Id.* § 31141(c)(5).

209. 49 C.F.R. § 355.25.

210. 49 U.S.C. § 140501. The statute also preempts certain laws and regulations on motor carriers of passengers (i.e., buses) and freight forwarders and brokers. *Id.*

211. *Id.* § 140501(c)(1). The statute exempts Hawaii. *Id.* § 140501(c)(4).

212. The difference in the statutory provisions is that the FAAA preemption uses the words "with respect to the transportation of property." Dan's City Used Cars, Inc. v. Pelkey, 569 U.S. 251, 261, 133 S. Ct. 1769, 185 L. Ed. 2d 909 (2013). "That phrase 'massively limits the scope of preemption' ordered by the FAAA." *Id.* (quoting City of Columbus v. Ours Garage & Wrecker Serv., Inc., 536 U.S. 424, 449, 122 S. Ct. 2226, 153 L. Ed. 2d 430 (2002) (Scalia, J., dissenting)).

213. Rowe v. N.H. Motor Transp. Ass'n, 552 U.S. 364, 370, 128 S. Ct. 989, 169 L. Ed. 2d 933 (2008).

214. *Id.* at 371 (quoting Morales v. Trans World Airlines, Inc., 504 U.S. 374, 378, 112 S. Ct. 2031, 119 L. Ed. 2d 157 (1992) (internal quotation marks omitted)).

215. *Id.* at 370–71 (internal citations omitted) (quoting *Morales*, 504 U.S. at 384, 386–87, 390).

216. *Id.* at 371 (quoting *Morales*, 504 U.S. at 390).

therein, insurance, intrastate transportation of household goods, certain tow truck operations, and certain uniform cargo rules and antitrust immunities.[217]

D. NTSB

The NTSB, though a transportation agency, is independent of the USDOT and investigates civil transportation crashes, advocates for safety improvements that can prevent crashes, and decides appeals for pilot and mariner certification requests. Congress has charged the NTSB with investigating all aircraft accidents, certain highway accidents selected in cooperation with states, railroad crashes, pipeline crashes, and marine casualties, as well as "any other accident related to the transportation of individuals or property when the Board decides (i) the accident is catastrophic; (ii) the accident involves problems of a recurring character; or (iii) the investigation of the accident would carry out this chapter."[218] Except for marine casualties, the NTSB has priority over any other United States agency, including the USDOT, in investigating accidents.[219] The NTSB maintains a docket that allows the public to review relevant documents for its analysis.

The NTSB has exercised its investigatory power over numerous crashes involving driving automation systems, including the Navya Arma automated shuttle crash on November 8, 2017, in Las Vegas,[220] several crashes involving Tesla Autopilot,[221] and the infamous crash involving an Uber automated vehicle in Tempe, Arizona.[222] The NTSB has shown that it is less friendly to the industry than NHTSA in its investigations and can be critical of NHTSA. In its final report on the Walter Huang Tesla Autopilot crash, the NTSB stated that manufacturers should geofence their driving automation systems and include driver monitoring systems.[223] The NTSB also criticized NHTSA's lack of oversight over driving automation systems' operational design domains and the foreseeable misuse of these systems.[224] Based on its history of investigating

217. 49 U.S.C. § 14501(c)(2), (3).

218. *Id.* § 1131(a)(1)(F).

219. *Id.* § 1131(2)(A).

220. Nat'l Transp. Safety Board, *Highway Accident Brief, Low-Speed Collision between Truck-Tractor and Autonomous Shuttle, Las Vegas, Nevada, November 8, 2017*, https://dms.ntsb.gov/public/62500-62999/62540/626197.pdf.

221. *E.g.*, Nat'l Transp. Safety Board, *Preliminary Report*, https://www.ntsb.gov/investigations/AccidentReports/Reports/HWY19FH008-preliminary.pdf (Tesla Model 3 crash on March 1, 2019); Nat'l Transp. Safety Board, *Preliminary Report*, https://ntsb.gov/investigations/AccidentReports/Reports/HWY18FH011-preliminary.pdf (Tesla Model X crash on March 23, 2018).

222. Nat'l Transp. Safety Board, *Preliminary Report*, https://www.ntsb.gov/investigations/AccidentReports/Reports/HWY18MH010-prelim.pdf (last visited May 4, 2020).

223. Nat'l Transp. Safety Board, *Abstract to Final Report, Collision between a Sport Utility Vehicle Operating with Partial Driving Automation and a Crash Attenuator* 2 (Feb. 25, 2020), https://www.ntsb.gov/news/events/Documents/2020-HWY18FH011-BMG-abstract.pdf.

224. *Id.* at 2–3.

crashes involving vehicles with automation features, the NTSB will likely have a significant role in the deployment of automated vehicles and crashes involving these vehicles.

III. NONTRANSPORTATION FEDERAL AGENCIES

The regulation of automated vehicles is not limited to the USDOT and NTSB. Nontransportation agencies will have responsibility over aspects of automated vehicles. Two agencies with prominent roles will be the Federal Communications Commission (FCC)[225] and the Federal Trade Commission (FTC).

A. FCC

Today's vehicles are no longer isolated and have become continually connected to other systems. Vehicles will use electromagnetic spectrum for, and will become more connected through, manufacturer software updates, infotainment systems, radars, GPSs, LiDARs, and vehicle-to-vehicle, vehicle-to-infrastructure, vehicle-to-pedestrian, and vehicle-to-everything communications. Spectrum is a finite, though nondepletable, resource. The use of telecommunications in automated vehicles means that the FCC will be involved in regulating automated vehicles.

The FCC regulates spectrum access, use, and rights.[226] One of the most important and potentially controversial allocations was the FCC's allocation to

225. The FCC is one of two agencies that has authority over electromagnetic spectrum allocation and use. The other agency is the National Telecommunications and Information Administration (NTIA). NTIA is, among other things, responsible for spectrum use by the federal government. The FCC, among other things, regulates all other spectrum use in the United States and its territories. Because this book is primarily concerned with automated vehicle use by nonfederal departments or agencies, this section will introduce the FCC's regulation of spectrum use and telecommunications.

226. *See* 47 U.S.C. § 303(y) ("Except as otherwise provided in this chapter, the Commission from time to time, as public convenience, interest, or necessity requires, shall . . . (y) Have authority to allocate electromagnetic spectrum so as to provide flexibility of use, if— (1) such use is consistent with international agreements to which the United States is a party; and (2) the Commission finds, after notice and an opportunity for public comment, that—(A) such an allocation would be in the public interest; (B) such use would not deter investment in communications services and systems, or technology development; and (C) such use would not result in harmful interference among users."). The FCC allocates spectrum bands by specified uses, such as specific applications like dedicated short-range communications or broad categories like aviation. Because the demand for spectrum is high, in many instances, multiple uses can be allocated to a given spectrum band. In these instances, the FCC allocates the priority and hierarchy among the competing demands for spectrum within the band. Robert B. Kelly & Mark D. Johnson, *Symposium, Defining a Stable, Protected and Secure Spectrum Environment for Autonomous Vehicles*, 52 Santa Clara L. Rev. 1271, 1288 (2012). The FCC can make a specific use primary, coprimary, secondary, or unlicensed. Primary has the highest priority, and secondary or unlicensed users cannot interfere with primary users and have to accept their interference. Coprimary users each have primary priority and must avoid causing interference to other primary users.

dedicated short-range communications (DSRC). In 1998, Congress directed the FCC to work with the USDOT to allocate spectrum for intelligent transportation systems, such as DSRC.[227] In 1999, the FCC allocated 75 megahertz of spectrum at 5.850–5.925 GHz for use by DSRC.[228] DSRC shares this band with military radar systems operated by the Department of Defense and the Fixed Satellite Service.[229] The FCC chairman announced plans in May 2019 to take a "fresh look" at the allocation to DSRC in the 5.9 GHz band.[230] A reallocation of the 5.9 GHz band could pit the FCC against the USDOT, as NHTSA believes "[p]reserving the 5.9 GHz band for transportation communications is essential to public safety today and in the future."[231] On December 12, 2019, the FCC adopted a notice of proposed rulemaking to consider the use of the 5.850–5.925 GHz band.[232]

Once the FCC has allocated spectrum for a specified use within a band, the FCC then assigns licenses to entities that seek to operate services within that band.[233] The FCC has some flexibility in determining how to award licenses. Each applicant must apply in writing for a license to use a band.[234] Before granting an application, the FCC must determine that the "public interest, convenience, and necessity will be served by the granting of such application."[235] The FCC uses frequency coordinators to help screen applicants and to examine applications.[236] If there are competing, qualifying applications for a similar license, the FCC can grant licenses on a shared basis, a first-come, first-served basis, randomly, or through competitive bidding.[237]

---

227. Transportation Equity Act for the 21st Century, Pub. L. No. 105–178, § 5206(f), 112 Stat. 107 (1998).

228. Fed. Commc'ns Comm'n, *Dedicated Short Range Communications on Intelligent Transportation Services*, 64 Fed. Reg. 66,405, 66,405 (Nov. 26, 1999).

229. *Id.*

230. Remarks of FCC Chairman Ajit Pai at the Wi-Fi World Congress 2019, Tysons Corner, Virginia (May 14, 2019), https://docs.fcc.gov/public/attachments/DOC-357456A1.pdf.

231. U.S. Department of Transportation's National Highway Traffic Safety Administration Issues Statement on Safety Value of 5.9 GHz Spectrum (Oct. 24, 2018), https://www.nhtsa.gov/press-releases/us-department-transportations-national-highway-traffic-safety-administration-issues.

232. Fed. Commc'ns Comm'n, *Notice of Proposed Rulemaking, In re Use of the 5.850-5.925 GHz Band* (Dec. 12, 2019), https://docs.fcc.gov/public/attachments/FCC-19-129A1.pdf.

233. *See* 47 U.S.C. § 307(e). Licensing by rule is permitted for the following types of radio services: "(A) the citizens band radio service; (B) the radio control service; (C) the aviation radio service for aircraft stations operated on domestic flights when such aircraft are not otherwise required to carry a radio station; and (D) the maritime radio service for ship stations navigated on domestic voyages when such ships are not otherwise required to carry a radio station." *Id.* In its ruling on DSRC, the FCC ruled that onboard units can be licensed by rule. *In the Matter of Amendment of the Commission's Rules Regarding Dedicated Short-Range Communication Services in the 5.850-5.925 GHz Band (5.9 GHz Band)*, Fed. Commc'ns Comm'n, 19 F.C.C. Rcd. 2458 ¶ 67 (Feb. 10, 2004).

234. 47 U.S.C. §§ 308, 309.

235. *Id.* § 309(a).

236. Kelly & Johnson, *supra* note 226, at 1308–09. Frequency coordinators that are certified by the FCC can review whether an applicant is qualified to hold a license. *Id.*

237. *See* 47 U.S.C. § 309.

Once licenses are issued within a band, the FCC generally oversees and manages the use to minimize interference.[238] Coordination can be effectuated by messaging or among licensees.[239] Spectrum can be managed through interference mitigations and resolution techniques, such as software review of applications to determine potential interference and conflict, channel registration, use of a third-party database manager, and access priority.[240]

B. FTC

The FTC is an independent federal agency that protects consumers and promotes competition. Section 5 of the Federal Trade Commission Act prohibits "unfair methods of competition in or affecting commerce" and "unfair or deceptive acts or practices in or affecting commerce."[241] An "unfair" method, practice, or act is one that "causes or is likely to cause substantial injury to consumers which is not reasonably avoidable by consumers themselves and not outweighed by countervailing benefits to consumers or to competition."[242] These two prongs of enforcement give the FTC a consumer protection role and a competition protection role. In its consumer protection role, the FTC, among other things, enforces laws that protect consumers from unfair, deceptive, or fraudulent practices and is the primary federal agency responsible for protecting consumers' privacy and personal information. The FTC has used both the unfair-method and the deceptive prongs of section 5 to address privacy and cybersecurity breaches. In its competition promotion role, the FTC enforces the antitrust laws, such as the Sherman Act and the Clayton Act, in the United States and reviews mergers and acquisitions to ensure that they do not substantially reduce competition.

The FTC could interact with the automated vehicle industry in either role. In its consumer protection role, the FTC could prohibit certain practices of automated vehicle manufacturers or service providers or take specific actions to protect data in automated vehicles.[243] The FTC, together with NHTSA, held a

---

238. Kelly & Johnson, *supra* note 226, at 1307.

239. *Id.* at 1309–13. The DSRC rules utilize a messaging priority system that prioritizes safety-of-life messages followed by public safety communication. *Id.* at 1310.

240. *Id.* at 1310–12.

241. 15 U.S.C. § 45(a)(1).

242. *Id.* § 45(n).

243. Nat'l Highway Traffic Safety Admin., *Vehicle Data Privacy*, U.S. Dep't of Transp., https://www.nhtsa.gov/technology-innovation/vehicle-data-privacy (last visited Sept. 2, 2019) ("The FTC has authority to bring actions against companies or individuals that engage in unfair or deceptive acts or practices, including those involving vehicle data privacy and security. The agency uses law enforcement, policy initiatives, and consumer and business education to accomplish its mission. In the motor vehicle context, for example, the FTC could use its enforcement authority in appropriate circumstances to bring an action against an automaker that uses a consumer's data in a way that violates the manufacturer's stated privacy policies."). In 2016, Mercedes released an advertisement that many believed misled the public about its 2017 E-Class's features and capabilities when it comes to automated driving, and the FTC was the agency these people filed complaints with about the Mer-

workshop on June 28, 2017, on privacy and security issues related to connected automated vehicles.[244] This role of the FTC is detailed in Chapter 16. In its competition protection role, the FTC could review and conceivably prevent consolidations, acquisitions, and vertical integrations that it deems anticompetitive.

IV. OTHER FEDERAL ACTS

Numerous stand-alone federal laws will have a significant impact on automated vehicles. This section, however, limits its focus to the Raechel and Jacqueline Houck Safe Rental Car Act of 2015 (the Safe Rental Car Act), the Magnuson-Moss Warranty Act, and the Americans with Disabilities Act of 1990.

A. The Safe Rental Car Act

Congress enacted the Safe Rental Car Act to prohibit rental companies from selling, leasing, or renting a motor vehicle or item of replacement equipment following receipt of a recall notice until the defect is fixed.[245] A rental company is defined as "a person who (A) is engaged in the business of renting covered rental vehicles; and (B) uses for rental purposes a motor vehicle fleet of 35 or more covered rental vehicles, on average, during the calendar year."[246] Covered motor vehicles are those that weigh under 10,000 pounds, that are rented for a term of less than four months, and that are part of a fleet of at least 35 motor vehicles of the rental company.[247]

In addition to car rental companies, the Safe Rental Car Act would likely cover an automated ridesharing company. A recall could effectively shut down an automated ridesharing operation. An automated ridesharing company that sources its vehicles from one manufacturer may be more particularly vulnerable than a company that sources its vehicles from multiple manufacturers.

B. The Magnuson-Moss Warranty Act

The Magnuson-Moss Warranty Act, codified at 15 U.S.C. § 2301 *et seq.*, governs written warranties for consumer products and authorizes the FTC to develop

---

cedes ad. Katie Reilly, *Mercedes Pulled a Self-Driving Car Ad Because It's Not Actually a Self-Driving Car*, TIME (July 30, 2016), https://time.com/4431956/mercedes-benz-ad-confusion-self-driving/.

244. *Connected Cars: Privacy, Security Issues Related to Connected, Automated Vehicles*, FED. TRADE COMM'N. (June 28, 2017), https://www.ftc.gov/news-events/events-calendar/2017/06/connected-cars-privacy-security-issues-related-connected.

245. 49 U.S.C. § 30120(i)(1). The Raechel and Jacqueline Houck Safe Rental Car Act of 2015, S. 1173, 114th Cong. (2015), is codified in various parts of title 49.

246. 49 U.S.C. § 30102(a)(11). State law may cover fleets of 34 vehicles or less. *See* California Consumer Automotive Recall Safety Act, codified at CAL. VEH. CODE §§ 11750-62.

247. *Id.* § 30102(a)(1).

rules on consumer product warranties.[248] The Magnuson-Moss Warranty Act covers vehicles,[249] and sales to consumers of automated vehicles with written warranties will be subject to its requirements. The Magnuson-Moss Warranty Act imposes certain duties and liabilities on suppliers who offer written warranties on their consumer products.[250] Generally, a warranty must explicitly state that it is a full warranty or a limited warranty.[251] A full warranty is one in which the warrantor (1) agrees to remedy defects within a reasonable time at no charge, (2) does not limit the duration of an implied warranty, (3) does not limit or exclude consequential damages unless such limitation or exclusion is conspicuously on the face of the warranty, and (4) provides a replacement or a refund for persistent defects.[252] The warranty and specific information concerning the product must be in a single document[253] and be available to the consumer before purchase.[254] One of the more significant provisions of the Magnuson-Moss Warranty Act is its prohibition on disclaiming or modifying implied warranties when a written warranty is given or when the seller enters into a service contract with the purchaser of the product within 90 days of the sale.[255] However, implied warranties may be limited in duration to the duration of the written warranty.[256] Another important component of the Magnuson-Moss Act is its prohibition on manufacturers' conditioning a warranty on specified replacement parts. Replacement parts may matter a great deal more for automated vehicles than they do for human-driven vehicles, given automated vehicles' dependence on their technology stacks.

V. AMERICANS WITH DISABILITIES ACT OF 1990

The Americans with Disabilities Act regulates certain types of automobile service providers and may apply to certain automated vehicle deployments. Title III of the Americans with Disabilities Act prohibits discrimination against persons

---

248. The FTC has promulgated its rules regarding the Magnuson-Moss Warranty Act in 16 C.F.R. subchapter G (§ 700 *et seq.*).

249. *See* 16 C.F.R. § 700.1 ("The Act applies to written warranties on tangible personal property which is normally used for personal, family, or household purposes.").

250. 16 C.F.R. § 700.3(a).

251. 15 U.S.C. § 2303. A warranty is full when it complies with the federal minimum standards of warranties, while a warranty is limited if it fails to comply with such minimum standards. *Id.*

252. *Id.* § 2304(a).

253. 16 C.F.R. § 701.3(a).

254. *Id.* § 702.3.

255. 15 U.S.C. § 2308(a).

256. *Id.* § 2308(b). For this exception to apply, the written warranty must be of "reasonable duration," conscionable, and "set forth in clear and unmistakable language and prominently displayed on the face of the warranty." *Id.*

with disabilities by providers of specified public transportation.[257] Specified public transportation, in turn, is "transportation by bus, rail, or any other conveyance (other than by aircraft) that provides the general public with general or special service (including charter service) on a regular and continuing basis."[258] The USDOT has stated that "[p]roviders of taxi service are subject to the requirements of this part for private entities primarily engaged in the business of transporting people which provide demand responsive service."[259] Taxicabs have historically avoided the need to have their fleets of vehicles be accessible by disabled persons by purchasing used vehicles.[260] The Americans with Disabilities Act imposes obligations on a service provider, including the need for an antidiscrimination policy for disabilities, prohibitions on special charges and preventing disabled persons from using services, and the need to make reasonable modifications and remove barriers to accommodate disabled persons.[261]

Uber and Lyft have faced lawsuits concerning whether they are required to comply with title III of the Americans with Disabilities Act by being either a public accommodation or a provider of specified public transportation.[262] In these cases, Uber and Lyft have attempted to dismiss these claims by, among other things, arguing they are not providers of specified public transportation because they are "technology companies that provide platforms for peer-to-peer sharing."[263] These courts have declined to dismiss these cases at the motion-to-dismiss stage.

The use of automated vehicles in public transportation will probably fall within the coverage of the Americans with Disabilities Act. Additionally, the operation of an automated ridesharing service where the service provider owns the fleet of automated vehicles will probably fall within the coverage of the act. When designing these services, transportation authorities and companies will need to review their obligations under the Americans with Disabilities Act.

---

257. 42 U.S.C. § 12184(a) ("No individual shall be discriminated against on the basis of disability in the full and equal enjoyment of specified public transportation services provided by a private entity that is primarily engaged in the business of transporting people and whose operations affect commerce.").

258. *Id.* § 12181(10).

259. 49 C.F.R. § 37.29(a). Taxi providers receive some relief in that they need not purchase accessible automobiles. *Id.* § 37.29(b).

260. Bryan Casey, *A Loophole Large Enough to Drive an Autonomous Vehicle Through: The ADA's "New Van" Provision and the Future of Access to Transportation*, 69 STAN. L. REV. ONLINE (Dec. 2016), https://www.stanfordlawreview.org/online/loophole-large-enough/ (describing the new van exception and how automated vehicles may make that exception inapplicable).

261. *See* 42 U.S.C. § 12184.

262. Rachel Reed, *Disability Rights in the Age of Uber: Applying the Americans with Disabilities Act of 1990 to Transportation Network Companies*, 53 GA. ST. U. L. REV. 517 (2017) (discussing cases filed against Uber and Lyft).

263. Ramos v. Uber Techs., Inc., No. SA-14-CA-502-XR, 2015 WL 758087, at *11 (W.D. Tex. Feb. 20, 2015); Nat'l Fed'n of the Blind of Cal. v. Uber Techs., Inc., 103 F. Supp. 3d 1073, 1084 (N.D. Cal. 2015).

Chapter 6

State Regulation of Motor Vehicles and Driving

The federal government's regulation of noncommercial vehicles is narrow in scope and focuses primarily on motor vehicle design. In contrast, states broadly regulate almost all components of driving and motor vehicles, including traffic laws, driver qualification and training, motor vehicle design, dealership laws, insurance, and transportation network companies (TNCs). This chapter canvasses how states exercise their regulatory authority and reviews potential complications created by automated vehicles. This chapter concludes by introducing the dormant Commerce Clause as a limitation on state authority to regulate the automated vehicle industry.

I. SURVEY OF STATE LAWS

Almost all state laws on motor vehicles and driving were enacted before the advent of automated vehicles and assume a human driver is necessary for the safe operation of the vehicle. Because of this fundamental assumption, some state laws create ambiguities when applied to driving automation systems, complicate or prohibit certain benefits of automated driving, lead to unnecessary costs, or are generally outdated. States need to conduct an audit and assessment of their laws, regulations, and rules to determine whether there are complications, barriers, or opportunities for automated vehicles and their use on the roadways. This chapter, with its space constraints, cannot perform a state-by-state audit. This section canvasses state laws that may create complications, opportunities, or barriers for automated vehicles.[1]

A. Rules of the Road

The rules of the road are the basic traffic laws within a jurisdiction. Though there is a uniform Model Vehicle Code,[2] traffic laws throughout the United States or even within individual states are not uniform.[3] Lack of uniformity among states is but one problem with these laws. As I have described elsewhere, "[s]ome seemingly bright line rules, such as speed limits, are not strictly enforced, or otherwise follow unwritten rules. Further complicating matters is that some rules of the road require subjective input of the operator of the vehicle. . . . And if those challenges did not make the algorithm writer's job difficult enough, drivers are sometimes required to follow directives of authorized individuals, even if such directives conflict with 'bright line rules,' like following traffic lights."[4] Additionally, localities have customs, such as the "Pittsburgh left," that, although not necessarily legal under the jurisdiction's rules of the road, drivers follow, the police ignore, and automated vehicles will encounter. Manufacturers will need to identify the rules of the road in each jurisdiction and geofence their programming to comply with those laws.

A hallmark of traffic laws is the use of the terms "driver," "operator," drive," and "operate." Many states use "driver" and "operator" interchangeably, and "drive" and "operate" are typically the controlling terms.[5] These concepts are relatively straightforward to apply when humans are driving or operating vehicles or for level 2 driving automation systems. These concepts create ambiguity and confusion when applied to automated driving systems that operate without the need of a human driver (or without human occupants).[6] Chapter 19 details these concepts, and states should clarify these concepts for automated driving systems.

---

1. For comprehensive analyses of state laws that could create impediments or barriers to automated vehicles, see generally Tammy E. Trimble, Lisa Loftus-Otway, Susanna Gallun, Sam Morrissey, Glenn Havinoviski, Betty Serian & Geoff Johnson, *Implications of Connected and Automated Driving Systems, Vol. 2: State Legal and Regulatory Audit*, NAT'L ACADS. OF SCIS., ENG'G, & MED. (2018), https://doi.org/10.17226/25294 [hereinafter Trimble et al.]; Bryant Walker Smith, *Automated Vehicles Are Probably Legal in the United States*, 1 TEX. A&M L. REV. 411 (2014). An example of a state-specific analysis is Bryan Gibson, *Analysis of Autonomous Vehicle Policies* (Research Report KTC-16-25/PL-26-1F), KY. TRANSP. CTR. (Mar. 2017), https://pdfs.semanticscholar.org/fd2f/1e0d7045049797f859974b8f71c29c32e367.pdf.

2. The last version of the uniform Model Vehicle Code was proposed in 2000.

3. Trimble et al., *supra* note 1, at 65. States allow their political subdivisions to enact certain types of regulations on the rules of the road. *See, e.g.*, CAL. VEH. CODE § 21100.

4. Jeffrey K. Gurney, *Driving into the Unknown: Examining the Crossroads of Criminal Law and Autonomous Vehicles*, 5 WAKE FOREST J.L. & POL'Y 393, 412–13 (2015).

5. Trimble et al., *supra* note 1, at 20.

6. *See id.* at 56.

B. Driver Education, Driver Tests, and Driver's Licenses

The road to driving, in almost all states, begins with successful completion of a driver education program, is followed by a satisfactory score on a driver's test, and concludes with the obtainment of a driver's license. Driving automation systems and advanced driver assistance systems (ADASs) require knowledge of components and skills that differ from conventional vehicles. Driver educational courses that focus only on conventional vehicles will not prepare future drivers with the knowledge and skills necessary to drive safely. States should consider incorporating advanced driving technologies into their education program requirements. These educational materials could instruct drivers on how to engage and disengage ADASs and driving automation systems, the limitations of these systems, and the roles and responsibilities of human drivers of level 1, level 2, and level 3 driving automation systems. States could also consider addressing the education of these features when already licensed vehicle owners seek to register vehicles equipped with such features.

The driver's examination typically involves a written test of basic competency of the rules of the road and general knowledge of driving laws and a skills test of the applicant's actual driving behavior.[7] In coordination with including advanced automation features in driver education courses, states should incorporate knowledge of the capabilities of automation features (including ADASs and driving automation systems) into their driver's tests.

An ancillary issue that may arise is whether states will have the right to condition registration of an automated vehicle or licensing of an automated vehicle on the vehicle passing a state-run, state-created examination. This requirement could conceivably apply to vehicles or manufacturers. This type of precertification would impose more significant requirements on manufacturers than those required under the Motor Vehicle Safety Act. States seeking to impose such a pretesting regime may find that the regime is preempted or violates the dormant Commerce Clause.

Every state requires a driver or operator of a motor vehicle to obtain a driver's license. Every state also prohibits classes or groups of people from obtaining driver's licenses. Typical prohibitions or limitations are based on age, physical and mental capabilities, and not being a habitual user of drugs or alcohol. Because these classes or groups of persons are prohibited from obtaining a driver's license, they currently rely on public transportation or others for transportation.

A license requirement for vehicles using level 4 or level 5 automated driving systems may be unnecessary and defeat important benefits of automated vehicles. Minors or individuals who suffer from physical or mental disabilities

---

7. *Id.*

or skill degeneration may not be able to avail themselves of the benefits of these vehicles if they need a traditional license.[8]

C. Registration of Motor Vehicles

States require vehicle owners to register their vehicles. Vehicle registration laws should not require any significant overhaul because they are based on ownership of the vehicle. However, the vehicle registration process provides opportunities for states to learn about automated vehicles on their roadways. This information could be useful to law enforcement to verify a vehicle's automated capabilities. Vehicle registration is a key component of the proposed Uniform Automated Operation of Vehicles Act described in Chapter 7.

D. Vehicle Inspection Laws

Some states require annual vehicle inspections, while other states require inspections but not annually.[9] In some of these states, the inspectors are police officers, while in others, they are private parties.[10] Vehicle inspection laws may become more or less appealing depending upon the capabilities of driving automation systems. On the one hand, inspections may be needed more when automated vehicles depend significantly upon the quality and effectiveness of the hardware and sensors on the vehicle. Automated vehicle inspections may require specialized expertise on sensors such as LiDARs, cameras, and radars. On the other hand, an automated vehicle should be able to detect when sensors or hardware components are not operating properly and affecting its ability to safely navigate the roadways, rendering vehicle inspections redundant or unnecessary.

E. Automobile Insurance

Although the federal government can conceivably regulate insurance using its Commerce Clause powers, Congress delegated the regulation of most types of insurance to states.[11] States have exercised this power to broadly regulate the automobile insurance industry, including mandating certain types of insurance coverage, setting or approving insurance rates, and mandating rating

8. *See id.* at 27 (stating that license requirements may mean that "many of the same users who hope to benefit from automated transportation may be disqualified").

9. *Id.* at 48.

10. *See, e.g.,* Cal. Veh. Code § 2814 (West).

11. In *United States v. S.-E. Underwriters Ass'n*, 322 U.S. 533, 64 S. Ct. 1162, 88 L. Ed. 1440 (1944), the United States Supreme Court held that the regulation of insurance fell within the Commerce Clause powers of the federal government. Congress reacted quickly in the McCarren-Ferguson Act of 1945 by delegating the power to regulate insurance to the states.

factors.[12] States will need to consider what types of insurance to require for different deployments of automated vehicles.

Some states explicitly outline rating factors.[13] These statutory or regulatory regimes may require insurance companies to consider factors that no longer make sense for highly automated vehicles. Rating factors based on driving history, credit score, age, gender, and marital status may no longer be indicative of risk when an automated vehicle does most of or all of the driving.

F. Motor Vehicle Design

States regulate motor vehicle design prospectively in the form of statutes and regulations and retrospectively in the form of product liability litigation. The Motor Vehicle Safety Act contains a preemption clause (the Preemption Clause) that limits prospective regulation by states of motor vehicle design:

> (1) When a motor vehicle safety standard is in effect under this chapter, a State or a political subdivision of a State may prescribe or continue in effect a standard applicable to the same aspect of performance of a motor vehicle or motor vehicle equipment only if the standard is identical to the standard prescribed under this chapter. However, the United States Government, a State, or a political subdivision of a State may prescribe a standard for a motor vehicle or motor vehicle equipment obtained for its own use that imposes a higher performance requirement than that required by the otherwise applicable standard under this chapter.
>
> (2) A State may enforce a standard that is identical to a standard prescribed under this chapter.[14]

Except for government-owned vehicles, states are permitted to enact identical safety standards. The Motor Vehicle Safety Act has a "savings clause" that purports to exclude common law liability claims, including products liability claims, from the Preemption Clause (the Savings Clause):

> Common law liability.—Compliance with a motor vehicle safety standard prescribed under this chapter does not exempt a person from liability at common law.[15]

The United States Supreme Court has, on three occasions, considered the interplay of the Preemption Clause and the Savings Clause for defective design

12. See generally Robert W. Peterson, *New Technology—Old Law: Autonomous Vehicles and California's Insurance Framework*, 52 SANTA CLARA L. REV. 1341 (2012), for a discussion of California insurance regulation and automated vehicles.

13. *See, e.g.*, CAL. CODE REGS. tit. 10, § 2632.5.

14. 49 U.S.C. § 30103(b).

15. *Id.* § 30103(e).

claims.[16] These cases and the Savings Clause are detailed in Chapter 8. This section focuses on the potential preemptive effect of guidance issued by the USDOT on automated vehicles.

The lack of a Safety Standard that directly addresses automated vehicles (other than human-driven vehicles more generically) does not mean that states can freely regulate automated vehicles and their designs without considering preemption. State regulations that create an obstacle to or conflict with federal law may be preempted.[17] Conflict preemption can arise "where it is impossible for a private party to comply with both state and federal requirements, or where state law stands as an obstacle to the accomplishment and execution of the full purposes and objectives of Congress."[18]

The USDOT has issued five policies on automated vehicles and a denial of a petition to commence rulemaking on automatic emergency braking that may be relevant to preemption. In these policies, the USDOT has explained its beliefs and goals in regulating—or more appropriately not regulating—automated vehicles. The USDOT and NHTSA have said that states should not regulate the design or performance of automated vehicles because this is a role reserved for NHTSA. The USDOT and NHTSA's position is not to directly regulate automated vehicles or their performance to allow for flexibility and innovation. NHTSA has explained that it is working with developers and manufacturers informally to "accomplish two goals: First, to make sure that new technologies are developed and deployed safely; and second, to leave room for flexibility and safety innovation."[19] State regulation on the design or performance of automated vehicles may interfere with NHTSA's policies and create an obstacle to NHTSA being able to "leave room for flexibility and safety innovation." State laws that interfere with these goals may be preempted, even in the absence of federal regulation. The Arizona Court of Appeals has held that a state products liability lawsuit based on failure to equip vehicles with automatic emergency braking was preempted for interfering with these goals.[20]

---

16. Freightliner Corp. v. Myrick, 514 U.S. 280, 115 S. Ct. 1483, 131 L. Ed. 2d 385 (1995); Geier v. Am. Honda Motor Co., 529 U.S. 861, 120 S. Ct. 1913, 146 L. Ed. 2d 914 (2000); Williamson v. Mazda Motor of Am., Inc., 562 U.S. 323, 131 S. Ct. 1131, 179 L. Ed. 2d 75 (2011).

17. *Freightliner Corp.*, 514 U.S. at 288–89 (holding that the existence of the Preemption Clause does not mean that implied conflict preemption cannot apply).

18. Sprietsma v. Mercury Marine, a Div. of Brunswick Corp., 537 U.S. 51, 64, 123 S. Ct. 518, 154 L. Ed. 2d 466 (2002) (internal quotation marks and citations omitted).

19. Nat'l Highway Traffic Safety Admin., *Federal Motor Vehicle Safety Standards; Automatic Emergency Braking*, 82 Fed. 8391, 8394 (Jan. 25, 2017).

20. Dashi v. Nissan N. Am., Inc., 445 P.3d 13 (Ariz. Ct. App. 2019).

G. Commercial Motor Vehicles and Their Drivers

Congress has not preempted state regulation of commercial motor vehicles and their drivers.[21] States have broadly regulated in this space. States are responsible for licensing commercial motor vehicle drivers and developing and administering the knowledge and skills tests.[22] States need to review their regulations of commercial motor vehicles and drivers to determine what revisions are needed to facilitate the adoption of automated commercial motor vehicles.

H. Lemon Laws

The mass production of motor vehicles inevitably will lead to defective parts and defective vehicles. Warranties on new vehicles are intended to provide vehicle owners with remedies for many of these defects. However, in some instances, consumers repeatedly need to have their vehicles repaired. After several repairs within a specified time, owners can require an automobile manufacturer to replace a defective vehicle or refund its purchase price.[23] These lemon laws supplement a manufacturer's warranty and a state's commercial code.[24]

Lemon laws are particularly well suited to hardware defects, but ambiguities may arise when applied to software defects. For example, if a software system is readily updated, is each update considered a "repair" of the software system? Also, if these updates address bugs in the software system and such bugs could cause an automated vehicle to crash and cause death or significant bodily injury,[25] is the vehicle subject to replacement (or refund) under a state lemon law? State courts may need to grapple with how many bugs are too many bugs before vehicles could be subject to remedies under state lemon laws.

---

21. Chapter 5 describes preemption of commercial motor vehicles.

22. *Commercial Drivers Licenses, States*, Fed. Motor Carrier Safety Admin., https://www.fmcsa.dot.gov/registration/commercial-drivers-license/states (last visited Dec. 1, 2019).

23. *See generally* Louis J. Sirico, Jr., *Automobile Lemon Laws: An Annotated Bibliography*, 8 Loy. Consumer L. Rev. 39 (1995).

24. *See generally* Elizabeth Vollmar, *Lemon Laws: Putting the Squeeze on Automobile Manufacturers*, 61 Wash. U. L. Q. 1125 (1984).

25. *See, e.g.*, Conn. Gen. Stat. § 42-179 ("If a motor vehicle has a nonconformity which results in a condition which is likely to cause death or serious bodily injury if the vehicle is driven, it shall be presumed that a reasonable number of attempts have been undertaken to conform such vehicle to the applicable express warranties if the nonconformity has been subject to repair at least twice by the manufacturer or its agents or authorized dealers within the express warranty term or during the period of one year following the date of the original delivery of the motor vehicle to a consumer, whichever period ends first, but such nonconformity continues to exist.").

I. Franchise Regulations

States regulate motor vehicle distribution to prohibit automakers from conducting specified activities, such as sales or service, directly with consumers.[26] These protectionist laws are intended to address a perceived lack of bargaining power that dealerships have with the automakers.[27] Every state regulates the relationship between automakers and automobile dealers. Some of these laws are more protectionist than others.[28] These statutes also regulate automakers' ability to terminate, cancel, or transfer franchise requirements.

There are at least four prominent ways these laws could interfere with automated vehicles. First, in the case of a massive automated ridesharing expansion

---

26. *See, e.g.*, S.D. Codified Laws § 32-6B-80 ("[N]o manufacturer or franchisor may directly or indirectly: (1) Own an interest in a vehicle dealer or dealership; (2) Operate or control a vehicle dealer or dealership; or (3) Act in the capacity of a vehicle dealer."); Va. Code Ann. § 46.2-1572 (West) ("It shall be unlawful for any motor vehicle manufacturer, factory branch, distributor, distributor branch, or subsidiary thereof, to own, operate, or control any motor vehicle dealership in the Commonwealth."); Wis. Stat. § 218.0121(2m) ("A [manufacturer] shall not, directly or indirectly, hold an ownership interest in or operate or control a motor vehicle dealership in this state."). The federal Automobile Dealer's Day in Court Act gives automobile dealers the right to sue automakers in federal court if automakers fail to perform or comply in good faith with the terms of a franchise agreement or in terminating, canceling, or declining to renew a franchise agreement. 15 U.S.C. § 1222. Good faith is defined to "mean the duty of each party to any franchise, and all officers, employees, or agents thereof to act in a fair and equitable manner toward each other so as to guarantee the one party freedom from coercion, intimidation, or threats of coercion or intimidation from the other party: *Provided,* That recommendation, endorsement, exposition, persuasion, urging or argument shall not be deemed to constitute a lack of good faith." 15 U.S.C. § 1221(e). Many courts have interpreted a breach of good faith to require coercion on the part of the automaker. *See, e.g.*, Ed Houser Enters., Inc. v. Gen. Motors Corp., 595 F.2d 366, 371 (7th Cir. 1978) ("The Act does not create a cause of action based on an alleged discriminatory practice without the requisite causal connection of coercion on the part of the defendant."); Hanley v. Chrysler Motors Corp., 433 F.2d 708, 712 (10th Cir. 1970) ("The lack of 'good faith' as used in the Act must be related in terms of coercion and intimidation on the part of the manufacturer."). Because of the need of the dealer to show coercion by the automaker, the Automobile Dealer's Day in Court Act has not been a panacea for dealers.

27. *See, e.g.*, Ford Motor Co. v. Texas Dep't of Transp., 106 F. Supp. 2d 905, 908 (W.D. Tex. 2000), *aff'd*, 264 F.3d 493 (5th Cir. 2001) ("Based on valid legislative findings, the Texas Legislature has enacted a statutory scheme that attempts to equalize the market power between manufacturers and dealers, and further the public interest of the citizens of Texas, by prohibiting manufacturers from acting in the capacity of a dealer."). This section will not discuss the underlying merits of these laws from an economic or social perspective. For discussions on those topics, see generally Daniel A. Crane, *Tesla, Dealer Franchise Laws, and the Politics of Crony Capitalism*, 101 Iowa L. Rev. 573 (2016); Francine Lafontaine & Fiona Scott Morton, *Markets: State Franchise Laws, Dealer Terminations, and the Auto Crisis*, 24 J. Econ. Persp. 233 (2010); Gerald R. Bodisch, Econ. Analysis Grp., *Economic Effects of State Bans on Direct Manufacturer Sales to Car Buyers*, U.S. Dep't of Justice (2009), https://www.justice.gov/atr/economic-effects-state-bans-direct-manufacturer-sales-car-buyers; Fiona Scott Morton & Ann McDermott, *Retail Auto Sales: Tesla v. State Vehicle Franchise Laws*, *in* The Antitrust Revolution: Economics, Competition, and Policy (John E. Kwoka Jr. & Lawrence J. White eds., 2017); Derek E. Empie, *Note, The Dormant Internet: Are State Regulations of Motor Vehicle Sales by Manufacturers on the Information Superhighway Obstructing Interstate and Internet Commerce?*, 18 Ga. St. U. L. Rev. 827 (2002).

28. Jessica Higashiyama, State Automobile Dealer Franchise Laws: Have They Become the Proverbial Snake in the Grass? 11 (unpublished manuscript version April 2009), https://papers.ssrn.com/sol3/papers.cfm?abstract_id=1394877.

where private ownership of vehicles declines rapidly, these laws will govern the potential closure of dealerships. Second, to the extent that automakers desire to have more control over private sales of automated vehicles, these laws may prohibit that closer role—even if the automakers claim that such direct contact at the point of sale could further safety. Third, automated vehicle manufacturers may desire to convey important information to consumers, such as the system's operational design domain, the means of activating and deactivating the system, and the system's underlying capabilities. However, these laws will prohibit this direct communication at the point of sale, and automakers will need to rely on independent dealerships to convey such information and perform such training. Fourth, many of these laws prohibit aftermarket service by automakers, meaning they cannot control service of their driving automation systems.

J. Transportation Network Company Laws

TNCs, like Uber and Lyft, rapidly transformed transportation throughout the country (and the world). As a result, states and local governments needed to react quickly. Every state regulates TNCs, and some local governments have added additional regulations. These laws have typically focused on driver age requirements, background checks, vehicle requirements, fees, insurance, and common carrier designation. Because companies are expected to offer automated ridesharing services that resemble TNC services, the first analysis is whether automated ridesharing companies are TNCs subject to these regulatory regimes.

The term "TNC" is narrowly defined to include entities that only provide certain activities. Qualification as a TNC can grant an entity certain benefits, privileges, or burdens. Except for states that have addressed automated ridesharing companies in their TNC laws,[29] this determination depends upon two

---

29. Florida's automated vehicle statute provides:

> An on-demand autonomous vehicle network shall operate pursuant to state laws governing the operation of transportation network companies and transportation network company vehicles as defined in s. 627.748, except that any provision of s. 627.748 that reasonably applies only to a human driver does not apply to the operation of a fully autonomous vehicle with the automated driving system engaged while logged on to an on-demand autonomous vehicle network. A fully autonomous vehicle with the automated driving system engaged while logged on to an on-demand autonomous vehicle network must meet the insurance requirements in s. 627.749.

FLA. STAT. ANN. § 316.85; *see also* UTAH CODE ANN. § 41-26-106 (West) ("Subject to Subsection (2), an on-demand autonomous vehicle network may only operate pursuant to state laws governing the operation of ground transportation for-hire under state law, including: (a) a transportation network company pursuant to Title 13, Chapter 51, Transportation Network Company Registration Act; (b) a public transit district as defined in Section 17B-2a-802; or (c) a private passenger carrier as defined in Section 53-3-102.").

broad variables: (1) the definition of TNC and (2) the services provided by the automated ridesharing provider.

The first inquiry necessitates a review of the state's definition of TNC. For example, in Texas, a TNC is defined as follows:

> (5) "Transportation network company" means a corporation, partnership, sole proprietorship, or other entity that, for compensation, enables a passenger to prearrange with a driver, exclusively through the entity's digital network, a digitally prearranged ride. The term does not include an entity that provides:
>
> (A) street-hail taxicab services;
>
> (B) limousine or other car services arranged by a method other than through a digital network;
>
> (C) shared expense carpool or vanpool arrangements; or
>
> (D) a type of ride service for which:
>
> (i) the fee received by the driver does not exceed the driver's costs of providing the ride; or
>
> (ii) the driver receives a fee that exceeds the driver's costs associated with providing the ride but makes not more than three roundtrips per day between the driver's or passenger's place of employment and the driver's or passenger's home.[30]

New York defines TNCs in a similar way:

> "Transportation network company" or "TNC" means a person, corporation, partnership, sole proprietorship, or other entity that is licensed pursuant to this article and is operating in New York state exclusively using a digital network to connect transportation network company passengers to transportation network company drivers who provide TNC prearranged trips.[31]

These definitions are relatively specific.

The second inquiry requires an analysis of the services provided by the automated ridesharing company. Broadly speaking, there are two distinct types of automated ridesharing companies: (1) one that uses third-party-owned automated vehicles to connect with passengers, and (2) one that uses its fleet of automated vehicles to connect with passengers. Conceivably, a given service could be a blend of these two categories; many automated ridesharing companies may divide their vehicle ownership and their ridesharing service into separate entities. Companies using the third-party-owned model have a better chance of qualifying as TNCs, since they connect riders with other people's

---

30. Tex. Occ. Code Ann. § 2402.001(5).
31. N.Y. Veh. & Traf. Law § 1691 (McKinney).

vehicles, similar to traditional TNCs. Companies using the fleet-owned model may struggle to qualify as TNCs because they are providing vehicles and more than a digital network. If fleet-owned automated ridesharing companies are not statutory TNCs, then they may be classified as some other type of transportation service[32] or not fall within a state classification at all. States that desire to treat automated ridesharing companies and TNCs similarly should amend their definition of TNCs.

II. DORMANT COMMERCE CLAUSE

The Commerce Clause of the United States Constitution grants Congress the power to "regulate commerce . . . among the several states."[33] The Supreme Court has interpreted the Commerce Clause to have both a positive part, i.e., when Congress directly regulates commerce, and a negative part, i.e., when Congress does not regulate a specific aspect of commerce, but a state regulation discriminates against or unduly burdens commerce. This negative part of the Commerce Clause is known as the dormant Commerce Clause and is a court-imposed doctrine that prevents states from interfering with interstate commerce.[34] A court engages in a two-step inquiry when reviewing whether a state law violates the dormant Commerce Clause. Under the first step, the court determines whether "a state statute directly regulates or discriminates against interstate commerce, or [whether] its effect is to favor in-state economic interests over out-of-state interests."[35] If a statute discriminates against interstate commerce or regulates extraterritorially, the statute is virtually per se invalid.[36] But if a "statute has only indirect effects on interstate commerce and regulates evenhandedly,"[37] then courts move to the second step, which is the balancing test outlined in *Pike v. Bruce Church, Inc.*, 397 U.S. 137 (1970). Under

32. There may not be a neat fit for classifying these types of entities. For example, a taxicab could be considered an alternative, but the definition of "taxicab" may specifically exclude these types of vehicles. *See, e.g.*, N.Y. Veh. & Traf. Law § 148-a (McKinney) ("Every motor vehicle, other than a bus, used in the business of transporting passengers for compensation, and operated in such business under a license or permit issued by a local authority. *However, it shall not include vehicles which are rented or leased without a driver.*" (emphasis added)).

33. U.S. Const. art. I, § 8, cl. 3.

34. S.-Cent. Timber Dev., Inc. v. Wunnicke, 467 U.S. 82, 87, 104 S. Ct. 2237, 81 L. Ed. 2d 71 (1984) ("Although the Commerce Clause is, by its text, an affirmative grant of power to Congress to regulate interstate and foreign commerce, the Commerce Clause has long been recognized as a self-executing limitation on the power of the States to enact laws imposing substantial burdens on such commerce.").

35. Brown-Forman Distillers Corp. v. N.Y. State Liquor Auth., 476 U.S. 573, 579, 106 S. Ct. 2080, 90 L. Ed. 2d 552 (1986).

36. *Id.*

37. *Id.*

the *Pike* balancing test, courts will strike down a state regulation if the "burden it imposes upon interstate commerce is clearly excessive in relation to the putative local benefits."[38] The roadways, whether interstate highways or local roads, have long been established within the Commerce Clause power of the federal government.

The dormant Commerce Clause may protect the automated vehicle industry from laggards and luddites alike. Laggards in the sense that if automated vehicles—especially automated trucking—caused a violation of state law, the dormant Commerce Clause may deem the application of that law to automated vehicles unconstitutional. Luddites in the sense that if a state legislature decided, for one reason or another, to prohibit or attempt to curtail automated vehicles, the dormant Commerce Clause may override such legislation.

38. Int'l Dairy Foods Ass'n v. Boggs, 622 F.3d 628, 644 (6th Cir. 2010) (quoting Pike v. Bruce Church, Inc., 397 U.S. 137, 90 S. Ct. 844, 25 L. Ed. 2d 174 (1970)).

Chapter 7

Automated Vehicle Guidance and Laws

Governments at every level—federal, state, and municipalities—have considered automated vehicle laws. This chapter begins by describing the two prominent federal automated vehicle bills that Congress has considered but failed to enact. The federal guidance has come in the form of policies, and this chapter outlines those policies. States have had more success in enacting automated vehicle laws. This chapter highlights those laws and identifies some unique provisions. In December 2019, the Uniform Law Commission proposed a uniform state law on automated vehicles. Historically, uniform laws have been influential on or enacted by states, so this chapter explores key provisions in the uniform act. This chapter concludes by considering whether automated vehicles are legal in states that have not adopted automated vehicle laws.

I. THE FEDERAL BILLS

There have been two principal automated vehicle laws introduced into Congress: the Safely Ensuring Lives Future Deployment and Research in Vehicle Evolution Act (the SELF DRIVE Act) and the American Vision for Safer Transportation through Advancement of Revolutionary Technologies Act (the AV START Act).[1] The SELF DRIVE Act is the House of Representatives' automated

1. Other bills have included the 21st Century Automated Vehicle Safety Act of 2018, H.R. 6058 (introduced on June 8, 2018, by Representative David Loebsack); the Transportation Workforce Modernization Act, H.R. 6200 (introduced on June 22, 2018, by Representative Eddie Bernice) (retraining of displaced employees); Early Warning Reporting System Improvement Act of 2018, H.R. 7192

vehicle law; the AV START Act is the Senate's automated vehicle law. The SELF DRIVE Act received a unanimous 54-0 vote in September 2017 in the House Energy and Commerce Committee and was approved by the whole House of Representatives on a voice vote. The SELF DRIVE Act has stalled in the Senate. The AV START Act has also stalled in the Senate. This section briefly describes the contents of each of these two bills.

A. SELF DRIVE Act

The SELF DRIVE Act would primarily regulate highly automated vehicles, defined as vehicles equipped with levels 3, 4, and 5 automated driving systems. The act would exclude commercial motor vehicles from the definition of highly automated vehicles.

Similar to the preemption language in the Motor Vehicle Safety Act, the act would broadly preempt state laws that regulate the "design, construction, or performance" of automated vehicles that are inconsistent with safety standards prescribed by NHTSA. The act would not preempt states and local governments from regulating "registration, licensing, driving education and training, insurance, law enforcement, crash investigation, safety and emissions inspections, congestion management of vehicles [in a state], or traffic" unless such regulation unreasonably restricts the design, construction, or performance of highly automated vehicles. States would retain their power to regulate the sale, distribution, repair, or service of highly automated vehicles. The act would include a savings clause that excludes "liability at common law" from preemption.

The act would require the USDOT to review existing Federal Motor Vehicle Safety Standards (the FMVSS or Safety Standards) and create new rules and Safety Standards to accommodate and regulate highly automated vehicles. The act would require the USDOT to issue a rule that requires automakers to submit safety assessment certifications to demonstrate how their vehicles maintain safety, function as intended, and contain fail-safes. The USDOT would be required to review and possibly update the rule every five years.

The act would require manufacturers of vehicles containing any level of automation to develop a cybersecurity plan and a privacy plan. In the cybersecurity plan, manufacturers would outline their practices for detecting and responding to cyberattacks. The privacy plan would provide information on the collection of nonanonymized information by the manufacturer and the use of such information.

(introduced on November 29, 2018, by Representative Matt Cartwright) and S. 344 (introduced on October 1, 2018, by Senator Edward J. Markey); and the Autonomous Vehicle Privacy Protection Act of 2015, H.R. 3876 (introduced on November 2, 2015, by Representative Grace Meng).

The act would increase the number of exempt vehicles under the exemptions that NHTSA can grant to manufacturers of highly automated vehicles from the particular application of a Safety Standard. Currently, NHTSA can allow manufacturers to produce up to 2,500 vehicles annually under an exemption. Under the act, this number would increase to 100,000.

If enacted, the act would direct the USDOT and the Federal Trade Commission (FTC) to conduct studies. The USDOT would be directed to study its Safety Standards, the terminology used to describe automated vehicles, and the methods of informing consumers about automated vehicles. The act would create a highly automated vehicle advisory council in NHTSA to collect information, analyze practices, and make recommendations concerning, among other topics, cybersecurity, privacy, employment effects of highly automated vehicles, environmental effects of automated vehicles, and independent verification of highly automated vehicles, as well as increased mobility opportunities for disabled persons, senior citizens, and persons underserved by public transportation. The act would direct the FTC to study data collection and privacy plans of manufacturers.

B. AV START Act

The AV START Act shares some common provisions with the SELF DRIVE Act. The AV START Act uses the SAE levels of automation and focuses primarily on levels 3, 4, and 5 automated vehicles. The AV START Act would broadly preempt states and their political subdivisions from regulating the "design, construction, or performance of" automated driving systems and includes similar exceptions and a savings clause as the SELF DRIVE Act. The act would create several committees, working groups, and collaborations to advance automated vehicles and to assist in creating rules, regulations, and guidance on automated vehicles.

The act would call for an expedited resolution of conflicts between highly automated vehicles and the FMVSS. The John A. Volpe National Transportation Systems Center of the Department of Transportation would be directed to prepare a report to the USDOT that identifies each "provision, requirement, specification, or procedure" in the FMVSS that is necessary for human drivers, specifies a location point or reference with respect to a human driver, or is intended to provide information to or receive input from a human driver while driving. For each part identified, the report would provide an alternative reference to an automated driving system or state that the applicable text is only necessary for human drivers or that such alternative reference would be impracticable. The USDOT would need to commence rulemaking within 90 days and issue final rules within one year of receiving the report from the Volpe Center.

The act would exempt certain testing of automated driving systems from the need to comply with the FMVSS, so long as those automated driving systems are introduced only for "the purposes of testing, evaluation, or demonstration"; such activities are only conducted by employees or agents of the tester or fleet management contractors; the manufacturer agrees not to sell, lease, or offer to sell the system or highly automated vehicle being tested, evaluated, or demonstrated at the conclusion of the time; and the manufacturer submits identification information prior to testing, evaluating, or demonstrating such systems.

The act would increase the number of vehicles that NHTSA could exempt from compliance with Safety Standards. Unlike the SELF DRIVE Act, which would immediately allow NHTSA to exempt 100,000 vehicles, the AV START Act would take a gradual approach by allowing NHTSA to exempt more vehicles by a manufacturer over time: 2,500 the first year, 15,000 the second year, 40,000 the third year, and 80,000 each year thereafter (and manufacturers could petition NHTSA for a higher number of vehicles after the fourth year).

The act would require manufacturers to submit safety evaluation reports each year to the USDOT. In its initial report, the manufacturer would identify how it addresses the safety of its automated driving system and describe its "documented assessment, testing, and validation process" for nine categories of information: (1) system safety, (2) data recording, (3) cybersecurity, (4) human-machine interface, (5) crashworthiness, (6) capabilities of the system or vehicle, (7) postcrash behavior of the vehicle or system if sensors or systems are damaged, (8) how the manufacturer is accounting for applicable traffic laws and rules of the road, and (9) the automation function (such as the operational design domain, the need for infrastructure, and the abilities of the vehicle to transition to a minimal risk condition). The annual report would require manufacturers to disclose significant changes to their automated driving systems or highly automated vehicles and provide aggregate results of safety deviations from expected performance disclosed in a prior report and a comparison of the safety of the systems or vehicles compared to human-driven vehicles. NHTSA could adopt a Safety Standard on a topic and provide that manufacturers no longer need to report on such topic.

The act would address cybersecurity risks of modern motor vehicles and automated vehicles. Manufacturers would need to develop and execute written plans for identifying and reducing cybersecurity risks to their systems and vehicles. The USDOT would be given authority to assist manufacturers in coordinating a vulnerability disclosure policy and practices to allow security researchers to disclose a vulnerability to a manufacturer privately.

II. AUTOMATED VEHICLE POLICIES

The USDOT has issued five policy statements on automated vehicles since 2013. The policy statements represent a belief by the USDOT that automated vehicles have significant safety benefits and can help the USDOT accomplish its goals of safe transportation and crash prevention. These informal policy statements are an effective use of soft law[2] for the introduction of automated vehicles. Although these policy statements are not binding law, NHTSA and the USDOT have regulatory tools to enforce the policy statements, and a court has given preemptive effect to these policy statements.[3]

A. 2013 Preliminary Statement of Policy Concerning Automated Vehicles

NHTSA issued a preliminary policy statement on automated vehicles in 2013 (Preliminary Statement),[4] in which it expressed its belief that automated vehicles can provide immense benefits to society, identified research areas, and made recommendations to states concerning automated vehicles. In this Preliminary Statement, NHTSA classified automation features based on five levels of automation—NHTSA and the USDOT would later abandon these five levels of automation for SAE's six levels of automation. NHTSA identified its three categories of research as human factors, electronic control systems safety, and system performance requirements. The most significant observation was NHTSA's recommendation that states not allow highly automated vehicles on their roadways other than for testing. NHTSA gave other recommendations concerning licensing drivers to operate highly automated vehicles for testing, as well as recommendations about the operating conditions and the basic principles for the use and safety of test vehicles on roadways. The Preliminary Statement said little on how NHTSA and the USDOT would regulate these vehicles.

B. *Federal Automated Vehicle Policy: Accelerating the Next Revolution in Roadway Safety*

The first major policy directive came in September 2016 when NHTSA and the USDOT issued their 112-page *Federal Automated Vehicle Policy: Accelerating the Next Revolution in Roadway Safety* (AV 1.0).[5] AV 1.0 provided guidance on vehicle performance, model state policy, an overview of NHTSA's regulatory tools, and potential new regulatory tools and authority. The guidance focused primar-

---

2. Soft laws are informal declarations or position statements that are not binding like traditional regulations or laws.

3. Dashi v. Nissan N. Am., Inc., 445 P.3d 13 (Ariz. Ct. App. 2019). *Dashi* is explored in detail in Chapter 8.

4. Nat'l Highway Traffic Safety Admin., *Preliminary Statement of Policy Concerning Automated Vehicles,* https://www.nhtsa.gov/staticfiles/rulemaking/pdf/Automated_Vehicles_Policy.pdf (last visited May 19, 2020).

5. U.S. Dep't of Transp. & Nat'l Highway Traffic Safety Admin., *Federal Automated Vehicles Policy: Accelerating the Next Revolution in Roadway Safety* (Sept. 2016), https://www.transportation.gov/sites/dot.gov/files/docs/AV%20policy%20guidance%20PDF.pdf.

ily on automated driving systems but also included level 1 and level 2 driving automation systems.

AV 1.0 targeted that guidance on vehicle performance toward manufacturers of driving automation systems and provided a framework for manufacturers to use in assessing the performance and safety of their automated vehicles. AV 1.0 requested (but did not require) manufacturers to provide reports on the safety of their automated driving systems that would be reviewed by NHTSA and available to the public. AV 1.0 requested that these safety assessment letters include information on system safety, cybersecurity, human-machine interface, privacy, data recording and sharing, ethical considerations, validation methods, and consumer education and training. The voluntary safety assessments are publicly available online.[6]

AV 1.0 was clear in reinforcing the USDOT and NHTSA's belief that the federal and state roles would remain unchanged for automated vehicles. The federal role involves setting and enforcing FMVSS, investigating and conducting recalls, educating the public on motor vehicle safety, and issuing guidance. The states' roles include licensing human drivers, registering motor vehicles, enacting and enforcing traffic laws, conducting safety inspections, regulating motor vehicle insurance, and imposing liability for crashes. The USDOT "strongly encouraged" states not to regulate the performance of automated driving systems and to leave such regulations to NHTSA. The USDOT also acknowledged that because NHTSA regulates motor vehicles and motor vehicle equipment and the states regulate human drivers, "[a]s motor vehicle equipment increasingly performs 'driving' tasks, DOT's exercise of its authority and responsibility to regulate the safety of such equipment will increasingly encompass tasks similar to 'licensing' of the non-human 'driver.'"

AV 1.0 also detailed recommendations for states on automated vehicle testing, administrative policies, the granting of permission to manufacturers to test on roadways, driver regulation, law enforcement, liability, and insurance. AV 1.0 suggested that states should work together to avoid inconsistent laws and to standardize their infrastructure.

AV 1.0 contained a discussion of additional regulatory authority and tools, some of which NHTSA has authority over and others for which it currently lacks authority (such as premarket approval). AV 1.0 discussed whether Congress should consider replacing the current self-certification method with a premarket approval method similar to the Federal Aviation Administration's

---

6. Nat'l Highway Traffic Safety Admin., *Voluntary Safety Self-Assessment*, U.S. Dep't of Transp., https://www.nhtsa.gov/automated-driving-systems/voluntary-safety-self-assessment (last visited Dec. 3, 2019). As of December 3, 2019, the following companies had submitted safety assessments: Apple, Aurora, AutoX, Ford, GM, Ike, Mercedes-Benz/Bosch L4-L5, Mercedes Benz L3, Navya, Nuro, Nvidia, Robomart, Startsky Robotics, TruSimple, Uber, Waymo, and Zoox.

premarket approval. AV 1.0 also considered whether Congress should give NHTSA a cease-and-desist authority to remedy conditions or practices that create emergencies. NHTSA also described how the use of replicable testing procedures for its Safety Standards could allow automated vehicles to "game" testing procedures. NHTSA suggested that it would need a clarifying amendment to the Motor Vehicle Safety Act to allow it to use varying testing procedures for automated vehicles.

C. *Automated Driving Systems 2.0: A Vision for Safety*

In 2017, the USDOT and NHTSA replaced and replicated much of AV 1.0 with *Automated Driving Systems 2.0: A Vision for Safety* (AV 2.0).[7] AV 2.0 contained a section on voluntary guidance for industry and a section on technical assistance to states. A common theme throughout AV 2.0 was the USDOT and NHTSA reiterating their belief that NHTSA alone should regulate the safety design and performance of automated driving systems. AV 2.0 made recommendations for manufacturers in addressing 12 automated driving system safety elements: system safety, operational design domain, object and event detection and response, fallback (minimal risk condition), validation methods, human-machine interface, vehicle cybersecurity, crashworthiness, postcrash automated driving system behavior, data recording, consumer education and training, and the system's accounting for applicable federal, state, and local laws. AV 2.0 encouraged manufacturers to submit voluntary safety self-assessments. AV 2.0 also updated its best practices for states on regulating automated vehicles and allowing testing and use of automated vehicles on their roadways.

D. *Automated Vehicles 3.0: Preparing for the Future of Transportation*

The USDOT issued *Automated Vehicles 3.0: Preparing for the Future of Transportation* (AV 3.0) in 2018.[8] Unlike AV 2.0, which superseded AV 1.0, AV 3.0 explicitly did not replace AV 2.0. By adding to AV 2.0, AV 3.0 was able to use its space to tread new ground and not merely repeat or replicate its predecessors, though some themes, such as the respective roles of the federal and state governments, were repeated and expounded. AV 3.0 also described the roles of the various administrations within the USDOT. The USDOT identified six principles that

7. U.S. Dep't of Transp. & Nat'l Highway Traffic Safety Admin., *Automated Driving Systems 2.0: A Vision for Safety* (Sept. 12, 2017), https://www.nhtsa.gov/sites/nhtsa.dot.gov/files/documents/13069a-ads2.0_090617_v9a_tag.pdf. The replacement of AV 1.0 almost certainly had nothing to do with a change in presidential administrations.

8. U.S. Dep't of Transp., *Automated Vehicles 3.0: Preparing for the Future of Transportation* (Oct. 4, 2018), https://www.transportation.gov/sites/dot.gov/files/docs/policy-initiatives/automated-vehicles/320711/preparing-future-transportation-automated-vehicle-30.pdf.

shape its automated vehicle policy: prioritizing safety, remaining neutral on technology, modernizing its regulations, encouraging a consistent regulatory and operational environment, preparing proactively for automation, and protecting and enhancing freedoms enjoyed by Americans.

AV 3.0 provided the most thorough position on the roles of the federal government, state governments, and local governments in regulating automated vehicles, transportation, and roadways. The USDOT also provided the respective roles of NHTSA, the Federal Motor Carrier Safety Administration (FMCSA), the Federal Highway Administration (FHWA), and the Federal Transit Administration (FTA). For NHTSA's authority, the USDOT referenced the preemptive effect that NHTSA's rulemaking and authority have on state law and tort claims that would create an inconsistency. The USDOT also acknowledged that NHTSA would consider revamping its approach to rulemaking and Safety Standards to use more performance-based, flexible Safety Standards that would allow NHTSA to test the capabilities of a system within and outside of its operational design domain.

The FMCSA provided its framework for analyzing the use of automated driving systems in commercial motor vehicles. The FMCSA will first analyze whether the commercial motor vehicle complies with the parts and accessory requirements. The FMCSA next considers whether the motor carrier complies with the operational requirements in the Federal Motor Carrier Safety Regulations (FMCSRs). The FMCSA stated that it would not apply the human-specific FMCSRs to automated driving systems that do not require a human operator. Importantly, the USDOT stated its position that the "FMCSA regulations will no longer assume that the CMV driver is always a human or that a human is necessarily present onboard a commercial vehicle during its operation." To the extent that the motor carrier cannot establish compliance with the operational requirements of the FMCSRs, then the motor carrier would need to seek an exemption. AV 3.0 also reminded states that the FMCSA has preemptive authority on inconsistent state laws. The FMCSA indicated that it would consider changes to its existing FMCSRs to account for differences between human drivers and computer drivers. The FMCSA also indicated the need to work with the Department of Labor to assess workplace concerns.

FHWA stated that it would pursue updating the Manual on Uniform Traffic Control Devices. FHWA, in coordination with stakeholders, will conduct research and experiments to gain a better understanding of the needs of automated vehicles. The FTA indicated that it is researching automated bus transit and working to provide flexibility to transit authorities in deciding whether to adopt automated transit.[9]

9. The FTA laid out its strategic research in a publication in January 2018. Fed. Transit Admin., *Strategic Transit Automation Research Plan*, U.S. Dep't of Transp. (Jan. 2018), https://www.transit.dot

The USDOT also provided additional guidance for states and their subdivisions in preparing for and regulating automated vehicles. In reflection of the broader scope of AV 3.0, this guidance was targeted to different aspects and parts of the state regulatory apparatus: state legislatures, highway safety officials, infrastructure owners and operators, commercial vehicle enforcement agencies, public transit, local government, and transportation sector cybersecurity.

The USDOT explained the role that industry has in building public acceptance of automated vehicles through transparency about their vehicle safety performance and public education campaigns. The USDOT reiterated the role of the voluntary safety self-assessments in building consumer acceptance. AV 3.0 gave specific advice to motor carriers and transit authorities that consider adopting automated driving systems. The USDOT also explained its safety risk management recommendations through the stages of development, testing, and deployment of automated vehicles.

E. *Automated Vehicles 4.0: Ensuring American Leadership in Automated Vehicle Technologies*

On January 8, 2020, the National Science & Technology Counsel from the Office of the President of the United States and the USDOT released *Automated Vehicles 4.0: Ensuring American Leadership in Automated Vehicle Technologies* (AV 4.0).[10] In AV 4.0, the federal government outlined its current or prior efforts to foster automated vehicles. The most significant takeaway from AV 4.0 is about the robustness of the federal government's efforts with respect to the automated vehicle industry. These efforts have included research on ensuring mobility for all Americans by the Department of Health and Human Services, the Department of the Interior, the Department of Justice, the National Council on Disability, the U.S. Access Board, and the USDOT; research on security and cybersecurity by the Department of Energy, the Department of Homeland Security, the Department of Justice, the USDOT, the National Institute of Standards and Technology, and the National Security Council; and research and guidance on spectrum and connectivity by the Department of Energy, the Department of Homeland Security, the USDOT, the Federal Communications Commission, the National Telecommunications and Information Administration, and the National Institute of Standards and Technology.

.gov/sites/fta.dot.gov/files/docs/research-innovation/114661/strategic-transit-automation-research-report-no-0116_0.pdf.

10. Nat'l Sci. & Tech. Council & U.S. Dep't of Transp., *Ensuring American Leadership in Automated Vehicle Technologies: Automated Vehicles 4.0* (Jan. 2020), https://www.transportation.gov/sites/dot.gov/files/2020-02/EnsuringAmericanLeadershipAVTech4.pdf.

III. STATE AUTOMATED VEHICLE LAWS

Nevada enacted the first automated vehicle law in 2011 to permit the testing of automated vehicles. Since then, states all over the country have enacted automated vehicle laws, and governors have issued executive orders on automated vehicles. Twenty-nine states[11] and the District of Columbia have enacted legislation relating to some aspect of automated vehicles. Ten governors have issued executive orders pertaining to automated vehicles.[12] In the interest of space, this book does not review each state's laws. Instead, this section explores common themes among the existing automated vehicle laws and unique or troublesome laws.

Many states specifically authorize the testing of automated vehicles. Some of these laws are more permissive than others. Some allow for testing without a human operator,[13] while others require a human operator.[14] Some states direct agencies or committees to help further the testing of automated vehicles within their states.[15] In some states, manufacturers are required to identify their test vehicles to the state department of motor vehicles before their operation on roadways.[16] Some states have allowed automated vehicle use without a human driver beyond testing.[17] A few states have explicitly authorized fully automated ridesharing or delivery vehicles.[18]

California uses a permit process for the testing and deployment of automated vehicles. One permit allows a manufacturer to test automated vehicles on public roadways with safety drivers.[19] A manufacturer can apply for a different

---

11. Alabama, Arkansas, California, Colorado, Connecticut, Florida, Georgia, Illinois, Indiana, Kentucky, Louisiana, Maine, Michigan, Mississippi, Nebraska, New York, Nevada, North Carolina, North Dakota, Oregon, Pennsylvania, South Carolina, Tennessee, Texas, Utah, Virginia, Vermont, Washington, and Wisconsin.

12. Arizona, Delaware, Hawaii, Idaho, Maine, Massachusetts, Minnesota, Ohio, Washington, and Wisconsin.

13. Ariz. Gov. Exec. Order 2018-04, Advancing Autonomous Vehicle Testing and Operating; Prioritizing Public Safety (Mar. 1, 2018) (autonomous vehicle pilot program); CAL. CODE REGS. tit. 13, § 227.38 (special permit for testing driverless automated vehicles); COLO. REV. STAT. § 42-4-242(1); MICH. COMP. LAWS ANN. § 257.665; NEV. REV. STAT. § 482A.070; Wash. Exec. Order 17-02, Autonomous Vehicle Testing & Technology in Washington State and Autonomous Vehicle Work Group (June 7, 2017).

14. Ill. Exec. Order 2018-13: Executive Order Establishing the Autonomous Illinois Initiative (Oct. 25, 2018); 2017 N.Y. S.B. 2005 (Jan. 23, 2017).

15. Haw. Exec. Order 17-07: Autonomous Vehicle Testing (Nov. 22, 2017) (directing certain agencies to assist in helping make Hawaii a test location); Idaho Exec. Order No. 2018-01, Creating the Autonomous and Connected Vehicle and Deployment Committee (Nov. 1, 2018).

16. CAL. CODE REGS. tit. 13, § 227.16.

17. GA. CODE § 408-11; IOWA CODE § 321.514 *et seq.*; NEB. REV. STAT. ANN. § 60-3302 (West); NEV. REV. STAT. § 482A.070; TENN. CODE ANN. § 55-30-103 (West); TEX. TRANSP. CODE ANN. § 545.453.

18. IOWA CODE § 321.518; MICH. COMP. LAWS ANN. § 257.665b (West); NEB. REV. STAT. ANN. § 60-3305 (West).

19. CAL. CODE REGS. tit. 13, § 227.04.

permit to test automated vehicles without safety drivers.[20] A manufacturer can obtain a separate permit to deploy automated vehicles beyond testing.[21] California's Public Utilities Commission (CPUC) also has a permit requirement for companies that desire to provide an automated ridesharing service.[22] CPUC does not allow companies to charge for their automated ridesharing services, perceiving somewhat perplexingly that there is a difference between charging a fee and not charging a fee. CPUC's rule is also shortsighted regarding the goal of the safe deployment of automated vehicles. California has a robust regulatory regime that could help foster safer development of automated vehicles, but CPUC has incentivized companies to operate in more lax regulatory states.

Some states have clarified the application of their motor vehicle codes to automated driving systems and human users of these systems.[23] Variations include carve-outs of specific code provisions[24] or general overrides of human driver-operator provisions.[25] States have differed in their approach about whether to require users of level 4 and level 5 automated vehicles to obtain a license. Texas provides explicitly that a licensed driver is not necessary to operate a motor vehicle while a level 4 or level 5 automated driving system is engaged.[26]

States have established advisory committees or commissions to study automated vehicles.[27] These advisory groups include members of the state government, representatives from automakers, and members of academia, with the purpose of reviewing state laws and the effects of automated vehicles in the state and making recommendations to the state government.

Some states have limited their laws thus far to automated commercial vehicles or laws that most seemingly benefit automated commercial vehicles by

---

20. *Id.* § 227.38.

21. *Id.* § 228.00 *et seq.*

22. *See* Cal. Pub. Util. Comm'n, *Autonomous Vehicle Passenger Service Pilot Programs*, CA.GOV, https://www.cpuc.ca.gov/avcpilotinfo/ (last visited Mar. 12, 2020).

23. *See, e.g.*, COLO. REV. STAT. § 42-4-110(2); IOWA CODE § 321.515(3); NEB. REV. STAT. ANN. § 60-3306 (West); NEV. REV. STAT. § 482A.200.

24. Fla. Ch. 2019-101: Motor Vehicles—Autonomous Vehicles (2019); Ga. House Bill 472; Mich. Pub. Act No. 332 (2016).

25. COLO. REV. STAT. § 42-4-110(2); IOWA CODE § 321.515(3); NEB. REV. STAT. ANN. § 60-3306 (West); NEV. REV. STAT. § 482A.200.

26. TEX. TRANSP. CODE ANN. § 545.453 (West).

27. Del. Exec. Order 14, Establishment of the Advisory Council on Connected and Autonomous Vehicles (Sept. 5, 2017); Idaho Exec. Order No. 2018-01, Creating the Autonomous and Connected Vehicle and Deployment Committee (Jan. 2, 2018); Maine H.P. 1204 (creating a commission on automated vehicles); MICH. COMP. LAWS ANN. § 257.665 (establishing a council on future mobility); Minn. Exec. Order 18-04 , Establishing the Governor's Council on Connected and Automated Vehicles (Apr. 1, 2019) (establishing an advisory council on connected and automated vehicles); Or. House Bill 4063 (task force on automated vehicles); 75 PA. CONS. STAT. § 8503 (advisory committee); Wash. Exec. Order 17-02, Autonomous Vehicle Testing & Technology in Washington State and Autonomous Vehicle Work Group (June 7, 2017).

authorizing platooning for commercial trucking.[28] States with more comprehensive automated vehicle laws allow platooning. Some of these statutes impose requirements on a commercial motor carrier before operating a platoon within the state.

Some states have expressly prohibited their political subdivisions from enacting or enforcing rules about certain parts of automated driving or vehicles.[29] Colorado's approach is to prohibit political subdivisions from enforcing different standards for automated driving systems and human-driven vehicles.[30] Texas's approach is to prohibit a "franchise or other regulation related to the operation of an automated motor vehicle or automated driving system."[31]

Some states have granted immunity to an automaker when a third party has converted its vehicles into automated vehicles.[32] This immunity, likely in response to Google's initial approach of installing its automated driving system on other automakers' vehicles, seemingly protects an automaker when its vehicles are converted into automated vehicles. Because the major nonautomaker automated driving system manufacturers, such as Waymo, Cruise, and Argo AI, are working in collaboration with or have the consent of the automaker, the effectiveness of this immunity may be limited.

Another unique law is Nevada's immunity for modified or altered automated vehicles:

> The original manufacturer or developer of an automated driving system that has been modified by an unauthorized third party is not liable for damages to any person injured due to a defect caused by the modification of the automated driving system by the third party unless the defect that caused the injury was present in the automated driving system as originally manufactured or developed.[33]

---

28. *See, e.g.*, Ind. Code § 9-21-22-1 *et seq.* (authorizing vehicle platoons); La. Rev. Stat. § 400.1 *et seq.* (same); Miss. Code Ann. § 63-3-103(k) (defining a platoon as a type of vehicle); N.C. Gen. Stat. § 20-152(c) (authorizing commercial motor vehicle platoons); S.C. Code Ann. § 56-5-1390 (authorizing platoons); Wis. Stat. Ann. § 346.14 (West).

29. Colo. Rev. Stat. § 42-4-110(6)(b); 625 Ill. Comp. Stat. Ann. 5/11-208 (e-10); Neb. Rev. Stat. Ann. § 60-3308 (West); Tenn. Code Ann. § 55-30-105 (West); Tex. Transp. Code Ann. § 545.452(b); Utah Code Ann. § 41-26-108.

30. Colo. Rev. Stat. § 42-4-110(6)(b).

31. Tex. Transp. Code Ann. § 545.452(b) (West).

32. D.C. Code Ann. § 50-2353 (West); Mich. Comp. Laws Ann. § 257.665a; Nev. Rev. Stat. § 482A.090. A lot of these immunity statutes state the obvious and seem unnecessary; if the automaker of the motor vehicle did not cause the crash, then the automaker of the motor vehicle is not liable for the underlying harm. Additionally, as described in Chapter 8, products liability law typically provides special protection to manufacturers of components when the components themselves are not defective.

33. Nev. Rev. Stat. § 482A.090.

This is a potentially pernicious provision. Three parties could modify or alter an automated vehicle after its original manufacture or development: (1) the manufacturer through software updates, (2) the vehicle owner or his or her agents through physical alterations or by jailbreaking the automated driving system, or (3) cyberattackers or others seeking to commit criminal acts. The immunity would not apply to manufacturer updates because the manufacturer is not an unauthorized third party. With respect to alterations or modifications made by the vehicle owner, these could be mere repairs to the system or other alterations or modifications. This statute does not seem to answer what happens if the vehicle owner repairs a part of the automated driving system, and a software update causes the fixed part not to interact correctly with the automated driving system and causes a crash. The manufacturer may argue that it receives the protection of this statute because the defect that caused the injury was not present in the automated driving system *"as originally manufactured or developed."* The owner may contend that his or her repairs or alterations were authorized. The third category, criminal actors, creates new problems. Is a vulnerability a defect, or was this statute intended to provide immunity to manufacturers from liability for harm caused by cyberattacks? If a vulnerability is a defect under this statute, what happens if a cyberattacker exploits a vulnerability that was created by a software update? Could the manufacturer claim that the system was not defective as originally manufactured or developed? This type of provision appears to create more confusion than solace, and rules like this will complicate an otherwise potentially straightforward liability analysis.

Almost all states avoid assigning or apportioning liability for automated vehicle crashes. Tennessee and Michigan are the exceptions to this general rule. Tennessee's automated vehicle law provides as follows:

> When the [automated driving system] is fully engaged, operated reasonably and in compliance with manufacturer instructions and warnings, the [automated driving system] shall be considered the driver or operator of the motor vehicle for purposes of determining: (1) Liability of the vehicle owner or lessee for alleged personal injury, death, or property damage in an incident involving the [automated driving system]-operated vehicle.[34]

If the automated driving system, and not the human driver, is treated as the driver or operator of the vehicle for determining liability while the automated driving system is engaged, then this statute attempts to shift liability from the human driver to the automated driving system. Since an automated driving system is incapable of paying a judgment, this statute seemingly places liability on the automated driving system manufacturer. Michigan speaks more clearly

---

34. Tenn. Code Ann. § 55-30-106(b)(1) (West).

and assigns liability to the manufacturer of the motor vehicle for crashes caused by automated driving systems used in ridesharing:

> [D]uring the time that an automated driving system is in control of a vehicle in the participating fleet, a motor vehicle manufacturer shall assume liability for each incident in which the automated driving system is at fault.[35]

Interestingly, the Michigan statute imposes liability on the motor vehicle manufacturer rather than the manufacturer of the automated driving system. Many of the key players developing automated driving systems are developing the automated driving system as a component added, at least in some cases, after the development of the vehicle.

California is the only state to have specifically addressed data collection by automated vehicles. California has specific requirements for manufacturers that collect personal information that is not necessary for the safe operation of automated vehicles.[36] The manufacturer must provide a written disclosure to the driver or, if driverless, the passengers of the automated vehicle that describes the personal information collected and how the manufacturer intends to use the information. Alternatively, the manufacturer can anonymize personal information. For automated vehicles sold or leased to consumers, to the extent that the personal information is not anonymized, the manufacturer must obtain written approval to collect the personal information. A manufacturer cannot condition sale on the written approval.

California has also been, albeit in a limited way, the first state to require specific training and education before the operation of automated vehicles. California requires manufacturers to maintain a training program for remote operators and to certify that the remote operators have received sufficient training.[37] Manufacturers must provide the Department of Motor Vehicles with a course outline and description of the program.

California is unique in its annual reporting requirements on disengagements for manufacturers testing automated vehicles.[38] A disengagement is defined as "a deactivation of the autonomous mode when a failure of the autonomous technology is detected or when the safe operation of the vehicle requires that the autonomous vehicle test driver disengage the autonomous mode and take immediate manual control of the vehicle, or in the case of driverless vehicles, when the safety of the vehicle, the occupants of the vehicle, or the public requires that the autonomous technology be deactivated."[39] The report must identify the location of each disengagement, whether the vehicle

35. Mich. Comp. Laws § 257.665b(4).
36. Cal. Code Regs. tit. 13, § 228.24.
37. *Id.* § 227.38.
38. *Id.* § 227.50.
39. *Id.*

had a driver at such time, a description of the cause of the disengagement, and who or what initiated the disengagement. The California Department of Motor Vehicles publishes the disengagement reports. They can be a helpful source of information on which manufacturers are testing automated vehicles in California, the size of their California fleet, their miles driven in California, the numbers of disengagements, and the ability to calculate each company's miles per disengagement. Miles per disengagement has been a popular but often criticized industry indicator of leaders and success of an automated driving system. In 2018, Waymo led with an average of 11,154.3 miles per disengagement, while Cruise had the second-most miles per disengagement at 5,204.9. The third-place company, Zoox, averaged 1,922.8 miles per disengagement. In 2019, Waymo's vehicles drove 1.45 million miles and had a disengagement every 13,219 miles. Cruise drove a total of 831,040 miles in 2019 and had a disengagement every 12,221 miles driven. Miles per disengagement is an imperfect statistic that suffers from at least two glaring flaws. It does not account for differences in company disengagement policies, and it does not consider the difficulty of the roads driven.

Pennsylvania has a data reporting requirement, too.[40] The Pennsylvania Department of Transportation (PennDOT) publishes an Automated Vehicle Testing Guidance manual that provides requirements of manufacturers seeking to test highly automated vehicles. PennDOT requires testers to complete a semiannual data collection form that identifies the miles traveled in Pennsylvania, the counties of testing, and whether the majority of testing occurred on limited-access roadways or arterial roadways, as well as information on the economic impact of testing.[41] Pennsylvania also requires crash reporting within six hours of any crash involving a highly automated vehicle.

IV. THE UNIFORM AUTOMATED OPERATION OF VEHICLES ACT

The Uniform Law Commission approved a Uniform Automated Operation of Vehicles Act (the Uniform Act) in December 2019.[42] The Uniform Act applies to

40. *Automated Vehicle Testing Guidance*, PA. DEP'T OF TRANSP. (July 23, 2018), https://www.penndot.gov/ProjectAndPrograms/ResearchandTesting/Autonomous%20_Vehicles/Documents/PennDOT%20HAV%20Testing%20Guidance.pdf.

41. *See Semi-Annual Data Collection Form*, PA. DEP'T OF TRANSP. (2018), https://www.penndot.gov/ProjectAndPrograms/ResearchandTesting/Autonomous%20_Vehicles/Documents/PennDOT%20HAV%20Semi-Annual%20Data%20Collection%20Form.pdf.

42. To see the Automated Operation of Vehicle Act Committee's actions, see https://www.uniformlaws.org/committees/community-home?communitykey=43730d6f-013d-4883-b2b7-98e66bda3239&tab=groupdetails. The final act can be found here: Unif. Automated Operation of

vehicles equipped with level 3, level 4, and level 5 automated driving systems.[43] This section explores the Uniform Act.

The concept of "automated driving provider" is at the heart of the Uniform Act. An automated driving provider is a person that (1) substantially participates in developing an automated driving system, (2) submits a safety self-assessment or equivalent report to NHTSA, to the extent such report is required or permitted, (3) is registered as a motor vehicle or motor vehicle equipment manufacturer, and (4) submits a declaration to a relevant state agency that it is an automated driving provider. Effectively, the Uniform Act requires companies to submit safety self-assessment reports to NHTSA—a requirement that the USDOT itself has declined to adopt, and a requirement with which companies seeking to become automated driving providers would need to comply.

The automated driving provider is required to designate its automated vehicles to the relevant state agency. An automated vehicle so designated is defined as an "associated automated vehicle." An associated automated vehicle remains as such until the automated driving provider's declaration is no longer recognized, the automated driving provider dissolves, or the automated driving provider dissociates the vehicle. An automated driving provider dissociates an automated vehicle by providing written notice to the relevant state agency.

Section 4 of the Uniform Act specifically addresses driver's licenses. The Uniform Act excuses individuals from possessing a driver's license for trips that are entirely operated by an automated driving system. Other than for those trips, drivers need a driver's license. The Uniform Act clarifies that an automated driving provider is not required to obtain a driver's license.

Section 5 of the Uniform Act provides rules on registering automated vehicles. The owner of a motor vehicle is required to register its vehicle as an automated vehicle. If a vehicle is retrofitted or becomes an automated vehicle, the owner must identify the vehicle as such upon obtaining a new registration. The Uniform Act only allows for the registration of automated vehicles that have been designated by an automated driving provider. The state agency responsible for registering motor vehicles can "decline, suspend, revoke, or decline to renew registration of an automated vehicle that is not an associated automated vehicle, not associated with an automated driving provider recognized by [such state agency], not properly maintained, not insured in accordance with applicable law, not compliant with a registration requirement, or otherwise not

---

Vehicles Act (2019), https://www.uniformlaws.org/HigherLogic/System/DownloadDocumentFile.ashx?DocumentFileKey=c9804af4-f1dd-9489-a116-136bab52d868&forceDialog=0.

43. Section 2 of the Uniform Act defines an "automated vehicle" as a "motor vehicle with an automated driving system." An "automated driving system" is defined as "the hardware and software collectively capable of performing the entire dynamic driving task on a sustained basis." This definition mirrors the definition in the Taxonomy for automated driving systems.

fit to be operated." For these vehicles, the agency can grant a temporary registration for nonautomated use of the vehicle.

The Uniform Act also clarifies certain rules that could complicate automated driving or deprive automated vehicles of certain benefits. Under these rules, prohibitions on using electronic devices, leaving vehicles unattended or abandoned, and following too closely do not apply to automated vehicles. The Uniform Act directs that the traffic laws should be interpreted to accommodate automated driving. It requires automated driving providers to "take reasonable steps" to comply with the rules of the road.

The Uniform Act provides that the "automated-driving provider is responsible for a violation of the state's rules of the road during automated operation of an associated automated vehicle." The Reporter's Comments expound upon this view:

> [A]n automated driving provider is responsible for a violation of the rules of the road by an associated automated vehicle under automated operation. This retrospective principle merely recognizes that the automated driving provider is the legal driver in these circumstances and is therefore subject to corresponding sanctions under the state's vehicle code. In other words, the automated driving provider should receive the speeding ticket when an associated automated vehicle under automated operation is caught speeding. At the same time, this section does not address the appropriate level of enforcement. It is expected that federal, state, and local authorities will continue to evaluate the role of various forms of automated enforcement (including self-reporting obligations) in improving road traffic safety.[44]

The decision to make the automated-driving provider responsible for violations of the rules of the road may be an indirect way to impose civil liability on the automated-driving provider. As discussed in Chapter 8, this decision may allow plaintiffs to assert negligence per se claims against automated-driving providers.

V. THE LEGALITY OF AUTOMATED VEHICLES IN STATES WITHOUT AUTOMATED VEHICLE LAWS

In those states that have not explicitly enacted laws, regulations, or pronouncements on automated vehicles, the question is whether these states prohibit personal or commercial use of automated vehicles. There has been scholarly debate and a divergence of opinions on whether automated vehicles could legally be

44. Unif. Automated Operation of Vehicles Act Reporter's Comments to Section 9.

operated without a human operator in the absence of specific authorization.[45] Legal scholar Bryant Walker Smith conducted the most comprehensive analysis of the legality of automated vehicles. Smith examined whether automated vehicles were prohibited by international, federal, and state law. After an exhaustive review of the three levels of regulations and based on the legal principle of nullum crimen sine lege (no crime without law), Smith concluded that states do not categorically prohibit automated vehicles on their roadways. Smith's conclusion was not devoid of nuance, and he noted that current law "may nonetheless discourage [automated vehicle] introduction or complicate their operation."[46] Some statutes could prohibit specific deployments and configurations of automated vehicles, as well as certain types of uses (such as distracted uses) of automated vehicles.

---

45. *Compare* Bryant Walker Smith, *Automated Vehicles Are Probably Legal in the United States*, 1 Tex. A&M L. Rev. 411 (2014) (suggesting that automated vehicles are probably legal because the driver concept can be broadly interpreted to include nonhuman drivers), *with* William J. Kohler & Alex Colbert-Taylor, *Current Law and Potential Legal Issues Pertaining to Automated, Autonomous and Connected Vehicles*, 31 Santa Clara High Tech. L.J. 99, 110–11 (2015) (stating their belief, in contrast to Smith's article, that courts would conservatively interpret the terms "driver" and "operator" to include only human drivers and operators).

46. Smith, *supra* note 45, at 516.

Part 3

Civil Liability

INTRODUCTION

Cars crash. Just as cars driven solely by humans crash, so too will cars driven by driving automation systems. When vehicles operated entirely or in part by driving automation systems crash, the aggrieved parties may seek to present new types of claims against other parties, such as the automaker or a government entity. Traditional at-fault parties, typically the human driver of the vehicle that caused the crash (or his or her insurer), may blame others for the crash. These new claims may complicate an otherwise straightforward tort claim for automobile crashes.

Historically, automobile crashes have distinctive features in comparison to other types of tort claims.[1] Approximately half of motor vehicle crash claims result in a formal or informal attempt to recover from another party—this number shrinks to 3 out of 100 for nonwork, non–motor vehicle incidents.[2] The overwhelming majority of claims center on who caused the crash. Only half of the claimants hire attorneys, and only a small portion of the claims end up in

1. Nora Freeman Engstrom, *When Cars Crash: The Automobile's Tort Law Legacy*, 53 Wake Forest L. Rev. 293, 299 (2018).

2. *Id.*

litigation.[3] Claimants are also likely to obtain recovery for their claims.[4] Automobile claims are ripe for exaggeration and fabrications of physical harm.[5]

The typical defendant is an individual who has no prior relationship with the plaintiff.[6] The typical defendant remains relatively uninvolved in most minor-injury, routine claims.[7] The claim process is typically handled through the plaintiff's and the defendant's insurance adjusters.[8] These minor injury claims are usually settled in a "fairly routinized manner—using claims categories, computer programs, going rates, or rules of thumb."[9] This process reduces the timing of claim resolution and the transaction costs.[10] If insurers are unable to resolve the claim, then the claim may end up in litigation.[11] Of the insubstantial number of claims that end up in litigation, only 3 percent to 6 percent of those cases are resolved by trial.[12] The trial itself is relatively straightforward without the need for expert witnesses.[13] Liability for most automobile crashes is relatively straightforward to resolve and has low transaction costs to those involved and to the judiciary.

Automated vehicles have the potential to upset the streamlined process that has developed over the past hundred years. Who or what should bear primary responsibility for crashes caused while a driving automation system is engaged is a question that has intrigued scholars, lawyers, engineers, software developers, and automakers. When a vehicle without any automation feature crashes, the presumption is that the driver caused the accident. There can then be a search for an intervening or superseding cause. When automated vehicles crash, there might well be an argument that the presumption is that the automated feature caused the accident.

This part explores civil liability issues that can arise from deploying automated vehicles. Chapter 8 describes broad-picture liability doctrines and their application to automated vehicles. In legal literature, there has been a lack of nuance in analyzing civil liability for automated vehicle crashes. This is a mistake. How litigants and courts analyze liability for automated vehicle crashes depends greatly upon the type of driving automation system and the method of deployment. Chapters 9, 10, and 11 analyze specific liability concerns for level 2, level 3, and levels 4 and 5 driving automation systems. The analyses in

3. *Id.* at 299–300.
4. *Id.* at 300.
5. *Id.*
6. *Id.* at 301–02.
7. *Id.* at 301.
8. *Id.*
9. *Id.*
10. *Id.*
11. *Id.* at 302.
12. *Id.*
13. *Id.*

these chapters focus primarily on privately owned automated vehicles. Immunity for level 4 and level 5 automated vehicle crashes has been a popular topic that is examined in Chapter 12. Chapters 13 and 14 explore liability when automated vehicles are rented or deployed in an automated ridesharing service. Chapter 15 considers government use of automated vehicles and additional liability risks of states and local governments in an automated vehicle world.

Before examining liability for automated vehicle crashes, it is essential to identify two caveats to the analysis. Responsibility for any crash is not an either-this-party-is-liable-or-this-other-party-is-liable determination. In most crashes (whether or not they have automation features), many parties may bear some sort of legal fault—and other parties may not bear legal fault but may have moral or technical fault for the underlying crash. Additionally, facts are an essential ingredient for any liability determination. Any legal analysis done in a vacuum without facts will never fully grasp the results in all cases.

It is important to distinguish among the three groups of persons who may be injured in a crash caused by an automated vehicle: (1) the vehicle owner (or vehicle lessee), (2) users and other passengers of the vehicle, and (3) everyone else.[14] The distinction among these groups of persons matters for a few reasons. Owners may receive instruction from the manufacturer—in the form of an owner's manual (that they of course read)—and the dealership[15]—perhaps in the form of in-person training or instruction from a sales representative. Users and passengers, if not the owner, may not have received that instruction and may need in-car instruction about the capabilities of the automation features in the vehicle. There should be a distinction between those who voluntarily use an automated vehicle and perhaps accept the risk—or misuse a given automation feature—and those on the road who unwittingly encounter multiton automated vehicles with or without supervision or assistance of the human users.

---

14. *See* W. Page Keeton et al., Prosser and Keeton on the Law of Torts 678 (5th ed. 1984) ("Currently those who suffer these different kinds of losses [from products liability claims] can be classified into three general groups as: (1) purchasers, (2) users who are not purchasers, and (3) non-users who are commonly referred to as bystanders when physical injuries are suffered."); F. Patrick Hubbard, *"Sophisticated Robots": Balancing Liability, Regulation, and Innovation*, 66 Fla. L. Rev. 1803, 1812 (2014) (stating that "injuries [caused by robots] will be borne primarily by three types of victims: (1) purchasers (owners) or lessees, (2) users (other than purchasers or lessees), and (3) other parties").

15. Most vehicle sales occur through dealerships that are owned independently of the vehicle manufacturer.

Chapter 8

Liability for Automated Vehicle Crashes

This chapter explores the principal liability doctrines that are used to impose fault for automobile crashes and how these doctrines apply to automated vehicles. Generally, there are two categories of claims that are asserted by litigants in automobile crash cases: negligence by the operator of a vehicle involved in the crash and products liability claims (whether in strict liability or negligence) alleging that a vehicle defect caused the crash or caused the injuries suffered in the crash to be more serious. The defendants of these cases usually have an array of defenses that they can use to shield themselves from liability, and this chapter discusses those defenses. The chapter concludes by exploring a few ancillary issues relating to postsale updates to driving automation systems.

I. NEGLIGENCE

Automobile crashes are overwhelmingly resolved through principles of negligence. Driver errors cause approximately 95 percent of current crashes, and most of these crashes are due to negligent behavior by drivers. Negligence is a tort law claim that imposes fault when people fail to act reasonably under the circumstances. There are four parts to a negligence claim: (1) duty, (2) breach, (3) causation, and (4) damages. Because every attorney is familiar with negligence,[1]

1. For those who need a refresher, the below discussion describes these elements. Duty is the "foundational element" of negligence. David G. Owen, Products Liability Law 62 (2nd ed. 2005). "An actor ordinarily has a duty to exercise reasonable care when the actor's conduct creates a risk of physical harm." Restatement (Third) of Torts: Prods. Liab. § 7 (1998). A duty can also be supplied from a statute, including from the state motor vehicle codes. Whether a defendant owed a duty to a plaintiff is

this section focuses on who has a duty to prevent automated vehicle crashes and how to define the reasonable person test.[2]

Who has a duty to exercise reasonable care for an automated vehicle crash may be a question of who is the driver or operator of the automated vehicle. Chapter 19 details who is the driver-operator of automated vehicles under state

---

typically a matter of law for a court to decide. Drivers have a number of duties when operating vehicles, including, but not limited to, duties to remain attentive and maintain a safe distance behind vehicles.

Breach "is the misconduct itself, the defendant's improper act or omission." David G. Owen, *The Five Elements of Negligence*, 35 Hofstra L. Rev. 1671, 1676 (2007). A breach can typically be determined by comparing the defendant's conduct to the conduct of a reasonable person. Another method of determining whether a defendant's conduct breached a duty is by using the "Hand Formula," which was developed by Judge Learned Hand in *United States v. Carroll Towing Co.*, 159 F.2d 169, 173 (2d Cir. 1947). Under this formulation, a defendant's duty of care to provide against resulting injuries is a function of three variables:

> (1) The probability [of injury]; (2) the gravity of the resulting injury . . .; (3) the burden of adequate precautions. Possibly it serves to bring this notion into relief to state it in algebraic terms: if the probability be called P; the injury, L; and the burden, B; liability depends upon whether B is less than L multiplied by P: i.e., whether B < PL.

A breach occurs under the Hand Formula when the burden is less than the probability of harm. Drivers breach their duties in a myriad of ways, including by becoming distracted, by speeding, or by tailgating.

Causation has two parts: (1) cause in fact and (2) proximate cause. Cause in fact is "but for" causation: "The defendant's conduct is a cause of the event if the event would not have occurred but for that conduct." W. Page Keeton et al., Prosser and Keeton on the Law of Torts 266 (5th ed. 1984). In some circumstances, such as when "two causes concur to bring about an event, and either one of them, operating alone, would have been sufficient to cause the identical result," courts determine cause in fact based on whether the defendant's conduct "was a material element and a substantial factor in bringing" about an event. *Id.* at 266–67. Cause in fact, in and of itself, is insufficient for the imposition of liability; the defendant's conduct must also have been the proximate cause in bringing about the harm. Proximate cause acts as a limitation to the potentially endless cause in fact of a given injury or given conduct. There are two principal tests for determining whether a defendant's conduct was the proximate cause of a plaintiff's harm: (1) directness and (2) foreseeability. "'The directness rule extends to all outcomes, even if not foreseeable, so long as they are the "direct" result of the tortious conduct.' Under the foreseeability test, the defendant is only responsible for foreseeable results of her action—not freakish and unforeseen outcomes." Jeffrey K. Gurney, *Crashing into the Unknown: An Examination of Crash-Optimization Algorithms through the Two Lanes of Ethics and Law*, 79 Alb. L. Rev. 183, 229 (2016) (quoting Jane Stapleton, *Legal Cause: Cause-in-Fact and the Scope of Liability for Consequences*, 54 Vand. L. Rev. 941, 996 (2001)).

Causation may be impacted by an intervening cause. "An intervening cause is one which comes into active operation in producing the result after the negligence of the defendant." Keeton et al., *supra*, at 301. Generally, a defendant is not excused from liability if an intervening cause "is one which the defendant has reason to anticipate under the particular circumstances." *Id.* at 303. This is true even when the intervening cause is another party's negligence or reasonably foreseeable intentional or criminal acts. When an intervening cause is not foreseeable, then the defendant is generally not responsible for the harm.

The final element—damages—is a requirement, for there can be no negligence without harm. This element requires actual loss or damage; nominal damages will not suffice. Damages in automobile crashes include physical human injuries (medical and lost wages), property damage, and injuries or harm to animals.

2. Other duties may be created by automated vehicles that do not relate to the driving behavior itself. For example, owners and operators of automated vehicles may have duties to update the vehicle's software on a regular basis.

traffic codes. Traffic codes can affect liability for crashes. Under negligence per se, a "violation of a statute is negligence in and of itself if it causes harm of the kind the statute was intended to avoid and to a person within the class of persons the statute was intended to protect."[3] Traffic codes are one of the more frequent uses of negligence per se because of the abundant traffic laws that are intended to make roads safer and prevent crashes. Thus, by designating which party is the operator or driver of an automated vehicle for purposes of traffic codes, state legislatures may be effectively establishing the party to which negligence per se claims could be asserted and create duties.

Most states have not legislatively determined who or what is the operator or driver of an automated vehicle. Under those state laws, the vehicle may have more than one operator or driver at any one time. With levels 1 and 2 driving automation systems, the human driver is required to maintain attention and control over the vehicle, so automation does not impact duties at those levels. Some states have made the following the driver-operator of highly automated vehicles: the automated driving system, the vehicle owner, the person sitting in the driver's seat, the automated vehicle's user, or, as the Uniform Automated Operation of Vehicles Act uses, the automated-driving provider. The decisions to legislatively define "operator" or "driver," as well as the decision not to, may have a significant effect on the continued application of negligence or negligence per se to automobile crashes.[4]

In applying negligence in the absence of a statutory duty, an important question will be whether to judge the conduct of the automated driving system against a reasonable human driver or a reasonable automated driving system. This is a complex determination that could have significant effects on the automated vehicle industry. Automated vehicles should be safer drivers

---

3. Dan B. Dobbs, Paul T. Hayden & Ellen M. Bublick, The Law of Torts § 148. *The Rule of Negligence Per Se* (2d ed. 2011).

4. The use of negligence to judge behavior for automated vehicle crashes has two main benefits. First, to the extent that plaintiffs begin asserting claims against manufacturers or entities that operate automated vehicles for these crashes, negligence is a more cost-efficient mechanism to judge the behavior of the automated driving system rather than products liability law and particularly design defect claims. Design defect claims under the risk-utility test would require multiple experts and the plaintiff to introduce a reasonable alternative design that could have prevented the crash. The costs of proving these claims may exceed the damages. For example, if a level 5 automated vehicle bumps into a lead car, the damages to the lead car may not justify the costs of proving a defective design of the level 5 automated vehicle. A legal system that creates artificial barriers to recovery, such as costs, seems unjust. Identifying the responsible party as the driver of the vehicle and applying negligence to that driver's behavior may be a simple way to avoid such injustices. The second justification for use of negligence is that juries more readily understand and comprehend whether the driver's behavior conformed to a "reasonable actor" standard than whether some alternative design proposed by the plaintiff is better than the design chosen by the manufacturer. *See generally* Jeffrey K. Gurney, *Imputing Driverhood: Applying a Reasonable Driver Standard to Accidents Caused by Autonomous Vehicles, in* Robot Ethics 2.0: The Ethical and Social Implications of Robotics 58–59 (Patrick Lin, Keith Abney & George A. Bekey eds., 2017).

than humans. Thus, using the "reasonable automated driving system" standard would hold the automated driving system driver to a higher standard than human drivers. This standard may mean that the deployment of automated vehicles, which are expected to reduce crashes, may be delayed because the system might be deemed to be not as effective as a "reasonable automated driving system." Such a standard could also make it difficult for newcomers in the industry.[5] Automated vehicles operate using artificial intelligence that improves over time, which means that automated vehicles should become better drivers over time and make it more difficult for industry newcomers to develop automated vehicles of the same caliber as existing companies. Therefore, newcomers may be deterred by the use of the "reasonable automated driving system" standard. Additionally, at least at first, juries will be unfamiliar with how a reasonable automated driving system may have acted in a similar situation, necessitating expert witnesses to explain the capabilities of automated driving systems to juries.

II. PRODUCTS LIABILITY LAW

The advent of the automobile and the way automobiles are distributed had a significant impact on the development of modern products liability law. Justice Cardozo's famous opinion abolishing privity for products liability law was in *MacPherson v. Buick Motor Company*.[6] Although automobiles had a significant role in shaping modern products liability law, products liability law currently plays a relatively small role in automobile crashes.[7] Given the shift in control over aspects of the dynamic driving task from the traditional driver to the driving automation system, many people believe that automated vehicle crashes will involve more products liability claims. This section introduces the underlying justifications of products liability law and then describes doctrines used to hold product manufacturers and other members of the supply chain liable for crashes.

---

5. Conversely, by testing their systems longer, assuming they could financially survive, their systems once ultimately released would be in a better position to compete than a system that would otherwise be rushed to market.

6. 217 N.Y. 382, 111 N.E. 1050 (1916).

7. Bryant Walker Smith, *Automated Driving and Product Liability*, 2017 MICH. ST. L. REV. 1, 36 (2017) (noting that "[a]bout 6% of the relevant verdicts in Westlaw's Case Evaluator involve a product liability claim").

A. Moral Underpinnings and Purposes of Products Liability Law

Products liability law is not some ironclad area of law that coldly examines whether products are not reasonably safe.[8] Products liability law has rich moral and societal foundations.[9] An analysis of the interplay between products liability law and new technologies, such as vehicle automation, necessarily begins with these principles. If liability were to shift from the conventional driver to the manufacturer as many have foretold, then the moral foundations and the purposes of products liability law would need to support that shift. This section briefly introduces these concepts.

1. Moral Foundations of Products Liability Law

Products liability law seeks to protect certain personal and societal interests.[10] The individual principles include freedom, truth, and equality.[11] The societal principles include community, utility, and sharing.[12] Freedom, the "first and most essential ideal," "is the first condition to protecting or advancing other values, such as equality, altruism, and communal welfare."[13] Truth, a component of autonomy, recognizes that autonomy is furthered when truth is promoted and that accidental harm is often the result "of the failure of at least one person, the actor or the victim, to expect the harm, to possess the truth concerning the things that caused the harm."[14] Equality draws the boundary among competing freedoms in a crowded world and treats each person's freedoms and rights equally.[15] For that reason, neither the victim's interests nor the actor's interests should be prioritized over the other.[16] Equality and freedom generally mean that there is not a moral basis "for a general rule of liability that holds actors strictly accountable for accidental harm."[17] The community interests "idealize[] the interests of the group" over the interests of the individuals that comprise the group.[18] Utility, a communal theory, seeks to aggregate the welfare of society as a whole.[19] The principle of utility finds accidental conduct justifiable "if

---

8. *See* David G. Owen, *The Moral Foundations of Products Liability Law: Toward First Principles*, 68 Notre Dame L. Rev. 427, 429–30 (1993).

9. *See id.*

10. *See id.*

11. *Id.* at 438–54.

12. *Id.* at 453–59.

13. *Id.* at 438–39.

14. *Id.* at 440–41.

15. *Id.* at 441–47.

16. *Id.* at 442.

17. *Id.* at 452–53. A principle of strict liability would value the interests of the victim over the interests of the actor. *Id.*

18. *Id.* at 453.

19. *Id.* at 455–56.

the social harm reasonably expected to result from the conduct is exceeded by the expected social benefit" from that conduct.[20] Sharing embodies the principles of risk-sharing that underlie the concept of insurance.[21] Efficiency also underlies products liability law and the assignment of liability upon the party that can most likely prevent an accident.[22]

2. Purposes of Products Liability Law

Products liability law has specific rationales that justify the imposition of liability on the manufacturer of a product. This section briefly introduces these concepts.

Consumer Expectations. A fundamental premise of products liability law is to protect the expectations of consumers.[23] As one scholar noted, "[T]he supplier, by placing the goods upon the market, represents to the public that they are suitable and safe for use; and by packaging, advertising or otherwise, he does everything that he can to induce that belief. He intends and expects that the product will be purchased and used in reliance upon this assurance of safety; and it is in fact so purchased and used."[24]

Cost Internalization. Strict liability for product defects requires the manufacturer to absorb the cost of harm and damages in the product's cost.[25] Internalizing costs means that the risk of defects is not being externalized to victims, and the product adequately reflects its cost and value. Cost internalization is related to risk-sharing and deterrence.

Insurer? One often-asserted rationale of products liability law is that "product sellers as business entities are in a better position than are individual users and consumers to insure against such losses."[26] This "loss spreading" rationale has not been widely embraced by commentators or courts.[27] The manufacturer's "deep pockets" are an inadequate reason to impose liability. Instead, for products liability to apply, "[i]t must be established that the manufacturer

---

20. *Id.* at 456. "The converse is also true: accidental conduct is improper if it may be reasonably expected to produce more harm than good." *Id.*

21. *Id.* at 458.

22. *Id.* at 457.

23. *Id.* at 290; Keeton et al., *supra* note 1, at 150 ("Frustration of consumer or user expectations [is] regarded as an important factor justifying . . . strict liability.").

24. William L. Prosser, *The Assault upon the Citadel (Strict Liability to the Consumer),* 69 Yale L.J. 1099, 1123 (1960).

25. Owen, *supra* note 1, at 292.

26. Restatement (Third) of Torts: Prods. Liab. § 2 cmt. a (1998).

27. Alvin S. Weinstein et al., Products Liability and the Reasonably Safe Product: A Guide for Management, Design, and Marketing 5 (1978).

violated a legal responsibility to the user or consumer."[28] There must be a defect in the product.[29]

Risk Sharing. By internalizing the costs of product defects, "manufacturers will relieve injured consumers of possibly crushing financial burdens and, at the same time, spread such losses widely among the consuming public," which means that manufacturers "provide all consumers, albeit indirectly, with mandatory crash insurance for harm from product defects."[30]

Deterrence. Products liability is imposed on manufacturers to force them to improve product safety.[31] As one scholar has noted, "Creating incentives for safety is one of the most important policies underlying strict products liability law."[32] In this respect, products liability law incentivizes manufacturers to design safer products and encourages manufacturers to withdraw defective products from the marketplace.[33]

Administrative Convenience. A justification of products liability is administrative convenience. When a product defect causes harm, the manufacturer's conduct was probably negligent, but the aggrieved party may have difficulty proving negligence.[34] As one scholar noted for manufacturing defects, courts are aware of "the great expense that plaintiffs would have to undergo to prove negligent manufacture of one out of a million products of the defendant. They may also realize that if plaintiff were sufficiently well-heeled he would probably succeed in a surprisingly large number of cases in proving that the manufacturer was indeed negligent in the ordinary sense of the word."[35]

B. Basis of Liability

Courts and legislatures achieve the purposes of products liability law by arming those harmed by products with numerous doctrines to assert against the product manufacturer and other members of the supply chain. A plaintiff

28. *Id.*

29. *Id.* at 43; *see, e.g.*, Segler v. Ford Motor Co., 438 So. 2d 297, 298 (Ala. 1983) (holding that a directed verdict should have been granted because "there was no evidence of any defect in the automobile which caused the injury"); Giordano v. Ford Motor Co., 165 Ga. App. 644, 645, 299 S.E.2d 897, 899 (1983) ("The existence of a defect is crucial, because a manufacturer is not an insurer against all risks of injury associated with its product."); Shoemake v. Omniquip Int'l, Inc., 152 S.W.3d 567, 573 (Tenn. Ct. App. 2003) ("As a general rule, an injury in and of itself is not proof of a defect and thereby does not raise any presumption of defectiveness. Consequently, it is not enough to show that the product caused the plaintiff's injury or was involved in it. Rather, the burden is upon the plaintiff to show that there is something wrong with the product." (internal quotation marks and citations omitted)).

30. Owen, *supra* note 1, at 295.

31. *Id.* at 294.

32. Gerald F. Tietz, *Strict Products Liability, Design Defects and Corporate Decision-Making: Greater Deterrence through Stricter Process*, 38 Vill. L. Rev. 1361, 1415–16 (1993).

33. *Id.* at 1417–22.

34. Owen, *supra* note 1, at 291.

35. Thomas A. Cowan, *Some Policy Bases of Products Liability*, 17 Stan. L. Rev. 1077, 1087 (1965).

must typically prove a product defect proximately caused harm to the plaintiff. This section focuses on the following defects: (1) negligence, (2) manufacturing defects, (3) design defects, (4) warning defects, and (5) crashworthiness.

1. Negligence-Based Products Liability Claims

Negligence can be asserted against a manufacturer for a product defect. Negligence-based products liability claims can mirror the strict liability claims. A plaintiff proceeding on a negligence theory for products liability typically must show a defect, whether in manufacture, design, or marketing, as well as the elements of negligence.[36] A product manufacturer has a duty "to exercise reasonable care to refrain from selling products that contain unreasonable risks of harm."[37] This duty is not limited to the purchaser of the product and extends to "foreseeable risks and foreseeable victims" without regard to privity.[38] Manufacturers are generally held to a standard as an "expert in the field" of their products.[39]

A products liability claim based on negligence, unlike strict liability, challenges the underlying conduct of the manufacturer.[40] These claims can be particularly helpful to obtain discovery on the inner thinking of the manufacturer or to challenge the testing of an automated vehicle. Negligence can be asserted in addition to the strict liability claims.[41]

2. "Strict" Liability

Modern products liability claims grew out of strict liability that courts began recognizing in the 1950s and 1960s. The three principal defects are manufacture, design, and warning. Over time, two of these claims—design defects and warnings—have begun to resemble negligence claims more than strict liability claims.

Many crashes caused by automated vehicles will be related to some problem within the software of the automated vehicle. An initial issue is whether the software within the automated driving system is a product that could be subject to products liability. The Products Liability Restatement states that

---

36. Owen, *supra* note 1, at 65.

37. *Id.* at 62.

38. *Id.*

39. *Id.* at 68; *see, e.g.*, Alcala v. Emhart Indus., Inc., 495 F.3d 360, 364 (7th Cir. 2007) ("[T]he jury may presume that the manufacturer has the skill and knowledge that other manufacturers at the time possessed."); O'Hare v. Merck & Co., 381 F.2d 286, 291 (8th Cir. 1967) ("A manufacturer is held to the skill of an expert in its particular field of endeavor, and is obligated to keep informed of scientific knowledge and discoveries concerning that field.").

40. *See* Keeton et al., *supra* note 1, at 695 ("In strict liability, the plaintiff is not required to impugn the conduct of the maker or other seller but he is required to impugn the product.").

41. *See, e.g.*, Connelly v. Hyundai Motor Co., 351 F.3d 535, 540 (1st Cir. 2003).

"[a] product is tangible personal property distributed commercially for use or consumption. Other items, such as real property and electricity, are products when the context of their distribution and use is sufficiently analogous to the distribution and use of tangible personal property that it is appropriate to apply the rules stated in this Restatement."[42] The driving automation system and the automated vehicle should be considered products.

a. Manufacturing Defect

The manufacturing defect doctrine imposes a basic obligation on a product manufacturer to produce its products according to its specifications. If a product manufacturer cannot meet its standards and that failure causes harm to someone, then the product manufacturer is strictly liable.[43] Manufacturing defects are today the only *strict* products liability claims.[44]

To assert a manufacturing defect, the plaintiff must introduce evidence of a defective product. A manufacturer's obligation to produce a product according to its specifications is not limited to defects that arise during the manufacturing process and includes defects that occur during the manufacturer's control over the product.[45] Irrelevant to whether a product was defectively manufactured are the extent of care exercised by the manufacturer in the production process and evidence of no prior claims.[46]

In most manufacturing defect cases, plaintiffs introduce direct evidence of a product defect. A plaintiff typically proves a manufacturing defect through expert testimony on the defective part of the product.[47] Occasionally, a plaintiff is entitled to prove a manufacturing defect with circumstantial

---

42. Restatement (Third) of Torts: Prods. Liab. § 19(a) (1998).

43. Richard A. Epstein, Modern Products Liability Law 68 (1980) ("In all [manufacturing defect] cases the plaintiff holds the defendant only to the standards that he has chosen to impose upon himself.").

44. Restatement (Third) of Torts: Prods. Liab. § 2 cmt. a (1998) ("In general, the rationale for imposing strict liability on manufacturers for harm caused by manufacturing defects does not apply in the context of imposing liability for defective design and defects based on inadequate instruction or warning."); Owen, *supra* note 1, at 449 ("[Q]uite unlike design and warning cases, the liability standards for manufacturing defects . . . are truly 'strict.'").

45. Mark A. Geistfeld, Products Liability Law 124 (2012) ("A construction or manufacturing defect is not limited to flaws in the construction or manufacture of the product. Such 'a defect may arise after manufacture, for example, during shipment or while in storage.'" (quoting Restatement (Third) of Torts: Prods. Liab. § 2 cmt. c (1998))).

46. *See* Restatement (Third) of Torts: Prods. Liab. § 2 (1998) (stating that "[a] product . . . contains a manufacturing defect when the product departs from its intended design even though all possible care was exercised in the preparation and marketing of the product"); Owen, *supra* note 1, at 455 (noting that manufacturers are generally strictly liable for manufacturing defects "even if [they] exercise due care"). However, this type of evidence may be relevant in certain types of cases "if it tends to show that the manufacturer is not responsible for causing the defect, to wit, that the defect (if any) probably arose after the product left the manufacturer's control." Owen, *supra* note 1, at 455.

47. Owen, *supra* note 1, at 453; Geistfeld, *supra* note 45, at 123.

evidence through what is known as the "malfunction doctrine."[48] "Under the malfunction doctrine, a plaintiff may establish a prima facie case of product defect by proving that the product failed in normal use under circumstances suggesting a product defect."[49] The plaintiff satisfies this burden by showing that "(1) the product malfunctioned, (2) the malfunction occurred during proper use, and (3) the product had not been altered or misused in a manner that probably caused the malfunction."[50] The malfunction doctrine does not excuse a plaintiff from proving that the product was defective; instead, it merely alleviates the burden of proving the specific defect.[51] Many scholars believe the malfunction doctrine may gain prominence (or at least be used by plaintiffs) in automated vehicle crashes.[52]

There are certain limitations to the malfunction doctrine that may hinder a plaintiff's ability to present a malfunction doctrine claim against a manufacturer.[53] One inherent limitation of the doctrine stems from the need to negate likely alternative causes for the malfunction, such as the vehicle's age, maintenance, or alterations.[54] In addition, not all courts have widely embraced the malfunction doctrine.[55]

A manufacturing defect, and conceivably the malfunction doctrine, may be asserted when an automated vehicle does not perform as intended by the

48. *See, e.g.*, Anderson v. Chrysler Corp., 184 W. Va. 641, 646, 403 S.E.2d 189, 194 (1991) ("[W]e hold that circumstantial evidence may be sufficient to make a prima facie case in a strict liability action, even though the precise nature of the defect cannot be identified, so long as the evidence shows that a malfunction in the product occurred that would not ordinarily happen in the absence of a defect."); Stackiewicz v. Nissan Motor Corp. in U.S.A., 100 Nev. 443, 448, 686 P.2d 925, 928 (1984) ("[P]roof of an unexpected, dangerous malfunction may suffice to establish a prima facie case for the plaintiff of the existence of a product defect."); Landahl v. Chrysler Corp., 144 A.D.2d 926, 927, 534 N.Y.S.2d 245, 246 (1988) ("A plaintiff in a products liability action need not establish the precise nature of the defect in order to make out a prima facie case. The existence of a defect may be inferred from the circumstances of the accident and from proof that the product did not perform as intended." (internal citations omitted)).

49. Owen, *supra* note 1, at 466.

50. *Id.*

51. *See, e.g.*, Sochanski v. Sears, Roebuck & Co., 689 F.2d 45, 50 (3d Cir. 1982) ("Evidence of a malfunction, then, is not a substitute for the need to establish that the product was defective. A malfunction is evidence that a defect existed and eliminates only the need to identify a specific failure.").

52. *See, e.g.*, Jeffrey K. Gurney, *Sue My Car Not Me: Products Liability and Accidents Involving Autonomous Vehicles*, 2013 U. Ill. J.L. Tech. & Pol'y 247, 259–60 (2013); Kevin Funkhouser, *Paving the Road Ahead: Autonomous Vehicles, Products Liability, and the Need for a New Approach*, 2013 Utah L. Rev. 437, 446–49 (2013); Sunghyo Kim, *Crashed Software: Assessing Product Liability for Software Defects in Automated Vehicles*, 16 Duke L. & Tech. Rev. 300, 317 (2018).

53. As a leading products liability law scholar has noted, the malfunction "doctrine presents a seductive but faulty shelter for plaintiffs with insufficient proof of defect and causation, and the law reports brim with decisions that recite the propriety of the doctrine as a general proposition but hold it inapplicable to the facts." Owen, *supra* note 1, at 469–70.

54. *Id.* at 470–71.

55. *See, e.g.*, Ford Motor Co. v. Ledesma, 242 S.W.3d 32, 42 (Tex. 2007) ("Texas law does not generally recognize a product failure or malfunction, standing alone, as sufficient proof of a product defect."); Dykes v. Husqvarna Outdoor Prod., N.A., Inc., 869 F. Supp. 2d 749, 758 (S.D. Miss. 2012) ("Proof of a malfunction of machinery in and of itself is not proof of a manufacturing defect under Mississippi law." (citation omitted)).

manufacturer's specifications.[56] Whether an automated vehicle performs as intended by the manufacturer's specifications and whether these are manufacturing defects are questions that may riddle litigants and courts. The inquiry into whether a defect is in manufacture becomes substantially more difficult when applied to intangible products, like software or the interactions of a complex product, instead of tangible products, like sensors, wheels, or transmissions. As one scholar has described, software development generally follows four phases: "(i) the design phase, (ii) the coding phase, (iii) the testing phase, and (iv) the replication and distribution phase."[57] The determination of whether a software defect is properly classifiable as a manufacturing defect or a design defect may depend upon the phase in which the defect arose. The two easy cases are the design phase and the replication and distribution phase—the former of which would be a design defect and the latter of which would be a manufacturing defect.[58] Disputes among plaintiffs and defendants will likely focus on whether defects that arise in coding or testing may be manufacturing defects or design defects.[59] Plaintiffs may claim that bugs in software and incorrect coding are like manufacturing defects in that manufacturers can prevent them with better oversight and testing, much like traditional manufacturing defects.[60] Defendants will contend that software bugs, incorrect coding, and testing mechanisms are issues that affect all driving automation systems and the entire product line.[61] Because software suffers from the same flaw,[62] these claims do not resemble manufacturing defects.[63]

---

56. *See, e.g.*, Michael I. Krauss, *Freedom from Control, Freedom from Choice? How Will Tort Law Deal with Autonomous Vehicles?*, 25 GEO. MASON L. REV. 20, 27 (2017).

57. Michael D. Scott, *Tort Liability for Vendors of Insecure Software: Has the Time Finally Come?*, 67 MD. L. REV. 425, 459 (2008).

58. *Id.*

59. *See id.* at 459–60; Ryan J. Duplechin, *The Emerging Intersection of Products Liability, Cybersecurity, and Autonomous Vehicles*, 85 TENN. L. REV. 803, 822–23 (2018).

60. *See* Peter A. Alces, *W(h)ither Warranty: The B(l)oom of Products Liability Theory in Cases of Deficient Software Design*, 87 CAL. L. REV. 269, 300–01 (1999).

61. *See* Duplechin, *supra* note 59.

62. As one court has noted in describing the distinction between manufacturing defects and design defects, "[m]anufacturing defect cases involve products which are flawed, i.e., which do not conform to the manufacturer's own specifications, and are not identical to their mass-produced siblings." Casey v. Toyota Motor Eng'g & Mfg. N. Am., Inc., 770 F.3d 322, 329 (5th Cir. 2014); *but see* Granillo v. FCA US LLC, No. CV 16-153 (FLW)(DEA), 2016 WL 9405772, at *14 (D.N.J. Aug. 29, 2016) (stating that "contrary to Defendant's assertions, it is plausible that a manufacturing defect might occur on a sufficiently wide scale to affect all, or nearly all of the Class Vehicles. A manufacturing defect is simply a defect that occurs when a product does not conform to the manufacturer's intended design. Although this type of defect is more likely to occur sporadically, it is entirely possible that some grave error in material or workmanship during assembly caused all Class Vehicles to deviate from Defendant's intended design" (internal quotation marks and citation omitted)).

63. *See, e.g.*, Jeffrey K. Gurney, *Sue My Car Not Me: Products Liability and Accidents Involving Autonomous Vehicles*, 2013 U. ILL. J.L. TECH. & POL'Y 247, 259 (2013); Mark A. Geistfeld, *A Roadmap for Autonomous Vehicles: State Tort Liability, Automobile Insurance, and Federal Safety Regulation*, 105 CAL. L. REV. 1611, 1633 (2017); Hughes v. Panasonic Consumer Elecs. Co., No. CIV.A. 10-846 SDW, 2011 WL 2976839, at *19 (D.N.J. July 21, 2011) ("A defect in the programming of the software which controlled the volt-

b. Design Defect

Unlike a manufacturing defect that alleges that a specific product had a defect, with a design defect, the plaintiff is challenging the underlying design decisions of an entire product line.[64] Thus, in a design defect case, the plaintiff seeks to condemn the whole product line by alleging that the manufacturer did not adequately consider the safety of the product. Liability for design defects is not "strict" as it is for manufacturing defects.[65] This section describes the two primary tests for challenging the adequacy of a product design, the consumer expectations test and the risk-utility test, as well as the combination of these tests and the application of the malfunction doctrine to design defects. This section concludes by exploring specific design defect issues for automated vehicles.

(i) The Consumer Expectations Test

The consumer expectations test was the first prominent theory to test the adequacy of a product design. Under the Restatement (Second) of Torts, "a product is defectively dangerous if it is dangerous to an extent beyond that which would be contemplated by the ordinary consumer who purchased it with ordinary knowledge common to the community as to the product's characteristics."[66] Many courts have substituted "ordinary user," "reasonable user," or "foreseeable user" for "ordinary consumer" to indicate that the application of the test is not limited to the consumer.[67]

From this statement of law, a few characteristics of the consumer expectations test can be deduced and its limitations revealed. First, "the consumer expectations test is an objective test based on the average, normal, or 'ordinary' expectations of the reasonable user or consumer."[68] This limitation created a significant hurdle for plaintiffs alleging a design defect claim for obvious dangers because "obvious dangers . . . are virtually always contemplated or expected by the user or consumer."[69] Thus, even where simple and inexpensive safety precautions could prevent an obvious significant danger, plaintiffs struggled to present claims with the consumer expectations test.[70] Second,

---

age adjustments is not akin to a defect in materials or substandard workmanship used by Panasonic in replacing or repairing parts under warranty.").

64. *See* Epstein, *supra* note 43, at 69 ("[Manufacturing] defects arise because of occasional miscarriages in the manufacturing process; design defects necessarily pervade an entire line of products whose production was exactly as intended.").

65. *See* Restatement (Third) of Torts: Prods. Liab. § 3 cmt. a (1998).

66. Keeton et al., *supra* note 1, at 698 (citing Restatement (Second) of Torts § 402A cmt. i (1965)).

67. *Id.*

68. Owen, *supra* note 1, at 505.

69. *Id.* at 506.

70. *Id.*

the test also limited the ability of plaintiffs to assert design defect claims for defects about which the reasonable purchaser "was adequately informed."[71] Third, the "test can result in the identification of products as being defectively dangerous that are clearly not."[72] Fourth, the consumer expectations test is vague.[73] The test does not explain what a reasonable consumer expects—all reasonable consumers likely know that products have adverse side effects.[74] Also, so long as manufacturers were to adequately describe the limitations of their driving automation systems and that the driving automation systems have error rates, this limitation may mean that the manufacturers of such systems may not be liable for otherwise avoidable harm.[75] Fifth, the focus on consumer expectations may prove helpful with noncomplex products, but, as courts have held, it seems unlikely that consumers generate expectations for complex or highly technical products.[76] Because driving automation systems are complex, many courts that still apply the consumer expectations test may not do so for automated vehicles. An additional problem with the consumer expectations test for automobiles is the question of whose expectations are at issue, given that numerous parties, including the car owner, the car driver, passengers, or bystanders, could have varying and diverse expectations about a given product.[77]

(ii) The Risk-Utility Test

Because of the shortcomings of the consumer expectations test, many courts and jurisdictions began to abandon the consumer expectations test in favor of the risk-utility test. The Products Liability Restatement expressly rejected the consumer expectations test for determining whether a product is defective and adopted the risk-utility test.[78] Many courts use the risk-utility test as the

---

71. Keeton et al., *supra* note 1, at 698.

72. *Id.* at 698–99. The example given by *Prosser & Keeton* is "when a new drug is a great boon to humanity but a few are victimized by a side effect or adverse reaction that was an unknowable risk." *Id.*

73. Owen, *supra* note 1, at 507; Keeton et al., *supra* note 1, at 699 ("The meaning is ambiguous and the test is very difficult of application to discrete problems.").

74. Keeton et al., *supra* note 1, at 699.

75. *See generally* Mark A. Geistfeld, *A Roadmap for Autonomous Vehicles: State Tort Liability, Automobile Insurance, and Federal Safety Regulation*, 105 Cal. L. Rev. 1611 (2017).

76. *See, e.g.*, Quintana-Ruiz v. Hyundai Motor Corp., 303 F.3d 62, 77 (1st Cir. 2002) (stating that "consumer expectations cannot be the basis of liability in a case involving complex technical matters"); Soule v. Gen. Motors Corp., 8 Cal. 4th 548, 882 P.2d 298, 308 (1994) (same); Izzarelli v. R.J. Reynolds Tobacco Co., 321 Conn. 172, 203, 136 A.3d 1232, 1249 (2016) (same).

77. *See* David G. Owen, *Design Defect Ghosts*, 74 Brook. L. Rev. 927, 942 (2009).

78. Restatement (Third) of Torts: Prods. Liab. § 2 cmt. g (1998) ("[C]onsumer expectations do not constitute an independent standard for judging the defectiveness of product designs."). Although the consumer expectations standard is not an independent standard, consumer expectations are relevant in determining whether a product is defective under the risk-utility test. *Id.*

sole test for determining whether a design is defective.[79] Unlike the consumer expectations test, which focuses on the expectations of consumers, the risk-utility test focuses on the foreseeable benefits and foreseeable costs of a particular safety feature that could have prevented the harm.[80]

The focus on the costs and benefits of a particular design tells a reader very little as to what inputs go into each factor. At a fundamental level, an essential initial determination is whether the risk-utility test balances "macro" or "micro" benefits and costs.[81] Some scholars have reached different conclusions on liability for automated vehicle crashes under the risk-utility test based upon which factors they include in their risk-utility analysis. Scholars who use macro factors, such as the overwhelming benefits that automated vehicles are projected to deliver, argue that plaintiffs will have a difficult time proving that a driving automation system is defective.[82] These macro formulations are typically a result of courts and scholars relying on the Dean John Wade factors for defining the risk-utility test.[83] In representing clients, an important task will be arguing for a broad or narrow formulation of the risk-utility test.

---

79. *See, e.g.*, Branham v. Ford Motor Co., 390 S.C. 203, 220, 701 S.E.2d 5, 14 (2010) ("We hold today that the exclusive test in a products liability design case is the risk-utility test with its requirement of showing a feasible alternative design."); Banks v. ICI Ams., Inc., 264 Ga. 732, 735, 450 S.E.2d 671, 674 (1994) ("[W]e conclude that the better approach is to evaluate design defectiveness under a test balancing the risks inherent in a product design against the utility of the product so designed. Hence, we hereby adopt the risk-utility analysis."); Uniroyal Goodrich Tire Co. v. Martinez, 977 S.W.2d 328, 337 (Tex. 1998) (declining to adopt the consumer expectations test); *but see, e.g.*, Aubin v. Union Carbide Corp., 177 So. 3d 489, 510 (Fla. 2015) ("In considering which approach is in line with our prior strict liability jurisprudence, we are in accord with those state supreme courts that have thoughtfully considered this issue and determined that the Third Restatement's new approach is inconsistent with the rationale behind the adoption of strict products liability.").

80. *See* Owen, *supra* note 1, at 511 ("A product's design is 'defective' under a risk-utility test if the costs of avoiding a particular hazard are foreseeably less than the resulting safety benefits.").

81. David G. Owen, *Toward a Proper Test for Design Defectiveness: "Micro-Balancing" Costs and Benefits*, 75 Tex. L. Rev. 1661 (1997); *see* Banks v. ICI Ams., Inc., 264 Ga. 732, 736, 450 S.E.2d 671, 675 (1994) ("We recognize that in setting forth a test under the risk-utility analysis for the determination whether a manufacturer should be liable for an entire product line, no finite set of factors can be considered comprehensive or applicable under every factual circumstance, since such matters must necessarily vary according to the unique facts of each case.").

82. *See* Matthew Blunt, *Highway to a Headache: Is Tort-Based Automotive Insurance on a Collision Course with Autonomous Vehicles?*, 53 Willamette L. Rev. 107, 127 (2017) ("Potential plaintiffs . . . face a significant hurdle in proving a design defect under the risk utility test, due to the safety benefits associated with autonomous vehicles.").

83. *See* Owen, *supra* note 1, at 516 ("Overbroad formulations of risk-utility analysis for design defect decisionmaking are traceable to a widely quoted set of liability factors proposed in an early, influential article written by Dean John Wade, *On the Nature of Strict Tort Liability for Products*.").

For those seeking to define the risk-utility test broadly, there is authority in many jurisdictions for such broad formulations.[84] For those seeking to define the risk-utility test narrowly, there is ample authority to define the test narrowly; indeed, a close examination of the cases that quote macro factors reveals that these courts actually apply micro factors in their analysis.[85] The inherent problem with a broad definition is that "[t]he great majority of products are beneficial at a macro-balance level, even if various risks may feasibly be designed away."[86]

The Products Liability Restatement defines the risk-utility test utilizing micro factors:

> A product . . . is defective in design when the foreseeable risks of harm posed by the product could have been reduced or avoided by the adoption of a reasonable alternative design by the seller or other distributor, or a predecessor in the commercial chain of distribution, and the omission of the alternative design renders the product not reasonably safe[.][87]

At its core, the risk-utility test seeks to ascertain "whether the aggregate dollar costs of adding some safety feature proposed by the plaintiff is or is not outweighed by the aggregate benefit of preventing foreseeable crashes like that which injured the plaintiff."[88] The risk-utility test weighs these costs: (1) the financial cost of the alternative design, (2) "any reduction in the product's usefulness," and (3) whether the "plaintiff's proposed alternative design . . . may

---

84. *See, e.g.*, Voss v. Black & Decker Mfg. Co., 59 N.Y.2d 102, 109, 450 N.E.2d 204, 208–09 (1983) ("In balancing the risks inherent in the product, as designed, against its utility and cost, the jury may consider several factors. Those factors may include the following: (1) the utility of the product to the public as a whole and to the individual user; (2) the nature of the product—that is, the likelihood that it will cause injury; (3) the availability of a safer design; (4) the potential for designing and manufacturing the product so that it is safer but remains functional and reasonably priced; (5) the ability of the plaintiff to have avoided injury by careful use of the product; (6) the degree of awareness of the potential danger of the product which reasonably can be attributed to the plaintiff; and (7) the manufacturer's ability to spread any cost related to improving the safety of the design." (internal citations omitted)); Calles v. Scripto-Tokai Corp., 224 Ill. 2d 247, 265, 864 N.E.2d 249, 261 (2007); Nunnally v. R.J. Reynolds Tobacco Co., 869 So. 2d 373, 379 (Miss. 2004); Banks v. ICI Ams., Inc., 264 Ga. 732, 450 S.E.2d 671, 675 n.6 (1994); Potter v. Chi. Pneumatic Tool Co., 241 Conn. 199, 694 A.2d 1319, 1333–34 (1997).

85. Owen, *supra* note 1, at 1674 (collecting cases).

86. *Id.* This type of macro balancing also is ill-suited for courts: "[T]he task is exceedingly complex, multi-layered, intrinsically 'polycentric' and 'legislative,' and hence especially inappropriate and prone to error in a court of law." *Id.* As one products liability scholar has noted in criticizing the macro test postulated in this way, "[A] fair macro-balance might well yield a finding that Pintos—even if their fuel-tank problems were simply and cheaply correctable—were not defectively designed because, on balance, they generated in the aggregate more social good than harm." *Id.* at 1679. For the benefits side of the equation, Professor Owen noted that for "[o]ver two decades, Pintos transported millions of Americans to work, to school, to church and synagogue, to visit friends and family, and to play at this and that. Surely the aggregate social 'benefits' or 'utility' of any such vehicle are enormous, amorphous, and practicably incalculable. . . ." *Id.*

87. Restatement (Third) of Torts: Prods. Liab. § 2(b) (1998).

88. Owen, *supra* note 1, at 524.

introduce new dangers into the product."[89] In these cases, a plaintiff pits his or her proposed alternative design against the manufacturer's chosen design, and most design defect cases boil down to whether the jury accepts the manufacturer's chosen design or the plaintiff's proposed alternative design.[90]

(iii) Combined Tests

Although the consumer expectations test and the risk-utility test seem like distinct tests, some courts have blended or merged the two.[91] These formulations have taken shape in either "(1) defining one test in terms of the other; or (2) establishing each as separate liability 'prongs,' either one of which may independently support a design defect finding."[92] Courts have followed the lead of the *Potter*[93] court in defining consumer expectations in terms of the risk-utility test.[94] For those jurisdictions that follow both tests, they tend to follow what is known as the *Barker-Soule*[95] approach. Under this approach, a plaintiff can prove a design defect under either the consumer expectations test or the risk-utility test, but the risk-utility test should be the only design defect test for complex products.[96] Courts may find that automated vehicles are complex; therefore, courts that follow the *Barker-Soule* approach will likely use the risk-utility test for design defects. A significant difference in the *Barker-Soule* approach adopted by California is that the manufacturer bears the burden under the risk-utility test after the plaintiff makes an initial showing.[97]

(iv) Malfunction Doctrine

Although the malfunction doctrine is used primarily for manufacturing defects, the Products Liability Restatement does not limit its application solely to manufacturing defect claims.[98] The Products Liability Restatement states that when a product fails during its intended performance parameters and

---

89. *Id.* at 525–27.

90. *Id.* at 530.

91. *Id.* at 532.

92. *Id.*

93. Potter v. Chi. Pneumatic Tool Co., 241 Conn. 199, 694 A.2d 1319 (1997).

94. David G. Owen & Mary J. Davis, Owen & Davis on Products Liability § 8:15 (4th ed. 2020).

95. Barker v. Lull Eng'g Co., 20 Cal. 3d 413, 573 P.2d 443 (1978); Soule v. Gen. Motors Corp., 8 Cal. 4th 548, 882 P.2d 298 (1994).

96. Soule v. Gen. Motors Corp., 882 P.2d 298 (Cal. 1994).

97. Kim v. Toyota Motor Corp., 6 Cal. 5th 21, 30, 424 P.3d 290, 296 (2018) ("The risk-benefit test requires the plaintiff to first 'demonstrate[ ] that the product's design proximately caused his injury.' If the plaintiff makes this initial showing, the defendant must then 'establish, in light of the relevant factors, that, on balance, the benefits of the challenged design outweigh the risk of danger inherent in such design.'" (quoting *Barker*, 573 P.2d 443)).

98. Restatement (Third) of Torts: Prods. Liab. § 3 cmt. b (1998); *see, e.g.*, Mullins v. Johnson & Johnson, 236 F. Supp. 3d 940, 944 (S.D.W. Va. 2017) ("West Virginia case law and the [pattern jury instructions] allow a plaintiff to prove his or her design defect strict liability case with circumstantial evidence.").

the plaintiff proves that the harm was not solely the result of other causes, "it should not be necessary for the plaintiff to incur the cost of proving whether the failure resulted from a manufacturing defect or from a defect in the design of the product."[99] But the Products Liability Restatement notes that these types of claims "are limited to situations in which a product fails to perform its manifestly intended function, thus supporting the conclusion that a defect of some kind is the most probable explanation."[100] Plaintiffs will likely assert malfunction doctrine claims to alleviate their burden under the risk-utility test when automated vehicles crash.

(v) Automated Vehicle Design Defects and Parties

(a) Categories of Design Defect Claims

There are three broad categories of design defect claims that may arise in crashes involving automated vehicles. The first are claims about the selection of the underlying technology that allows an automated vehicle to perceive, sense, and analyze the world.[101] This type of claim could also involve the integration of the technologies within the driving automation system as a whole. Fundamental questions about design and choice of technology and the integration of those technologies will inevitably be defective design cases. For example, Tesla Motors has been outspoken in its belief that light detection and radar (LiDAR) is unnecessary for highly automated vehicles, while almost every other manufacturer disagrees with Tesla Motors on this point.[102] If a Tesla Motors automated vehicle is involved in a crash that could have been prevented by LiDAR, then those harmed may assert that Tesla Motors's automated vehicle is defective for failing to use LiDAR. Tesla Motors, for its part, will need to defend its decision to forgo LiDAR and demonstrate how its LiDAR-less driving automation systems are reasonably safe.

The second broad category of design defect claims will be bugs and problems within the actual driving automation system itself, a software-type defect.

---

99. Restatement (Third) of Torts: Prods. Liab. § 3 cmt. b (1998). The Products Liability Restatement acknowledges the primary difference between design defect and manufacturing defect claims with respect to the malfunction doctrine: namely, for design defects all products suffer from the alleged design defect and the plaintiff can analyze the design of those products, while for manufacturing defects other products may not have the specific manufacturing flaw that the alleged defective product had. *Id.*

100. *Id.*

101. There are several cases where courts have forced a manufacturer to adopt certain technologies. *See, e.g.,* The T.J. Hooper, 60 F.2d 737 (2d Cir. 1932) (a precursor of modern products liability but finding that the lack of a radio system in tugboats to allow captains to communicate about fast-moving storms was negligence).

102. *See, e.g.,* Andrew J. Hawkins, *Elon Musk Still Doesn't Think LIDAR Is Necessary for Fully Driverless Cars*, Verge (Feb. 7, 2018), https://www.theverge.com/2018/2/7/16988628/elon-musk-lidar-self-driving-car-tesla.

In these cases, litigants will be arguing about how the plaintiff can prove the reasonable alternative design that the manufacturer should have adopted that could have prevented the crash. A fundamental question that courts will need to resolve will be whether the plaintiff is required to introduce an alternative "algorithm" or software that could have prevented the crash. In high-exposure cases, the plaintiff may be able to afford a software opinion, but the software opinion may be cost-prohibitive in other cases. There is limited authority on whether a plaintiff has to prove an alternative design by introducing an algorithm that could have prevented the crash. One court has held that requiring presentation of an alternative algorithm imposes a nonsensical evidentiary burden.[103]

The third category of design defect claims will surround inadequate testing of the driving automation system. Broadly, plaintiffs may contend that the driving automation system should have been tested further before being deployed on roads. Although this argument may be appealing to plaintiffs, they may struggle to establish causation for this type of claim; how much further testing would have been needed to prevent the crash? Narrowly, plaintiffs may contend that the driving automation system should have been tested in the specific spot of the crash or tested more in such a spot. This claim is more directed to the particular causation of the crash but may, if taken to its furthest conclusion, lead to extraordinary costs by requiring manufacturers to test their systems on every road or close to every road in the United States (and continue to repeat that process).

(b) Parties

An ancillary issue that can arise in the design defect cases is the proper defendant. There are three ways that highly automated vehicles could be developed: (1) an automaker designs and develops its own automated driving system and vehicle; (2) an automated driving system manufacturer procures vehicles manufactured by an automaker and installs its automated driving system in such vehicles; or (3) a traditional automaker acquires an automated driving system from a supplier and installs the automated driving system in its vehicles. In the first variation, the proper defendant is the only defendant—the automaker. In the second and third variations, the difference between the two examples is the identity of the component manufacturer.

Under products liability law, component manufacturers are generally not liable when harm is caused by the final product when the underlying

103. Wickersham v. Ford Motor Co., 194 F. Supp. 3d 434, 439 (D.S.C. 2016) ("To the extent Ford argues that plaintiff can only prevail if it provides an actual algorithm that Ford could have used in the 2010 Escape, Ford seeks to impose an evidentiary burden well above any sensible interpretation [of the case law].").

component is not defective.[104] Generally, a supplier is only liable when its component is defective or when the supplier "substantially participates in the integration of the component into the design of the product," the integration causes the product to be defective, and the defect causes harm.[105]

The general rules for component suppliers may lead to different results between variations two and three. Although some states have granted immunity to the automaker in the second variation,[106] it appears that such immunity may be unnecessary or redundant to existing laws. In crashes involving automated vehicles, it seems unlikely that the underlying cause of a crash will be a defect in the vehicle but not the automated driving system. Thus, so long as the conventional vehicle did not suffer from any defect, the vehicle manufacturer in the second variation may not be liable. However, this conclusion may be limited to the extent that the vehicle manufacturer is a partner in or assists with the development of the automated vehicle.

The manufacturer of the automated driving system in the third variation may not be shielded like the automaker in the second example. Important questions for supplier liability would be whether the supplier was involved in designing the positioning of sensors on the automated vehicle or whether it merely supplied the technology. If the supplier was involved in the design of the vehicle, the supplier might not be able to avoid liability if that design was faulty. If the supplier was not involved in the process, the supplier might still be responsible in addition to the automaker if the plaintiff can show that there was a defect in the automated driving system, such as faulty software.

---

104. *See, e.g.*, Cipollone v. Yale Indus. Prod., Inc., 202 F.3d 376, 379 (1st Cir. 2000) ("When a component of an integrated product is not itself defective, the maker of the component is not liable for injury that results from a defect in the integrated product."); Sperry v. Bauermeister, Inc., 4 F.3d 596, 598 (8th Cir. 1993) ("A supplier is not liable unless the product is defective at the time it leaves the supplier's control and enters the stream of commerce."); Woods v. Graham Eng'g Corp., 183 Ill. App. 3d 337, 342, 539 N.E.2d 316, 320 (1989) ("[T]he manufacturer is to be held liable only when it is the manufacturer that is responsible for the design of the final product or the component part of the product that caused the injury."); Trevino v. Yamaha Motor Corp., U.S.A., 882 F.2d 182, 186 (5th Cir. 1989), *abrogated on other grounds by* Little v. Liquid Air Corp., 37 F.3d 1069 (5th Cir. 1994); Zaza v. Marquess & Nell, Inc., 144 N.J. 34, 51, 675 A.2d 620, 628 (1996); Noonan v. Texaco, Inc., 713 P.2d 160, 164 (Wyo. 1986); Mitchell v. Sky Climber, Inc., 396 Mass. 629, 631, 487 N.E.2d 1374, 1376 (1986). Traditional automakers have also been considered component manufacturers in some cases; this typically arises in the optional safety device cases. *See, e.g.*, Shaw v. Gen. Motors Corp., 727 P.2d 387, 389 (Colo. App. 1986); Verge v. Ford Motor Co., 581 F.2d 384 (3d Cir. 1978).

105. Restatement (Third) of Torts: Prods. Liab. § 5 (1998).

106. *See* D.C. Code Ann. § 50-2353 (West); Mich. Comp. Laws Ann. § 257.665a; Nev. Rev. Stat. § 482A.090.

c. Warning Defects

A manufacturer has an obligation to provide information to consumers about its products.[107] The Products Liability Restatement provides the following rule on warning defects:

> A product . . . is defective because of inadequate instructions or warnings when the foreseeable risks of harm posed by the product could have been reduced or avoided by the provision of reasonable instructions or warnings by the seller or other distributor, or a predecessor in the commercial chain of distribution, and the omission of the instructions or warnings renders the product not reasonably safe.[108]

A manufacturer's informational obligations impose two duties: (1) a duty to warn of latent (hidden) dangers, and (2) a duty to instruct on how to use a product safely.

(i) Duty to Warn of Latent Dangers

A manufacturer must provide adequate warnings on material, foreseeable dangers.[109] If there is no warning, the principal question is whether the manufacturer has to warn of that danger. If there is a warning, the inquiry focuses on whether the warning is adequate. A warning is adequate if it "can reasonably be expected to catch the attention of the consumer," is "comprehensible and give[s] a fair indication of the specific risks involved with the product, and" is "of an intensity justified by the magnitude of the risk."[110] Although, from a risk-utility standpoint, the obligation to provide a warning appears so inexpensive that almost every risk would conceivably mandate a warning, "warning pollution" is a significant risk of overwarning.[111]

In addition to usual warnings, the main concern for manufacturers will be making sure that they adequately explain a driving automation system's inherent limitations. For level 2 and level 3 driving automation systems, these limitations include the systems' operational design domains and the respective

---

107. Restatement (Third) of Torts: Prods. Liab. § 2 cmt. i (1998) ("Commercial product sellers must provide reasonable instructions and warnings about risks of injury posed by products.").

108. *Id.* § 2(c).

109. Owen, *supra* note 1, at 590 ("A product with inadequate information about product dangers and how to avoid them is 'defective.'"); Richard M. Goodman et al., Automobile Design Liability § 1:3. *Failure to Warn* (3d ed. 2016) (Courts have held "automobile manufacturers liable in strict liability actions absent a specific defect where manufacturers knew of a problem and their failure to warn rendered the product unreasonably dangerous.").

110. Owen, *supra* note 1, at 594.

111. *Id.*; *see also* Keeton et al., *supra* note 1, at 686 ("Those who argue for warning as the judicial solution to latent design defects labor under a naïve belief that one can warn against all significant risks. Too much detail can be counterproductive. A warning to be effective must be read and understood.").

roles of the human driver and the driving automation. For level 4 automated driving systems, a manufacturer will need to describe the operational design domain for the vehicle. For level 5 automated driving systems, although there is not an operational design domain, the classification of a system as level 5 does not mean that the system can operate everywhere in every weather condition—the manufacturer will want to clarify whatever these limitations may be.[112] The duty to warn may extend to warning or explaining to consumers the effects that certain repairs and unmaintained or unclean sensors may have on their driving automation systems.

Plaintiffs may also seek recourse against dealers for failing to warn of the inherent limitations. Under this type of argument, plaintiffs may allege that the sales representative or the dealer had a duty to warn of the inherent limitations of automated vehicles. In these cases, dealers have two arguments for why they should not be responsible. First, the dealer may contend that it lacked knowledge of the underlying inherent limitation and had no duty to warn of the defect.[113] Second, the dealer can argue that it does not have a duty to warn to the extent that the automaker adequately provided warnings.[114]

Finally, plaintiffs may begin to challenge the adequacy of warnings that are merely provided in an owner's manual. Plaintiffs may contend that manufacturers can convey information concerning their driving automation systems and their limitations in infotainment systems. Plaintiffs may also argue that manufacturers must use information they possess to provide adequate warnings. For example, if the driving automation system knows that the weather is snowy and the driving automation system in the vehicle cannot be used in snow, plaintiffs may contend that the system should specifically warn of such risk at that time (or, alternatively, that the system is defectively designed by

112. For example, consumers may believe that the superior driving of a level 5 automated vehicle extends to extreme weather conditions, such as a blizzard. Manufacturers may want to clarify under what weather conditions the automated vehicle may struggle to operate.

113. *See, e.g.*, Sanns v. Butterfield Ford, 2004 UT App. 203, 94 P.3d 301, 304 (2004), *overruled on other grounds by* Bylsma v. R.C. Willey, 2017 UT 85, 416 P.3d 595 (2017) ("Butterfield Ford, as a passive retailer and not the van's manufacturer, did not owe a duty to its customers to warn them of a manufacturing defect that it did not know of itself."); Bren-Tex Tractor Co. v. Massey-Ferguson, Inc., 97 S.W.3d 155 (Tex. App. 2002).

114. *See, e.g.*, Ford Motor Co. v. Rushford, 868 N.E.2d 806, 811 (Ind. 2007) ("[A]bsent special circumstances, if the manufacturer provides adequate warnings of the danger of its product and the seller passes this warning along to the buyer or consumer, then the seller has no obligation to provide additional warnings."); Abramowitz v. Maghsoudlou, 295 A.D.2d 548, 548, 745 N.Y.S.2d 36 (2002) (finding that a car dealership had no duty to warn against the dangers associated with driving a vehicle with different-sized tires and, even if it did, the duty was not breached because the owner's manual contained such warnings); Farmer v. Brannan Auto Parts, Inc., 231 Ga. App. 353, 498 S.E.2d 583, 585 (1998) (holding that a part distributor "had no duty to communicate to users a danger already clearly listed on the product itself").

allowing its operation in these circumstances).[115] Manufacturers, for their part, will contend that they do not have a duty to provide the best type of warnings possible.[116]

(ii) Duty to Instruct

Instructions are different than warning of latent defects, and instructions, in and of themselves, cannot satisfy the duty to warn of latent defects.[117] "Instructions inform persons how to use and consume products safely."[118] The duty to instruct does not require manufacturers to train people on how to use the product.[119]

With crashes caused by driving automation systems, especially those that have at least one inherent limitation (levels 1–4) and rely on humans, initial lawsuits may involve whether the automakers adequately instructed drivers on those inherent limitations and the role of the driver in the safe navigation of the vehicle. A lack of uniformity among driving automation systems of a single manufacturer or among manufacturers may create additional responsibilities for the automaker to adequately instruct the driver on how to use the system at hand. With human-driven vehicles that conform to a set design, such as a round steering wheel, the throttle on the right, and the brake pedal on the left, drivers do not necessarily need instructions on how to safely drive different vehicles produced by the same manufacturer or by other manufacturers. Experienced human drivers can enter many vehicles and immediately know how to use each one. The same may not be true for automated vehicles, and plaintiffs should frame their arguments, especially for a defense of misuse or comparative fault, about whether the drivers were instructed correctly about how to use the driving automation system.

(iii) The Interplay between Warnings and Designs

One important question is whether a manufacturer can warn about a design limitation instead of adopting a safer alternative design. The Products Liability Restatement answers that question in the negative: "In general, when a safer design can reasonably be implemented and risks can reasonably be designed out of a product, adoption of the safer design is required over a warning that

115. See Bryant Walker Smith, *Proximity-Driven Liability*, 102 Geo. L.J. 1777 (2014), for an argument of this type of liability.

116. *See, e.g.*, Gurley *ex rel.* Gurley v. Am. Honda Motor Co., 505 So. 2d 358, 361 (Ala. 1987) ("Where a warning is necessary, the warning need only be one that is reasonable under the circumstances and it need not be the best possible warning."); Nolan v. Dillon, 261 Md. 516, 276 A.2d 36, 40 (1971) (same).

117. Owen, *supra* note 1, at 597–98.

118. Restatement (Third) of Torts: Prods. Liab. § 2 cmt. i (1998).

119. *See, e.g.*, Glorvigen v. Cirrus Design Corp., 816 N.W.2d 572, 583 (Minn. 2012).

leaves a significant residuum of such risks."[120] The Products Liability Restatement justifies this rule because "instructions and warnings may be ineffective because users of the product may not be adequately reached, may be likely to be inattentive, or may be insufficiently motivated to follow the instructions or heed the warnings."[121] Warnings and instructions may prevent a product from being potentially defective if the alternative design cannot be readily adopted.[122] An example is that courts may require a driver monitoring system in level 2 automated vehicles and not allow a manufacturer to rely upon warnings and instructions to drivers to supervise the drive.

d. Crashworthiness

Congress defines crashworthiness as "the protection a passenger motor vehicle gives its passengers against personal injury or death from a motor vehicle accident."[123] Crashworthiness is about the enhanced injuries that result from the "second collision" of the occupants with the interior of the vehicle.[124] Under a crashworthiness claim, a manufacturer of a vehicle could be liable even though the manufacturer was not at fault for the underlying crash. Traditional design defect claims and crashworthiness differ because traditional design defect claims seek to determine whether a defect caused the initial crash.

Crashworthiness is not a requirement of absolute safety: "'[C]rashworthy' means optimal—not perfect—automotive safety."[125] Crashworthiness is generally analyzed under a risk-utility analysis: the plaintiff needs to introduce

---

120. RESTATEMENT (THIRD) OF TORTS: PRODS. LIAB. § 2 cmt. l (1998).

121. *Id.*

122. *Id.*

123. 49 U.S.C. § 32301(1). As a liability concept, crashworthiness is not limited to passenger vehicles. OWEN, *supra* note 1, at 1132. Crashworthiness is also not limited to merely the protection of vehicle passengers and may apply to bystanders. *See* 1 AUTO. DESIGN LIABILITY § 1:4. *Crashworthiness* (3d ed. 2016) ("Although crashworthiness is usually associated with the occupants of automobiles, the doctrine has also been held to apply to non-occupants.").

124. *See* GEISTFELD, *supra* note 45, at 395 ("These defects do not cause the initial accident, and so the plaintiff would have suffered some injury from the crash even if the vehicle has not been defective in this respect."); GOODMAN ET AL., *supra* note 109, at § 1:4. *Crashworthiness* (stating that under the crashworthiness doctrine "a supplier is liable for injuries sustained in a vehicular accident because of a defect that was not the cause of the accident, but that caused or enhanced the degree of injuries suffered"); Larsen v. Gen. Motors Corp., 391 F.2d 495, 502 (8th Cir. 1968); *but see* Holiday Motor Corp. v. Walters, 292 Va. 461, 477, 790 S.E.2d 447, 455 (2016) ("In Virginia, there is no duty on the part of vehicle manufacturers to design or supply a crashworthy vehicle.").

125. OWEN, *supra* note 1, at 1135; *see, e.g.*, Roberts v. May, 41 Colo. App. 82, 85, 583 P.2d 305, 308 (1978) ("Manufacturers are not required to produce automobiles with the strength and crash-damage resistance features of an M-2 Army tank." (internal quotation marks and citation omitted)); Kern v. Gen. Motors Corp., 724 P.2d 1365, 1367 (Colo. App. 1986) ("In designing and manufacturing its automobiles, defendant must take into account the possibility that these vehicles may be involved in accidents. However, this does not imply a duty to prevent all injuries in all accidents.").

evidence of a reasonable, feasible alternative design that would have prevented the enhanced injuries, and the costs of that reasonable, feasible alternative design cannot outweigh its benefits.[126] One significant cost is the impact that the alternative design would have on the safety of the vehicle as a whole. In essence, a vehicle is not defective if the plaintiff's alternative design creates equal or greater risks of harm than the ones suffered by the plaintiff.[127]

For automated vehicles to maximize the happiness of occupants, vehicle interiors should be redesigned.[128] Forward-facing seats organized as they currently are in vehicles would not maximize happiness for occupants. Vehicles have been designed this way to maximize safety. From a products liability standpoint, manufacturers can redesign automated vehicle interiors, but they need to balance the risks created by such designs to ensure that their vehicles are crashworthy.

C. Limitations to Manufacturer Liability

This section addresses what many call affirmative defenses to products liability claims. Some of these doctrines may be affirmative defenses, such as comparative fault, or they may be part of the plaintiff's case-in-chief in proving that the product was defective or that a given defect was the cause of the plaintiff's harm.[129] In some cases, these limitations may mean that the plaintiff fails to satisfy his or her burden. In other cases, the plaintiff may satisfy his or her

---

126. Owen, *supra* note 1, at 1136–38, 1140–42.

127. *Id.* at 1138 ("Because of the different ways in which accidents can occur, automotive engineers are forced to balance a vehicle's safety in one type of crash against its safety in other types of crashes; making it safer for one type of accident may make it more dangerous for another type of accident."); Restatement (Third) of Torts: Prods. Liab. § 2 cmt. f (1998) ("When evaluating the reasonableness of a design alternative, the overall safety of the product must be considered. It is not sufficient that the alternative design would have reduced or prevented the harm suffered by the plaintiff if it would also have introduced into the product other dangers of equal or greater magnitude.").

128. *See, e.g.*, Eric A. Taub, *Envisioning the Car of the Future as a Living Room on Wheels*, N.Y. Times (June 15, 2017), https://www.nytimes.com/2017/06/15/automobiles/wheels/driverless-cars-interior.html (discussing various methods of designing the interior of an automated vehicle); Hod Lipson & Melba Kurman, Driverless: Intelligent Cars and the Road Ahead 266 (2016) ("We predict that driverless cars will look different, both inside and outside. The steering wheel will disappear, the dashboard will become flexible workspace, and the car's cabin will contain whatever people need for their on-board leisure and work activities.").

129. *See* Restatement (Third) of Torts: Prods. Liab. § 2 cmt. p (1998); William J. McNichols, *The Relevance of the Plaintiff's Misconduct in Strict Tort Products Liability, the Advent of Comparative Responsibility, and the Proposed Restatement (Third) of Torts*, 47 Okla. L. Rev. 201, 207 (1994) ("The most important distinction to make in analyzing the plaintiff's conduct issues is whether the evidence about the plaintiff's behavior relates to an issue on which the plaintiff bears the burden of proof, such as the existence of defect, or whether evidence about the plaintiff's conduct is being used to establish one of the strict tort affirmative defenses, such as unreasonable assumption of risk, or unforeseeable misuse or contributory negligence in those states which recognize the latter two as defenses."); *but see* Peter Zablotsky, *The Appropriate Role of Plaintiff Misuse in Products Liability Causes of Action*, 10 Touro L. Rev. 183, 188–89 (1993) (arguing that misuse should not be part of the plaintiff's case-in-chief).

burden of proving a defect, but the plaintiff's conduct also contributed to that harm and these doctrines limit or possibly bar the plaintiff's recovery.

One last area of confusion, especially with automobiles, occurs when a bystander is harmed as a result of a product user's use of a product. In these cases, the plaintiff is not the person whose conduct may have caused the harm, and these doctrines would seemingly be inapplicable against the plaintiff.[130] Instead, the proper analysis for these types of claims is whether the manufacturer's conduct was the proximate cause of the plaintiff's harm.[131] In many respects, a product user can be a superseding cause of the harm.[132]

1. Contributory Negligence/Comparative Negligence

Contributory negligence and comparative negligence are two methods that courts use to limit a plaintiff's recovery when the plaintiff's conduct was a cause of the plaintiff's harm. In states that follow contributory negligence,[133] if the plaintiff was a cause of his or her own harm, then the plaintiff cannot recover absent an exception. Given the harshness of the contributory negligence rule, the Restatement (Second) of Torts rejected it as a defense to strict liability products liability claims,[134] and most states have abandoned it altogether for comparative negligence regimes. Contributory negligence and comparative negligence examine the same underlying issue: whether the plaintiff's conduct was negligent.

States apply different rules when it comes to the effect of the plaintiff's negligence. Some states follow "pure" comparative negligence under which a plaintiff's damages are reduced by his or her fault.[135] Most states have adopted one of the two types of modified comparative negligence systems.[136] "Under the 'equal fault bar' approach, the plaintiff cannot recover anything if his fault is equal to or greater than that of the defendant; he is allowed to recover, in other words, only if his negligence is less than that of the defendant."[137] "Under

---

130. McNichols, *supra* note 129, at 208.

131. *Id.*

132. *Id.*

133. Contributory negligence remains the rule in Alabama, the District of Columbia, Maryland, North Carolina, and Virginia.

134. Restatement (Second) of Torts § 402A cmt. n (1965). The Restatement did provide that contributory negligence can be an applicable defense when the contributory negligence is based on assumption of risk. *See id.*; Zablotsky, *supra* note 129, at 186 ("[C]omment n to section 402A of the Restatement (Second) states that contributory negligence (i.e., the failure to discover a defect in a product) is not a defense in a products liability action, while assumption of risk (i.e., continuing to make use of a product after having discovered its defect and becoming aware of its danger) is a defense.").

135. Keeton et al., *supra* note 1, at 472.

136. *Id.* at 473.

137. *Id.*

the 'greater fault bar' system, the plaintiff is prevented from all recovery only if his fault exceeds the defendant's; he is therefore allowed to recover if his negligence is less than that of the defendant."[138] In either the "equal fault bar" system or the "greater fault bar" system, the plaintiff still cannot recover for his or her portion of the fault.[139]

2. *Misuse*

Product misuse can be a defense to a products liability claim. Misuse, in this context, is different than its everyday meaning. In this context, misuse means that manufacturers are not liable for unforeseeable product uses.[140] Generally, "[i]f the manufacturer or other product seller reasonably should have contemplated and guarded against [a] risk, the defendant is subject to liability for harm; if the plaintiff put the product to an unforeseeable, unfair use, the defendant simply is not liable."[141]

Some product misuse cases involve failures by the product user to follow the instructions or heed warnings about a given product.[142] Comment j to section 402A of the Restatement (Second) of Torts provided that "[w]here warning is given, the seller may reasonably assume that it will be read and heeded; and a product bearing such a warning, which is safe for use if it is followed, is not in a defective condition, nor is it unreasonably dangerous."[143] The Products Liability Restatement expressly rejected comment j, and courts typically do

---

138. *Id.*

139. *Id.*

140. OWEN, *supra* note 1, at 885; COLO. REV. STAT. ANN. § 13-21-402.5 (West) ("A product liability action may not be commenced or maintained against a manufacturer or seller of a product that caused injury, death, or property damage if, at the time the injury, death, or property damage occurred, the product was used in a manner or for a purpose other than that which was intended and which could not reasonably have been expected, and such misuse of the product was a cause of the injury, death, or property damage."); IND. CODE ANN. § 34-20-6-4 (West) ("It is a defense to an action under this article (or IC 33-1-1.5 before its repeal) that a cause of the physical harm is a misuse of the product by the claimant or any other person not reasonably expected by the seller at the time the seller sold or otherwise conveyed the product to another party."); MICH. COMP. LAWS ANN. § 600.2947 (West) ("A manufacturer or seller is not liable in a product liability action for harm caused by misuse of a product unless the misuse was reasonably foreseeable. Whether there was misuse of a product and whether misuse was reasonably foreseeable are legal issues to be resolved by the court."); N.C. GEN. STAT. ANN. § 99B-4 ("No manufacturer or seller shall be held liable in any product liability action if: (1) The use of the product giving rise to the product liability action was contrary to any express and adequate instructions or warnings delivered with, appearing on, or attached to the product or on its original container or wrapping, if the user knew or with the exercise of reasonable and diligent care should have known of such instructions or warning. . . .").

141. OWEN, *supra* note 1, at 895.

142. *Id.*

143. RESTATEMENT (SECOND) OF TORTS § 402A cmt. j (1965).

not absolve product sellers that fail to address misuse through feasible engineering solutions.[144]

Misuse will be asserted in various ways by manufacturers sued for crashes caused by automated vehicles. For level 2 driving automation systems, the misuse defense will focus primarily on the driver not supervising the road. Chapter 9 explores misuse of level 2 driving automation systems. Misuse of level 3 automated driving systems will involve the user not fulfilling his or her role as the dynamic driving task fallback-ready user, a topic described in Chapter 10. Misuse of level 4 automated driving systems will likely involve users who alter the system or use the system outside of its operational design domain. Users may misuse level 5 automated driving systems by using them where they cannot safely operate, such as off-road or in inclement weather, or by altering the systems.

3. Assumption of Risk

Assumption of risk is an affirmative defense for manufacturers to products liability claims.[145] There are two elements to assumption of risk: "[F]irst, the plaintiff must know and understand the risk; and, second, the plaintiff's choice to encounter it must be free and voluntary."[146] Knowledge requires the plaintiff

---

144. Restatement (Third) of Torts: Prods. Liab. § 2 cmt. l (1998) ("Reasonable designs and instructions or warnings both play important roles in the production and distribution of reasonably safe products. In general, when a safer design can reasonably be implemented and risks can reasonably be designed out of a product, adoption of the safer design is required over a warning that leaves a significant residuum of such risks. For example, instructions and warnings may be ineffective because users of the product may not be adequately reached, may be likely to be inattentive, or may be insufficiently motivated to follow the instructions or heed the warnings. However, when an alternative design to avoid risks cannot reasonably be implemented, adequate instructions and warnings will normally be sufficient to render the product reasonably safe. Compare Comment e. Warnings are not, however, a substitute for the provision of a reasonably safe design."); *see, e.g.*, Lewis v. Am. Cyanamid Co., 155 N.J. 544, 564, 715 A.2d 967, 977 (1998) ("Allowing such a warning to defeat a design-defect claim would not encourage manufacturers to place such warnings on their products. Indeed, it would frustrate the imposition of liability when a product's design fails to take into account an injured party's objectively foreseeable misuse of the product."); Delaney v. Deere & Co., 268 Kan. 769, 786, 999 P.2d 930, 942–43 (2000) ("[W]e decline to adopt that portion of Comment j which holds that a product bearing a warning which is safe for use if the warning is followed is not in a defective condition or unreasonably dangerous."); Uniroyal Goodrich Tire Co. v. Martinez, 977 S.W.2d 328, 336–37 (Tex. 1998) (rejecting comment j); Uloth v. City Tank Corp., 376 Mass. 874, 879–80, 384 N.E.2d 1188, 1192 (1978) ("An adequate warning may reduce the likelihood of injury to the user of a product in some cases. We decline, however, to adopt any rule which permits a manufacturer or designer to discharge its total responsibility to workers by simply warning of the dangers of a product. Whether or not adequate warnings are given is a factor to be considered on the issue of negligence, but warnings cannot absolve the manufacturer or designer of all responsibility for the safety of the product.").

145. *See* Restatement (Second) of Torts § 402A cmt. j (1965) (providing that assumption of risk is a defense to products liability claims).

146. Owen, *supra* note 1, at 863.

to "understand with some particularity how the product may cause an injury, so that he or she is able to evaluate the likelihood and seriousness of potential injury and thereby make an informed decision on whether or not to engage the risk."[147] A plaintiff is not required to understand the specific technical mechanisms that may cause harm or the precise nature of the product defect.[148] The "free and voluntary" element of assumption of risk typically requires that

> [t]he plaintiff make[] a true and meaningful *choice* to engage a particular risk, to expose himself to a particular risk of harm, presumably to advance an interest (even mere convenience) that the plaintiff considers more valuable than avoidance of the risk. "Choice" in this context means, first, that the plaintiff understands the nature of the risk encountered; second, that the plaintiff has available one or more alternative courses of action by which he or she may reasonably avoid the risk; and third, that the plaintiff then decides that his or her interests will best be served by encountering the danger.[149]

Assumption of risk is a subjective inquiry into the knowledge, voluntariness, and appreciation of risk for the specific plaintiff and not a "reasonable plaintiff."[150] Assumption of risk may become an issue when drivers misuse level 2 or level 3 driving automation systems or use automated vehicles in inclement weather or outside their operational design domains.

4. State of the Art

State of the art is a limitation to the plaintiff's case-in-chief and a defense asserted by manufacturers. The meaning of "state of the art" differs among jurisdictions, with legislatures and courts considering such concepts as industry custom and scientific and technological capabilities.[151] At its core, state of the art stands for the proposition "that manufacturers should be held respon-

---

147. *Id.* at 865.

148. *Id.* at 866.

149. David Owen, *Products Liability: User Misconduct Defenses*, 52 S.C. L. Rev. 1, 31 (2000).

150. *Id.* at 25.

151. Owen, *supra* note 1, at 712; *see, e.g.*, Ariz. Rev. Stat. Ann. § 12-681(10) ("'State of the art' means the technical, mechanical and scientific knowledge of manufacturing, designing, testing or labeling the same or similar products that was in existence and reasonably feasible for use at the time of manufacture."); Mo. Ann. Stat. § 537.764 (West) ("'[S]tate of the art' means that the dangerous nature of the product was not known and could not reasonably be discovered at the time the product was placed into the stream of commerce."); N.H. Rev. Stat. Ann. § 507:8-g ("In product liability actions brought by or in consequence of harm to a user, it is an affirmative defense that the risks complained of by the plaintiff were not discoverable using prevailing research and scientific techniques under the state of the art and were not discoverable using procedures required by federal or state regulatory authorities charged with supervision or licensing of the product in question. Discoverability of risk shall be measured as of the time the manufacturer parted with possession and control of, or sold the product in question, whichever occurred last.").

sible only for what they can foresee and reasonably prevent" at the time of production.[152] As one court noted, "A consumer would not expect a Model T to have the safety features which are incorporated in automobiles made today."[153]

For some components, such as the hardware comprising the driving automation systems, the state-of-the-art doctrine should apply in a relatively straightforward way. A significant debate will concern the role of the state-of-the-art doctrine for software components of driving automation systems. Over-the-air updates and software updates may mean that the design of the driving automation system, or at least its software, is never finished or at least that plaintiffs are not limited to introducing alternative software designs that existed at the point of sale.

With warning defects, "the state-of-the-art issue involves a manufacturer's ability to foresee a product danger to enable it to warn consumers of its presence."[154] A manufacturer should not—and most courts find that the manufacturer does not—have a duty to warn against unforeseeable dangers.[155]

Although state of the art can be a useful limitation to product manufacturers for claims of design defects and warning defects, product manufacturers have not been able to use state of the art as a defense to manufacturing defects.[156] The manufacturer must be capable of producing its products according to its own specifications.

5. Wear and Tear

Vehicles, like every other product, suffer from wear and tear, causing them to become more dangerous over time absent replacement of parts.[157] A manufacturer does not have an obligation to produce a product that will not wear out over time.[158] Additionally, "manufacturers generally are not responsible for dangers that naturally arise from the normal wear and tear of products that eventually do wear out."[159] Therefore, "as products age, responsibility for

---

152. Owen, *supra* note 1, at 737.
153. Bruce v. Martin-Marietta Corp., 544 F.2d 442, 447 (10th Cir. 1976).
154. Owen, *supra* note 1, at 707.
155. *Id.* at 733.
156. *Id.* at 449 n.18.
157. *Id.* at 744–45.
158. *See, e.g.*, Auld v. Sears, Roebuck & Co., 261 A.D. 918, 918, 25 N.Y.S.2d 491, 493 (App. Div. 1941), *aff'd*, 288 N.Y. 515, 41 N.E.2d 927 (1942) ("There is no duty upon a manufacturer to furnish a machine that will not wear out."); *see also* Tenn. Code Ann. § 29-28-108 (West) ("If a product is not unreasonably dangerous at the time it leaves the control of the manufacturer or seller but was made unreasonably dangerous by subsequent unforeseeable alteration, change, improper maintenance or abnormal use, the manufacturer or seller is not liable."); Mitchell v. Ford Motor Co., 533 F.2d 19, 21 (1st Cir. 1976) (stating that "a manufacturer does not have to anticipate that maintenance will be neglected").
159. Owen, *supra* note 1, at 746.

accident prevention shifts incrementally from the manufacturer to the user."[160] However, courts have held that product age, in and of itself, does not excuse a manufacturer from liability if the product's defect did not manifest itself for years.[161]

6. Government Standards

As described in Part 2, Congress is unlikely to explicitly preempt all products liability claims from being asserted against manufacturers of automated vehicles. However, NHTSA likely will exercise its regulatory authority to regulate automated vehicle design and impose certain testing requirements for automated driving systems before their release into the marketplace. Many manufacturers may seek refuge in complying with these design or testing requirements. Manufacturers may argue that these federal design or testing requirements preempt state products liability claims or, in any event, that compliance should be conclusive on defectiveness.

a. Preemption

Preemption is a powerful defense that can block products liability claims. The Motor Vehicle Safety Act guides preemption in the automotive industry. The Motor Vehicle Safety Act broadly preempts state law:

> When a motor vehicle safety standard is in effect under this chapter, a State or a political subdivision of a State may prescribe or continue in effect a standard applicable to the same aspect of performance of a motor vehicle or motor vehicle equipment only if the standard is identical to the standard prescribed under this chapter. However, the United States Government, a State, or a political subdivision of a State may prescribe a standard for a motor vehicle or motor vehicle equipment obtained for its own use that imposes a higher performance requirement than that required by the otherwise applicable standard under this chapter.[162]

---

160. *Id.*

161. *See, e.g.,* Pryor v. Lee C. Moore Corp., 262 F.2d 673, 674 (10th Cir. 1958) (holding that the trial court erred when it "directed a verdict and entered judgment for the defendant-manufacturer, on the grounds that fifteen years of safe use for the purpose for which the equipment was intended foreclosed any probability that it was defectively or negligently made"); Mickle v. Blackmon, 252 S.C. 202, 240, 166 S.E.2d 173, 190 (1969) ("We readily concede that the passage of thirteen years between the marketing of a product and its injury-producing failure is a formidable obstacle to fastening liability upon the manufacturer. However, it may reasonably be inferred in this case that the advanced age of the ball was coincidental with its failure rather than the cause of it, and that the knob would have shattered upon a comparable impact had it occurred much earlier in the life of the car. The important inquiry is not how long the knob lasted but what caused its failure. Mere passage of time should not excuse Ford if its negligence was the cause.").

162. 49 U.S.C. § 30103(b).

The Motor Vehicle Safety Act includes a "savings clause" that purports to exclude common law liability claims, including products liability claims, from the preemption clause: "Compliance with a motor vehicle safety standard prescribed under this chapter does not exempt a person from liability at common law."[163] The United States Supreme Court has considered whether a products liability claim asserted against an automobile manufacturer is preempted.[164] Federal law can either expressly or impliedly preempt a state law cause of action.

Express preemption is relatively straightforward; an act of Congress expressly precludes a state law cause of action.[165] The Supreme Court has held that Congress's inclusion of the savings clause in the Motor Vehicle Safety Act defeats any express preemption arguments.[166] The automated vehicle bills that Congress has considered have also included a savings clause for common law claims.[167] Therefore, it is unlikely that express preemption will be involved in tort lawsuits involving automated vehicle crashes.

Implied preemption arises with field preemption or conflict preemption. Field preemption occurs when "[t]he scheme of federal regulation [is] so pervasive as to make reasonable the inference that Congress left no room for the States to supplement it, . . . [or when] the federal interest is so dominant that the federal system will be assumed to preclude enforcement of state laws on the same subject."[168] A finding of field preemption over a tort lawsuit is rare because states have historically regulated public health and safety.[169] Conflict preemption can arise "where it is impossible for a private party to comply with both state and federal requirements, or where state law stands as an obstacle to the accomplishment and execution of the full purposes and objectives of Congress."[170] Implied preemption does not require an actual regulation—the absence of regulation may be an important federal policy to leave an industry unregulated. The Supreme Court has held that "a federal decision to forgo regulation in a given area may imply an authoritative federal determination that

---

163. *Id.* § 30103(e).

164. Freightliner Corp. v. Myrick, 514 U.S. 280, 115 S. Ct. 1483, 131 L. Ed. 2d 385 (1995); Geier v. Am. Honda Motor Co., 529 U.S. 861, 120 S. Ct. 1913, 146 L. Ed. 2d 914 (2000); Williamson v. Mazda Motor of Am., Inc., 562 U.S. 323, 131 S. Ct. 1131, 179 L. Ed. 2d 75 (2011).

165. See *Cipollone v. Liggett Grp., Inc.*, 505 U.S. 504, 112 S. Ct. 2608, 120 L. Ed. 2d 407 (1992), for an example of express preemption.

166. *Geier*, 529 U.S. at 868.

167. See Chapter 7 for a discussion of these bills.

168. Rice v. Santa Fe Elevator Corp., 331 U.S. 218, 230, 67 S. Ct. 1146, 91 L. Ed. 1447 (1947) (internal citations omitted).

169. Owen & Davis, *supra* note 94, at § 15:11 ("Field preemption has not been as important a category of preemption for state law product liability claims because of the traditional operation of state law in the area of public health and safety.").

170. Sprietsma v. Mercury Marine, a Div. of Brunswick Corp., 537 U.S. 51, 64, 123 S. Ct. 518, 154 L. Ed. 2d 466 (2002) (internal quotation marks and citations omitted).

the area is best left unregulated, and in that event would have as much preemptive force as a decision to regulate."[171] This preemptive lack of regulation can be important for automated vehicles because the United States Department of Transportation (USDOT) has not yet explicitly regulated driving automation systems or other advanced driver-assistance systems for consumer vehicles.[172]

The United States Supreme Court's first foray into the preemptive effect of the Motor Vehicle Safety Act occurred in *Freightliner Corp. v. Myrick*.[173] This case involved two nearly identical crashes where a tractor-trailer jackknifed into oncoming traffic after the driver attempted to brake suddenly. In one crash, the vehicle struck and killed the driver, and in the other, the driver was left permanently paraplegic and brain damaged. The plaintiff, in each case, alleged that the tractor-trailers were defective by failing to utilize an antilock braking system (ABS). The defendant, Freightliner Corporation, alleged that this type of design defect claim was barred by the Motor Vehicle Safety Act and, in particular, Safety Standard 121.

By way of background, NHTSA had adopted Safety Standard 121 in 1971 to "impose[] stopping distances and vehicle stability requirements for trucks." Manufacturers claimed that the standard essentially required that they incorporate ABSs in their trucks. After NHTSA refused to relent on Safety Standard 121, several manufacturers and trade associations challenged it. The Ninth Circuit Court of Appeals found that Safety Standard 121 was not reasonable or practicable when put into effect and remanded the case. After that, NHTSA amended Safety Standard 121 to suspend the application of the new requirement to trucks and trailers.

In a unanimous decision, the Supreme Court easily dispensed with Freightliner Corporation's preemption argument. The Court ruled that the preemption clause of the Motor Vehicle Safety Act only expressly preempts state law when there is a Safety Standard in effect. Because NHTSA had suspended the requirement in Safety Standard 121, "[t]here is simply no minimum, objective standard stated at all." The Supreme Court then found that there was not implied preemption because (i) "there is simply no federal standard for a private party to comply with" and (ii) the requirement could not conflict with or frustrate the objectives of Congress because the Motor Vehicle Safety Act and Safety Standard 121 do not address ABSs, nor had NHTSA ordered or required manufacturers to refrain from using these devices.

In *Geier v. American Honda Motor Co., Inc.*, the Supreme Court ruled that federal law preempted a products liability claim based on a failure to equip

---

171. Ark. Elec. Co-op. Corp. v. Ark. Pub. Serv. Comm'n, 461 U.S. 375, 384, 103 S. Ct. 1905, 76 L. Ed. 2d 1 (1983); *accord Sprietsma*, 537 U.S. at 66.

172. See Chapter 7 for a discussion on the efforts by the USDOT.

173. 514 U.S. 280 (1995).

a vehicle with airbags.[174] Geier was seriously injured when his 1987 Honda Accord collided with a tree. The Accord had a three-point seat belt, which Geier was using at the time of the crash, but the Accord did not have airbags. Geier alleged that the Accord was defective by failing to have a driver's side airbag.

American Honda claimed that the Motor Vehicle Safety Act preempted this claim. Safety Standard 208 did not require manufacturers to equip their vehicles with airbags. Instead, it gave manufacturers options for the protection of front-seat passengers: a manufacturer needed to equip the vehicle with one of the following: (1) passive protection from frontal and angular collisions; (2) passive protection from head-on collisions, supplemented by seat belts and a belt warning system; or (3) lap and shoulder belts, plus a belt warning system.[175] Because Safety Standard 208 gave manufacturers an option to protect front-seat passengers without the use of an airbag, American Honda claimed that a design defect claim predicated on an airbag requirement was preempted.

The Supreme Court first considered the interplay between the preemption clause and the savings clause in the Motor Vehicle Safety Act. The Supreme Court ruled that the existence of the savings clause makes it such that the preemption clause cannot expressly preempt common law actions, including an action based on a defective design. The Supreme Court then shifted its focus to the effect that the savings clause has on implied preemption claims. The Supreme Court ruled that the existence of the savings clause does not preclude a finding that federal law preempts, through conflict preemption, a products liability action.

The Supreme Court conducted an exhaustive review of the background of Safety Standard 208 and the purposes and intent of NHTSA in providing manufacturers with an option to equip their vehicles with airbags. Its review led the Supreme Court to find that NHTSA intended to give manufacturers a choice of passive restraints and not to require airbags. Because the plaintiff's products liability claim—based on a failure to equip the vehicle with airbags—would frustrate that choice, the plaintiff's cause of action was preempted for creating an obstacle to the federal policy.

Geier led to a potentially expansive application of preemption to other Federal Motor Vehicle Safety Standards that provided manufacturers with a choice. This expansive reading of *Geier* found its way back to the Supreme Court in 2011 in *Williamson v. Mazda Motor of America, Inc.*[176] *Williamson* also involved Safety Standard 208, but this time regarding a provision that gave manufacturers a choice about whether to install lap-and-shoulder belts or simple lap belts for rear inner seats, including middle seats or those next to an

174. 529 U.S. 861 (2000).
175. 49 C.F.R. § 571.208.
176. 562 U.S. 323 (2011).

aisle. The Williamson family was involved in a crash, and the occupants using lap-and-shoulder belts survived the crash. The person sitting in the rear aisle seat of the minivan, which only had a simple lap belt, died from the crash. The estate claimed that the vehicle was defective because Mazda did not install lap-and-shoulder belts on the rear aisle seat. Mazda argued that, under *Geier*, the choice in Safety Standard 208 preempts the plaintiff's failure to equip.

The Supreme Court stated that *Geier* stands for the proposition that implied preemption applies when a products liability action would conflict with a "significant objective of the federal regulation." Thus, according to the Supreme Court, it was not the mere conflict with a choice that mandated preemption; rather, the conflict must be with a choice that is a significant objective of NHTSA. The Supreme Court conducted a review of the background and purposes of this provision in Safety Standard 208 and found that the choice between simple lap belts and shoulder-and-lap belts was not a significant objective of federal regulation. Thus, even though the plaintiff's design defect claim, if proven, would deprive manufacturers of that choice, the claim is not preempted.

Following *Williamson*, implied preemption in the "choice cases" is a three-pronged inquiry. First, the regulation must leave manufacturers with a choice. Second, the products liability claim, if proven, would need to restrict that choice. Third, there must be a finding that the choice itself "is a significant regulatory objective." The first two inquiries can be discerned by reading the applicable Safety Standard and asking whether the plaintiff's claim would restrict the choice. The third inquiry requires a significant review of the legislative and regulatory history and the guidance given by NHTSA.[177] NHTSA's contemporaneous opinion on the significance of the choice is also important.[178]

This three-pronged inquiry was recently applied in *Dashi v. Nissan North America, Inc.*[179] In *Dashi*, the Arizona Court of Appeals considered whether federal law preempts a design defect claim based on Nissan's failure to equip its 2008 Nissan Rogue with automatic emergency braking (AEB). Dashi was significantly injured when the driver of the Nissan Rogue crashed into her vehicle when she was conducting a U-turn and was perpendicular to the Nissan Rogue. The underlying basis of her claim against Nissan was that the crash could have been prevented if Nissan had installed AEB as standard equipment and that its failure to do so rendered the vehicle defective. Nissan argued that federal law preempts Dashi's design defect claim because, if proven, the

---

177. *See* Williamson v. Mazda Motor of Am., Inc., 562 U.S. 323 (2011) (conducting an exhaustive review of the history behind the Safety Standard); Geier v. Am. Honda Motor Co., Inc., 529 U.S. 861 (2000) (same).

178. *Williamson*, 562 U.S. at 332 (giving "weight to the Solicitor General's view" provided as amicus).

179. 247 Ariz. 56, 445 P.3d 13 (Ct. App. 2019), *review denied* (Jan. 7, 2020).

claim would defeat a significant objective of NHTSA in not mandating that AEB be standard equipment in vehicles. At the heart of Nissan's argument was a denial of a petition by NHTSA to mandate AEB as standard equipment and the automated vehicle policies.[180] No Safety Standard provided manufacturers with a choice on whether to install AEB as standard equipment.

The Arizona Court of Appeals conducted a review of the regulatory history and NHTSA's guidance on AEB to distill whether NHTSA gave manufacturers a choice and whether that choice was a significant regulatory objective. The court considered the denial of the petition on AEB, the USDOT's and NHTSA's amicus briefs in *Williamson*, and the federal automated vehicle policies to glean NHTSA's policies and objectives concerning AEB. The court ruled that NHTSA explicitly desired to give manufacturers a choice to innovate with advanced technologies like AEB and that the choice was a substantial federal objective. Based on that finding, the court ruled that Dashi's claim that AEB should be standard equipment would frustrate that objective and be an obstacle. Therefore, federal law preempts her design defect claim.

Dashi is significant because it interpreted the preemptive effect of the automated vehicle policies. Many of the advanced driver assistance systems (ADASs) are safety-critical systems that invite design defect claims when manufacturers do not include them as standard equipment in their motor vehicles. No automaker currently offers every one of its ADASs as standard equipment. Automakers offer some ADASs as standard equipment and others as optional equipment on more expensive trims. *Dashi*, its reasoning, and its arguments can stand as an impediment to failure-to-equip claims that seek to mandate ADASs as standard equipment.

Dashi also lends significant credibility to the effects of the automated vehicle policies and other informal guidance of NHTSA on ADASs and automated vehicles. Under the reasoning in *Dashi*, NHTSA's statements included in its guidance on automated vehicles need to be carefully reviewed to determine NHTSA's intent concerning preemption and automated vehicles. To the extent that NHTSA has established a significant policy objective that the plaintiff's claim, if proven, would frustrate, preemption may prohibit such claims.

In January 2020, a Kansas district court judge reached the opposite conclusion from *Dashi* in *Butler v. Daimler Trucks North America, LLC*.[181] In *Butler*, the plaintiffs sued, claiming that a freightliner was defective for not being equipped with forward collision warning (FCW) and AEB. The freightliner was involved in a crash that killed five people. Although NHTSA denied a petition that sought to commence rulemaking on AEB for consumer vehicles,

---

180. Nat'l Highway Traffic Safety Admin., *Federal Motor Vehicle Safety Standards; Automatic Emergency Braking*, 82 Fed. 8391 (Jan. 25, 2017) (denying a consumer vehicle petition).

181. No. 2:19-CV-2377-JAR-JPO, 2020 WL 128055 (D. Kan. Jan. 10, 2020).

NHTSA had granted a petition in 2015 to commence rulemaking on heavy vehicles. NHTSA has not finished its rulemaking process. The *Butler* court considered the history behind AEB and FCW and found that federal law did not preempt the claims because NHTSA did "not make any authoritative statement of a federal policy against FCW/AEB systems."

On May 5, 2020, a different panel of the Arizona Court of Appeals narrowly read *Dashi* in *Varela v. FCA US LLC.*[182] In this case, Varela and her daughter were killed when a 2014 Jeep Grand Cherokee rear-ended her car while traveling at a high speed. The plaintiff claimed that the Jeep was defective by not being equipped with Chrysler's existing automated collision-avoidance technology, which, according to the plaintiff, would have prevented the crash. The automated collision-avoidance technology was available on different Jeep Cherokee trims and other Chrysler models. The trial court granted Chrysler's motion to dismiss based on obstacle preemption, and on appeal Chrysler argued that *Dashi* required the court to affirm the dismissal. The plaintiff argued that *Dashi* applies when the manufacturer does not offer an AEB because that claim invariably would require a plaintiff to establish a performance standard. In contrast, the plaintiff in *Varela* was arguing that Chrysler should have installed its existing AEB technology used on other trims or models, and the plaintiff was not trying to change the AEB used by Chrysler. The court of appeals disagreed with Chrysler and ruled that the petition denial by NHTSA on AEB did not create obstacle preemption in this case. The court of appeals narrowly read *Dashi* to apply only when the plaintiff's claim would require a specific performance standard on AEB. Under *Varela*, an AEB design defect claim based on the existing technology offered by a manufacturer would not be preempted.

The *Varela* opinion seems conflicted with the broad opinion of *Dashi* that found obstacle preemption, and the two panels read the same guidance from NHTSA and interpreted the guidance differently. The net result of *Varela* and *Dashi* may give automakers an incredible choice on AEB: offer AEB on all models and trims to avoid facing design defect claims on trims and models that do not offer AEB, or do not offer AEB at all and be protected from all design defect cases by federal preemption. It also rewards manufacturers with worse AEB, since the worse AEB would be less likely to prevent crashes, and a plaintiff, even after *Varela*, could not impose a performance standard on a manufacturer by mandating that the manufacturer use better AEB technology.

b. Compliance as a Defense

When the statute or regulation does not preempt the cause of action, what impact, if any, does compliance with the statute have in the products liability

---

182. Varela v. FCA US LLC, No. 1 CA-CV 19-0209, 2020 WL 2123281 (Ariz. Ct. App. May 5, 2020).

action? Generally, compliance with safety statutes or regulations is not a bar to a products liability action, though evidence of compliance is relevant to determining whether a product is defective.[183] Thus, compliance with federal statutes and regulations is generally admissible as evidence to show that the product was not defective.

7. Compliance with Industry Custom

Industry custom and compliance (or lack thereof) can be relevant in a products liability action, but it is not dispositive.[184] Industry custom can be relevant to show state of the art and whether the omission of the plaintiff's proposed design rendered the vehicle not reasonably safe.[185] Juries also find industry customs persuasive.

8. Nonmanufacturer Distributors

Although under general products liability rules, retailers, suppliers, and other nonmanufacturers could be liable for a defect in a product,[186] many states have granted partial immunity to nonmanufacturer distributors and retailers in certain instances.[187] These statutes condition immunity on a lack of fault by

---

183. RESTATEMENT (THIRD) OF TORTS: PRODS. LIAB. § 4(b) (1998). Some states give more conclusive effect to compliance with state statutes. *See, e.g.,* Beatty v. Trailmaster Prod., Inc., 330 Md. 726, 625 A.2d 1005, 1014 (1993) ("[C]ompliance with a statute does not necessarily preclude a finding of negligence or product defectiveness where a reasonable person would take precautions beyond the statutorily required measure. But where no special circumstances require extra caution, a court may find that conformity to the statutory standard amounts to due care as a matter of law."). Some courts may not allow compliance with a government standard to be introduced as evidence. *See, e.g.,* Gaudio v. Ford Motor Co., 2009 Pa. Super. 102, 976 A.2d 524, 543–44 (2009) (holding that compliance with a Federal Motor Vehicle Safety Standard is not admissible in a crashworthiness case).

184. *See, e.g.,* The T.J. Hooper, 60 F.2d 737, 740 (2d Cir. 1932) ("Indeed in most cases reasonable prudence is in fact common prudence; but strictly it is never its measure; a whole calling may have unduly lagged in the adoption of new and available devices. It never may set its own tests, however persuasive be its usages."); Brooks v. Beech Aircraft Corp., 1995-NMSC-043, ¶ 40, 120 N.M. 372, 382, 902 P.2d 54, 64 (1995) ("We hesitate to embrace a standard that would allow an industry to set its own standard of reasonable care and to determine how much product-related risk is reasonable.").

185. RESTATEMENT (THIRD) OF TORTS: PRODS. LIAB. § 2 cmt. d (1998).

186. Section 1 of the Products Liability Restatement provides that "[o]ne engaged in the business of selling or otherwise distributing products who sells or distributes a defective product is subject to liability for harm to persons or property caused by the defect." Comment e elaborates that "[t]he rule stated in this Section provides that all commercial sellers and distributors of products, including nonmanufacturing sellers and distributors such as wholesalers and retailers, are subject to liability for selling products that are defective."

187. *See, e.g.,* COLO. REV. STAT. § 13-21-402; DEL. CODE ANN. tit. 18, § 7001 (West); IDAHO CODE § 6-1407; 735 ILL. COMP. STAT. ANN. 5/2-621; IOWA CODE § 613.18; KAN. STAT. ANN. § 60-3306 (West); KY. REV. STAT. ANN. § 411.340 (West); MD. CODE ANN., CTS. & JUD. PROC. § 5-405 (West); MINN. STAT. § 544.41; MO. REV. STAT. § 537.762; N.C. GEN. STAT. § 99B-2; N.D. CENT. CODE § 28-01.3-05; N.J. STAT. ANN. § 2A:58C-9 (West); OHIO REV. CODE ANN. § 2307.78 (West); S.D. CODIFIED LAWS § 20-9-9; TENN. CODE ANN. § 29-28-106 (West); WASH. REV. CODE § 7.72.040.

the nonmanufacturer distributor, and many statutes condition immunity on a solvent manufacturer. Three states have laws that require automobile manufacturers to indemnify dealers explicitly.[188] Because states have unique provisions in their statutes, litigants need to review these statutes to determine their applicability. These statutes may shield distributors and retailers of automated vehicles from products liability lawsuits.

D. Updates to Driving Automation Systems

Driving automation systems will need to be updated by manufacturers for various reasons, including to correct or update maps, install new security patches, or correct buggy software. As described earlier, these updates may complicate a design defectiveness analysis and the state-of-the-art defense. This section describes two other areas where updates to software may be implicated in products liability actions. The first concerns whether the update itself is admissible in a products liability action. The second concerns whether the update, especially an over-the-air update, can invite class action lawsuits or otherwise limit them.

1. *Subsequent Remedial Measures*

After a crash caused by an automated vehicle, manufacturers and regulators will want to determine the cause of the crash. If a software bug or defect caused the crash, the manufacturer may—and indeed may be required to[189]—update the driving automation system to correct the software bug or defect. Tesla Motors has shown it can update vehicle software through over-the-air updates.[190] People injured by automated vehicles may desire to use software corrections as evidence that the computer code had a defect or that the manufacturer is liable for the harm.

In federal court, this type of evidence would be inadmissible to show a product defect.[191] Federal Rule of Evidence 407 provides as follows:

> When measures are taken that would have made an earlier injury or harm less likely to occur, evidence of the subsequent measures is not admissible to prove:

---

188. Del. Code Ann. tit. 6, § 4905 (West); Idaho Code § 49-1622; Me. Stat. tit. 10, § 1175.

189. See Chapter 5 for a discussion on NHTSA's recall power.

190. Sean O'Kane, *Tesla Can Change So Much with Over-the-Air Updates That It's Messing with Some Owners' Heads*, Verge (June 2, 2018), https://www.theverge.com/2018/6/2/17413732/tesla-over-the-air-software-updates-brakes.

191. Fed. R. Evid. 407. The justifications for this rule are that the fix is not necessarily an admission of fault and may be consistent with a mere accident, and that the rule incentivizes defendants to increase safety. Fed. R. Evid. 407 Advisory Committee Notes, 1972 Proposed Rules.

- negligence;
- culpable conduct;
- a defect in a product or its design; or
- a need for a warning or instruction.

> But the court may admit this evidence for another purpose, such as impeachment or—if disputed—proving ownership, control, or the feasibility of precautionary measures.

Rule 407 could be an impediment to using subsequent updates to show a product was defective.

Plaintiffs will have more success in state courts. Some states have allowed plaintiffs to use this type of evidence for products liability claims.[192] Because plaintiffs can utilize this evidence in some state courts, plaintiffs will want to review the applicable rulings to determine whether such evidence is admissible in state court. In such cases, more pressure may be placed on defeating diversity jurisdiction to keep the action in state court.

2. *Class Action Lawsuits*

Courts will need to wrestle with the appropriateness of class certification for design defect claims based on software defects that are corrected by over-the-air updates. Court decisions may have significant ripple effects for the use of over-the-air updates and security patches. A potentially troublesome outcome would be that automakers face class action lawsuits for every security patch or algorithm update that increases the safety of their vehicles.[193] In these cases, putative classes may attempt to use the benefit-of-the-bargain theory to prove that they have suffered economic damages by purchasing, leasing, or renting an automated vehicle that had an unmanifested defect or vulnerability. Under the benefit-of-the-bargain test, plaintiffs claim that they suffered economic harm through overpayment, loss in value, or loss of usefulness by purchasing

---

192. *See, e.g.*, Haw. Rev. Stat. Ann. § 626-1, Rule 407 (West) ("When, after an event, measures are taken which, if taken previously, would have made the event less likely to occur, evidence of the subsequent measures is not admissible to prove negligence or culpable conduct in connection with the event. This rule does not require the exclusion of evidence of subsequent measures when offered for another purpose, such as proving dangerous defect in products liability cases, ownership, control, or feasibility of precautionary measures, if controverted, or impeachment."); Iowa R. Civ. P. 5.407 ("When measures are taken that would have made an earlier injury or harm less likely to occur, evidence of the subsequent measures is not admissible to prove negligence or culpable conduct. But the court may admit this evidence when offered on a manufacturing defect claim based on strict liability in tort, breach of warranty, or when offered for another purpose, such as impeachment or—if disputed—proving ownership, control, or feasibility of precautionary measures.").

193. *See In re* Apple Processor Litig., No. 5:18-CV-00147-EJD, 2019 WL 3533876 (N.D. Cal. Aug. 2, 2019) (class action lawsuit that followed a security patch by Apple to fix a vulnerability in its products).

a vehicle with the alleged defect.[194] The benefit-of-the-bargain test, if not limited, could open the floodgates in automated vehicle cases. An example may help provide flesh to the issue:

> ABC Motors manufactures automated vehicles. While conducting a safety review of its automated driving system's code, its team learns of a bug in the code that may have caused a crash had one of its automated vehicles encountered a particular roadway. The bug existed in every vehicle and had the potential to cause a crash. ABC Motors fixes the bug with an over-the-air update prior to any crash.

Conceivably, plaintiffs may claim that that they suffered economic harm because they purchased a vehicle marketed as safe. Had they known that ABC Motors' vehicles suffered from this defect, they would have either paid less or purchased an automated vehicle from another automaker. These economic benefit claims are brought under consumer protection statutes and are not products liability actions.[195]

In recent years, courts have considered several unmanifested injury cases involving automobiles and other products.[196] In these unmanifested injury cases, courts have grappled with whether a putative class has standing to assert claims against the defendants. A plaintiff invoking federal jurisdiction must establish standing by showing fulfillment of the following requirements:

> First, the plaintiff must have suffered an "injury in fact"—an invasion of a legally protected interest which is (a) concrete and particularized, and (b) actual or imminent, not "conjectural" or "hypothetical." Second, there must be a causal connection between the injury and the conduct complained of—the injury has to be fairly traceable to the challenged action of the defendant, and not the result of the independent action of some third

---

194. *In re* Toyota Motor Corp. Unintended Acceleration Mktg., Sales Practices & Prods. Liab. Litig., 754 F. Supp. 2d 1145, 1162 (C.D. Cal. 2010); *see also* Mazza v. Am. Honda Motor Co., 666 F.3d 581, 595 (9th Cir. 2012) (holding that plaintiffs satisfy the injury-in-fact element when they "contend that class members paid more for the [collision mitigation braking system] than they otherwise would have paid, or bought it when they otherwise would not have done so, because Honda made deceptive claims and failed to disclose the system's limitations"); Cole v. Gen. Motors Corp., 484 F.3d 717, 723 (5th Cir. 2007) (finding that the putative class has suffered an injury in fact where they based their claim on "recovery for their actual economic harm (e.g., overpayment, loss in value, or loss of usefulness) emanating from the loss of their benefit of the bargain").

195. Products liability law is unavailable for strictly economic damages.

196. *See, e.g.*, Cahen v. Toyota Motor Corp., 147 F. Supp. 3d 955 (N.D. Cal. 2015), *aff'd*, 717 F. App'x 720 (9th Cir. 2017); *In re* Toyota Motor Corp., 790 F. Supp. 2d 1152 (C.D. Cal. 2011); *In re Toyota Motor Corp. Unintended Acceleration Mktg., Sales Practices & Prods. Liab. Litig.*, 754 F. Supp. 2d 1145; Contreras v. Toyota Motor Sales USA, Inc., No. C 09-06024 JSW, 2010 WL 2528844 (N.D. Cal. June 18, 2010), *aff'd in part, rev'd in part sub nom.* Contreras v. Toyota Motor Sales U.S.A. Inc., 484 F. App'x 116 (9th Cir. 2012); *Cole*, 484 F.3d 717.

> party not before the court. Third, it must be "likely," as opposed to merely "speculative," that the injury will be redressed by a favorable decision.[197]

Putative classes have struggled proving an injury in fact. The named plaintiffs in a class action must establish that they satisfy the standing requirements and cannot rely on harm caused to other persons.[198] Chapter 18 details standing.

One of the most popular unmanifested injury class actions for automobiles was the sudden unintended acceleration (SUA) cases against Toyota.[199] In these cases, a putative class sued Toyota, claiming that they suffered harm "for diminution in the market value of their vehicles in light of acknowledged and/or perceived defects." Toyota claimed that the putative class could not prove that its members had suffered an injury in fact. Toyota first focused on how numerous members had not experienced SUA and that the benefit-of-the-bargain test, if applicable, only applies when a product manifests a defect. The court ruled that the plaintiffs did not obtain the benefit of their bargain and suffered an injury in fact: "Plaintiffs bargained for safe, defect-free vehicles, but instead received unsafe, defective vehicles. A vehicle with a defect is worth less than one without a defect. The overpayment for the defective, unsafe vehicle constitutes the economic-loss injury that is sufficient to confer standing."[200]

In *Contreras v. Toyota Motor Sales USA, Inc.*,[201] the putative class filed suit against Toyota based on a defect in the vacuum-assisted brake system that, in sustained low temperatures, can freeze, causing the brakes to become harder to apply. Toyota conducted a limited recall of the affected vehicles in 19 states and of vehicles that could be operated in the weather conditions that led to the recall. The putative class did not include vehicles that manifested the defect, persons who would likely operate their vehicles in the 19 states, or persons who would operate their vehicles in the sustained low temperatures that cause the defect. The class's alleged injury was that their vehicles were worth less due to the defect or that they would not have purchased their vehicles had

---

197. Lujan v. Defs. of Wildlife, 504 U.S. 555, 560–61, 112 S. Ct. 2130, 119 L. Ed. 2d 351 (1992) (internal quotation marks and citations omitted).

198. *E.g.*, Warth v. Seldin, 422 U.S. 490, 502, 95 S. Ct. 2197, 45 L. Ed. 2d 343 (1975) (stating that the named plaintiffs "must allege and show that they personally have been injured, not that injury has been suffered by other, unidentified members of the class to which they belong and which they purport to represent. Unless these petitioners can thus demonstrate the requisite case or controversy between themselves personally and respondents, none may seek relief on behalf of himself or any other member of the class." (internal quotation marks and citations omitted)).

199. *In re Toyota Motor Corp.*, 790 F. Supp. 2d 1152; *In re Toyota Motor Corp. Unintended Acceleration Mktg., Sales Practices & Prods. Liab. Litig.*, 754 F. Supp. 2d 1145. Standing was considered in both opinions.

200. *In re Toyota Motor Corp.*, 790 F. Supp. 2d at 1163.

201. No. C 09-06024 JSW, 2010 WL 2528844 (N.D. Cal. June 18, 2010), *aff'd in part, rev'd in part on other grounds in* Contreras v. Toyota Motor Sales U.S.A. Inc., 484 F. App'x 116 (9th Cir. 2012).

they known of the defect. The court held that the putative class lacked standing because, among other reasons, their vehicles had not manifested the defect, they did not allege they were reasonably likely to use their vehicles in conditions that could cause the defect to manifest, they did not allege any out-of-pocket damages or that they were forced to repair their vehicles, and their allegations about diminished value were conclusory.

In *Beyer v. Symantec Corp.*,[202] a putative class sued Symantec for vulnerabilities in Symantec's AntiVirus Decomposer Engine. Symantec received notice of the vulnerabilities from a cybersecurity team at Google. Symantec notified the public of the vulnerabilities and issued security patches to fix the vulnerabilities on the same date. The court considered whether the putative class had standing under the overpayment theory—that they would not have purchased the product or would have paid less for the product had they known of the defect—and whether actual harm was caused by their computer systems being infiltrated.[203] As to the overpayment theory, the court found that the plaintiffs failed to establish an injury in fact for that claim. The court focused on the speculative nature of the claims, stating that they were based "on a purported *past* risk that has never been alleged to manifest and presumably never will" after the security patches.

Standing could be a significant threshold for putative classes that seek to bring consumer protection–type claims after significant over-the-air updates that repair vulnerabilities or correct the software. As shown in *Beyer*, this could be a significant threshold when the problem is fixed through a software patch before there is harm. This threshold, in light of the desire for frequent use of over-the-air updates, seems warranted, especially in cases where the over-the-air update predates any actual harm or exploitation of the vulnerability.

Even if a putative class can survive a challenge to standing in these cases, the next question is whether a relatively quick update to a driving automation system can defeat class certification. Two prominent circuit court cases, *Winzler v. Toyota Motor Sales U.S.A., Inc.* and *In re Aqua Dots Products Liability Litigation*, lend credence to the manufacturers' ability to curtail class certification, at least for software defects, by fixing the software defect with an over-the-air update. In *Winzler v. Toyota Motor Sales U.S.A., Inc.*, the 10th Circuit, in an opinion written by now-Justice Gorsuch, considered whether a voluntary recall by Toyota mooted a potential class certification by a plaintiff who owned one of the voluntarily recalled Toyotas.[204] The plaintiff contended that the 2006 Toyota Corolla

---

202. No. 18-CV-02006-EMC, 2019 WL 935135 (N.D. Cal. Feb. 26, 2019).

203. There was no proof of actual infiltration and harm at the time of the court's order. However, the court granted limited discovery on this issue to allow plaintiffs an opportunity to establish standing.

204. 681 F.3d 1208, 1209 (10th Cir. 2012).

and the 2006 Toyota Matrix had a defective engine control module. She sought an order requiring Toyota "to notify all relevant owners of the defect and then to create and coordinate an equitable fund to pay for repairs." The district court dismissed her complaint for failing to state a claim; she appealed, but while the case was on appeal, Toyota voluntarily recalled the vehicles at issue to fix the engine control module. On appeal, Toyota claimed its voluntary recall mooted the plaintiff's case and the appeal. After discussing mootness, Justice Gorsuch stated that a case can be "prudentially mooted" "if events so overtake a lawsuit that the anticipated benefits of a remedial decree no longer justify the trouble of deciding the case on the merits [then] equity may demand not decision but dismissal." The opinion focused principally on the statutory mechanism of the voluntary recall and the fact that the executive branch, through NHTSA, would ensure that Toyota followed through with its voluntary recall. The court dismissed the case on prudential mootness grounds.

In re Aqua Dots Products Liability Litigation, decided by the Seventh Circuit, is an example of another court declining to certify a class where a manufacturer conducted a voluntary recall.[205] Aqua Dots, a toy for small children, was created using a substance that, when consumed, caused medical problems and could lead to death. After learning of the defect, the distributor recalled the product and offered to replace it with a properly produced product, a comparative children's toy, or to allow consumers to return their products to the retailer. The recall appeared to be successful. The plaintiffs purchased Aqua Dots for their children, but their children were fortunately not harmed by the product. The plaintiffs did not seek a refund or challenge the recall program. The district court denied class certification because the plaintiffs would be better off using the method provided by the distributor than any that the court could order. The Seventh Circuit disagreed with the district court. However, the Seventh Circuit held that class certification should be denied because the plaintiffs could not be said to represent the class's interests when they were seeking a remedy with a high transaction cost (a class action suit) when a low transaction cost was already available (using the manufacturer's recall mechanism).

205. 654 F.3d 748 (2011).

Chapter 9

Level 2 Driving Automation Systems

Level 2 driving automation systems, such as Tesla's Autopilot or General Motors's Super Cruise, are currently used on our roadways. Level 2 driving automation systems automate the longitudinal vehicle motion control subtask and the lateral vehicle motion control subtask of the dynamic driving task. They are essentially adaptive cruise control combined with lane assistance. Under the Taxonomy, drivers must supervise the roadways while using level 2 driving automation systems. This chapter explores whether the increased control by the driving automation system causes its manufacturer to bear liability (or a share thereof) when a driver crashes a level 2 automated vehicle.

I. LIABILITY FOR DESIGN LIMITATIONS OF A LEVEL 2 DRIVING AUTOMATION SYSTEM

Manufacturers could face claims from drivers or bystanders. Drivers are responsible for supervising their drives and being ready to regain control over their vehicles at any time. A common theme involving level 2 automated vehicle crashes has been that the drivers were distracted. Joshua Brown tragically died in May 2016 when his Tesla Model S crashed into a tractor-trailer while Autopilot was engaged.[1] According to the NTSB, Brown was inattentive at the

1. Aarian Marshall, *Tesla Bears Some Blame for Self-Driving Crash Death, Feds Say*, Wired (Sept. 13, 2017), https://www.wired.com/story/tesla-ntsb-autopilot-crash-death/.

time of the crash.[2] Other crashes have occurred with inattentive drivers.[3] Drivers (and their estates) will have a more difficult time presenting claims against manufacturers because distracted drivers bear fault for the underlying crash, or they have assumed the risk of the crash.[4]

In contrast, bystanders bear no fault for drivers becoming distracted and ignoring warnings and instructions about the need for drivers to supervise their drives and be vigilant. Bystanders, especially those who have damage in excess of the driver's insurance policy coverage, may allege that the manufacturers are responsible for the crashes. Manufacturers will likely make numerous arguments to prevent the levees from breaking at level 2. First, manufacturers will argue that the responsibility for safe operation of level 2 driving automation systems remains with the driver and that they have adequately warned and instructed the driver on the limitations and the safe use of their systems. Manufacturers will likely analogize their level 2 driving automation systems to cruise control or autopilot for airplanes. When the driver is not supervising the drive, manufacturers will also likely claim that the driver caused the crash or misused the system.

A level 2 driving automation system does not replace the driver in any meaningful way. The driver is responsible for supervising the drive and is primarily responsible for the object and event detection and response subtasks of the dynamic driving task. The driver is also responsible for the dynamic driving task fallback. A level 2 driving automation system is more reminiscent of a feature that assists the driver, such as cruise control, rather than a feature that replaces the driver.

Because of the limited capabilities of a level 2 driving automation system, not every crash that occurs will be caused by a defect in the level 2 driving automation system.[5] Many of these crashes will result from design limitations of the level 2 driving automation system and the driver not supervising the drive or executing his or her role. A manufacturer is not obliged to forgo designing a feature with inherent limitations, so long as the manufacturer adequately apprises the consumer of those limitations.[6] If a crash occurs and the driving automation system did not malfunction or fail to operate properly, liability for

---

2. Alex Davies, *Tesla's Latest Autopilot Death Looks Just Like a Prior Crash*, WIRED (May 16, 2019), https://www.wired.com/story/teslas-latest-autopilot-death-looks-like-prior-crash/.

3. *Id.*

4. See Chapter 8 for a discussion on comparative fault, product misuse, and assumption of the risk.

5. See Chapter 8 for a discussion on the need for a defect.

6. James A. Henderson, Jr. & Aaron D. Twerski, *Closing the American Products Liability Frontier: The Rejection of Liability without Defect*, 66 N.Y.U. L. REV. 1263, 1329–30 (1991) ("[M]ost courts that have considered product-category liability claims have rejected them out of hand. And of the very few decisions that have embraced the notion, each has been reversed by its respective state legislature."). See Chapter 10 for a discussion of categorical liability.

the crash should be principally borne by the at-fault driver. Imposing liability on drivers also reinforces to them and the public that drivers are blameworthy for these crashes.

Driver liability could increase safety and prevent crashes by incentivizing people to execute their roles within the shared driving experience. Drivers are primarily responsible for the safe operation of the level 2 automated vehicle, and a liability system that makes manufacturers liable for crashes could remove part of the incentive of drivers to supervise their drives.

Another justification for driver liability is that manufacturer liability would allow less scrupulous drivers to force safer drivers to bear part of the cost of their behavior. A principal purpose of products liability law is to force manufacturers to internalize the costs of a product defect into the purchase price of the product so that the product reflects its true costs to consumers. This internalization occurs by imposing liability on the manufacturer for harm caused by product defects. When a manufacturer bears liability for harm not caused by a product defect, manufacturers would still internalize the liability into the product's price. However, the increased price does not reflect the product's cost—the price reflects compelled insurance and higher costs for less scrupulous drivers.

Cases involving cruise control may help determine liability risks for manufacturers. The driver is generally responsible for crashes caused while the driver uses cruise control. The driver has been able to shift responsibility for a cruise control crash to the manufacturer only when there are allegations "that the cruise control systems caused the cars to unexpectedly accelerate and fail to respond to braking."[7]

Additionally, airplane autopilot cases may help forecast how a court may apply liability at level 2.[8] There have been numerous airplane crashes where plaintiffs have brought products liability claims against the manufacturer

---

7. Gary E. Marchant & Rachel A. Lindor, *The Coming Collision between Autonomous Vehicles and the Liability System*, 52 Santa Clara L. Rev. 1321, 1325 (2012). A representative case is *Cole v. Ford Motor Co.*, 136 Or. App. 45, 900 P.2d 1059, 1060 (1995). In *Cole*, the plaintiff, a passenger, was riding in a Ford Aerostar van while her husband drove the van. The husband activated the cruise control system. The husband pressed the brake pedal to disengage the cruise control system, but the van "just took off wide open," causing the vehicle to accelerate through a stop sign and ultimately crash into a guardrail. The plaintiff filed a strict products liability suit against Ford, alleging that the cruise control system was defective due to a malfunction. At trial, the plaintiff's experts testified that the crash probably occurred due to a malfunction, while Ford's experts testified that the cruise control system was free from defects. The jury found for the plaintiff.

8. *See, e.g.*, K. Krasnow Waterman & Matthew T. Henshon, *Imagine the Ram-If-Ications: Assessing Liability for Robotics-Based Car Accidents*, 5 No. 4 A.B.A. SciTech Law 14, 14–16 (Spring 2009); Marchant & Lindor, *supra* note 7 ("[A]irplanes capable of flying on 'autopilot' (while also manned by a live pilot) provide a close analogy to autonomous vehicles.").

alleging a defective autopilot system.[9] A common theme in many of these cases is a defense verdict for the manufacturer, with the pilot being found responsible in one way or another for the underlying accident, unless the plaintiff can show that the autopilot was defective.[10] In one case, the court stated that "[t]he obligation of those in charge of a plane under robot control to keep a proper and constant lookout is unavoidable. It is particularly clear when the plane is in flight across a recognized and established airway."[11]

Courts should impose the duty of reasonable lookout and supervision by human drivers when they use level 2 driving automation systems. Roadways are too busy to allow drivers to use these systems without imposing such a duty. Many of these level 2 driving automation systems will likely collect substantial data on vehicle performance and on the behavior of the driver. Vehicle data should help courts learn the cause of the crash and whether the driver was fulfilling his or her duty.

II. THE PROBLEM OF THE FORESEEABLE MISUSE

Plaintiffs will counter manufacturers' contentions about the human driver's responsibility for safe driving by arguing that their level 2 driving automation systems are defective because of their effects on drivers. A manufacturer generally must design its products to address foreseeable misuses.[12] This section describes how the misuse of a level 2 driving automation system is foreseeable and how design defect claims may be framed at level 2.

A. The Foreseeability of the Misuse

Level 2 driving automation systems create greater opportunities for drivers to neglect their duties. Human-driven vehicles cannot be safely operated without full attention from drivers, and traffic laws require drivers to always pay

9. *See, e.g.*, Ferguson v. Bombardier Servs. Corp., 244 F. App'x 944 (11th Cir. 2007); Sussman v. Cessna Aircraft Co., No. 2007-20336 CINS (Vousia Cnty. Cir. Ct. Fla. Apr. 10, 2014); Long v. Muncie Aviation Co., No. 2006L006992 (Cook Cnty. Cir. Ct. of Ill. Nov. 7, 2016); *In re* Jacoby Airplane Crash Litig., No. 99-cv-6073 (N.Y.D.C. Nov. 2007); Moe v. Avions Marcel Dassault-Breguet Aviation, 727 F.2d 917 (10th Cir. 1984); Brouse v. United States, 83 F. Supp. 373 (N.D. Ohio 1949).

10. In one case, Lee and Johnnie Arrendale died while Lee was flying a plane. 15 No. 10 VERDICTS, SETTLEMENTS & TACTICS 373 (Oct. 1995). The plane crashed after an uncontrolled dive. The National Transportation Safety Board found that the crash was due to pilot error. But the plaintiffs, the Arrendales' estates, filed suit against King, the manufacturer, because of the "unusual nature of the radar track, the manner in which the plane crashed, and because there had been other reported incidents of a similar nature involving crashes or near crashes of aircraft equipped with King KFC-200 autopilot systems (the autopilot system installed in the Arrendale aircraft), or its similar systems (the King KFC-150 and KFC-250)." *Id.* at 373. The case settled confidentially.

11. *Brouse*, 83 F. Supp. at 374–75.

12. See Chapter 8 for a discussion of product misuse.

attention. However, drivers continuously risk their safety and the safety of others by distracting themselves or driving under or in nonoptimal conditions (such as while sleepy or intoxicated or in inclement weather). Intuitively, people are probably going to pay less attention to the roadway as level 2 driving automation systems assume greater roles in the operation of vehicles.

Countless human factors studies support this commonsense observation.[13] Human factors is the study of how humans behave in particular environments. For level 2 driving automation systems, human factors is concerned with the human-machine interaction and the effects that automation has on drivers.[14]

Level 2 driving automation systems adversely affect situational awareness and driver performance. Drivers need situational awareness to operate vehicles safely. Situational awareness is "a person's 'perception of the elements of the environment within a volume of time and space, the comprehension of their meaning and the projection of their status in the near future.'"[15] Studies have shown that drivers get bored watching the roadway while using level 2 driving automation systems.[16] Drivers alleviate their boredom by diverting their attention from the roadway and engaging in other activities. Studies have also shown that drivers using level 2 driving automation systems suffer from passive fatigue, which results when "the workload on the driver is too little during periods of automation."[17] Passive fatigue degrades driver performance through reduced vigilance and inattentiveness.[18] Other studies show drivers suffer from automation bias and automation complacency.[19] Automation bias is the realization that humans tend to trust automation quickly. People are less apt to trust a driving automation system that continually needs human oversight and correction. Herein lies the paradox: The better a driving automation system is at driving, the more dangerous the vehicle may become.[20]

---

13. *See, e.g.*, Robert E. Llaneras, Jeremy Salinger & Charles A. Green, Human Factors Issues Associated with Limited Ability Autonomous Driving Systems: Drivers' Allocation of Visual Attention to the Forward Roadway, at the Seventh International Driving Symposium on Human Factors in Driver Assessment, Training, and Vehicle Design (June 17–20, 2013); Catherine Neubauer, Gerald Matthews, Lisa Langheim & Dyani Saxby, *Fatigue and Voluntary Utilization of Automation in Simulated Driving*, 54 HUMAN FACTORS 734–46 (2012) (finding that automation can cause fatigue).

14. Mitchell Cunningham & Michael A. Regan, Autonomous Vehicles: Human Factors Issues and Future Issues, Conference: Australasian College of Road Safety Conference 2015 (Sept. 2015) ("[F]rom a human factors perspective, [a fundamental issue] is how to design automation so that drivers understand fully the capabilities and limitations of the vehicle, and maintain situational awareness of what the vehicle is doing and when manual intervention is needed.").

15. Mica R. Endsley, *From Here to Autonomy: Lessons Learned from Human-Automation Research*, 59 No. 1 HUMAN FACTORS 5, 8 (Feb. 2017).

16. *Id.*

17. Cunningham & Regan, *supra* note 14, at 2.

18. *Id.*

19. *Id.*

20. Endsley, *supra* note 15 (noting a "fundamental automation conundrum: The more automation is added to a system, and the more reliable and robust that automation is, the less likely that human

B. Design Defect Claims

The significant products liability question for manufacturers of level 2 driving automation systems is how to design a driving automation system that reasonably engages the driver and helps the driver retain situational awareness. The most common approach has been to utilize a driver monitoring system that prohibits the use of the level 2 driving automation system if the system determines the driver is not supervising the drive. Some companies, like Tesla Motors, use torque sensors on the steering wheel to ensure that drivers keep their hands on the steering wheel while using Autopilot.[21] General Motors's Super Cruise utilizes an infrared camera to track the driver's eyes.[22] A driver monitoring system can be coupled with other components to keep drivers engaged, such as how the system conveys its actions to the driver.[23]

A driver monitoring system does not need to be perfect. It need not work with every driver all the time; otherwise, less scrupulous drivers may be violating the equality of other consumers by making them pay higher prices for their negligent behavior. A reasonably designed driving automation system that adequately attempts to resolve the human factors problems at level 2 should be sufficient to defeat these design defect claims.

A manufacturer that does not address the human factors problems of level 2 driving automation systems and instead relies merely upon instructions or warnings to the driver may have a difficult time preventing claims from reaching juries.[24] Currently, technologies exist to address human factors (with potentially varying degrees of success) while a level 2 driving automation system is engaged. The failure to utilize that technology to address a seemingly foreseeable misuse potentially places bystanders on the roadway—whether other drivers, pedestrians, bicyclists, or fire trucks—at a serious risk of harm and deprives these third parties of their equality[25] and freedom on the roadway.

---

operators overseeing the automation will be aware of critical information and able to take over manual control when needed"); Oliver Casten et al., *Control Task Substitution in Semiautomated Driving: Does It Matter What Aspects Are Automated*, 54 Human Factors 747, 749 (2012) ("The more capable the automation of vehicle control, the more reliance drivers will be willing to place in that automation and the more willing they will therefore be to engage in non-driving-related tasks.").

21. Alex Davies, *A New Camera Turns Its Gaze to the Driver—And Everyone Else*, Wired (Jan. 11, 2019), https://www.wired.com/story/guardian-optical-cabin-control-driver-monitoring/.

22. *Id.*

23. See generally Autonocast Podcast, *Bonus Episode: Alex and Kelly Funkhouser of Consumer Reports Can't Stop Talking about Automated Driving and HMI* (Sept. 26, 2019), for a discussion of this topic.

24. *See* Mark A. Geistfeld, *A Roadmap for Autonomous Vehicles: State Tort Liability, Automobile Insurance, and Federal Safety Regulation*, 105 Cal. L. Rev. 1611, 1627 (2017) ("A manufacturer that does not adopt a reasonably safe, fault-tolerant design is subject to tort liability for the resultant physical harms.").

25. The product seller is valuing price—whether to provide a cheaper product to induce consumers to purchase its goods or to increase profits to shareholders—over safety and depriving third parties of their equal worth.

Under a risk-utility test, the harm is significant when the level 2 feature malfunctions because a multiton automobile can cause significant injury, death, and property damage. The probability of harm may be considerable (depending upon the competence of the level 2 driving automation system) because of the foreseeability that drivers will be derelict in their duties. Given the serious risk of harm, a manufacturer likely has to install some sort of driver monitoring system.[26]

26. *See* Geistfeld, *supra* note 24, at 1628 ("[M]anufacturers cannot merely warn drivers to stay 'alert' in order to take over the driving responsibilities when prompted by [driver-assistance systems]; they must instead adopt fault-tolerant designs for [driver-assistance systems] whenever doing so would be a cost-effective method for reducing the risk of driver error."). A manufacturer that does not equip its systems with a driver monitoring system may find itself sharing some of the responsibility for the crash with the driver.

Chapter 10

Level 3 Automated Driving Systems

Level 3 automated driving systems represent the first significant automation system that fundamentally alters the role and responsibilities of the human driver. The user of a level 3 automated driving system is no longer required to supervise the drive while the feature is engaged; instead, the user must remain receptive to a request to intervene and intervene when requested. In an ironic twist, the driver becomes the advanced driver assistance system (ADAS) for the level 3 automated driving system. The Taxonomy recognizes this fundamental role change by declining to call people who use these features "drivers" and instead referring to them as "dynamic driving task fallback-ready users" (or, as used herein, "users"). This chapter explores whether this fundamental role shift also causes a fundamental shift in liability by considering liability for users and manufacturers.

The role reversal of the driver and the vehicle also reflects an ongoing trend that is occurring in automobile product liability claims. Historically, a manufacturer had an obligation to produce a product that did not cause harm. Recent automobile products liability claims are challenging this basic premise by asserting that automobile manufacturers have a duty not only not to cause crashes but also to prevent crashes. This is the theme of most cases involving ADAS failure-to-equip claims.[1] This philosophy change will become even more pronounced with level 3 automated driving systems where the automaker has

1. *See, e.g.,* Dashi v. Nissan N. Am., Inc., 247 Ariz. 56, 445 P.3d 13 (Ct. App. 2019), *review denied* (Jan. 7, 2020); Butler v. Daimler Trucks N. Am., LLC, No. 2:19-CV-2377-JAR-JPO, 2020 WL 128055 (D. Kan. Jan. 10, 2020).

affirmatively allowed a driver to become distracted when the automated driving system is engaged.

The role reversal of the traditional driver and the automated driving system invites claims by users of these systems. With the level 2 driving automation system, comparative fault, misuse, and assumption of the risk are significant hurdles for drivers claiming manufacturers should bear responsibility for crashes. These hurdles may be smaller or nonexistent for level 3 automated driving systems. Users of these systems are allowed to engage in other activities, and thus their distraction, in and of itself, should not be a defense for manufacturers.

From a liability standpoint, a principal new design concern created by level 3 automated driving systems is the intervention requests to the users. At its core, the level 3 automated driving system must (1) provide notice to the user that the automated driving system needs intervention and (2) provide such notice to the user with sufficient time to retake control safely.

Based on these two core functions of the notification system, there are two general categories of crashes of level 3 automated vehicles. The first is the no-notification cases. The car crashes, and the automated driving system is just as surprised as the human occupants in the vehicle. The second is the notification cases. The level 3 automated driving system provides notice to the user to intervene, and a crash results. For purposes of this analysis, the first category of crashes will be called System Failures, and the second category of crashes will be called Handoff Failures.

To nonoccupants harmed by level 3 automated driving systems, a crash results regardless of whether there was a System Failure or a Handoff Failure. These people will likely sue the user and the manufacturer. The identity of the failure as a System Failure or a Handoff Failure may matter in identifying the appropriate target.

I. SYSTEM FAILURES

In System Failure cases, third parties harmed will likely assert claims against the user, the vehicle manufacturer, and, if different, the manufacturer of the level 3 automated driving system. The user and other occupants of the vehicle will likely sue the vehicle manufacturer and the automated driving system manufacturer. Manufacturers will presumably defend these cases by alleging that their systems are not defective or by challenging the causation of the crashes.

A significant question for these System Failure cases is whether the user can be distracted and whether the manufacturer should be liable for these crashes. At level 3, the user is allowed to rely on the technology and engage in other activities. For distracted users, there may not be an opportunity for them to execute the dynamic driving task fallback without a notification from the system that intervention is needed. Without the notification, the user has been deprived of the opportunity to execute the dynamic driving task fallback. The deprivation of this opportunity would be the basis for imposing liability on the manufacturer for a crash caused while the level 3 automated driving system is engaged.

Although most users of level 3 automated driving systems may be distracted, there will be times when the users are not distracted and are paying attention. Liability for these crashes may not be so neatly allocated between the manufacturer and the attentive user. However, these users were denied truth and had their expectations frustrated by the System Failure. In essence, they purchased a vehicle with a level 3 automated driving system that was supposed to notify them when intervention was needed. Unless the system was obviously malfunctioning, an attentive user might not know he or she needs to intervene. Thus, seemingly, there may be some justification for holding manufacturers liable for System Failures even for attentive users. However, in other circumstances, some System Failures may be so apparent that an attentive user should be obliged to intervene to prevent the crash.

In System Failure cases, an important question will be the type of products liability defect. Given the focus on notifications and warnings, one could mistakenly believe that these are warning defects. The automated driving system did not provide the warning, and the crash resulted. However, these claims do not fit neatly within the general framework of warning defects, which primarily addresses latent defects or product instructions.[2] These claims fit better as manufacturing or design defects.

Whether this type of claim constitutes a manufacturing or design defect will depend upon the facts of the case. The classification of the defect as being in manufacture or design may significantly affect the amount of proof the parties would need to demonstrate.[3] A situation where the System Failure results from the system failing to detect an object or the need for a takeover seems more analogous to design defect claims. However, plaintiffs undoubtedly will contend that the automated driving system did not perform as intended by failing to alert the driver of the need to intervene. Chapter 8's discussion about whether software defects are manufacturing defects or design defects is relevant to identifying the proper claim.

---

2. See Chapter 8 for a discussion of warning defect claims.

3. Design defect claims generally require proof of an alternative design that could have prevented the crash. See Chapter 8 for a discussion of design defect claims.

II. HANDOFF FAILURES

Moral and legal responsibility for Handoff Failures is more ambiguous than for System Failures. For level 3 automated driving systems, the user must be "receptive to [automated driving system]-issued requests to intervene and to evident [dynamic driving task] performance-relevant system failures in the vehicle compelling him or her to perform the [dynamic driving task] fallback."[4] The user may also be summoned when the level 3 automated driving system is exiting its operational design domain.

There are two categories of Handoff Failure crashes. The first Handoff Failure crash category is when the automated driving system notifies the user of the need to intervene, but the user does not become a driver before the crash. One reason may be that the user is not receptive to a request for intervention. The user could be asleep or listening to music on his or her headphones. Another reason may be insufficient time between the request and the crash. The second Handoff Failure crash category occurs when the automated driving system notifies the user, the user becomes the driver, and the driver is unable to prevent or otherwise avoid the crash or causes a new type of crash (perhaps out of panic). These crashes may stem from insufficient time for the user to safely regain situational awareness and avoid the crash or from poor decision making. The insufficient-time claims for both categories of crashes are where liability and moral responsibility could become significantly ambiguous and will be the focus of the remaining part of the chapter.

Human factors studies have shown that allowing the user to engage in other behavior "can lead to a contingently serious deterioration of situation[al] awareness."[5] "If the system requires a take-over, lost situation[al] awareness has to be regained in order to perform safely and comfortably."[6] For a user to safely intervene, the user will need "to comprehend the current scene" and regain situational awareness.[7] Numerous factors impact the ability of a user to intervene and regain situational awareness safely.[8] These factors include the current traffic situation, the individual user's abilities, and the distracting

---

4. *Taxonomy and Definitions for Terms Related to On-Road Motor Vehicle Automated Driving Systems*, J3016, SAE Int'l 17 (June 18, 2018).

5. Jonas Radlmayr et al., *How Traffic Situations and Non-Driving Related Tasks Affect the Take-Over Quality in Highly Automated Driving*, Proceedings of the Human Factors and Ergonomics Society 58th Annual Meeting 2063–2067 (2014).

6. *Id.*

7. Christian Gold, Daniel Dambock, Lutz Lorenz & Klaus Bengler, *"Take Over!" How Long Does It Take to Get the Driver Back into the Loop?*, Proceedings of the Human Factors and Ergonomics Society 57th Annual Meeting 1938–42 (2013), *available at* https://pdfs.semanticscholar.org/2e72/498ec95d07f6af362ab1edf543bc543eea6a.pdf.

8. *See* Radlmayr et al., *supra* note 5, at 2063 (discussing a number of factors that can impact the time needed to safely intervene).

activities of the user.[9] The problem created by the handoff has been described as "the most challenging problem" for an automated vehicle.[10]

Because of these difficulties, companies have forgone developing level 3 automated driving systems. As Erik Coelingh, Volvo's head of safety and driver-assist technologies, stated, "That problem's [the handoff problem] just too difficult."[11] In its monthly report for October 2015, Google's Self-Driving Car Project explained why it ceased producing automated vehicles that relied on humans.[12] In the fall of 2012, Google employees tested the self-driving technology on freeways during their commutes. The Google team instructed volunteers to pay attention to the road at all times because of the early stage of the technology. The volunteers also signed forms promising to pay attention at all times and understood that cameras would record them. After the test, the volunteers gave positive feedback about the technology. The Google team then watched the film. Many volunteers failed to pay attention as they promised. As the report noted, "We saw human nature at work: people trust technology very quickly once they see it works. As a result, it's difficult for them to dip in and out of the task of driving when they are encouraged to switch off and relax."

For companies that pursue level 3, the most important new design consideration is how to design the system to ensure that the user has sufficient time to intervene and take control of the vehicle safely.[13] Studies have shown that with shorter times, people "come to a decision more quickly, reacting faster, but the quality is generally worse."[14] Given the severity of any crash, from a design defect perspective, the potential liability could be significant for any system that does not provide sufficient time to drivers.

Because some manufacturers of vehicle automation have famously eschewed level 3 and because of the aforementioned difficulties with the handoff, an important question is whether those vehicle manufacturers who

---

9. *See, e.g.*, Kathrin Zeeb, Axel Buchner & Michael Schrauf, *Is Take-Over Time All That Matters? The Impact of Visual-Cognitive Load on Driver Take-Over Quality after Conditionally Automated Driver*, 92 Accident Analysis & Prevention 230, 231 (2016); Alexander Eriksson & Neville A. Stanton, *Take-over Time in Highly Automated Vehicles: Noncritical Transitions to and from Manual Control*, 59 Human Factors 689, 701 (June 2017) (noting that "the range of time in which drivers resume control from the automated driving system was between 1.9 and 25.7 [seconds] depending on task engagement").

10. David A. Mindell, Our Robots, Ourselves: Robotics and the Myths of Autonomy 201 (2015).

11. Alex Davies, *The Very Human Problem Blocking the Path to Self-Driving Cars*, Wired (Jan. 1, 2017), https://www.wired.com/2017/01/human-problem-blocking-path-self-driving-cars/.

12. Google Self-Driving Car Project Monthly Report (Oct. 2015).

13. Gold et al., *supra* note 7, at 1938 ("It is fundamental that the driver is made aware of th[e] transition [from the automated driving system to the driver] early enough to avoid potentially dangerous situations and to ensure a comfortable take-over process.").

14. *Id.* at 1942. As this study noted, as the time to react decreases, "the gazes in mirrors and shoulder checks decrease, the accelerations increase, and the brake is used excessively. If other road users had been in the blind spot, the risk of collision would have been high." *Id.*

incorporate level 3 automated driving systems into their vehicles or those who produce such systems should be liable for selling a per se defective product. Initially, the reasons for manufacturers forgoing level 3 are not necessarily an admission by those manufacturers that level 3 is too dangerous. The justifications are more nuanced than that. It may be that these manufacturers have forgone level 3 because the difficulty and costs of designing a level 3 automated driving system are more significant than for higher levels of automation (such as a level 4 or level 5 system). In other words, these manufacturers may believe it is easier to develop higher levels of automation than to address the handoff problem. Of course, this is the type of decision manufacturers routinely make when deciding whether to develop a product and should not, in and of itself, be proof of a defect.[15]

More importantly, courts have routinely rejected the idea of categorical liability.[16] The risk-utility analysis is generally limited to incremental improvements within a given category of products, while the individual consumer

---

15. Industry custom or practice may be relevant evidence for the jury. See Chapter 8 for a discussion on industry custom.

16. *See, e.g.*, Moore v. R.G. Indus., Inc., 789 F.2d 1326, 1327 (9th Cir. 1986) (California law) (declining to find a handgun defective because "[i]t performed as it was intended"); Martin v. Harrington & Richardson, Inc., 743 F.2d 1200, 1204 (7th Cir. 1984) (Illinois law) (declining to impose strict liability on handgun manufacturer when the gun is not defective because, among other reasons, it would require manufacturers to "act as insurers against all damages produced by their products"); Baughn v. Honda Motor Co., 107 Wash. 2d 127, 727 P.2d 655 (1986) (declining to impose absolute liability on an accident involving a mini-trail bike); James A. Henderson, Jr. & Aaron D. Twerski, *Closing the American Products Liability Frontier: The Rejection of Liability without Defect*, 66 N.Y.U. L. Rev. 1263, 1329–30 (1991) ("[M]ost courts that have considered product-category liability claims have rejected them out of hand. And of the very few decisions that have embraced the notion, each has been reversed by its respective state legislature."). Example 10 to section 2 of the Products Liability Restatement states as follows:

> ABC Armour Co. manufactures bullet-proof vests for use by police and security personnel. ABC offers several models, some providing front and back protection only, and others providing wrap-around protection. State A highway patrol officials chose to purchase the model that provides front and back protection only. They did so because that model is less expensive, allows greater flexibility of movement, and is more comfortable. Robert, a state trooper, was shot and killed while making a routine traffic arrest. The bullet entered the side of his body, where the vest did not provide protection. Robert's legal representative argues that the design of the vest is defective because it does not provide wrap-around coverage. Robert has not established a defect within the meaning of Subsection (b). The mere fact of the availability of a wrap-around vest does not render defective the vest that provides only front-and-back protection. Although the wrap-around design is somewhat safer, it is also more costly to buy and use. Moreover, the differences in advantages and disadvantages are sufficiently understood by consumers that omission of the wrap-around feature does not render the front-and-back design not reasonably safe. To subject sellers to liability based on that design would unduly restrict the range of consumer choice among products.

Restatement (Third) of Torts: Prod. Liab. § 2 (1998). Example 10 is based on *Linegar v. Armour of America, Inc.*, 909 F.2d 1150, 1154 (8th Cir. 1990).

reserves the determination of whether to purchase a product among categories.[17] Therefore, as one legal scholar has noted, a "plaintiff . . . cannot claim that the design of a [vehicle with a level 3 automated driving system] is defective because its reliance on a human driver makes it unreasonably dangerous as compared to a fully automated vehicle."[18] The manufacturer of a vehicle equipped with a level 3 automated driving system is obligated to design a reasonably safe level 3 automated driving system, warn about latent defects and the system's limitations, and instruct users on how to be a user. The manufacturer is not obligated to forgo designing that system altogether.

Although level 3 automated driving systems may prove to be riskier than other levels of automation, allowing consumers to purchase level 3 automated driving systems furthers their autonomy.[19] Autonomy means that "consumers ought to be able to purchase a riskier product even though a safer model might be available."[20] Because a consumer who has been adequately informed of the capabilities of a level 3 automated driving system voluntarily purchases that feature to further the consumer's own autonomy, the consumer should accept the responsibility for his or her autonomy by being responsible for harm caused when he or she fails to reasonably act as the dynamic driving task fallback-ready user. If a reasonably diligent dynamic driving task fallback-ready user could have prevented the crash, then the user should be liable for the crash. This inquiry is highly factual and will almost certainly require expert witnesses from the parties to establish whether the time and notification were sufficient.

---

17. Mark A. Geistfeld, Products Liability Law 190 (2012).

18. Mark A. Geistfeld, *A Roadmap for Autonomous Vehicles: State Tort Liability, Automobile Insurance, and Federal Safety Regulation*, 105 Cal. L. Rev. 1611, 1628 (2017).

19. *See* Richard C. Ausness, *Risky Business: Liability of Product Sellers Who Offer Safety Devices as Optional Equipment*, 39 Hofstra L. Rev. 807, 810 (2011) (discussing how "allowing consumers to bear product-related risks is based principally on the concept of personal autonomy").

20. *Id.* at 812.

Chapter 11

Level 4 and Level 5 Automated Driving Systems

Level 4 and level 5 automated driving systems represent significant steps from level 3 automated driving systems because they remove humans from the loop during their use. The user of the level 4 or level 5 automated driving system is not required to execute any part of the dynamic driving task or the dynamic driving task fallback. The difference between a level 4 automated driving system and a level 5 automated driving system is that the former has an operational design domain. While operating within its operational design domain, a level 4 automated driving system should be as proficient as a level 5 automated driving system.[1] Thus, the difference is not one of capability, but rather it is about where the capability can be used.

Many scholars have argued that the removal of the human from the loop significantly changes the risk calculus such that manufacturers increase their share of liability for crashes caused by their automated vehicles.[2] This chapter

1. This statement, of course, is not always going to be the case. Some level 4 systems may be more proficient than level 5 systems and vice versa—much of the difference may stem from the capabilities of the underlying manufacturer and its system.

2. *See, e.g.*, Jack Boeglin, *The Costs of Self-Driving Cars: Reconciling Freedom and Privacy with Tort Liability in Autonomous Vehicle Regulation*, 17 Yale J.L. & Tech. 171 (2015); Dorothy J. Glancy, *Sharing the Road: Smart Transportation Infrastructure*, 41 Fordham Urb. L.J. 1617 (2014); Kyle Graham, *Of Frightened Horses and Autonomous Vehicles: Tort Law and Its Assimilation of Innovations*, 52 Santa Clara L. Rev. 1241 (2012); F. Patrick Hubbard, *"Sophisticated Robots": Balancing Liability, Regulation, and Innovation*, 66 Fla. L. Rev. 1803 (2014); Gary E. Marchant & Rachel A. Lindor, *The Coming Collision between Autonomous Vehicles and the Liability System*, 52 Santa Clara L. Rev. 1321 (2012); David C. Vladeck, *Machines without Principals: Liability Rules and Artificial Intelligence*, 89 Wash. L. Rev. 117 (2014); Kyle Colonna, *Autonomous Cars and Tort Liability*, 4 Case W. Reserve J.L. Tech. & Internet 81 (2012); Kevin Funkhouser, *Note, Kevin Funkhouser, Paving the Road Ahead: Autonomous Vehicles, Products Liability, and*

initially presents the arguments for why the vehicle manufacturer (or the manufacturer of the automated driving system) should bear a greater share of liability for crashes caused by highly automated vehicles. This chapter then explains the reasons why courts should consider not making manufacturers bear a greater share of liability. This dual approach to responsibility for these crashes is intended to fully present the issue, explore potential arguments, and allow the reader to assess whether the manufacturer, the user, or the vehicle owner should bear liability.

Although the issue of manufacturers bearing a larger share of liability for crashes of these vehicles is presented in a binary discussion, liability, of course, is not binary. No crash is identical, and each crash will have separate facts that shape its liability analysis. Important facts include the specific manufacturer and what it did or did not do, the specific system and its capabilities, marketing, owner or user instruction, roadway conditions, other drivers' behavior, user's and occupants' behavior, the vehicle's condition, and vehicle maintenance.

In deciding which party should bear liability for crashes caused by level 4 and level 5 automated vehicles, the initial inquiry should be into the causes of such crashes. Crashes may result from four broad categories of errors: (1) faulty software, sensors, or interaction among software and sensors, including software bugs, errors, sensor fusion, or the failure of the system to anticipate a driving situation and react properly (System Errors); (2) lack of vehicle maintenance, use of old parts, or uncleanliness of sensors (Maintenance Errors); (3) alterations to parts or software (Alteration Errors); or (4) use of level 4 automated driving systems outside of their operational design domains or use of level 4 or level 5 automated driving systems in dangerous conditions (Use Errors).[3] Crashes caused by cyberattacks are described in Chapter 18. Maintenance

---

the Need for a New Approach, 2013 UTAH L. REV. 437 (2013); Andrew P. Garza, *"Look Ma, No Hands!" Wrinkles and Wrecks in the Age of Autonomous Vehicles*, 46 NEW ENG. L. REV. 581 (2012); Jeffrey K. Gurney, *Sue My Car Not Me: Products Liability and Accidents Involving Autonomous Vehicles*, 2013 U. ILL. J.L. TECH. & POL'Y 247 (2013); Dylan LeValley, *Autonomous Vehicle Liability—Application of Common Carrier Liability*, 36 SEATTLE U.L. REV. 5 (2013); Eric Cheng, *Obamacars: Applying an Insurance Mandate to Manufacturers of Fully Autonomous Vehicles*, 87 GEO. WASH. L. REV. 973 (2019); Ryan Abbott, *The Reasonable Computer: Disrupting the Paradigm of Tort Liability*, 86 GEO. WASH. L. REV. 1 (2018); Orly Ravid, *Don't Sue Me, I Was Just Lawfully Texting & Drunk When My Autonomous Car Crashed into You*, 44 SW. L. REV. 175 (2014); John W. Zipp, *The Road Will Never Be the Same: A Reexamination of Tort Liability for Autonomous Vehicles*, 43 TRANSP. L.J. 137 (2016).

3. People have considered liability that depends upon certain decisions made by the automated driving system. These decisions may include path choices or choices on who or what to collide with. *See, e.g.*, Jeffrey K. Gurney, *Crashing into the Unknown: An Examination of Crash-Optimization Algorithms through the Two Lanes of Ethics and Law*, 79 ALB. L. REV. 183 (2016); Alfred R. Cowger, Jr., *Liability Considerations When Autonomous Vehicles Choose the Accident Victim*, 19 J. HIGH TECH. L. 1, 43 (2018); Bryant Walker Smith, *The Trolley and the Pinto: Cost-Benefit Analysis in Automated Driving and Other Cyber-Physical Systems*, 4 TEX. A&M L. REV. 197 (2017).

Errors, Alteration Errors, and Use Errors may cause System Errors. For example, a dirty sensor may cause a sensor not to interact properly.

I. THE CASE FOR MANUFACTURERS BEARING MORE LIABILITY

Maintenance Errors, Alteration Errors, and Use Errors seem, at least at first, to fall within the province of the automated vehicle owner or the user. The manufacturer has long relinquished control over its automated vehicles and cannot usually interact with customers at the point of sale.[4] The manufacturer has no power or control over whether vehicle owners or users maintain or alter their vehicles or whether they use their vehicles outside the system's operational design domain or in dangerous conditions. These types of errors seemingly belong in the category of vehicle owner or user responsibility.[5]

A deeper dive may collapse Maintenance Errors, Alteration Errors, and Use Errors into System Errors. These errors foreseeably cause automated driving systems not to operate properly or otherwise cause crashes. Foreseeability, the linchpin of duty, invites courts to determine whether an alternative design to the automated vehicle could have prevented the harm. An engineering solution may exist for all of these claims. For Maintenance Errors, automated driving systems could prevent their use when replacement parts are needed or sensors are dirty. An automated driving system could detect alterations to the vehicle that affect its safety or attempts to jailbreak the software. Level 4 automated vehicles could be geofenced to their operational design domain, and level 5 automated vehicles could be designed not to operate in inclement weather or

4. See Chapter 6 for a discussion of how vehicle manufacturers generally are prohibited from selling motor vehicles directly to consumers.

5. With respect to Alteration Errors, see, for example, *Kobza v. General Motors Corp.*, 63 Ohio App. 3d 742, 745, 580 N.E.2d 47, 48–49, *dismissed*, 47 Ohio St. 3d 706, 547 N.E.2d 987 (1989) (stating that "a claim for strict liability in tort cannot be maintained if there has been a material alteration to the defective product once it has left the manufacturer's control, and if this material alteration significantly contributed to the plaintiff's injury"); *Allstate Insurance Co. v. Volkswagen of America*, 2004 WI App 109, ¶ 5, 273 Wis. 2d 786, 680 N.W.2d 833 ("To prove a strict product liability claim, the plaintiff must prove, among other things, that the product has not undergone a substantial, material change since it left the manufacturer's or seller's control."); *Cox v. General Motors Corp.*, 514 S.W.2d 197, 197 (Ky. 1974); see also MICH. COMP. LAWS ANN. § 600.2947 (West) ("A manufacturer or seller is not liable in a product liability action for harm caused by an alteration of the product unless the alteration was reasonably foreseeable. Whether there was an alteration of a product and whether an alteration was reasonably foreseeable are legal issues to be resolved by the court."). With respect to Maintenance Errors, see, for example, TENN. CODE ANN. § 29-28-108 (West) ("If a product is not unreasonably dangerous at the time it leaves the control of the manufacturer or seller but was made unreasonably dangerous by subsequent unforeseeable alteration, change, improper maintenance or abnormal use, the manufacturer or seller is not liable."). Chapter 8 describes defenses relating to Maintenance Errors and Use Errors.

other dangerous conditions. The "proximity," as one scholar has called it, of vehicle manufacturers to their vehicles may necessitate new duties and obligations that historically have been too burdensome or nearly impossible for manufacturers to fulfill.[6] For example, a system that cannot operate safely in snow and can detect that it is snowing may be defective by not connecting the dots.

System Errors not caused by Maintenance Errors, Alteration Errors, or Use Errors have been the focus of scholarly attention. Liability for crashes caused by System Errors could be borne by the vehicle owner, the vehicle user, the vehicle manufacturer, and/or the automated driving system manufacturer. The cornerstone of civil liability is duty. With the automated vehicle owner and user, the fundamental question is whether he or she has a duty for these System Errors. From a technical standpoint, the answer is simple: there is no duty.

Legal duties are established by courts and legislatures, not by the Society of Automotive Engineers. One duty that courts could impose on users of level 4 and level 5 automated vehicles, at least when they are occupied and have human-driving capabilities, is a duty to monitor the drive and intervene to prevent System Errors. Under the Hand Formula, a duty should be imposed where the cost of a safety precaution is less than the expected harm that would result from not having the safety precaution. Litigants and courts need to weigh the costs and benefits.

The costs of a duty to monitor and intervene for level 4 and level 5 automated vehicles are significant. For consumers, a primary benefit of highly automated vehicles is to engage in other activities and be able to delegate the driving responsibilities to automated driving systems. Other consumers may purchase these vehicles because they cannot otherwise drive vehicles due to disabilities or a degradation of their mental and physical facilities. This duty would defeat these purposes of highly automated vehicles. The duty may also dissuade people from purchasing and using automated vehicles and lead to more dangerous roads.[7]

Even if consumers were willing to purchase automated vehicles with a duty to monitor, the duty might increase the risk of harm.[8] Automated driving systems and what they are perceiving (or not perceiving) will likely be opaque

---

6. Bryant Walker Smith, *Proximity-Driven Liability*, 102 GEO. L.J. 1777 (2014).

7. See Chapter 4 for a discussion of the expected safety benefits of automated vehicles.

8. In weighing costs and benefits, the risk of harm created by a duty or an alternative design is relevant. *See, e.g.*, Phatak v. United Chair Co., 756 A.2d 690, 695 (Pa. Super. Ct. 2000) (noting that if "an automobile were made to resemble a tank[, i]t might make its occupants safer, but if in so doing it creates an unacceptable hazard to other motorists or pedestrians, the risk-utility is negative and the product design feature should be thought of as a negative, not a positive"); Crespo v. Chrysler Corp., 75 F. Supp. 2d 225, 228 (S.D.N.Y. 1999) (noting that the design defect test "must mean, not simply that the manufacturer could have designed the product so that it would not have caused the victim's injuries, but also that doing so would not have rendered the product more-than-offsettingly unsafe for other relevant users").

to their users.[9] What may seem like a potential crash to a human may be avoidable by a level 4 or level 5 automated driving system. Intervention in such cases by potentially panicked users scared of breaching such a duty could lead to more crashes than if the users had been permitted to get distracted.

Another cost of this duty would be that drivers will, as they cede significant aspects of their driving to automated vehicles, become unpracticed in driving and suffer from an erosion of their skills.[10] Young drivers may never acquire driving experience to become skilled. These situations would require drivers to use skills they no longer possess or never developed.

The risk of System Errors occurring where human intervention is needed also seems low, or at least it should be low for a safe and competent automated driving system. Level 4 and level 5 automated driving systems perform the entire dynamic driving task fallback and transition the vehicle to a minimum risk condition when there is a system failure. Human intervention is only required when there is both a System Error that could cause a crash and the failure of the automated driving system to reach a minimum risk condition to avoid the crash. This duty might create more harm by incentivizing overly cautious (and perhaps panicking) users to intervene at inopportune and unnecessary times.

The benefits of this duty seem minimal. Although car crashes do cause significant harm to persons and damage to property, as is evidenced by the considerable appeal of automated vehicles, the assertion that drivers could minimize this harm by monitoring and intervening seems naïve at best. Human users will almost certainly be distracted, regardless of the duty.[11] In some cases, a human user may be able to avoid crashes caused by System Errors. In most cases, the human user may increase the risk of harm by making panicked decisions before regaining situational awareness.

Courts could also consider equating the user of a level 4 or level 5 automated driving system to a passenger of the vehicle. Passengers are generally

---

9. *See* Mica R. Endsley, *From Here to Autonomy: Lessons Learned from Human-Automation Research*, 59 Human Factors 5, 7 (2017) ("Many automation interfaces do not provide needed information to the operator on the state of the automation and, often, little feedback on the state of the system it is controlling.").

10. Hod Lipson & Melba Kurman, Driverless: Intelligent Cars and the Road Ahead 62 (2016); Mitchell Cunningham & Michael A. Regan, Autonomous Vehicles: Human Factors Issues and Future Issues, Conference: Australasian College of Road Safety Conference 2015 (Sept. 2015) ("Drivers that over-rely on highly automated driving systems may fail to use their manual driving skills over long periods of time. The neglect of manual driving skills may, in turn, degrade both the psychomotor dexterity and cognitive skills required to manually complete a task successfully and safely" (internal citations omitted).).

11. See Chapter 9 for a discussion of the misuse of level 2 driving automation systems. See Chapter 10 for a discussion of the time necessary for a distracted driver to regain situational awareness.

not liable for crashes caused by drivers.[12] "Generally speaking, a motor vehicle occupant, other than the driver, who does not actively interfere with the operation of the vehicle is not liable for harm caused to third parties on account of the driver's negligence or wilful misconduct in the absence of evidence that the occupant had some control over the driver, that the driver was in the occupant's employ or was his agent, or that the driver and the occupant were engaged in a joint enterprise."[13] The user of a level 4 or level 5 automated driving system lacks control over the operation of the software driving the vehicle. The user's control is limited to inputting destinations, which, at least for passengers of human-driven vehicles, has been insufficient to impose liability on the passenger.[14] An agency relationship has not been formed, at least not in the traditional formulation of an agency relationship.[15] Automated driving systems cannot consent to be the agent of the user, and the user likely does not have the means to control the agency. And finally, the automated driving system and the user could not be said to be in a joint enterprise—a relationship similar to that of a partnership whereby each actor has "an equal voice in directing the conduct of the enterprise."[16] Therefore, the aspects of the law where liability would traditionally be imposed on a passenger appear inapplicable.

A third avenue of imposing liability for these crashes could be to use doctrines, such as negligent entrustment of the family purpose doctrine, that make vehicle owners strictly liable for crashes caused by permissible users of their vehicles.[17] Courts justify this liability on two principal considerations. First, there are financial reasons for imposing liability on the vehicle owner in the absence of negligence: "Bluntly put, it is felt that, since automobiles are

12. Gregory G. Sarno, *Liability of Motor Vehicle Passenger for Accident*, 50 Am. Jur. Proof of Facts 2d 677 (1988) (providing that "as a general rule, a person in a motor vehicle other than the driver is not liable for an injury to or the death of a third party due to the negligent or wilful misconduct of the driver").

13. *Id.; see, e.g.*, Warren *ex rel.* Warren v. Glascoe, 852 So. 2d 634, 640 (Miss. Ct. App. 2003), *aff'd*, 880 So. 2d 1034 (Miss. 2004) ("Mississippi case law does recognize a duty of care for passengers in a vehicle. 'A guest or passenger in a vehicle is not chargeable with the negligence of the driver over whom the passenger has no control, until the guest or passenger has become aware of that negligence and of the probable danger thereof and there upon fails reasonably to remonstrate or to escape, if escape be safely or reasonably possible.'" (quoting Miss. Cent. v. Roberts, 173 Miss. 487, 504, 160 So. 604, 606 (1935))).

14. *See, e.g.*, Desuza v. Andersack, 63 Cal. App. 3d 694, 699, 133 Cal. Rptr. 920, 924 (Ct. App. 1976) (stating that "it is settled that if one person is riding in another's automobile, the right to control may not be inferred merely because the trip is undertaken for the passenger's benefit and the destination is selected by him").

15. "Agency is the fiduciary relationship that arises when one person (a 'principal') manifests assent to another person (an 'agent') that the agent shall act on the principal's behalf and subject to the principal's control, and the agent manifests assent or otherwise consents so to act." Restatement (Third) of Agency § 1.01 (2006).

16. W. Page Keeton et al., Prosser & Keeton on the Law of Torts 517 (5th ed. 1984).

17. *See, e.g.*, Jacob D. Walpert, *Carpooling Liability? Applying Tort Principles to the Joint Emergence of Self-Driving Automobiles and Transportation Network Companies*, 85 Fordham L. Rev. 1863, 1883 (2017).

expensive, the owner is more likely to be able to pay any damages than the driver, who may be entirely impecunious; and that the owner is the obvious person to carry the necessary insurance to cover the risk, and so to distribute any losses among motorists as a class."[18] The second justification is that the vehicle owner can control the risk by determining who can use the vehicle.[19]

The underlying purposes of these rules seem limited for automated vehicles. In any of these cases, the driver who causes the crash is responsible for his or her negligence. The strict liability doctrines are used to ensure that those harmed have an avenue for recovery. In considering the potential alternative liable party for these crashes, it seems that the manufacturer is almost always going to have "deeper pockets" than the vehicle owner. To the extent that courts base these claims on owner control, vehicle owners lack control over the automated driving system. The owner's only control is over whether to purchase and use an automated vehicle. The inherent risks of an automated driving system may be unquantifiable at the time of purchase. After-sale actions or inactions may affect the safety of an automated vehicle. Software that is not updated to reflect roadway changes or does not receive security patches may become dangerous postpurchase.

In any case, under these types of doctrines, owner or user liability for these types of crashes would be strict. At its core, the battle of liability for highly automated vehicles is a question of who should bear liability for these crashes. On one side of the battlefield, courts could impose strict liability on the owner or user for System Errors. On the other side of the battlefield, courts could impose products liability on the manufacturer. This battle should be resolved primarily on whether the imposition of liability on the user/owner, on the one hand, or the manufacturer, on the other hand, furthers the primary purposes of tort law and products liability law.

More likely than not, System Errors will be the result of a defect in the design and programming of the automated driving system. Manufacturers are responsible for designing and programming their automated driving systems and will, in most cases, have complete control over the coding and software of their systems. Because System Errors are within the control of manufacturers, when these systems malfunction and cause crashes, the manufacturers are probably negligent or at least at fault for the crash.[20]

Justice Oliver Wendell Holmes, Jr.'s cogent observation that "the safest way to secure care is to throw the risk upon the person who decides what

---

18. KEETON ET AL., *supra* note 16, at 522.

19. *Id.* at 522–23.

20. See Chapter 8 for a discussion of the purposes of products liability law and that one of the reasons for strict products liability is administrative convenience.

precautions shall be taken"[21] leans heavily in favor of manufacturer liability for any crash caused by System Errors. For these crashes, the vehicle owner or user, unable to fix the software, is incapable of preventing the crash from recurring, unless he or she monitors the drive with its potential pitfalls identified above or forgoes using automated vehicles.

In contrast, the manufacturer can update its software to prevent the crash from recurring. The manufacturer can also update the software of its other automated vehicles to prevent those vehicles from causing similar or identical crashes. The ability to prevent the crash from occurring in other vehicles manufactured by the company is a decisive difference between conventional vehicles and automated vehicles. With conventional vehicles, a driver may commit an almost identical crash or crashes similar to ones that he or she previously committed or that others committed. Still, those crashes are more of the exception than the rule. With some System Errors, similar crashes may be the rule and not the exception, and only the manufacturer can prevent these crashes.

Another justification for manufacturer liability is that it internalizes the riskiness of the automated driving system into its cost.[22] This internalization of the costs of System Errors into the price of the vehicle should mean that safer automated vehicles will be less expensive and, thus, more appealing to consumers. With vehicle owner or user liability, internalization becomes more convoluted and may create inefficiencies. If courts impose strict liability on vehicle owners, they may be unable to determine the riskiness of an automated vehicle. When automated vehicles are initially released, insurers may have insufficient data about their risks to price insurance properly. Purchasers may have insufficient information about automated vehicles and the manufacturer to appreciate their long-term insurance risks. An automated vehicle that is safe today may be unsafe or riskier tomorrow. With conventional vehicles, vehicle owners can purchase a vehicle with the expectation that it will work as intended, and their insurance risk is dependent primarily on their driving behavior. Insurance companies can generally ascertain the riskiness of a driver based on objective (and some subjective) factors about the driver. When the insurance risk becomes about the capabilities of an automated driving system, that risk calculus changes significantly.

Strict liability imposed on vehicle users instead of vehicle owners creates its own inefficiencies in the insurance markets. Automated vehicles will have different risk profiles based on the effectiveness of their automated driving systems. Given that people may ride in and use automated vehicles manufactured

---

21. Oliver Wendell Holmes, Jr., The Common Law 117 (1881).

22. *See* Vladeck, *supra* note 2, at 146 (stating that, among other reasons, "a strict liability regime is warranted because, in contrast to the injured party, the vehicle's creators are in a position to either absorb the costs, or through pricing decisions, to spread the burden of loss widely").

by various companies, determining the actual risk that an insured person poses may be more trying. The inefficiency in this system may make those who purchase a safer automated vehicle pay an additional premium based on the existence of less safe automated vehicles. Imagine a scenario where two college roommates, Andrew and Nick, each own an automated vehicle. Nick, who is more cautious, purchases a safer automated vehicle from a more reputable manufacturer than his cost-cutting roommate. If courts impose strict liability on the vehicle owner, Nick's decision should result in a lower insurance premium than the premium Andrew should pay. However, if courts impose strict liability on vehicle users, a sophisticated insurer would let Andrew benefit from living with Nick and require Nick to pay extra for living with Andrew because there is a chance that Andrew will use Nick's automated vehicle and vice versa.

Additionally, frustrated expectations may be another basis for imposing liability on the manufacturer instead of the vehicle owner or user. Level 4 and level 5 automated vehicles are explicitly or implicitly, depending upon marketing, sold on the basis that they are capable of safely driving within their operational design domains without the need for humans to monitor drives or intervene. Thus, the essential function of the product is that it can drive safely without human supervision, monitoring, or intervention. When automated driving systems fail to perform as intended, the trust upon which the relationship between the manufacturer and vehicle owner (or user) depends has been violated.

Vehicle owner or user liability may disincentivize people to purchase and use level 4 and level 5 automated vehicles, even if these vehicles are safer than their driving. Many people irrationally believe that they are excellent or very good drivers and believe other drivers are not as capable.[23] For these people, car crashes occur because of other people's poor habits and skills. Vehicle owner or user liability presents people with the choice of trusting their "excellent" driving skills or trusting an automated driving system they may not fully comprehend.

User liability also suffers from some additional practical flaws. If more than one person occupies an automated vehicle, who is liable? Perhaps if there is a steering wheel and traditional driver seat, then the person who sits in that seat could be liable, but that begs the question as to why anyone would voluntarily subject themselves to such risks. If instead all passengers share liability, would

---

23. According to a study commissioned by Allstate Insurance Company in 2011, 64 percent of those surveyed rate themselves "excellent" or "very good" drivers, while they only considered 29 percent of their close friends "excellent" or "very good" drivers and 22 percent of people their age "excellent" or "very good." *New Allstate Survey Shows Americans Think They Are Great Drivers—Habits Tell a Different Story*, PR Newswire (Aug. 2, 2011), https://www.prnewswire.com/news-releases/new-allstate-survey-shows-americans-think-they-are-great-drivers---habits-tell-a-different-story-126563103.html.

such liability be joint and several to avoid problems created by judgment-proof passengers (such as minors or perhaps the uninsured)? Given that car crashes can result in damages in excess of insurance limits, passengers who own any assets would be incurring substantial risk in riding in automated vehicles.

The foregoing should establish that liability for crashes caused by System Errors should fall primarily on the vehicle manufacturer or automated driving system manufacturer. By imposing liability on the manufacturer, courts would properly incentivize safe designs, leading to more efficient pricing of automated vehicles, and avoid complications imposed by vehicle owner or user liability.

II. THE CASE AGAINST MANUFACTURERS BEARING GREATER LIABILITY

The prior section argued that manufacturers should bear a greater share of responsibility for crashes caused by level 4 and level 5 automated vehicles. This section reexamines manufacturer liability for System Errors from the other side of the liability battlefield and argues why courts should resist manufacturer liability. The prior section, for all of its theoretical reasoning, was devoid of the practical realities and problems that can arise from manufacturer liability. Manufacturers come armed not just with their practical complaints about manufacturer liability but also with the historical treatment courts have given to new products. Courts have historically granted manufacturers of new technologies a grace period before imposing potentially enormous liability on them.[24] The automobile industry itself received such a honeymoon period when vehicles, many of which were defective, first entered the marketplace.[25]

A. Manufacturer Liability Would Make Manufacturers Insurers

A fundamental premise of products liability law is that manufacturers are not insurers of their products. As one court stated:

---

24. Graham, *supra* note 2, at 127 (stating that "tort law sometimes (though not always) blesses innovations with a 'honeymoon' period"); *see also* Jerry L. Mashaw & David L. Harfst, The Struggle for Auto Safety 41 n.23 (1990); Keeton et al., *supra* note 16, at 25 ("[C]ourts frequently have been reluctant to saddle an industry with the entire burden of the harm it may cause, for fear that it may prove ruinously heavy. . . . It is . . . likely to be true as to a new industry, which may be unduly hampered in its development. . . .").

25. Graham, *supra* note 2 (describing the initial lawsuits involving motor vehicles and how manufacturers were not blamed for defective motor vehicles).

> Although a manufacturer has a duty to design a product that is reasonably safe for its foreseeable use, the manufacturer is not required to design "the best possible product." That a product is inherently dangerous, or that it creates a risk of harm, or that it causes injury does not, alone, make it defective. Some products, no matter how carefully manufactured, are inherently dangerous, and virtually every product is capable of causing injury. As is often stated, manufacturers are not insurers against all injury involving their products. Similarly, proof that technology existed, which if implemented could feasibly have avoided a dangerous condition, does not alone establish a defect. Thus, for example, automobiles in the 1950's were manufactured without seat belts, even though the technology was available for their installation. Yet, as counsel for [the plaintiff] conceded at oral argument, we would not now impose liability on the manufacturer of those automobiles on the basis that they were defective for not having seat belts. Government and industry standards did not require them and consumers did not, at the time, demand or expect them.[26]

Unless courts require plaintiffs to identify and show a defect, manufacturer liability would make manufacturers insurers of their automated vehicles.[27] Liability should not be imposed on manufacturers unless plaintiffs can introduce some defect that caused the crash—and not merely that a crash resulted.

B. Lack of Uniformity

Absent a federal liability statute, there will be no uniformity in liability and responsibility among the 50 states.[28] Lack of uniformity among courts can create significant costs for victims of these crashes and create uncertainties for users, owners, manufacturers, and insurers. This uncertainty may deter manufacturers from entering the automated vehicle industry or delay the introduction of automated vehicles. Additionally, it may create unnecessary costs for consumers and users of automated vehicles. For example, if South Carolina courts find that the manufacturer of an automated vehicle is primarily responsible for crashes caused by System Errors, but Georgia courts find that the users of the automated vehicles are liable, automated vehicles and insurance may be

---

26. Sexton *ex rel.* Sexton v. Bell Helmets, Inc., 926 F.2d 331, 336 (4th Cir. 1991) (internal citations and quotation marks omitted); *accord, e.g.,* Robinson v. Brandtjen & Kluge, Inc., 500 F.3d 691, 696 (8th Cir. 2007).

27. See Chapter 8 for a discussion of the need of plaintiffs to prove a defect.

28. As one scholar noted, the complexity and novelty of many of the issues involving an automated vehicle crash means that "there will be some degree of trial and error (no pun intended) and jurisprudential inconsistency among jurisdictions before the law becomes more settled." Tracy Hresko Pearl, *Compensation at the Crossroads: Autonomous Vehicles & Alternative Victim Compensation Schemes*, 60 WM. & MARY L. REV. 1827, 1863 (2019).

priced artificially higher. Because occupants may still be liable for the crash in Georgia, their insurers may still charge a significant premium for insuring them, and, because manufacturers are liable in South Carolina, they may price liability into their vehicles. Manufacturer liability may, at least at first, create significant inefficiencies.

C. Deterrence

A major reason why scholars have opposed a liability shift from the traditional driver to the manufacturer is that the shift may deter innovation and the development of automated vehicles. The deterrence argument seems overstated, given that some manufacturers have appeared to accept responsibility for crashes they cause[29] and given that manufacturers are racing to develop automated vehicles notwithstanding the manufacturer liability drum beating louder each year.[30] Although deterrence may be overstated, manufacturer liability will likely cause them to spend more time developing, testing, and validating their automated driving systems.

A delay in deploying automated vehicles will result in safer automated vehicles and may ultimately lead consumers to trust automated vehicles. However, the delay comes with an immense cost. Human driving imposes significant costs, including over 36,000 people dying and millions more crashes on American roadways each year, lost productivity through driving and congestion, and limited autonomy for people who suffer from physical or mental disabilities.[31] If highly automated vehicles are equally as safe as human driving, then they would be better for society because productivity could be increased, and persons who suffer from disabilities could use these vehicles. If highly automated vehicles are twice as safe as human driving, then society would also benefit from reduced deaths, injuries, and crashes. However, if automated vehicles could halve the number of deaths and injuries, the damages a manufacturer could face for crashes could still be devastating.

---

29. *See, e.g.*, Michael Ballaban, *Mercedes, Google, Volvo to Accept Liability When Their Autonomous Cars Screw Up*, JALOPNIK (Oct. 7, 2015), https://jalopnik.com/mercedes-google-volvo-to-accept-liability-when-their-1735170893.

30. *See* Kenneth S. Abraham & Robert L. Rabin, *Automated Vehicles and Manufacturer Responsibility for Accidents: A New Legal Regime for a New Era*, 105 VA. L. REV. 127, 154 (2019) (stating that they believe the deterrent "effect is likely to be minimal. [Highly automated vehicle] technology promises transportation and safety advances of such magnitude, and profits that will inevitably accompany these advances, that the threat of financial responsibility for [highly automated vehicle]-related losses is unlikely to deter research and development or the marketing of [highly automated vehicles] in any significant way.").

31. See Chapter 4 for a discussion on the benefits that automated vehicles are expected to deliver.

The USDOT provides recommended monetized values of crashes.[32] Each death is valued at $9,400,000, each critical injury is valued at $5,574,200, each severe injury $2,500,400, each serious injury $987,000, each moderate injury $441,800, and each minor injury $28,200. USDOT uses a monetary value of $3,927 per vehicle in property-damage-only crashes. If automated vehicles were to halve the number of fatalities from the 36,560 people who died in 2018 to 18,280 people, in deaths alone, the amount of potential liability that manufacturers would be incurring (or insuring for) would be $171.832 billion—even though the manufacturers themselves saved society that much by introducing the technology. There were approximately 2,746,000 people injured in car crashes and another 4,530,000 crashes that resulted in only property damage in 2017.[33] If automated vehicles halve the number to 1,373,000 injuries and 2,265,000 property-damage crashes, and assuming that each injury results in $75,000 of damages and each property-damage-only crash averages $5,000,[34] then the potential liability for the nonfatality crashes is approximately another $102.975 billion for crashes involving injuries and $11.325 billion for property-damage-only crashes. The total crash liability risk for manufacturers would be $286.132 billion. If they were able to spread the liability cost among 20,000,000 vehicles, the increased price for the liability risk would be $14,306.60 per vehicle per year. If manufacturers were able to spread these costs among 50,000,000 automated vehicles, then the average liability cost per year would be $5,722.64 a vehicle.

This simplistic, unscientific calculation should show that the liability risks for manufacturers imposed by such a regime are not insignificant. Society would undoubtedly be better off collectively if automated vehicles could halve crashes and fatalities. Yet liability for those crashes may cause manufacturers to delay releasing their vehicles until the costs become more manageable. Imposing liability on manufacturers preemptively or manufacturers believing that such liability will be imposed could cause manufacturers to delay their release of automated vehicles, potentially causing thousands more people to die and millions more crashes by human drivers.

---

32. *Tiger Benefit-Cost Analysis (BCA) Resource Guide*, U.S. Dep't of Transp., https://www.transportation.gov/policy-initiatives/tiger/tiger-benefit-cost-analysis-bca-resource-guide (last updated Apr. 2, 2015).

33. Nat'l Highway Traffic Safety Admin., *Traffic Safety Facts 2017, 2017 National Statistics*, U.S. Dep't of Transp., https://crashstats.nhtsa.dot.gov/Api/Public/ViewPublication/812806 (last visited May 11, 2020).

34. According to NHTSA, 39.2 percent of crashes involve a collision with another motor vehicle in transit, 29.9 percent involve a collision with a fixed object, 21.8 percent involve a collision with a nonfixed object (such as a pedestrian, parked motor vehicle, or animal), and 9.1 percent are noncollision crashes (such as a rollover). *Id.* at Table 29. If NHTSA values the average damage of each vehicle in a property-damage-only crash as $3,927, then we would expect that 39.2 percent of the crashes to have average damages of at least $7,854 ($3,927 x 2). There was no science in picking $5,000 except that it is probably a conservative number and works for the purposes of showing the amount of costs.

Additionally, even when manufacturers decide to deploy automated vehicles notwithstanding their liability risks, consumers may not appreciate that the higher price they are paying for automated vehicles includes coverage for crashes. This liability risk will make the sticker price for automated vehicles artificially higher than the sticker price for conventional vehicles, incentivizing purchases of (less safe) conventional vehicles. A liability shift would not eliminate the need for insurance, as insurance coverage extends beyond liability risk coverage to, among other things, uninsured motorist coverage and underinsured motorist coverage.

D. Manufacturers Are "Being Punished" for Saving Lives

Automated vehicles are expected to reduce the number of crashes significantly and, as a consequence, the number of deaths and injuries on our roadways. Level 4 and level 5 automated vehicles should be safer than level 0, level 1, level 2, and level 3 vehicles. The benefits will be mostly positive externalities that manufacturers are providing to society by innovating. However, manufacturer liability imposes a high cost and risk on manufacturers. There is an inherent unfairness to imposing significant liability on manufacturers after they, through innovation, provided substantial savings and benefits to society.

E. Limitations in Design

Manufacturer liability may slow down innovation and lead to overly conservative designs of automated vehicles. Manufacturers may seek to assert more control over motor vehicles and not allow consumers to modify or alter their vehicles. The lack of customization of vehicles may make innovation entirely dependent upon the automakers and their suppliers. Automated vehicles may also be designed not to speed and to be very deferential to other drivers and pedestrians. Ultraconservative driving behaviors may seem like a benefit of manufacturer liability, but they may reduce demand for automated vehicles and increase demand for conventional vehicles.

Manufacturer liability also creates significant barriers to entry into the automated vehicle industry and favors large companies. Because manufacturer liability may only be manageable with significant sales numbers, this regime may create insurmountable barriers for small manufacturers or new entrants. Manufacturer liability may also result in industry consolidation. Competition breeds innovation, and a liability regime that limits competition will hamper innovation.

F. Future Causes May Not Be Known or Static

Projecting the causes of future crashes is a difficult and trying endeavor. Many people believe that the primary cause of these automated vehicle crashes will be System Errors, but these beliefs may not be true. Other causes for automated vehicle crashes may be more pronounced than scholars suspect, and System Errors may be less significant than was otherwise thought. Crashes caused by inadequate maintenance, alterations, or shortcomings with infrastructure may become more pronounced. The primary cause could also be cyberattacks. Thus, to the extent that legislatures impose strict or absolute liability on automated vehicle manufacturers for what they perceive may be the cause of crashes, liability may eventually be imposed on the wrong party.

G. Manufacturer Liability Is Unnecessary for Safety

One of the main themes in favor of manufacturer liability is that such a regime is necessary and desirable to ensure that manufacturers only release safe automated driving systems and continue to update and monitor these systems. This underlying basis seems overstated as there are plenty of other, more significant reasons why manufacturers would release safe automated driving systems. Safety is important for manufacturers to establish their goodwill and reputation. Additionally, to the extent that the automated driving system is defective, manufacturers are under a duty to report such defects to NHTSA, and NHTSA has the power to force a manufacturer to recall its systems.[35]

H. When Does the Duty End?

One potential problem with manufacturer liability is determining when the manufacturer's obligations to maintain, patch, and update the automated vehicle software end. People currently own their vehicles for a long time. The average age of a vehicle on the road in 2016 was 11.6 years.[36] There were 264 million cars and light trucks (SUVs included) owned in 2016.[37] Sixty-two million of those vehicles were 16 years old or older, and IHS Markit predicted that number would grow to 81 million vehicles in 2021, with 20 million of those 81 million vehicles being more than 25 years old.[38] Given that many people predict automated vehicles—with their ability to smoothly brake and accelerate—to increase the mileage that a vehicle can drive, the longevity of ownership may also increase.[39]

---

35. See Chapter 5 for a thorough discussion of manufacturers' responsibilities with respect to safety defects and NHTSA's recall power.

36. *See Vehicles Getting Older: Average Age of Light Cars and Trucks in U.S. Rises Again in 2016 to 11.6 Years, IHS Markit Says* (Nov. 22, 2016), http://news.ihsmarkit.com/press-release/automotive/vehicles-getting-older-average-age-light-cars-and-trucks-us-rises-again-201.

37. *Id.*

38. *Id.*

39. Conversely, if motor vehicle use increases, vehicle age may decrease. See Chapter 4 for a discussion on the effects that automated vehicles will have on vehicle miles traveled.

Old automated vehicles create two problems for manufacturers. The first problem is that the automated vehicle hardware stack—the LiDAR, radars, sensors, GPSs, and other technology—may become outdated, or better, presently unknown technology may be developed. There is no effective way for manufacturers to compel updates to or replacements of the automated vehicle hardware stacks if they sell their vehicles to consumers. Although products liability law historically looks to the capabilities of hardware and technology as they existed at the point of sale, the use of software updates may muddy an otherwise straightforward analysis, and the outdated hardware may complicate software updates to the system. Additionally, manufacturers may desire to cycle out old hardware in favor of newer and better technology.

The second problem is the longevity of software support for a 30-year-old automated vehicle. Software developers have traditionally discontinued support for older software models. In 2014, Microsoft discontinued its immensely popular 12-year-old Windows XP operating system when more than 400 million computers worldwide still used Windows XP.[40] Other companies have discontinued software or ceased providing software updates for older versions of their products.[41]

With automated driving systems, a failure to protect software from new methods of cyberattacks or to fix bugs may result in significant bodily harm to the users and others on the roadway. The risk of this harm raises an important question: should manufacturers be able to discontinue software updates for their vehicles? The short answer is yes; they should be able to stop updating software. The more complicated answer is that it depends upon the marketing and warranties provided at the point of sale.

One benefit of allowing manufacturers to disable their automated driving system is that the requirement to update and maintain old software continually may create significant inefficiencies in the market. When many vehicles are operating under a similar system based on the design of the vehicle, the company can spread the costs of maintaining the software among many purchasers. If, for example, a great many purchasers only intend to use their vehicles for a certain period of years, say ten years, before they purchase an upgrade, and only a small number of users, say fewer than 1 percent, intend to use their vehicles for 30 years, then all of the users may pay, when purchasing the vehicle, for continued updating of services potentially forever, even though most people do not want to purchase the service for that long.

---

40. *How Can I Use Windows XP Safely Now It's No Longer Supported?*, GUARDIAN (Apr. 11, 2014), https://www.theguardian.com/technology/askjack/2014/apr/11/how-can-i-use-windows-xp-safely-now-its-no-longer-supported.

41. *Apple's iOS 11 Update Ends Support for iPhone 5 and 5C*, BBC (June 6, 2017), https://www.bbc.com/news/technology-40172019.

I. Statutes of Repose

A legal impediment in some states to a perpetual liability on manufacturers for crashes caused by level 4 and level 5 automated vehicles may be a statute of repose—a bar to a cause of action. Products liability statutes of repose are of two flavors.[42] Some states, approximately 17, bar causes of action after a period following the sale.[43] For example, Indiana's statute of repose provides that "a product liability action must be commenced . . . within ten (10) years after the delivery of the product to the initial user or consumer."[44] Some states, approximately seven, bar causes of action based on the product's useful life.[45] An example of the useful-life statute is Minnesota's useful-life statute, which provides that "it is a defense to a claim against a designer, manufacturer, distributor or seller of the product or a part thereof, that the injury was sustained following the expiration of the ordinary useful life of the product."[46] In those states that impose a time-based statute of repose, manufacturer liability may be prohibited beyond the statute of repose and result in a shift of liability back to the vehicle user or owner.

J. Lack of Control over Sales, Repairs, and Alterations

Almost all states generally prohibit traditional manufacturers from selling vehicles directly to consumers.[47] The Magnuson-Moss Act and state law constrain manufacturers' abilities to control repairs and parts used to repair automated vehicles. Under the Magnuson-Moss Act, manufacturers cannot condition coverage under a warranty on the consumer's use of specific parts or services, unless they freely provide such parts or services.[48] Practically

---

42. David G. Owen & Mary J. Davis, Owen & Davis on Products Liability § 14:13 *Statutes of Limitations and Repose—Types of Repose Statutes* (4th ed. 2020).

43. *Id.*

44. Ind. Code Ann. § 34-20-3-1(b). The statute of repose applies to products liability actions that sound in negligence or strict liability. *Id.* § 34-20-3-1(a).

45. Owen & Davis, *supra* note 42, at § 14:13.

46. Minn. Stat. Ann. § 604.03. The statute provides the following factors that juries and courts consider in determining the useful life of products:

> The useful life of a product is not necessarily the life inherent in the product, but is the period during which with reasonable safety the product should be useful to the user. This period shall be determined by reference to the experience of users of similar products, taking into account present conditions and past developments, including but not limited to (1) wear and tear or deterioration from natural causes, (2) the progress of the art, economic changes, inventions and developments within the industry, (3) the climatic and other local conditions peculiar to the user, (4) the policy of the user and similar users as to repairs, renewals and replacements, (5) the useful life as stated by the designer, manufacturer, distributor, or seller of the product in brochures or pamphlets furnished with the product or in a notice attached to the product, and (6) any modification of the product by the user.

47. See Chapter 6 for a discussion on franchise laws.

48. 16 C.F.R. § 700.10. The Magnuson-Moss Act is described in Chapter 5.

speaking, there may not be a feasible way for manufacturers to control repairs and maintenance for vehicles owned by consumers. If automated vehicles' safety depends upon specific parts or services, it seems inherently unfair and unreasonable to impose liability on a manufacturer that cannot control those parts or services.

III. AUTOMATED VEHICLES AS A SUBSCRIPTION, LEASE, OR LICENSE

Given the complications that exist when manufacturers bear an increased share of liability, it seems that manufacturers may seek to avoid those negative effects by changing their personal-ownership model for motor vehicles. Simply stated, problems created by a lack of control over modifications, alterations, repairs, and maintenance of automated vehicles; outdated hardware and software; and potentially limitless support obligations could seemingly be limited or avoided if manufacturers decline to sell automated vehicles outright to consumers.

Manufacturers could gain control over the vehicle by offering their automated vehicles or the automated driving system (in dual-use vehicles) as a subscription, lease, or license. Under this model, the manufacturer would be able to contractually obligate certain maintenance and repairs, prohibit modifications and alterations, and control their fleet of automated vehicles. Once the term of the agreement expires, the manufacturer could reobtain possession of the automated vehicle. The manufacturer could then assess (or upgrade) the technology stack and the software before releasing the automated vehicle for a new subscription, lease, or license. Manufacturers could also build insurance coverage into the cost of the subscription, lease, or license.

Maintaining this type of control over automated vehicles would make roadways safer. Manufacturers could replace outdated automated vehicle stacks frequently so that automated vehicles utilize the best technology. Another advantage is that by removing older motor vehicles from the roadway periodically, worn-out equipment could also be repaired and under the control of the manufacturer. Many crashes occur each year due to worn-out equipment—most commonly tires.[49]

---

49. According to NHTSA, in 2017, 738 motor vehicle traffic fatalities resulted from tire-related crashes. *See* Nat'l Highway Traffic Safety Admin., *Tires*, U.S. Dep't of Transp., https://www.nhtsa.gov/equipment/tires (last visited Jan. 24, 2020).

Chapter 12

Immunity for Level 4 and Level 5 Crashes

The potential problems and complications with manufacturer liability, as well as the ambiguity of finding an at-fault party and the costs incurred in litigating these cases, have caused some scholars to suggest an immunity and compensation system for crashes caused by highly automated vehicles. The potential problems answer the question of *why* an immunity and compensation system may be needed; this chapter explores the question of *how* to structure an immunity and compensation system and canvasses examples suggested in scholarship. An immunity and compensation system broadly consists of four main determinants: (1) which level of government should grant immunity, (2) when the immunity and compensation system should be enacted, (3) the coverage of the immunity and compensation system, and (4) compensation.

Scholars who have considered an immunity system for automated vehicles have argued that an immunity system is only viable if created by the federal government.[1] Immunity in one state but not in other states would mean that society could not reap the benefits of the immunity system. Additionally, any immunity and compensation system may be cost prohibitive for an individual state, especially states with a smaller pool of persons who pay into the system. A state-run system would also need to address the "free rider" problem of persons who use the state's roads but may not live in the state or otherwise pay into the tax base.

1. *See, e.g.*, Kenneth S. Abraham & Robert L. Rabin, *Automated Vehicles and Manufacturer Responsibility for Accidents: A New Legal Regime for a New Era*, 105 VA. L. REV. 127, 148–49 (2019).

Scholars generally agree that governments should only create an immunity and compensation system for crashes involving levels 4 and 5 automated vehicles.[2] Within this category, the immunity proposed has focused on the potential liabilities for the manufacturers of automated vehicles and automated driving systems. Vehicle owner or vehicle user immunity has not been popular, although lawsuits involving highly automated vehicles may involve these parties, too. Some scholars have carved out certain claims against manufacturers that should not be subject to immunity, including crashworthiness,[3] warning defects,[4] and punitive damage allegations.[5] Scholars have differed on what injuries or harms should be compensable by their immunity and compensation system. Some proposals exclude property damage and suggest that property damage—by far the most common type of claim for automobile crashes—could be handled through other forms of insurance.[6]

Another important question for an immunity and compensation system is at what point in the lifecycle of the automated vehicle industry the federal government should intervene to create the system. Most scholars who advocate for immunity argue that the federal government should preemptively intervene before automated vehicles become available for commercial use and sale.[7] Another method has been to suggest implementing an immunity and compensation system when automated vehicles reach a certain market share, such as 25 percent of registered vehicles.[8] These scholars do not consider that the actual need for an immunity system may not depend upon preemptive action or market share but rather whether the industry, at such time, needs an immunity and compensation system. As described below, Congress enacted existing immunity and compensation systems to address market instability or to foster market entry in the face of unknown or unquantifiable risks. Given that highly automated vehicles have not yet been deployed for commercial sale and use at any meaningful level, whether the horrors of liability or the problems addressed in Chapter 11 become a reality is a story still being written. Additionally, the risks of automobile crashes are known and quantifiable, and

2. *See, e.g.*, Tracy Hresko Pearl, *Compensation at the Crossroads: Autonomous Vehicles & Alternative Victim Compensation Schemes*, 60 Wm. & Mary L. Rev. 1827, 1878 (2019); Antonio Davola, *A Model for Tort Liability in a World of Driverless Cars: Establishing a Framework for the Upcoming Technology*, 54 Idaho L. Rev. 591, 596 (2018).

3. Matthew Blunt, *Highway to a Headache: Is Tort-Based Automotive Insurance on a Collision Course with Autonomous Vehicles?*, 53 Willamette L. Rev. 107, 132 (2017).

4. Kevin Funkhouser, *Paving the Road Ahead: Autonomous Vehicles, Products Liability, and the Need for a New Approach*, 2013 Utah L. Rev. 437, 461–62 (2013).

5. Abraham & Rabin, *supra* note 1, at 159.

6. *See, e.g.*, Pearl, *supra* note 2, at 1878–79; Abraham & Rabin, *supra* note 1, at 152.

7. *See, e.g.*, Pearl, *supra* note 2, at 1891 ("Given that the existing tort system will always remain a fallback option, there appear to be few reasons not to give a victim compensation fund a chance and many reasons to design and implement one before fully autonomous cars come to market.").

8. Abraham & Rabin, *supra* note 1, at 132.

the insurance industry has already shown that it is capable of dealing with millions of these crashes each year. Without a concrete idea of when manufacturers will sell or use significant numbers of automated vehicles commercially, it may be too early to preemptively call for and design immunity and compensation systems for problems that have not yet manifested.[9]

This chapter describes numerous proposals scholars have suggested for immunity and compensation systems. It discusses five different categories of proposals for an immunity and compensation system. Scholars mostly propose immunity and compensation systems based on existing immunity systems, such as no-fault insurance, the National Vaccine Injury Compensation Program (the NVICP), and the Price-Anderson Nuclear Industries Indemnity Act (the Price-Anderson Act). Other scholars have used the Federal Insurance Contribution Act (FICA) as a model or created a new form of immunity, such as the manufacturer enterprise responsibility system.

I. NO-FAULT INSURANCE

No-fault insurance, with its current use for automobile crashes in some states, makes sense to consider as an option for an immunity and compensation system for highly automated vehicles.[10] There are generally three variations of no-fault insurance: pure, hybrid, and choice.[11] In a pure system, those harmed by an automobile crash must seek recovery under their insurance policies. No jurisdiction in the United States utilizes a pure no-fault system. Most no-fault states in the United States use a hybrid no-fault regime. In a hybrid regime, those harmed by an automobile crash recover under their no-fault insurance policies and can, if damages exceed a certain threshold, seek recovery in tort. In some respects, the ability to recover in tort may be curtailed by statute, such as Michigan's limitation on the ability of litigants to recover noneconomic damages except in some instances.[12] The choice system merely gives motorists the ability to purchase no-fault insurance in addition to proceeding through the conventional tort system.

No-fault has been discussed as a potentially popular choice to handle automated vehicle crashes. One proposal suggests a no-fault immunity to protect manufacturers from the risk of liability for crashes caused by automated

---

9. *See generally* Jeffrey K. Gurney, *Imputing Driverhood: Applying a Reasonable Driver Standard to Accidents Caused by Autonomous Vehicles*, *in* ROBOT ETHICS 2.0: THE ETHICAL AND SOCIAL IMPLICATIONS OF ROBOTICS 58–59 (Patrick Lin, Keith Abney & George A. Bekey eds., 2017).

10. Blunt, *supra* note 3.

11. See generally Gary T. Schwartz, *Auto No-Fault and First-Party Insurance: Advantages and Problems*, 73 S. CAL. L. REV. 611 (2000), for a discussion of the types of no-fault insurance.

12. *See* MICH. COMP. LAWS § 500.3135.

vehicles.[13] Under this proposal, those harmed by automated vehicle crashes would seek recovery under a federal public insurance corporation instead of private insurers.

II. THE NVICP MODEL

The NVICP model has been popular among those who support an immunity and compensation system.[14] Congress enacted the National Childhood Vaccine Injury Act (the Vaccine Act) in 1986 to address concerns over potentially ruinous lawsuits that vaccine manufacturers were facing. These lawsuits caused some manufacturers to threaten to stop producing vaccines and leave the industry. The Vaccine Act created the NVICP as an alternative to medical malpractice and products liability lawsuits when vaccines harm people.

The NVICP does not provide broad immunity to vaccine manufacturers and only covers a limited number of vaccines. Litigants must first present their claims in the NVICP before resorting to traditional litigation. The injured party files a petition with the United States Court of Federal Claims. The plaintiff must identify his or her claim as a table claim—an injury listed on the Vaccine Injury Table—or an off-table claim—one that is not on the Vaccine Injury Table. Off-table claims require a higher standard of proof and evidence of causation. Claims are paid from the Vaccine Trust Fund, which is funded by a 75-cent tax imposed on the purchase of covered vaccinations. A plaintiff who is dissatisfied with the results in the NVICP can bring a traditional lawsuit, but manufacturers receive immunity from causes of action based on the "unavoidable side effects" of vaccinations (if they provide proper directions and warnings) and from failing to provide direct warnings.[15]

An immunity and compensation fund comparable to the NVICP was proposed by Professor Tracy Hresko Pearl to cover damages for human injuries and fatalities, but not property damage, caused by level 4 and level 5 automated vehicles. The system would be funded by a tax imposed on manufacturers and consumers of these automated vehicles—Professor Pearl suggests a tax of approximately $1,000 to be split between the manufacturer and the purchaser. NHTSA would administer the compensation fund. Victims of highly automated vehicle crashes would need to present their claims to the fund before resorting to the traditional tort system. Following NHTSA's resolution of the claim, the victim would have a choice to accept the offer and waive rights to litigate the case or reject the offer and litigate the case in the traditional

13. Blunt, *supra* note 3, at 132.
14. *E.g.*, Pearl, *supra* note 2; Funkhouser, *supra* note 4.
15. 42 U.S.C. § 300aa-22(b).

tort system. In exchange for the protections offered by the immunity and compensation system, manufacturers would need to "participate in a data-sharing and design improvement program administered by NHTSA." This participation would help NHTSA in regulating automated vehicles and help the system incentivize safety.

The analogy of automated vehicles to the NVICP has two potential shortfalls. First, there is no indication that the automated vehicle industry will face the ruinous liability of the vaccine industry. Every vaccine has unavoidable side effects that can create significant liability risks, as the vaccine industry *actually* experienced. Automated vehicles will, of course, have bugs and software errors, but not all of those bugs and software errors will result in harm before manufacturers patch them, and a single crash due to a bug or error will not bring ruin to any significant manufacturer and will likely be covered by insurance. Second, society has no alternative to vaccines. If automated vehicles create ruinous liability, there are alternative means of transportation, including human-driven vehicles, public transit, or bicycles. Society is not forced to accept particularly risky automated vehicles that would lead to runaway liability.

III. THE PRICE-ANDERSON ACT MODEL

Scholars have proposed a Price-Anderson Act model for granting immunity to highly automated vehicle crashes.[16] Congress enacted the Price-Anderson Act in 1957 to address a liability "roadblock" for the private development of nuclear power plants and to ensure that those who are harmed by nuclear power plants would have a means to recovery.[17] The Price-Anderson Act effectuates its dual purpose by requiring nuclear reactors to obtain a first-tier private insurance policy at an amount set by the Atomic Energy Commission.[18] The Atomic Energy Commission has currently set this amount at $450 million.[19] In the event of an accident that has damages over $450 million, a second tier applies to provide damages. The second tier requires each nuclear reactor to be assessed a prorated share of the excess up to $131.056 million per reactor.

16. *See* Kyle Colonna, *Autonomous Cars and Tort Liability*, 4 CASE W. RESERVE J.L. TECH. & INTERNET 81 (2012) (arguing that Congress should model immunity of automated vehicle liability on the Price-Anderson Act).

17. See Harold P. Green, *Nuclear Power: Risk, Liability, and Indemnity*, 71 MICH. L. REV. 479 (1973), for an excellent overview of the Price-Anderson Act.

18. 42 U.S.C. § 2210.

19. *Backgrounder on Nuclear Insurance and Disaster Relief*, U.S. Nuclear Regulatory Comm'n, https://www.nrc.gov/reading-rm/doc-collections/fact-sheets/nuclear-insurance.html (last updated May 3, 2019).

As of April 2019, there were 98 reactors, so this second tier provides coverage up to $12.9 billion.[20]

Under this proposal, manufacturers would obtain a base liability insurance policy that covers crashes. Under one proposal, each highly automated vehicle manufacturer and technology manufacturer would fund the second tier by paying a pro rata share based "on what the manufacturer produces, its predicted revenue, and the amount of risk it is predicted to incur in the future."[21] Manufacturers of highly automated vehicles and highly automated technologies would register with the government to become eligible for the protections.[22]

There are two critical differences between the nuclear reactor industry and the automated vehicle industry that make the Price-Anderson Act more appropriate for the former industry than for the latter: frequency and catastrophe. Nuclear power plant incidents are rare, and when they do occur, they are catastrophic and can potentially, from an insurance standpoint, create unquantifiable risks. Automated vehicle crashes could be a daily or weekly occurrence, but the people impacted are known and quantifiable for insurance companies. Additionally, an immunity that would require safer automated vehicle companies to backstop unsafe automated vehicle companies seems misplaced and could lead to a decrease in overall safety. The second and more significant problem is that any one nuclear meltdown can cripple a company, while any one automated vehicle crash should not cripple the manufacturer or an insurer. Because of the noncatastrophic nature of an automated vehicle crash in comparison to a nuclear meltdown, the risk of an automated vehicle crash should be insurable, while the risk of a nuclear meltdown was not fully insurable.

IV. NATIONAL CAR INSURANCE FUND SIMILAR TO FICA

FICA is the funding mechanism for Medicare and Social Security. Taxpayers, along with their employers, make contributions (i.e., pay a tax) each pay period to the Medicare and Social Security trust funds. The trust funds are available to qualifying recipients. The Social Security Administration oversees Social Security claims and Social Security disability insurance claims, while the Department of Health and Human Services oversees Medicare.

---

20. *Id.*
21. Colonna, *supra* note 16, at 125.
22. *Id.* at 125–26.

One scholar has suggested that a federal insurance fund similar to FICA be used to compensate those harmed by automated vehicles.[23] Under this proposal, NHTSA would operate a National Car Insurance Fund whereby NHTSA would receive claims, assess the claims, and set the tax rate to be assessed to fund the National Car Insurance Fund. The National Car Insurance Fund would be financed by taxes assessed on users and owners of highly automated vehicles and manufacturers. NHTSA would set the tax rate for the parties based on their safety record.

V. MANUFACTURER ENTERPRISE RESPONSIBILITY

Professors Kenneth Abraham and Robert Rabin suggested a system they call Manufacturer Enterprise Responsibility or MER.[24] They suggest that the federal government should implement the MER system when automated vehicles account for 25 percent of the vehicles on the road—though they are flexible on that percentage. MER is a "manufacturer-financed, strict responsibility bodily injury compensation system, administered by a fund created through assessments levied on [level 4 and level 5 automated vehicle] manufacturers."[25]

The MER system would compensate persons for bodily injuries that "arise out of the operation" of level 4 and level 5 automated vehicles.[26] Although Professors Abraham and Rabin do not believe that their numbers are "set in stone," they suggest that the MER system provide unlimited medical benefits, $1 million in compensation for wage losses, and an additional $500,000 for specified long-term or permanent injuries.[27] The MER system would cover any defects in highly automated vehicles, such as defective brakes or an uncrashworthy vehicle.[28] The MER system would exclude property damage suffered by any party.[29] Professors Abraham and Rabin suggest that the exclusion of property damage would not create a "hardship or unfairness" because anyone could purchase insurance that covers property damage.[30] Thus, under the MER system, recovery for property damage would resemble a no-fault system, and persons harmed by automated vehicles would need to ensure that they have insurance policies that cover no-fault harm. The MER system would

23. *See* Carrie Schroll, *Splitting the Bill: Creating a National Car Insurance Fund to Pay for Accidents in Autonomous Vehicles*, 109 Nw. U. L. Rev. 803 (2015).
24. Abraham & Rabin, *supra* note 1.
25. *Id.* at 147.
26. *Id.* at 151.
27. *Id.* at 149, 160–61.
28. *Id.* at 156.
29. *Id.* at 152.
30. *Id.*

also not cover the negligence of the vehicle owner and conceivably users of automated vehicles.[31]

Manufacturers of level 4 and level 5 automated vehicles would finance the MER system.[32] Initially, each manufacturer's pro rata share of financing would be based on market share until adequate data can be collected about the relevant frequency and amount of payouts by manufacturers. Eventually, funding of the MER system would be targeted according to the frequency and amount of payouts by manufacturers.

The MER system would be an almost exclusive remedy for all crashes arising out of the operation of level 4 and level 5 automated vehicles.[33] The only three exceptions that Professors Abraham and Rabin suggest to this exclusive remedy are claims against vehicle manufacturers and component manufacturers for punitive damages, third-party harm (such as through cyberattacks), or crashes involving owner-modified automated vehicles.[34] Professors Abraham and Rabin do not describe how the punitive damage exception would apply in practice, leaving a lot of important questions unanswered. Can those harmed seek recovery under the MER system and then file a products liability suit alleging punitive damages, or do they need to choose the traditional liability system or the MER system? If the former, can persons harmed recover in whole through the civil liability system, including for property damages and bodily injury recovery in excess of the MER system? How will courts determine whether the punitive damage award is excessive if the person receives only limited compensation under the MER system? Is pleading punitive damages in a complaint sufficient to seek recovery outside of the MER system?

Professors Abraham and Rabin also discuss mixed crashes—those involving level 4 or level 5 automated vehicles and all other vehicles. For these crashes, they argue that the MER system should be applicable both when the automated vehicle causes harm and when a human driver causes harm using a non–level 4 or level 5 automated vehicle.[35] The abolishment of the tort system for crashes caused by human drivers calls for absolute liability on manufacturers of automated vehicles. This absolute (or categorical) liability for manufacturers seems strange considering the exclusion of property damages from the MER system—Professors Abraham and Rabin excuse manufacturers from one type of harm where fault is seemingly more applicable and replace it with an absolute liability unseen in tort law.

---

31. *Id.* at 151–52, n.52.
32. *Id.* at 169.
33. *Id.* at 156–58.
34. *Id.* at 159–60.
35. *Id.* at 165–69.

Chapter 13

Automated Rental Vehicles

Companies like Hertz, Enterprise, and Budget are expected to rent automated vehicles. This chapter explores new liability risks created by renting automated vehicles. The major liability protection for car rental companies is the Graves Amendment, and this chapter details Graves Amendment immunity. The chapter concludes by reviewing whether automated car-sharing or ridesharing companies can claim Graves Amendment protection.

Before August 2005, some states imposed liability on car rental companies for their customers' negligence under doctrines such as dangerous instrumentality or vicarious liability. These doctrines impose liability on a vehicle owner for the negligence of the permissible user of the vehicle. In August 2005, within the Safe, Accountable, Flexible, Efficient Transportation Equity Act: A Legacy for Users, Congress enacted an amendment proposed by Representative Samuel Graves to preempt state law common law claims that impose liability on car rental companies and lessors of motor vehicles for the negligence of their customers or lessees.[1] This so-called Graves Amendment provides, in pertinent part, as follows:

1. "The legislative history of the Graves Amendment indicates that it was intended to 'protect the vehicle rental and leasing industry against claims for vicarious liability where the leasing or rental company's only relation to the claim was that it was the technical owner of the vehicle.'" Cioffi v. S.M. Foods, Inc., 129 A.D.3d 888, 893, 10 N.Y.S.3d 620 (N.Y. App. Div. 2015) (quoting Rein v. CAB E. LLC, No. 08 CIV. 2899PAC, 2009 WL 1748905, at *1 (S.D.N.Y. June 22, 2009)); Jacob D. Walpert, *Carpooling Liability?: Applying Tort Law Principles to the Joint Emergence of Self-Driving Automobiles and Transportation Network Companies*, 85 Fordham L. Rev. 1863, 1876 (2017).

> (a) In general.—An owner of a motor vehicle that rents or leases the vehicle to a person (or an affiliate of the owner) shall not be liable under the law of any State or political subdivision thereof, by reason of being the owner of the vehicle (or an affiliate of the owner), for harm to persons or property that results or arises out of the use, operation, or possession of the vehicle during the period of the rental or lease, if—
>
> (1) the owner (or an affiliate of the owner) is engaged in the trade or business of renting or leasing motor vehicles; and
>
> (2) there is no negligence or criminal wrongdoing on the part of the owner (or an affiliate of the owner).[2]

A defendant seeking Graves Amendment protection must show that it (1) owns the vehicle or is an affiliate of the owner, (2) is in the business of renting or leasing motor vehicles, and (3) is not (and its affiliates are not) negligent or has not committed criminal wrongdoing. For traditional car rental companies, they easily satisfy the first two elements, in that they are the owners of their vehicles and are involved in the trade or business of renting or leasing motor vehicles. These two elements warrant further inquiry for car-sharing companies or automated ridesharing companies. This inquiry is undertaken at the end of this chapter. The question of immunity for traditional car rental companies essentially boils down to whether they are free from negligence or criminal wrongdoing.

I. CLAIMS AGAINST TRADITIONAL CAR RENTAL COMPANIES

Congress intended the Graves Amendment to protect car rental companies from a specific perceived problem: liability imposed on car rental companies when their customers caused the underlying harm. The user's responsibility for an automated vehicle crash depends, in part, on the level of driving automation system. For levels 1, 2, and 3 driving automation systems, the driver (or user of the level 3 automated driving system) can be negligent in a crash when the automation feature is engaged.[3] For crashes involving level 4 and level 5 automated driving systems, the user of the vehicle may not be responsible for causing the crash, a topic described at length in Chapter 11. Level 4 and level 5 automated vehicle crashes may put pressure on the applicability of Graves Amendment immunity because litigants would not be seeking to impose

2. 49 U.S.C. § 30106(a).

3. See Chapters 9 and 10 for a discussion of liability for level 2 and level 3 driving automation systems.

liability on car rental companies based on the vicarious liability of their customers. This section discusses whether the Graves Amendment would shield car rental companies from three broad categories of claims that litigants may assert: software defects or bugs, maintenance, and instructions.

A. Software Defects or Bugs

In the first category of claims, litigants may seek to hold car rental companies liable when their vehicles have software defects or bugs that cause the crashes. This type of claim would be a products liability claim. The underlying question becomes whether car rental companies can be liable in products liability and, if so, whether the Graves Amendment shields them from such liability.

Courts have long held that car rental companies can be subject to products liability for harm caused by defects in vehicles they rent to consumers.[4] The Products Liability Restatement provides the following general rule for the defendants liable for product defects: "One engaged in the business of selling or otherwise distributing products who sells or distributes a defective product is subject to liability for harm to persons or property caused by the defect."[5] Section 20 of the Products Liability Restatement specifically includes lessors: "One otherwise distributes a product when, in a commercial transaction other than a sale, [it] provides the product to another either for use or consumption or as a preliminary step leading to ultimate use or consumption. Commercial nonsale product distributors include, but are not limited to, lessors, bailors, and those who provide products to others as a means of promoting either the use or consumption of such products or some other commercial activity."[6] Although plaintiffs can assert products liability actions against car rental companies, some state statutes may limit or reduce that exposure.[7]

The second question is whether the Graves Amendment could shield these car rental companies from liability for products liability actions. Courts have held that the Graves Amendment does not act as a shield against products

---

4. *See* American Law of Products Liability 3d Treatise, § 36:19. *Applicability of Doctrine to Leases and Bailments* (Feb. 2020) ("Most jurisdictions that have adopted the doctrine of strict liability in products liability have also, when faced with the question, applied the doctrine to commercial lessors or bailors of products, or to lease transactions which are commercial in nature and are not isolated cases."); *see, e.g.*, Cintrone v. Hertz Truck Leasing & Rental Serv., 45 N.J. 434, 212 A.2d 769 (1965); Coleman v. Hertz Corp., 1975 OK CIV APP 5, 534 P.2d 940 (Okla. Ct. App. 1975); Stewart v. Budget Rent-A-Car Corp., 52 Haw. 71, 470 P.2d 240 (1970); Stang v. Hertz Corp., 83 N.M. 730, 497 P.2d 732 (1972), *abrogated on other grounds by* Livingston v. Begay, 98 N.M. 712, 652 P.2d 734 (1982); Price v. Shell Oil Co., 2 Cal. 3d 245, 466 P.2d 722 (1970).

5. Restatement (Third) of Torts: Prods. Liab. § 1 (1998).

6. *Id.* § 20(b).

7. See Chapter 8 for a discussion of state law immunity granted to nonmanufacturer distributors.

liability claims.[8] Thus, if highly automated vehicles create a shift from driver negligence actions to products liability actions, the Graves Amendment may no longer be a protective shield for car rental companies.

B. Negligent Maintenance

The second category of claims will be ones where a failure to maintain the vehicle caused the crash. These claims could be related to basic mechanical maintenance, dirty sensors, or a failure to update vehicle software. Since the enactment of the Graves Amendment, negligent maintenance claims have been a popular means used by plaintiffs to sue car rental companies.[9]

Generally, in canvassing the applicability of these types of claims, some courts have begun their analysis with the lease or rental agreement.[10] For example, one court, after noting that the savings clause in the Graves Amendment permits claims of negligent maintenance, held that the plaintiffs, in that case, could not assert a negligent maintenance claim against the lessor because the lease agreement required the lessee, not the lessor, to maintain the vehicle.[11] Thus, at least for one court, the applicability of a negligent maintenance claim may be within the control of the lessor, and the lessor can weigh whether, for business or other reasons, it makes sense to limit its maintenance duties in the contract.

C. Failure to Instruct

The final broad category of claims involves those that relate to a car rental company's failure to instruct its customers on how to properly use the driving

---

8. *See, e.g.*, Hack v. SAI Rockville L, LLC, No. CIV. WDQ-14-1985, 2015 WL 795853, at *4 (D. Md. Feb. 24, 2015); Hagen v. U-Haul Co. of Tenn., 613 F. Supp. 2d 986, 1000 (W.D. Tenn. 2009).

9. *See, e.g.*, Novovic v. Greyhound Lines, Inc., No. CV-08-3190 (CPS), 2008 WL 5000228, at *3 (E.D.N.Y. Nov. 19, 2008) ("Plaintiffs . . . allege that the injuries suffered by Novovic were due in whole or in part to the negligence of defendant MCI in the maintenance of the Greyhound bus. This claim renders the Graves Amendment inapplicable."); Dubose v. Transp. Enter. Leasing, LLC, No. 608-CV-385-ORL-31DAB, 2009 WL 210724, at *4 (M.D. Fla. Jan. 27, 2009) (noting that "§ 30106(a)(2) only appears to apply to claims predicated on criminal wrongdoing and negligent maintenance claims. . . ."); Colon v. Bernabe, No. 07 CIV. 3369 AJP, 2007 WL 2068093, at *4 (S.D.N.Y. July 19, 2007) (a claim based on negligent maintenance is not frivolous); Berkan v. Penske Truck Leasing Canada, Inc., 535 F. Supp. 2d 341, 345 (W.D.N.Y. 2008) (considering a negligent maintenance claim but finding that no evidence was adduced to prove negligent maintenance); *Hagen*, 613 F. Supp. 2d at 1000 ("The Plaintiffs assert theories of direct liability against the Defendants, such as negligence, failure to warn, negligent repair, etc. Such claims would qualify as 'negligence . . . on the part of the owner (or an affiliate of the owner)' and are explicitly exempted under the statute." (quoting 49 U.S.C. § 30106(a)(2))).

10. *See, e.g.*, *Dubose*, 2009 WL 210724, at *1; *Colon*, 2007 WL 2068093, at *4; Cukoviq v. Iftikhar, 169 A.D.3d 766, 768, 93 N.Y.S.3d 710, 712 (N.Y. App. Div. 2019).

11. *Dubose*, 2009 WL 210724, at *5.

automation system and a failure to warn of the system's limitations.[12] Historically, a car rental company could feel comfortable that its customers know how to use vehicles rented to them. Cars are relatively standard in design and operation. A customer who owns a vehicle manufactured by General Motors can drive a vehicle made by Ford.

Automation features affect that uniformity in a meaningful way. For vehicles equipped with level 2 driving automation systems, driver monitoring systems and operational design domains may differ among manufacturers (or even among different brands of the same manufacturer). For level 3 automated vehicles, notification systems, reengagement methods, and operational design domains may differ. For level 4 automated vehicles, reengagement methods and operational design domains may differ. Level 5 automated vehicles may have different internal configurations, such as for seating or sleeping, that require additional instructions or have different means for a user to become a driver. Consumers may also not intuitively understand their maintenance requirements concerning the general cleanliness of sensors and other necessary hardware. The differences between conventional vehicles and automated vehicles, as well as the differences among automation features and different vehicle models, create new duties and obligations for car rental companies. Indeed, these obligations may be more significant in the infancy of the technology than when the technology becomes ubiquitous.

II. APPLICABILITY OF THE GRAVES AMENDMENT TO CAR-SHARING AND RIDESHARING

Companies have deployed two relatively new styles of rental companies: (1) car-sharing companies and (2) transportation network companies (TNCs). This section describes the applicability of the Graves Amendment to these companies.

12. Though one of the big problems with liability here would be whether customers would even read the warnings or instructions. Additionally, if these warnings and instructions are provided by the manufacturer, an additional issue is whether courts would impose a redundant obligation on the car rental company. Courts have historically found that dealers do not owe a duty to provide more warnings and instructions than are provided in the owner's manual. *See, e.g.*, Ford Motor Co. v. Rushford, 868 N.E.2d 806, 811 (Ind. 2007) ("[A]bsent special circumstances, if the manufacturer provides adequate warnings of the danger of its product and the seller passes this warning along to the buyer or consumer, then the seller has no obligation to provide additional warnings."); Abramowitz v. Maghsoudlou, 295 A.D.2d 548, 548, 745 N.Y.S.2d 36 (2002) (finding that a car dealership had no duty to warn against the dangers associated with driving a vehicle with different-sized tires and, even if it did, the duty was not breached because the owner's manual contained such warnings); Farmer v. Brannan Auto Parts, Inc., 231 Ga. App. 353, 498 S.E.2d 583, 585 (1998) (holding that a part distributor "had no duty to communicate to users a danger already clearly listed on the product itself").

A. Car-Sharing Companies

Car-sharing companies, such as ZipCar, Car2Go, and Turo, offer their members an opportunity to rent a vehicle by the hour, by the day, or for an extended time. For example, a ZipCar membership costs $7 per month and entitles the member to use vehicles located throughout specific cities for a fee.[13] According to ZipCar's website, rental of a Honda Civic costs approximately $9 per hour and $74 per day through ZipCar. The rental fee includes insurance and gas for the trip. Because car-sharing companies resemble car rental companies, courts have held that these car-sharing companies are entitled to protection under the Graves Amendment.[14] This conclusion should not be changed for automated vehicles.

B. Automated Ridesharing Companies

Currently, the main distinction between a car rental company and a TNC is that the car rental company allows a driver to borrow a vehicle to drive for a period of time, while a TNC connects a person who needs a ride with a driver for a specific trip.[15] In the future, the line between car rental companies and TNCs may begin to blur for highly automated vehicles. Companies could deploy automated ridesharing services under one of two models that necessitate separate legal analyses. A ridesharing company could own a fleet of automated vehicles and use those vehicles to operate its ridesharing services. For these companies, the only distinction between a traditional car rental company and the ridesharing company will (likely) be the length of use of the vehicle—a distinction that may not matter legally. The Graves Amendment provides a protective shield for owners of vehicles that are "engaged in the trade or business of renting or leasing motor vehicles."[16] A ridesharing company that owns a fleet of automated vehicles and rents its vehicles for short-term use should be able to claim Graves Amendment immunity.[17]

The other approach would be an automated ridesharing company that does not own the automated vehicles but merely connects riders with other people's vehicles. This approach would be similar to the traditional TNC. Under this alternative model, the Graves Amendment analysis becomes murkier, and two different defendants exist: the ridesharing company and the car owner.

---

13. *Zipcar Membership Plans*, ZIPCAR, https://www.zipcar.com/pricing (last visited Nov. 7, 2019).

14. *See, e.g.*, Moreau v. Josaphat, 42 Misc. 3d 345, 975 N.Y.S.2d 851, 859 (Sup. Ct. 2013).

15. *See* Agnieszka A. McPeak, *Sharing Tort Liability in the New Sharing Economy*, 49 CONN. L. REV. 171, 181 (2016).

16. 49 U.S.C. § 30106(a)(1).

17. *See* Vargas v. Enter. Leasing Co., 60 So. 3d 1037, 1042 (Fla. 2011) (holding that the Graves Amendment protects short-term lessors).

The term "owner" is defined broadly for purposes of the Graves Amendment to mean a person who is

> (A) a record or beneficial owner, holder of title, lessor, or lessee of a motor vehicle;
> (B) entitled to the use and possession of a motor vehicle subject to a security interest in another person; or
> (C) a lessor, lessee, or a bailee of a motor vehicle, in the trade or business of renting or leasing motor vehicles, having the use or possession thereof, under a lease, bailment, or otherwise.[18]

Under this broad definition, a ridesharing company would be an "owner" so long as it has an agreement with the vehicle owner that grants the company the right to use or possess the automated vehicle. The automated vehicle owner, as the holder of the title, would be an owner, too.

The second element of Graves Amendment immunity requires the owner to be "engaged in the trade or business of renting or leasing motor vehicles."[19] This issue has rarely been litigated.[20] Typically, to dismiss a case based on the Graves Amendment, the entity claiming its protection must submit sworn affidavits attesting to this element.[21] The automated ridesharing company should be able to establish that it is in the trade or business of renting or leasing motor vehicles. Car owners, on the other hand, may struggle to establish this element when they passively allow companies to use their automated vehicles; they may need to rely on contractual provisions, like indemnification from the ridesharing companies.

18. 49 U.S.C. § 30106(d)(2).
19. *Id.* § 30106(a)(1).
20. An example of a case where a motion to dismiss was not granted on this element is *Murphy v. Pontillo*, 12 Misc. 3d 1146, 1147, 820 N.Y.S.2d 743, 745 (Sup. Ct. 2006).
21. *See, e.g.*, *Murphy*, 820 N.Y.S.2d at 745; *Moreau*, 975 N.Y.S.2d at 856.

Chapter 14

Automated Ridesharing Vehicles

An automated ridesharing service has a few unique aspects from a liability standpoint that are not considered when analyzing privately owned automated vehicles or automated rental vehicles. Chapter 6 described how states have enacted rules and laws governing transportation network companies (TNCs), and whether an automated ridesharing company would fall within the definition of a TNC depends, in part, upon its activities and ownership of the fleet. Qualification as a TNC matters—at least until state legislatures amend their definitions of TNCs—for liability purposes because some states have legislatively declared that TNCs are not common carriers. Chapter 13 dealt with an automated ridesharing company's liability in terms of its resemblance to a traditional car rental company.

Torts and civil liability abound in the automated ridesharing context. Bystanders, as well as those who use automated ridesharing services, can be harmed by automated ridesharing vehicle crashes. Users of automated ridesharing vehicles could be harmed by the second collision between them and the interior of the automated vehicles. Additionally, users could be harmed by persons who "hijack"—whether digitally or physically—the automated ridesharing vehicle or otherwise interfere with its use, or by other users within shared automated vehicles.

I. CRASHES CAUSED BY AUTOMATED RIDESHARING VEHICLES

When automated ridesharing vehicles cause crashes, those harmed by the vehicles will likely sue automated ridesharing companies among other parties (such as the users or the manufacturers).[1] Ridesharing users could sue these companies, even when other vehicles cause crashes.[2] This section considers whether the automated ridesharing company can be liable for crashes.

A. Products Liability Law

When these crashes occur, the aggrieved parties may assert products liability claims. Certain types of automated ridesharing companies—the ones that do not own the fleet of vehicles—may contend that they merely provide services.[3] Products liability law does not typically apply to services, even if those services involve the use of a product.[4] TNCs have generally been able to avoid products liability law by contending that the application that connects drivers to customers is merely a service.[5] Currently, the role of a TNC to the consumer is likely that of a service provider.[6] The TNC provides an application-based platform that connects people who need a ride with drivers.[7] A defining feature of the current system is that the TNCs do not own or control the underlying vehicles. In the non-fleet-owned model, these automated ridesharing companies would have a more persuasive argument to avoid products liability law. The fleet-owned companies may begin to resemble car rental companies and could be treated similarly.[8]

B. Negligence

Automated ridesharing companies will owe passengers and bystanders numerous duties; breaches of any one of these duties that proximately cause harm to others can result in liability for negligence. The duties owed by these companies will be similar to the ones described in Chapter 8. A paramount issue when

1. User and manufacturer liability are described in depth in Chapter 11.

2. These claims will be more prominent when the at-fault party has insufficient insurance coverage or assets to make the users whole.

3. Indeed, companies may further protect against risk, to the extent permissible under local law, by having their ridesharing service and the vehicle ownership further divided between subsidiary entities. Under this structure, the vehicle owner and the service provider would be separated. However, this may make it easier to establish that the vehicle owner is subject to products liability law because it would not be providing any services but would be solely in the business of renting its fleet of automated vehicles.

4. *See* William C. Powers, Jr., *Distinguishing between Products and Services in Strict Liability*, 62 N.C. L. Rev. 415, 415 (1984) (noting that "most courts have held that 'pure' services are beyond the scope of strict products liability because they are not 'product sales'").

5. *See, e.g.*, Liu v. Uber Techs., Inc., No. CGC-14-536979 (Cal. Super. Ct. filed Jan. 27, 2014).

6. *See* Jacob D. Walpert, *Carpooling Liability?: Applying Tort Law Principles to the Joint Emergence of Self-Driving Automobiles and Transportation Network Companies*, 85 Fordham L. Rev. 1863, 1876 (2017).

7. *See id.*

8. See Chapter 13 for a discussion on the application of products liability to car rental companies.

the plaintiff is a passenger[9] is whether tort law will treat these companies as common carriers or private carriers. A common carrier has heightened duties (typically imposed by statute), including affirmative duties, to its passengers.[10] At least for tort liability, designation as a common carrier concerns the duty the carrier owes to its passengers.[11] An automated ridesharing company could be considered a common carrier.

Generally, "[a] common carrier is one who undertakes to, or does, transport all persons indiscriminately, or at least contracts to transport all persons within a definite class."[12] Courts distinguish between a common carrier and a private carrier, but the terms "common" and "private" have nothing to do with whether the carrier is publicly owned or privately owned. A private carrier is generally one who "undertakes by special agreement, in a particular instance only, to transport persons or property from one place to another either gratuitously or for hire."[13] In determining whether a carrier is a common carrier, courts have considered "whether: (1) the defendant maintains a regular place of business for the purpose of transportation; (2) the defendant advertises its services to the general public; and (3) the defendant charges standard fees for its services."[14] Common carriers are required to "use great caution to protect passengers entrusted to their care; and this has been described as 'the utmost caution characteristic of very careful prudent men,' or 'the highest degree of

---

9. The heightened duty for common carriers generally only applies to passengers or to-be passengers and not bystanders. DAN B. DOBBS, PAUL T. HAYDEN & ELLEN M. BUBLICK, THE LAW OF TORTS, § 263 *Scope of the Duty; Who Counts as a Common Carrier or Passenger* (2d ed. 2019) ("The carrier's duty of utmost care is owed only to those who count as passengers, not to strangers such as pedestrians or drivers of other vehicles.").

10. See Dylan LeValley, *Autonomous Vehicle Liability—Application of Common Carrier Liability*, 36 SEATTLE U.L. REV. 5 (2013), for a discussion of common carrier liability for automated vehicles.

11. The determination of an entity as a common carrier can invite a whole host of far more significant regulatory obligations on the entity.

12. DOBBS, HAYDEN & BUBLICK, *supra* note 9, at § 263; *see also* Doe v. Rockdale Sch. Dist., No. 84, 287 Ill. App. 3d 791, 794, 679 N.E.2d 771, 773 (1997) ("A common carrier undertakes for hire to carry all persons indifferently, who may apply for passage so long as there is room and there is no legal excuse for refusal." (citation omitted)); Emp'rs Mut. Cas. Co. v. Chicago & N.W. Transp. Co., 521 N.W.2d 692, 695 (Iowa 1994) ("A 'common carrier' is one who undertakes to transport, indiscriminately, persons and property from one place to another." (citation omitted)); Van Maanen v. Youth With a Mission-Bishop, 852 F. Supp. 2d 1232, 1245 (E.D. Cal. 2012), *aff'd sub nom.* Van Maanen v. Univ. of the Nations, Inc., 542 F. App'x 581 (9th Cir. 2013) ("Under California law, '[e]very one who offers to the public to carry persons, property[,] or messages, excepting only telegraphic messages, is a common carrier of whatever he thus offers to carry.'" (quoting CAL. CIV. CODE § 2168)).

13. James Buchwalter et al., *§ 471 Nature of Carriers of Passengers*, 13 C.J.S. CARRIERS (Mar. 2020).

14. *Van Maanen*, 852 F. Supp. 2d at 1246; Jackson v. Stancil, 253 N.C. 291, 301, 116 S.E.2d 817, 824 (1960) (in an airplane case, noting that "the following are generally considered important factors in the operation of common carriers: An established place of business. Engaging in the operation as a regular business and not merely as a casual or occasional undertaking. Regular schedule of charges." (internal citations omitted)).

vigilance, care, and precaution.'"[15] This heightened duty may be created by statute[16] or case law.[17]

Although part of the legal maneuvering of the TNCs, such as Uber and Lyft, has been to argue that they are not common carriers, courts have considered and in one instance refused to dismiss a claim that sought to treat them as "common carriers."[18] Numerous states have legislatively stated that TNCs and the drivers for such services are not common carriers.[19] In these states, it is important to consider whether an automated ridesharing company would be considered a TNC for purposes of the statutory regime. The conclusion depends, in part, upon the type of services, the underlying structure of the automated ridesharing company, and the definition of TNC under the applicable regulation.[20] For those automated ridesharing companies that can claim that they are a TNC for purposes of these statutes, they should be able to avoid designation as a common carrier as a matter of legislative grace.

Regardless of the classification of the automated ridesharing company as a common carrier or a private carrier, these companies may face claims that they were negligent in selecting their automated vehicles. This claim presumes a duty—namely that these companies have an obligation to the users (or the public) to exercise reasonable care in choosing the automated vehicles used in its owned fleet or nonowned fleet. For automated vehicle ridesharing companies that utilize their own fleet, this obligation may be relatively straightforward

---

15. W. Page Keeton et al., Prosser & Keeton on the Law of Torts 209 (5th ed. 1984); *see, e.g., Jackson*, 116 S.E.2d at 823 ("The rule of 'utmost care' does not merely require the degree of care usual among ordinarily prudent and competent carriers. It requires the degree of care to be expected of an unusually prudent and competent carrier. It is then something more stringent than the rule of 'ordinary' care under all the circumstances." (internal quotation marks and citation omitted)).

16. Cal. Civ. Code § 2100 (West); N.D. Cent. Code § 8-02-02; Okla. Stat. tit. 13, § 32.

17. Pa. Co. v. Roy, 102 U.S. 451, 456, 26 L. Ed. 141 (1880) ("These and many other adjudged cases, cited with approval in elementary treatises of acknowledged authority, show that the carrier is required, as to passengers, to observe the utmost caution characteristic of very careful, prudent men. He is responsible for injuries received by passengers in the course of their transportation which might have been avoided or guarded against by the exercise upon his part of extraordinary vigilance, aided by the highest skill.").

18. Doe v. Uber Techs., Inc., 184 F. Supp. 3d 774, 787 (N.D. Cal. 2016) (refusing to dismiss a claim that Uber is a common carrier); *see also* Diva Limousine, Ltd. v. Uber Techs., Inc., 392 F. Supp. 3d 1074 (N.D. Cal. 2019) (stating that "[t]he parties are in accord that Uber is a 'public utility corporation' [because] '[p]ublic utility' includes every common carrier . . .").

19. *See, e.g.*, Fla. Stat. § 627.748 (2017) ("A TNC or TNC driver is not a common carrier, contract carrier, or motor carrier and does not provide taxicab or for-hire vehicle service. In addition, a TNC driver is not required to register the vehicle that the TNC driver uses to provide prearranged rides as a commercial motor vehicle or a for-hire vehicle."); N.Y. Veh. & Traf. Law § 1692 (McKinney) (2017) ("A TNC or a TNC driver shall not be deemed a common carrier . . ."); 66 Pa. Cons. Stat. Ann. § 102 (2016) (explicitly providing that "[t]he term [common carrier] does not include a transportation network company or a transportation network company driver"); Tex. Occ. Code Ann. § 2402.002 (West) (2017) ("Transportation network companies and drivers logged in to the company's digital network are not common carriers, contract carriers, or motor carriers.").

20. See Chapter 6 for a discussion of whether automated ridesharing companies are TNCs.

to satisfy. Avoiding claims of negligence requires reasonable care—but may require heightened care under the common carrier standard—and so long as the company made a reasonable, informed choice of the supplier of its vehicles, these claims may not be successful. For automated vehicle ridesharing companies that utilize privately owned automated vehicles, there may be a greater duty to conduct "background checks" on the vehicle owner and the vehicle owner's maintenance of the vehicles. These companies probably cannot blindly trust that the owners have properly maintained their vehicles. Automated ridesharing companies will also be sued for harms that they could prevent. For those companies that own their vehicles, they will need to maintain the vehicles properly and to avoid altering the vehicle in a way that could affect performance.

C. Joint Enterprise Theory

Some scholars have argued that courts should treat current TNCs as being in a joint enterprise with their drivers.[21] The joint enterprise theory is another way to impose liability on an actor for the negligence or wrongful acts of another. The comments to the Restatement (Second) of Torts provide the following statement about a business enterprise with a vehicle:

> When the journey on which the plaintiff and driver of the vehicle are participating is itself a part of a business enterprise in which the parties are mutually interested, the two are engaged in a joint enterprise. It is immaterial that the particular journey is a single transaction and is not a part of the general course of a business in which they are associated as partners or otherwise. It is also immaterial whether the car is owned or hired by the one or the other or by both jointly; the use of the car as a part of a common business enterprise makes each responsible for the manner in which it is operated both as defendant and plaintiff.[22]

Courts have used joint enterprise theories in transportation contexts,[23] such as when taxicab companies do not own their cabs but have the authority to operate the cabs[24] and for third-party logistic providers in the trucking industry.[25] Non-fleet-owned automated ridesharing companies may face these types of claims.

21. *See, e.g.*, Agnieszka A. McPeak, *Regulating Ridesharing Platforms through Tort Law*, 39 U. Haw. L. Rev. 357, 378–80 (2017).

22. Restatement (Second) of Torts § 491 cmt. g (1965).

23. McPeak, *supra* note 21, at 385–89.

24. *See, e.g.*, Callas v. Indep. Taxi Owners Assoc., 66 F.2d 192, 193 (D.C. Cir. 1933); Am. Ass'n of Cab Cos., Inc. v. Parham, 291 Ga. App. 33, 661 S.E.2d 161 (2008).

25. *See, e.g.*, Johnson v. Pac. Intermountain Exp. Co., 662 S.W.2d 237 (Mo. 1983).

D. Claims Relating to the Interior of the Vehicles

Unique designs of the interior of automated vehicles may invite crashworthiness claims. There is nothing unique about crashworthiness claims for automated ridesharing companies with one exception. Automated ridesharing companies may seek to differentiate themselves by developing their own interior designs. Many automated ridesharing companies may not be vehicle manufacturers; so, third-party manufacturers will likely produce these vehicles based on specifications provided by automated ridesharing companies. If a plaintiff is harmed by a defective design of the interior, who is liable for the "uncrashworthy" design?

The interior customization by these automated ridesharing companies increases their liability risks and allows the manufacturer to use the contract specifications defense. "The contract specifications defense shields a manufacturer from liability for injuries caused by a design defect in products it manufactures pursuant to plans and specifications supplied by the purchaser, unless the design is obviously defective."[26] The contract specifications defense is rooted in the common law rule for independent contractor liability for the materials and specifications provided by an employer.[27] The vast majority of courts recognize the contract specifications defense.[28] The contract specifications defense may mean that an automated ridesharing company is principally liable for crashworthiness claims.[29]

II. CRIMINAL ACTS OF THIRD PARTIES

In today's TNCs, criminal acts to passengers occur.[30] Many of these criminal acts are committed by or with compliance of the driver of the vehicle.[31] In the automated ridesharing vehicles of tomorrow, there may not be a driver in the

---

26. David G. Owen, Products Liability Law 922 (2nd ed. 2005). Manufacturers would still have an obligation to ensure that the vehicles meet the Federal Motor Vehicle Safety Standards and would have an obligation to make the certification under federal law.

27. *Id.* at 922–23.

28. *See, e.g.*, Garrison v. Rohm & Haas Co., 492 F.2d 346, 351 (6th Cir. 1974) ("To hold [the manufacturer] liable for defective design would amount to holding a non-designer liable for design defect. Logic forbids any such result."); Bloemer v. Art Welding Co., 884 S.W.2d 55, 58–59 (Mo. Ct. App. 1994); *but see* Janusz v. Symmetry Med. Inc., 256 F. Supp. 3d 995, 1007 (E.D. Wis. 2017) ("The contract specification defense significantly undermines the policies underlying strict liability. Therefore, the court concludes that the defense does not exist under Wisconsin law.").

29. However, it seems unlikely that courts would insulate a traditional automaker that blindly produces vehicles according to specifications provided by third parties, no matter how flawed those specifications are.

30. Emily L. Dyer, *Need a Ride? Uber: The Trendy Choice That Could Turn Threatening*, 17 Nev. L.J. 239, 252–54 (2016) (detailing criminal incidents).

31. *Id.*

vehicle. Depending upon how automated ridesharing vehicles are deployed, the lack of a driver may or may not increase the safety of users. For automated ridesharing vehicles that only pick up one person, the removal of the driver may reduce the risk of harm to the single passenger but may increase the risk that third parties may attempt to physically intercept the vehicle. For automated ridesharing vehicles that pick up multiple groups of persons, the removal of the driver—who, customarily, wore many hats, including driving the vehicle, ensuring that the proper people enter the vehicle, and maintaining decorum within the vehicle—may increase the risk of harm among passenger groups. In either case, the lack of a driver to monitor the interactions among passengers and to control the vehicle may invite mischief and criminal misconduct in these vehicles.

Automated ridesharing companies will have duties to protect their passengers from criminal acts. The Restatement (Second) of Torts noted that an actor can be responsible for criminal acts of another "[w]here the actor stands in such a relation to the other that he is under a duty to protect him against such misconduct."[32] The duty owed to passengers may depend upon whether automated ridesharing companies are considered common carriers. If courts find that automated ridesharing companies are common carriers, they will have little trouble finding that automated ridesharing companies have to protect their passengers from reasonably foreseeable criminal attacks.[33] This duty to protect passengers is not absolute, as a common carrier "is not an insurer of the safety of its passengers, nor is it obliged by law to foresee and to guard against unlikely dangers and improbable harms."[34] Courts (and perhaps legislatures) will need to define and delineate what this duty entails.

---

32. Restatement (Second) of Torts § 302B cmt. e (1965).

33. Sonja A. Soehnel, *Liability of Land Carrier to Passenger Who Becomes Victim of Third Party's Assault on or about Carrier's Vehicle or Premises*, 34 A.L.R.4th 1054 (1984) ("The courts are generally in agreement that a carrier has a duty to protect passengers from assaults at the hands of third parties and is liable to a passenger where it or its employees knew or ought to have known of the assault and failed to protect the passenger."); Stuart M. Speiser et al., *§ 9:19. Duty to Prevent Criminal Acts of Third Persons; Liability to Victim Harmed by Criminal Acts of Third Persons—Liability of Persons in Special Relationship Owing Duty to Protect against Criminal Misconduct; Public Agencies or Bodies*, 2A American Law of Torts (2020) ("So long as the exposure to and incidence of a criminal attack by a third person on a passenger is reasonably foreseeable and likely to be anticipated, the carrier may be subjected to liability for breach of its duty to exercise the degree of care required by the circumstances to protect its passengers."); *see, e.g.*, Jones v. Yellow Cab & Baggage Co., 176 Kan. 558, 271 P.2d 249 (1954); Bullock v. Tamiami Trail Tours, Inc., 266 F.2d 326 (5th Cir. 1959) (Florida law); Citizens Coach Co. v. Wright, 228 Ark. 1143, 313 S.W.2d 94 (1958); Hines v. Garrett, 131 Va. 125, 108 S.E. 690 (1921).

34. Quigley v. Wilson Line of Mass., Inc., 338 Mass. 125, 154 N.E.2d 77, 79 (1958).

Chapter 15

State and Local Government Liability Risks

State and local governments will be heavily involved in the automated vehicle industry as regulators, participants, and facilitators. Governments may make policymaking decisions about automated vehicles, such as whether to permit them on their roadways and in what fashion; may own and operate automated vehicles; may implement intelligent transportation systems (ITSs) that interact with automated vehicles; and may make other customary decisions, such as road maintenance, that significantly impact automated vehicles. In many of these circumstances, people who are harmed by automated vehicles (or, conceivably, human-driven vehicles) may blame the state or local government for their harms.

The automated vehicle industry will impact and increase the risk profile for state and local governments, regardless of whether they use automated vehicles or ITSs. Automated vehicles will rely upon infrastructure far more than human drivers, and poor or unmaintained infrastructure may be fatal. Plaintiffs have already sued state and local governments following two prominent automated vehicle crashes.[1] In *Huang v. Tesla, Inc.*, Walter Huang died on March 23, 2018, when he crashed into a median while using Autopilot on his Tesla Model X. Huang's estate sued Tesla and the state of California. Huang's estate claimed, in relevant part, that the California Department of Transportation (Caltrans) created a dangerous condition by being negligent and careless in its ownership, maintenance, inspection, repair, and control of the highway

1. Huang v. Tesla Inc., No. 19-cv-346663 (Cal. Cnty. Santa Clara filed Apr. 26, 2019); Wood v. Arizona, No. CV2019-090948 (Ariz. Maricopa Cnty. filed Mar. 20, 2019).

where the crash resulted. The central allegation against Caltrans was that a safety device called a crash attenuator guard, which is intended to dissipate crash energy, was not on the median. Huang's estate alleged that the crash attenuator guard had been damaged in a crash 11 days earlier, and Caltrans had not yet replaced the guard. In *Wood v. Arizona*, Elaine Herzberg died when an Uber automated vehicle crashed into her on March 18, 2018, in Tempe, Arizona. Herzberg's estate settled with Uber, so this lawsuit was solely against Arizona. One claim was for inadequate oversight by the Arizona Department of Transportation (ADOT) and the City of Tempe over the testing of automated vehicles on the roadways and for failing to keep the roadways safe. Another claim was against the City of Tempe for negligent median design by having brick-paved walkways that lead to the roadway without crosswalks or lighting for pedestrian safety. These two cases foreshadow the types of claims that state and local governments may encounter with automated vehicles.

In determining whether a governmental entity could bear liability for a particular harm, the two initial questions are whether the governmental entity is at fault for such harm and, even if the governmental entity is at fault, whether immunity shields the government from liability. The resolution of the first question, at least for motor vehicles and roadways, is typically based on negligence. Did the government owe a duty for the harm, and did the government's breach of that duty proximately cause the plaintiff's harm? This inquiry, though necessary, is not particularly unique to governmental entities. The uniqueness of government entity liability is whether an at-fault government is immune.

Historically, states and municipalities were completely immune from tort suits. Almost all states have abolished the absolute immunity that traditionally protected them. States have taken three broad approaches to waive their immunity.[2] The most common approach—adopted by approximately 30 states—has been to abolish immunity but preserve it in specified situations.[3] The less common approach has been to retain sovereign immunity but abolish it for specific types of actions.[4] Alabama, Wisconsin, and South Dakota retain a comprehensive immunity.[5] States impose specific procedural requirements, such as presentment to an administrative body or use of a specified court, when plaintiffs file claims against governments.[6] The approach taken by states can affect the underlying analysis of liability for government action or inaction.

In any jurisdiction, states are immune for discretionary or policymaking decisions. "The test for discretionary acts asks whether the act encompasses an

---

2. Dan B. Dobbs, Paul T. Hayden & Ellen M. Bublick, The Law of Torts, § 342. *State Sovereign Immunity and Its Waiver* (2d ed. 2019).

3. *Id.*

4. *Id.*

5. *Id.*

6. *Id.*

exercise in judgment and choice about what is proper and just under the circumstances, based on public policy considerations, or whether the conduct at issue entails policy-making or policy choice involving social, political, or economic considerations."[7] This immunity extends to bad policy decisions.[8] States are generally not immune for ministerial decisions or those decisions that require them to carry out the policymaking decisions.[9] Additionally, motor vehicles and roadway safety are areas where states have commonly waived immunity.[10]

This chapter explores five different situations where plaintiffs could sue governments: (1) the decision to allow automated vehicles on their roadways, (2) the use of automated vehicles by governments, (3) road maintenance, (4) the procurement and use of ITSs, and (5) data collection.

I. THE DECISION TO ALLOW AUTOMATED VEHICLES ON ROADWAYS

The decision to allow testing or deployment of automated vehicles is inherently the type of discretionary or policymaking decision for which immunity is intended.[11] So long as governments make an informed decision, governments can likely decide whether to allow automated vehicle testing or the

---

7. George Blum et al., *§ 65 Definition of Discretionary Act of Function*, 57 AM. JUR. 2D MUNICIPAL, ETC., TORT LIABILITY (Feb. 2020).

8. *See* Evangelical United Brethren Church of Adna v. State, 67 Wash. 2d 246, 407 P.2d 440, 444 (1965) ("Practically all jurisdictions that have broken varying amounts of ground in the abdication of governmental immunity from tort liability have judicially, if not statutorily, recognized that the legislative, judicial, and purely executive processes of government, including as well the essential quasi-legislative and quasi-judicial or discretionary acts and decisions within the framework of such processes, cannot and should not, from the standpoint of public policy and the maintenance of the integrity of our system of government, be characterized as tortious however unwise, unpopular, mistaken, or neglectful a particular decision or act might be.").

9. *E.g.*, *Evangelical United Brethren Church of Adna*, 407 P.2d at 444–45 (identifying that some courts distinguish between "planning" and "operational" decisions, others between "discretionary" and "ministerial"); Wainscott v. State, 642 P.2d 1355, 1356 (Alaska 1982) (stating "the distinction is between basic policy formulation, which is immune, and the execution or implementation of that basic policy, which is not immune" (internal quotation marks and citation omitted)).

10. For motor vehicles, see, for example, OHIO REV. CODE ANN. § 2744.02 (West) ("Except as otherwise provided in this division, political subdivisions are liable for injury, death, or loss to person or property caused by the negligent operation of any motor vehicle by their employees when the employees are engaged within the scope of their employment and authority."). For roadways, see, for example, MICH. COMP. LAWS § 691.1402 ("Each governmental agency having jurisdiction over a highway shall maintain the highway in reasonable repair so that it is reasonably safe and convenient for public travel. A person who sustains bodily injury or damage to his or her property by reason of failure of a governmental agency to keep a highway under its jurisdiction in reasonable repair and in a condition reasonably safe and fit for travel may recover the damages suffered by him or her from the governmental agency."); CONN. GEN. STAT. § 13a-144.

11. *See* Dorothy J. Glancy, *Autonomous and Automated and Connected Cars—Oh My! First Generation Autonomous Cars in the Legal Ecosystem*, 16 MINN. J.L. SCI. & TECH. 619, 660–61 (2015).

deployment of automated vehicles on their roads without facing tort liability for that decision.

A critical question in these cases is at what point does the discretionary decision-making end such that the governmental authority responsible for allowing automated vehicles on roadways could be liable for their negligent or reckless deployment? Typically, the protective shield disappears when governments implement and operate their policymaking decisions; the line between policymaking and operational decisions is often challenging to draw.[12] Generally speaking, the government is immune when establishing policy but not when implementing that policy.[13] Thus, to the extent that governments are negligent in implementing their automated vehicle policies, they may lose their protective shield and be subject to liability for harm foreseeably caused by that negligence.

II. GOVERNMENT-OWNED LEVEL 4 AND LEVEL 5 AUTOMATED VEHICLES

Governments will own and use level 4 and level 5 automated vehicles, and inevitably those vehicles will be involved in crashes. Motor vehicle use and operation by government employees are one of the more significant exceptions

---

12. *See, e.g.*, Commercial Carrier Corp. v. Indian River Cnty., 371 So. 2d 1010, 1020 (Fla. 1979) (noting the difficulty of identifying "discretionary functions, certain policymaking, planning, or judgmental government functions").

13. One court has described the initial inquiry in these cases as a consideration of four questions:

> Whatever the suitable characterization or label might be, it would appear that any determination of a line of demarcation between truly discretionary and other executive and administrative processes, so far as susceptibility to potential sovereign tort liability be concerned, would necessitate a posing of at least the following four preliminary questions: (1) Does the challenged act, omission, or decision necessarily involve a basic governmental policy, program, or objective? (2) Is the questioned act, omission, or decision essential to the realization or accomplishment of that policy, program, or objective as opposed to one which would not change the course or direction of the policy, program, or objective? (3) Does the act, omission, or decision require the exercise of basic policy evaluation, judgment, and expertise on the part of the governmental agency involved? (4) Does the governmental agency involved possess the requisite constitutional, statutory, or lawful authority and duty to do or make the challenged act, omission, or decision? If these preliminary questions can be clearly and unequivocally answered in the affirmative, then the challenged act, omission, or decision can, with a reasonable degree of assurance, be classified as a discretionary governmental process and nontortious, regardless of its unwisdom. If, however, one or more of the questions call for or suggest a negative answer, then further inquiry may well become necessary, depending upon the facts and circumstances involved.

Evangelical United Brethren Church of Adna, 407 P.2d at 445.

to sovereign immunity.[14] Questions and disputes may arise about the applicability of sovereign immunity depending upon how the state has waived it (broad retention with carve-outs or broad abolishment with specific retentions). In states that broadly retained sovereign immunity and carved out motor vehicle crashes, the immunity analysis depends upon the language of the carve-outs. For example, some states base their waiver of statutory immunity on whether the government authority is using a motor vehicle, which they seemingly would with level 4 and level 5 automated vehicles; in contrast, other states base their waivers on whether a government employee negligently operated the motor vehicle.[15] With human-driven vehicles, there is no practical difference between these two types of waivers—immunity only applies when the government authority would otherwise be liable, and in the realm of employee use of motor vehicles, such liability would be based in negligence. With level 4 and level 5 automated vehicles, emphasis may be placed on the language of the waiver because the user may no longer be negligent.[16]

III. ROAD MAINTENANCE

Automated vehicles may put new pressure on government entities to maintain roadways. Automated vehicles will probably depend more on our roads and infrastructure than human drivers. For example, properly painted fog lines may be needed. Generally, the governmental entity responsible for roads is not required to maintain roads in a perfect condition; instead, the governmental entity's responsibility is limited to maintaining roads in a reasonably safe condition.[17] A governmental entity is generally not responsible unless it had actual or constructive knowledge of the dangerous roadway condition.[18] Some states give the governmental entity time to fix a dangerous roadway condition before

---

14. *See, e.g.*, Ohio Rev. Code Ann. § 2744.02(B)(1) (West); Tex. Civ. Prac. & Rem. Code Ann. § 101.021 (West).

15. *Compare* Tex. Civ. Prac. & Rem. Code Ann. § 101.021 (West) ("A governmental unit in the state is liable for: (1) property damage, personal injury, and death proximately caused by the wrongful act or omission or the negligence of an employee acting within his scope of employment if: (A) the property damage, personal injury, or death arises from the operation or use of a motor-driven vehicle or motor-driven equipment") *with* Ohio Rev. Code Ann. § 2744.02 (West) ("Except as otherwise provided in this division, political subdivisions are liable for injury, death, or loss to person or property caused by the negligent operation of any motor vehicle by their employees when the employees are engaged within the scope of their employment and authority.").

16. See Chapter 11 for a discussion of liability for level 4 and level 5 automated vehicle crashes.

17. Mark S. Dennison, *Liability of Public Authority for Injury Caused by Pothole in Surface of Highway or Street*, 92 Am. Jur. Proof of Facts 3d 461 (2006).

18. *Id.*

liability can attach.[19] Other states will not allow liability until after the governmental entity received written notice of the condition.[20] The driving automation system may also better detect and determine when the cause of a crash is due to inadequate or defective infrastructure. The combination of the greater need to rely on well-maintained roads and the driving automation system's ability to pinpoint the exact cause of a crash may place inordinate pressure on governments to maintain roads.

IV. ITSS

The decision whether to use or not use ITSs and whether to place certain traffic devices on a particular roadway are seemingly discretionary functions for which the governmental entity should have immunity.[21] Once a government decides to use ITSs, the government entity responsible for implementing that decision needs to exercise reasonable care and avoid being negligent.[22] Certain

---

19. *See, e.g.*, MICH. COMP. LAWS § 691.1403 (providing that "knowledge of the defect and time to repair the same shall be conclusively presumed when the defect existed so as to be readily apparent to an ordinarily observant person for a period of 30 days or longer before the injury took place").

20. *See, e.g.*, Betzold v. Town of Babylon, 18 A.D.3d 787, 796 N.Y.S.2d 680, 680 (2005).

21. *See* Felix V. Lapine, *Liability of Highway Authorities Arising out of Motor Vehicle Accident Allegedly Caused by Failure to Erect or Properly Maintain Traffic Control Device at Intersection*, 34 A.L.R.3D 1008 (1970) ("As to the initial decision of whether or not to place a traffic control device at a particular intersection, most courts, in the absence of statute, have either assumed or ruled that highway authorities are exercising a governmental function in determining whether such traffic control is necessary, and they are therefore held immune from liability for accidents claimed to have been caused by their failure to erect a traffic control device."); *see also* Kohl v. City of Phoenix, 215 Ariz. 291, 160 P.3d 170 (2007) (holding that the decision to prioritize signalization is fundamental policymaking that is entitled to government immunity). Some states legislatively provide that sovereign immunity does not apply to these types of decisions; for example, Texas law provides the following rule:

> This chapter [waiving sovereign immunity] does not apply to a claim arising from:
>
> (1) the failure of a governmental unit initially to place a traffic or road sign, signal, or warning device if the failure is a result of discretionary action of the governmental unit;
>
> (2) the absence, condition, or malfunction of a traffic or road sign, signal, or warning device unless the absence, condition, or malfunction is not corrected by the responsible governmental unit within a reasonable time after notice; or
>
> (3) the removal or destruction of a traffic or road sign, signal, or warning device by a third person unless the governmental unit fails to correct the removal or destruction within a reasonable time after actual notice.

TEX. CIV. PRAC. & REM. CODE ANN. § 101.060 (West).

22. *See* Indian Towing Co. v. United States, 350 U.S. 61, 69, 76 S. Ct. 122, 100 L. Ed. 48 (1955) ("The Coast Guard need not undertake the lighthouse service. But once it exercised its discretion to operate a light on Chandeleur Island and engendered reliance on the guidance afforded by the light, it was obligated to use due care to make certain that the light was kept in good working order; and, if the light did become extinguished, then the Coast Guard was further obligated to use due care to discover this fact and to repair the light or give warning that it was not functioning. If the Coast Guard

functions, such as the maintenance and repair of the ITSs, are analogous to road maintenance and are seemingly the type of ministerial and operational functions that are not protected by the policymaking immunity.[23] Other functions, such as the type of ITS to use and street locations of ITSs, may be more difficult to identify as operational or policymaking.[24] The actual procurement of the ITS may be an operational decision that is unprotected by immunity.[25]

V. GOVERNMENT DATA

Data sharing among the automated vehicle industry and states may create new risks for governments that collect and transmit data. The decision as to whether to share data and receive data from industry (or others, such as road maintenance workers) would be the type of policymaking decision that would be protected by immunity. However, a government that undertakes sharing and receiving data may open itself up to new liability risks. One risk may be that harm may be caused, at least in part, by the government's failure to provide timely data. An example would be a government that collects construction data from construction workers and transmits that information to automated vehicle companies to update their maps. If the government collects data but does not timely transmit data and a crash results, could the government incur liability for these types of crashes? A second risk would be the collection and transmission of faulty data. An example would be a government that collects data, converts data to a new format, and transmits data for use. In this case, if an error results in the conversion of data that causes harm, is the government liable?

---

failed in its duty and damage was thereby caused to petitioners, the United States is liable under the Tort Claims Act.").

23. *See, e.g.,* City of Midland v. Sullivan, 33 S.W.3d 1, 11 (Tex. App. 2000) (holding that "the visibility of the pavement markers due to disrepair and the failure of the school zone sign to activate the school zone thirty minutes prior to class" are not policymaking functions that are entitled to immunity); *but see* Town of Fort Oglethorpe v. Phillips, 224 Ga. 834, 165 S.E.2d 141, 143 (1968) ("The operation and maintenance of traffic lights and other traffic control devices is a governmental function conducted on behalf of the public safety and for the negligent performance of which municipal corporations are not liable. . . . We are of the opinion that the maintenance and operation of the traffic light by the Town of Fort Oglethorpe was a governmental function and the municipality is not liable for negligent performance." (internal citations and quotation marks omitted)).

24. *See* Wainscott v. State, 642 P.2d 1355, 1357 (Alaska 1982) (holding that "the decision to install flashing red and yellow lights in lieu of a sequential traffic signal constituted a planning level decision").

25. *See* Dep't of Transp. v. Brown, 267 Ga. 6, 471 S.E.2d 849, 851 (1996) ("Whether to buy copier paper from a particular vendor, and in which colors, are decisions that might be affected by all three factors, but they are not policy decisions.").

There have been cases where plaintiffs have sued governments for inaccuracies in data they provide.[26] These cases have involved the provision of defective data directly and indirectly to consumers. In *Reminga v. United States*,[27] a direct case, the Federal Aviation Administration (FAA) was sued after a plane crashed into a guy wire attached to a TV tower. The FAA issued a sectional chart that was allegedly being used by the pilot—the chart inaccurately displayed the location of a TV tower. The Second Circuit stated that "[t]hough not required by law to do so, when the FAA arranges for the publication of aeronautical navigation charts and engenders reliance on them, it is required to use due care to see that they accurately depict what they purport to show. Failure to show the location of the tower accurately rendered the United States liable for injury to those who relied upon the chart." In *Brocklesby v. United States*,[28] a plane crashed into a mountain in Alaska. The heirs and the airplane owner sued the United States and Jeppesen, claiming that "the accident was caused by defects in an instrument approach developed by the Government and published by Jeppesen." The FAA produced compilations of data in tabular form, and Jeppesen portrayed that data in graphic form. The FAA settled with the plaintiffs for $5 million after the jury returned a verdict against Jeppesen for $11,630,000.[29]

Under these two lines of authority, governments may incur more significant liability risks when they become data providers or aggregators. To protect against these risks, government providers of data should disclaim liability for companies and people who use data.[30] Depending upon the circumstances, these disclaimers should be coupled with indemnification provisions for recipients and users of data.

---

26. *See, e.g.*, Reminga v. United States, 631 F.2d 449 (6th Cir. 1980); Brocklesby v. United States, 767 F.2d 1288 (9th Cir. 1985).

27. 631 F.2d 449 (6th Cir. 1980).

28. 767 F.2d 1288 (9th Cir. 1985).

29. The issue on appeal was whether Jeppesen could be held liable in products liability for the defective information provided by the FAA. The Ninth Circuit ruled that it could. *See also* Aetna Cas. & Sur. Co. v. Jeppesen & Co., 642 F.2d 339, 343 (9th Cir. 1981).

30. *See* Rosenstein v. Standard & Poor's Corp., 264 Ill. App. 3d 818, 636 N.E.2d 665, 671–72 (1993).

Part 4

Data Privacy and Cybersecurity

INTRODUCTION

Automated vehicles will collect troves of data and personal information. One prediction is that automated vehicles will generate approximately four terabytes of data every hour and a half of operation.[1] Sensors that help automated vehicles perceive the world will collect data about their surroundings. Sensors within vehicles, such as driver monitoring systems, will collect data about vehicle occupants. GPSs will collect location data about the vehicle. Connected vehicles will transmit data between or among other vehicles, infrastructure, or pedestrians. Telematics will monitor the vehicle and driving behavior. Onboard entertainment systems will collect additional data points about occupants of automated vehicles. All of these data will reveal, to quote Justice Sotomayor, a person's "familial, political, professional, religious, and sexual habits."[2]

Data collected by automated and connected vehicles will be valuable to commercial entities for marketing, law enforcement for crime prevention and

1. Kathy Winter, *For Self-Driving Cars, There's Big Meaning behind One Big Number: 4 Terabytes*, Intel Newsroom (Apr. 14, 2017), https://newsroom.intel.com/editorials/self-driving-cars-big-meaning-behind-one-number-4-terabytes/#gs.pmoas9.

2. United States v. Jones, 565 U.S. 400, 415, 132 S. Ct. 945, 181 L. Ed. 2d 911 (2012) (Sotomayor, J., concurring).

prosecutions, and individuals for their personal privacy. Legislatures may need to decide who owns data recorded by automated or connected vehicles. Currently, Congress and some legislatures have provided certainty as to who owns a limited type of data collected by vehicles: data collected by an event data recorder, the black box of a vehicle. Other than data collected by an event data recorder, Congress and state legislatures have not tried to determine who owns all other data collected by automated vehicles.

This part tackles the dual angles of regulatory and societal concern with the mass collection of data. The first angle is about data privacy and ensuring that those who properly collect data do not misuse that data and utilize it only for proper purposes. The mass collection of data by automated vehicles on both users and the world at large will significantly affect personal privacy. Privacy loss may be consensual, such as when an automated vehicle user consents to such loss of privacy. For many people, the privacy loss may be nonconsensual, such as when automated vehicles capture personal information about bystanders and the public. Examining the issue with respect to automated vehicles requires an inquiry into the existing federal and state regulatory regimes. Chapters 16 and 17 explore existing federal and state privacy regulations that may apply directly to automated vehicles.

The second angle is ensuring that third parties do not access that data without permission—in other words, the prevention of data access by cyberattacks or hackers. Chapters 16 and 17 examine federal and state laws that address cybersecurity, and Chapter 18 addresses civil liability for cybersecurity vulnerabilities and breaches. Chapter 20 in Part 5 explores federal and state criminal laws for cyberattacks.

Chapter 16

Federal Regulation of Data Privacy and Security

The United States does not have an omnibus privacy act that governs the privacy rights of its citizens or an omnibus data security act. The approach in the United States has been to enact industry-specific privacy or data acts. This approach leaves significant gaps in coverage on privacy and data security that federal agencies have sought to fill. This chapter begins by describing industry-specific privacy and data security laws that may apply to automated vehicles. The chapter then considers how agencies have tried to fill the regulatory gap, focusing primarily on how the National Highway Traffic Safety Administration (NHTSA) and the Federal Trade Commission (FTC) have regulated privacy and cybersecurity.

I. FEDERAL ACTS

Even with the proliferation of the Internet of things and mass data collection by private companies, Congress has done little in recent years to address consumer privacy or cybersecurity. Congress was not always this reluctant. In 1986, before the Internet became what it is today, Congress enacted the Stored Communications Act (the SCA) to regulate data privacy in wire and electronic communications. One recent act that Congress enacted was the Cybersecurity Act of 2015, which provides for information sharing among industry members and between industry and government for cybersecurity risks. Congress has also enacted two acts entitled the Driver Privacy Protection Act of 1994 and the Driver Privacy Protection Act of 2015. These two acts do not broadly regulate driver privacy; they regulate two different aspects of privacy. The final act that this section describes is the Privacy Act of 1974, which applies to the collection of data by federal agencies.

A. Section 2702 of the Stored Communications Act

The SCA, codified at 18 U.S.C. § 2701 *et seq.*, is a privacy act that provides safeguards and protections of privacy for networks and communications. Section 2701—described in Chapter 20—addresses unlawful access to communications, and section 2703—described in Chapter 22—limits government access to communications. Section 2702, discussed here, governs voluntary disclosure of data by persons or entities providing electronic communication services or remote computing services. Generally, subject to several exceptions, a provider of electronic communication services or remote computing services to the public cannot knowingly divulge the contents of a communication or a record or other information about a subscriber or customer.[1]

In determining whether section 2702 applies, the first inquiry is whether the entity is a provider of electronic communication service or remote computing service to the public. "Electronic communication service" is defined as "any service which provides to users thereof the ability to send or receive wire or electronic communications."[2] A "remote computing service" is "the provision to the public of computer storage or processing services by means of an electronic communications system."[3] Courts have generally found that traditional sellers of products or services do not provide an electronic communication service.[4] Instead, section 2702 is generally limited to Internet service providers or

1. 18 U.S.C. § 2702.
2. *Id.* § 2510(15).
3. *Id.* § 2711(2).
4. *See, e.g.,* Dyer v. Nw. Airlines Corp., 334 F. Supp. 2d 1196, 1199 (D.N.D. 2004) ("Courts have concluded that 'electronic communication service' encompasses internet service providers as well as telecommunications companies whose lines carry internet traffic, but does not encompass businesses selling traditional products or services online."); *In re* Jetblue Airways Corp. Privacy Litig., 379 F. Supp. 2d 299, 307 (E.D.N.Y. 2005) (stating that "a company such as JetBlue does not become an 'electronic communication service' provider simply because it maintains a website that allows for the transmission of electronic communications between itself and its customers"); Copeland v. Nw. Airlines Corp., No. 04-2156 M1/V, 2005 WL 2365255, at *2 (W.D. Tenn. Feb. 28, 2005) ("Courts have found that the term 'electronic communication service' refers to entities that provide internet access to customers, such as internet service providers ('ISP's'), as opposed to merchants and businesses that sell their products or services over the internet."). A notable outlier was that the Ninth Circuit held that American Airlines, through the use of its computerized travel reservation program called SABRE, "is a provider of wire or electronic communication service." United States v. Mullins, 992 F.2d 1472, 1478 (9th Cir. 1993). Because the Ninth Circuit examined the question within the context of the wire or electronic communication interception statute and whether American Airlines employees violated the defendant's Fourth Amendment rights, courts within the Ninth Circuit have not applied *Mullins* to section 2702. Pica v. Delta Air Lines, Inc., No. CV 18-2876-MWF (EX), 2019 WL 1598761, at *7 (C.D. Cal. Feb. 14, 2019).

other entities that resemble communication companies.[5] Because automakers and driving automation system manufacturers would be traditional sellers of products that do not resemble companies in the business of providing communications, section 2702 probably would not apply to them. Thus, data collection activities by these companies with automated vehicles should fall outside the scope of section 2702.

Wire and electronic communications collected by communication providers in the automated vehicle industry would be protected. A wire communication is "any aural transfer made in whole or in part through the use of facilities for the transmission of communications by the aid of wire, cable, or other like connection between the point of origin and the point of reception (including the use of such connection in a switching station) furnished or operated by any person engaged in providing or operating such facilities for the transmission of interstate or foreign communications or communications affecting interstate or foreign commerce."[6] An electronic communication is "any transfer of signs, signals, writing, images, sounds, data, or intelligence of any nature transmitted in whole or in part by a wire, radio, electromagnetic, photoelectronic or photooptical system that affects interstate or foreign commerce."[7] Given the broad definition of electronic communication as any transfer of data, any transmission of data to or from an automated vehicle would be considered an electronic communication. This data may be text messages and e-mails sent through the automated vehicle's infotainment system or other data, such as geolocation or telematics data, that is transferred from the automated vehicle to the manufacturer or a service provider of the manufacturer.

B. The Cybersecurity Act of 2015

In 2015, Congress enacted the Cybersecurity Act of 2015 (the Cybersecurity Act). One part of the Cybersecurity Act was the Cybersecurity Information Sharing Act of 2015 (the CISA).[8] The CISA expires on September 30, 2025, unless Congress elects to extend it. The CISA allows and instructs relevant federal agencies to develop practices that promote the sharing of cybersecurity threat indicators or cybersecurity threats with federal entities and nonfederal entities (depending upon the classification of the information).[9]

---

5. *See supra* note 4; Crispin v. Christian Audigier, Inc., 717 F. Supp. 2d 965, 982 & n.35 (C.D. Cal. 2010) (concluding that Facebook and MySpace are providers of electronic communication services).
6. 18 U.S.C. § 2510(1).
7. *Id.* § 2510(11).
8. The CISA is codified at 6 U.S.C. § 1501 *et seq.*
9. *Id.* § 1502.

The CISA authorizes a "private entity"[10] to, for cybersecurity purposes,[11] "monitor (A) an information system of such private entity; (B) an information system of another non-Federal entity, upon the authorization and written consent of such other entity; (C) an information system of a Federal entity, upon the authorization and written consent of an authorized representative of the Federal entity; and (D) information that is stored on, processed by, or transiting an information system monitored by the private entity under this paragraph."[12] A private entity is also authorized to operate a defensive measure.[13] Nonfederal entities are authorized to share and receive cyber threat indicators or defensive measures with federal entities and nonfederal entities.[14] The sharer can impose lawful restrictions on the cyber threat indicator or defensive measure.[15] Before sharing a cyber threat indicator, the nonfederal entity must review the indicator and remove any personal information of a specific individual that is not directly related to a cybersecurity threat.[16]

The CISA overrides inconsistent laws, including the SCA.[17] The CISA excludes industry coordination permitted under the act from antitrust enforcement.[18] The CISA also preempts and requires a prompt dismissal of causes of action based on a private entity monitoring an information system or information under the CISA or based on the sharing or receipt of a cyber threat indicator or defensive measure under the CISA.[19] The immunity granted under the CISA is limited and would not shield a company from liability based on a data security breach itself. The CISA also prevents plaintiffs from using the act to create tort law duties.[20]

---

10. A private entity is defined to include true private entities and state, tribal, or local governments that perform utility services. *Id.* § 1501(15).

11. Cybersecurity purpose is "the purpose of protecting an information system or information that is stored on, processed by, or transiting an information system from a cybersecurity threat or security vulnerability." *Id.* § 1501(4).

12. *Id.* § 1503(a)(1).

13. A defensive measure is "an action, device, procedure, signature, technique, or other measure applied to an information system or information that is stored on, processed by, or transiting an information system that detects, prevents, or mitigates a known or suspected cybersecurity threat or security vulnerability." *Id.* § 1501.

14. *Id.* § 1503(c)(1).

15. *Id.* § 1503(c)(2).

16. *Id.* § 1503(d)(2).

17. *Id.* § 1503(a)(1).

18. *Id.* § 1503(e)(1) ("[I]t shall not be considered a violation of any provision of antitrust laws for 2 or more private entities to exchange or provide a cyber threat indicator or defensive measure, or assistance relating to the prevention, investigation, or mitigation of a cybersecurity threat, for cybersecurity purposes under this subchapter."). There is a limited carve-out if the CISA is used for traditional antitrust behavior, like market allocation, price fixing, or boycotts. *See id.* § 1507(e).

19. *Id.* § 1505(a), (b).

20. *Id.* § 1505(c).

The CISA presents opportunities for the automated vehicle industry to share cybersecurity risks. The automobile industry has created an industry information-sharing mechanism for cybersecurity threats and vulnerabilities, the Automotive Information Sharing and Analysis Center (Auto-ISAC)—the Auto-ISAC was not created pursuant to the CISA.[21] Auto-ISAC launched in August 2015, and membership includes automakers, suppliers, and members of the commercial motor vehicle industry. Auto-ISAC conducts workshops, summits, and information exchange events; produces best practices; and creates intelligence reports based on cybersecurity risks facing connected vehicles throughout the world.

C. Driver Privacy Protection Act of 1994

The Driver Privacy Protection Act of 1994 (the 1994 Act) governs the release and use of personal information by state departments of motor vehicles (DMVs).[22] The 1994 Act applies to data collected in a "motor vehicle record," which is "any record that pertains to a motor vehicle operator's permit, motor vehicle title, motor vehicle registration, or identification card issued by a department of motor vehicles."[23] The 1994 Act distinguishes between "personal information" and "highly restricted personal information." Personal information is "information that identifies an individual, including an individual's photograph, social security number, driver identification number, name, address (but not the 5-digit zip code), telephone number, and medical or disability information, but does not include information on vehicular accidents, driving violations, and driver's status."[24] Highly restricted personal information is "an individual's photograph or image, social security number, [and] medical or disability information."[25]

Generally, a DMV or its employees or contractors cannot disclose personal information except for specified reasons, and highly restricted personal information can only be disclosed in very limited circumstances.[26] There are 14 exceptions that allow the use of personal information, including use by government agencies and their personnel in carrying out their functions; "use in connection with matters of motor vehicle or driver safety and theft"; recalls; motor vehicle emissions; motor vehicle research activities; performance monitoring

---

21. The Auto-ISAC was created pursuant to Presidential Policy Directive-21 (PPD-21): Critical Infrastructure Security and Resilience (signed in February 2013), the National Infrastructure Plan of 2013, and Presidential Decision Directive 63 in 1998. *See Frequently Asked Questions, About Us, What Is an ISAC?*, AUTO-ISAC, https://www.automotiveisac.com/faqs/ (last visited Jan. 18, 2020).

22. 18 U.S.C. § 2721 *et seq.*

23. *Id.* § 2725(1).

24. *Id.* § 2725(3).

25. *Id.* § 2725(4).

26. *Id.* § 2721(a).

of motor vehicles; use in court proceedings; use in research activities so long as the personal information is redacted; use pursuant to written consent by the individual; and "for any other use specifically authorized under the law of the State that holds the record, if such use is related to the operation of a motor vehicle or public safety."[27]

The 1994 Act imposes criminal fines on violators.[28] The 1994 Act also creates a private cause of action for people harmed to sue disclosers and those who obtain or use personal information in violation of the 1994 Act.[29] The 1994 Act allows for the recovery of actual damages and attorney's fees, as well as punitive damages upon a showing of "willful or reckless disregard of the law" and other equitable relief.[30]

Because the 1994 Act is limited to information contained in or acquired in connection with motor vehicle records, it will have limited application to the troves of data that automated or connected vehicles will collect. For example, the 1994 Act would not apply to personal information collected by a state or local government by using connected infrastructure or connected or automated vehicles. Indeed, it likely would not provide any more protection than it does for conventional vehicles.

D. Driver Privacy Protection Act of 2015

While the 1994 Act addressed driver privacy for data shared with state DMVs, the Driver Privacy Protection Act of 2015 (the 2015 Act) addressed concerns over data privacy and ownership of event data recorders (EDRs), the black boxes of vehicles.[31] Congress firmly established that the vehicle owner or the vehicle lessee is the owner of data recorded by an EDR:

> Any data retained by an event data recorder (as defined in section 563.5 of title 49, Code of Federal Regulations), regardless of when the motor vehicle in which it is installed was manufactured, is the property of the owner, or, in the case of a leased vehicle, the lessee of the motor vehicle in which the event data recorder is installed.[32]

The 2015 Act prohibits access by other persons to EDR data except in limited circumstances. These circumstances are (1) a court order authorizing the retrieval of EDR data, (2) consent by the vehicle owner or lessee for any reason, (3) investigations by federal agencies under sections 1131(a) or 30166 of title 49,

---

27. The exceptions are set forth in subsection (b).
28. 18 U.S.C. § 2723.
29. *Id.* § 2724(a).
30. *Id.* § 2724(b).
31. Driver Privacy Protection Act of 2015, S. 766, 114th Congress.
32. *Id.* § 2.

(4) as needed to facilitate emergency medical response following a crash, and (5) traffic safety research provided that personally identifiable information is not disclosed.

Thus, under the 2015 Act, the owner or lessee of an automated vehicle would be the owner of data collected by an EDR. This ownership may help establish who has the right to consent to a search of that data or how the owner of the vehicle can obtain EDR data to reconstruct a crash in a civil lawsuit. An EDR is defined to mean "a device or function in a vehicle that records the vehicle's dynamic time-series data during the time period just prior to a crash event (e.g., vehicle speed vs. time) or during a crash event (e.g., delta–V vs. time), intended for retrieval after the crash event. For the purposes of this definition, the event data do not include audio and video data."[33] Basically, EDR data includes the precrash information about the vehicle. NHTSA's regulations on EDRs are outlined in 49 C.F.R. § 563.1 *et seq.*

The 2015 Act does not answer who or what owns the most private data of an automated or connected vehicle: non-EDR data. There will be a great quantity of this non-EDR data, and it will contain sensitive personal information that will be sought for commercial purposes. This data will clearly be more valuable, especially when coupled with people's other data from the Internet and other sources or other people's data. Accordingly, although the 2015 Act provides some clarity as to one aspect of data collection, by limiting itself to data collected by an EDR, the 2015 Act does not give the needed clarity on the wealth of other types of data collected by modern vehicles, connected vehicles, or automated vehicles.

E. The Privacy Act of 1974

The Privacy Act of 1974 (the Privacy Act), codified at 5 U.S.C. § 552a, provides rules and limitations on federal agency collection of personal information. Because federal agencies may own or have access to data collected by federal-government-owned automated vehicles or intelligent transportation systems, the Privacy Act governs the collection, possession, use, and dissemination of that data. However, as will be seen below, the Privacy Act provides little protection in the context of data collection by automated vehicles or intelligent transportation systems.

Under the Privacy Act, unless one of 12 exceptions applies, "[n]o agency shall disclose any record which is contained in a system of records by any means of communication to any person, or to another agency, except pursuant to a written request by, or with the prior written consent of, the individual to

33. 49 C.F.R. § 563.5.

whom the record pertains."[34] The Privacy Act also includes specific requirements on agencies that maintain a "system of records." For example, the agency must grant an individual access to the system of records and the right to correct incorrect information within the records. The Privacy Act imposes rules on the collection of records in a system of records. The Privacy Act includes both criminal and civil remedies. For an individual to seek recovery for a wrongful disclosure of a record on such person, he or she would need to show an adverse effect and a willful and intentional disclosure. The Privacy Act requires actual damages, and the Supreme Court has interpreted this to exclude emotional distress damages.[35]

An agency is defined in the same manner as it is under the Federal Freedom of Information Act.[36] The Privacy Act does not cover state and local governments and private entities.[37] A record means any "information about an individual that is maintained by an agency."[38] A "system of records" is "a group of any records under the control of any agency from which information is retrieved by the name of the individual or by some identifying number, symbol, or other identifying particular assigned to the individual."[39]

Based on the definitions of "records" and "systems of records," data collected by federal automated vehicles or an ITS that does not include identifying information would be excluded from the calculation. For example, a LiDAR sensor may collect data on a pedestrian's walking mannerisms but may not be able to identify the individual and thus would not be a record. Even when sensors capture identifying information, such as images of an individual, those records may not be considered a system of records unless organized to be a system of records. The United States Court of Appeals for the District of Columbia examined the collection of data by the Transportation Security Administration (the TSA) in *Electronic Privacy Information Center v. United States Department of Homeland Security*.[40] The case was about the TSA's Advanced Imaging Technology (AIT) that it began deploying in airports in 2007 as a more effective means to detect contraband on passengers than the traditional magnetometer. The AIT produces a crude, unclothed image of each passenger that is only viewable to a person in a back room; the TSA took specific actions to protect the privacy of passengers, including filtering facial features and prohibiting the use of cell

34. 5 U.S.C. § 552a(b).

35. F.A.A. v. Cooper, 566 U.S. 284, 287, 132 S. Ct. 1441, 182 L. Ed. 2d 497 (2012).

36. 5 U.S.C. § 552a(a)(1).

37. There is a limited caveat to this statement because Pub. L. No. 93-579, 88 Stat. 1896 (Dec. 31, 1974), added a note that provides Social Security usage restrictions on state and local government agencies.

38. 5 U.S.C. § 552a(a)(4).

39. *Id.* § 552a(a)(5).

40. 653 F.3d 1 (D.C. Cir. 2011).

phones or cameras in the room where the images are reviewed. The plaintiff claimed that these images were records and systems of records for purposes of the Privacy Act. The D.C. Circuit rejected that contention because the TSA did "not maintain data from AIT scanners in a 'system of records' linked to names or any other identifier. Even if, as the petitioners speculate, the TSA has the ability to combine various sources of information and then to link names to the images produced using AIT, their Privacy Act claim still fails because they offer no reason to believe the TSA has in fact done that."[41] Accordingly, unless federal agencies maintain data collected by their automated vehicles or ITSs in an individually identifiable system of records, a Privacy Act claim would likely be unsuccessful.

II. FEDERAL AGENCY ACTIVITIES

Federal agencies have sought to fill the void created by congressional inaction. Different federal agencies have issued guidance on privacy and cybersecurity. Informal guidance from an agency is not hard law, but that does not mean that organizations should avoid reading these publications or ignore their recommendations. These publications provide valuable information and strategies for all organizations that collect or use data, and they often reflect industry standards on the topics.[42] Most agencies cooperate with industry and academia through workshops, meetings, and comments in developing their cybersecurity and privacy guidance. This cooperation helps establish baseline cybersecurity and privacy practices that can be used by litigants in a civil lawsuit to show that an organization failed to exercise reasonable care or by a government agency, such as NHTSA or the FTC, in prosecuting violators.

This section initially introduces the authority of the Department of Homeland Security (the DHS) in cybersecurity, followed by a discussion of the efforts of two federal bodies, the National Institute of Standards and Technology (NIST) and the National Telecommunications and Information Administration (NTIA), to establish baselines in cybersecurity and privacy. The remainder of this section focuses on the efforts of NHTSA and the FTC in providing guidance and their enforcement mechanisms for cybersecurity and privacy.

---

41. *Id.* at 8.

42. Another effective source of industry standards and guidance comes from industry groups, such as the Society of Automotive Engineers International (SAE), the International Organization for Standardization, and the Institute of Electrical and Electronics Engineers. SAE's comprehensive cybersecurity publication is J3061, and SAE has subcommittees dedicated to cybersecurity of automobiles.

A. The DHS

The DHS is the primary federal agency that oversees the national efforts to protect critical infrastructure from cyberattacks and vulnerabilities. The DHS's principal efforts are housed in the Cybersecurity and Infrastructure Security Agency. The Cybersecurity and Infrastructure Security Agency leads the national effort on cybersecurity by, among other things, providing 24/7 "cyber situational awareness, analysis, incident response and cyber defense capabilities" to all levels of government, private industry, and international partners; coordinating resilience for industry; and operating the National Risk Management Center.[43] The DHS's Science and Technology Directorate has been researching potential utility and vulnerabilities of automated vehicles and operates a testbed, spoofs protection for global navigation satellite systems, provides analytics for evaluating the performance of advanced driver assistance systems, and uses AI and machine learning for automated systems.

B. NIST and NTIA

NIST is a nonregulatory agency within the Department of Commerce that helps foster industry innovation and competitiveness through research and standard setting. NIST has issued and continually updates guidance on cybersecurity. NIST's publications have included "Framework for Improving Critical Infrastructure Cybersecurity"[44] and draft publications entitled "NIST Privacy Framework: A Tool for Improving Privacy through Enterprise Risk Management"[45] and "Recommendations for IoT Device Manufacturers: Foundational Activities and Core Device Cybersecurity Capability Baseline."[46] NIST's Cybersecurity Framework is widely considered the industry baseline for cybersecurity practices. NHTSA has issued a report on applying NIST's Cybersecurity Framework to automobiles.[47] NIST also operates the National

---

43. *What Does CISA Do?*, Cybersecurity & Infrastructure Security Agency, https://www.cisa.gov/ (last visited Apr. 4, 2020).

44. *Framework for Improving Critical Infrastructure Cybersecurity*, Nat'l Inst. Stds. & Tech. (Apr. 16, 2018), https://nvlpubs.nist.gov/nistpubs/CSWP/NIST.CSWP.04162018.pdf.

45. *NIST Privacy Framework: A Tool for Improving Privacy through Enterprise Risk Management*, Preliminary Draft, Nat'l Inst. Stds. & Tech. (Sept. 6, 2019), https://www.nist.gov/system/files/documents/2019/09/09/nist_privacy_framework_preliminary_draft.pdf.

46. Michael Fagan, Katerina N. Megas, Karen Scarfone & Matthew Smith, *Recommendations for IoT Device Manufacturers: Foundational Activities and Core Device Cybersecurity Capability Baseline*, Second Draft, Nat'l Inst. Stds. & Tech. (Jan. 2020), https://nvlpubs.nist.gov/nistpubs/ir/2020/NIST.IR.8259-draft2.pdf.

47. Charlie McCarthy & Kevin Harnett, *National Institute of Standards and Technology Cybersecurity Risk Management Framework Applied to Modern Vehicles*, Nat'l Highway Traffic & Safety Admin. (Oct. 2014), https://www.nhtsa.gov/sites/nhtsa.dot.gov/files/812073_natlinstitstandardstechcyber.pdf.

Cybersecurity Center for Excellence, a collaboration among industry, government, and academia to conduct security research and establish cybersecurity standards.

NTIA is an agency within the Department of Commerce that is responsible for advising the United States president on issues relating to telecommunications and information. NTIA has used this role to issue guidance and collaborate on privacy and cybersecurity. NTIA issued guidance in September 2018 on what it believes are core privacy principles and objectives for federal action on privacy.[48] In addition to its focus on privacy, NTIA has been working with stakeholders to enhance security vulnerability disclosures.[49]

C. NHTSA

NHTSA has taken both a proactive and reactive role in cybersecurity. NHTSA has broad authority over the design and safety of motor vehicles and motor vehicle equipment.[50] This authority includes the ability to set Federal Motor Vehicle Safety Standards (Safety Standards) and to conduct recalls when a vehicle or item of motor vehicle equipment fails to comply with the Safety Standards or suffers from a defect related to motor vehicle safety. Because NHTSA's statutory authority concerns motor vehicle safety and design, it may have limited power over privacy and data collection of the automobile industry.

In setting Safety Standards, NHTSA establishes requirements for manufacturers of motor vehicles and motor vehicle equipment. In 2012, Congress required NHTSA to study and submit a report on "the need for safety standards with regard to electronic systems in passenger motor vehicles. In conducting this examination, the Secretary shall consider the electronic components, the interaction of electronic components, the security needs for those electronic systems to prevent unauthorized access, and the effect of surrounding environments on the electronic systems."[51] NHTSA has not proposed a Safety Standard on cybersecurity.[52] This approach seems prudent given the need for a Safety Standard to be "practicable, meet the need for motor vehicle safety, and be stated in objective terms,"[53] as well as the amount of time it takes NHTSA to

---

48. Nat'l Telecomms. & Info. Admin., *Developing the Administration's Approach to Consumer Privacy*, 83 Fed. Reg. 48,600-01 (Sept. 26, 2018).

49. *See* Angela Simpson, *Improving Cybersecurity through Enhanced Vulnerability Disclosure*, NAT'L TELECOMMS. & INFO. ADMIN. (Dec. 15, 2016), https://www.ntia.gov/blog/2016/improving-cybersecurity-through-enhanced-vulnerability-disclosure.

50. Chapter 5 details NHTSA's authority.

51. Moving Ahead for Progress in the 21st Century Act, Pub. L. No. 112-141, § 31402, 126 Stat. 405 (2012).

52. Nat'l Highway Traffic Safety Admin., *Report to Congress: "Electronic Systems Performance in Passenger Motor Vehicles"*, U.S. DEP'T OF TRANSP. (Dec. 2015), https://www.nhtsa.gov/sites/nhtsa.dot.gov/files/electronic-systems-performance-in-motor20vehicles.pdf.

53. 49 U.S.C. § 30111(a). See Chapter 5 for a discussion of these elements.

develop a Safety Standard. Cybersecurity best practices and vulnerabilities are dynamic and are constantly changing. NHTSA's efforts, though not through the promulgation of a Safety Standard, reflect that its best role in cybersecurity is to proactively issue guidance and work with industry to establish best practices.

In this regard, NHTSA has been busy providing substantial guidance on cybersecurity of motor vehicles and automated vehicles. This guidance has included issuing guides and reports, conducting workshops and roundtables on cybersecurity, researching the need for a Safety Standard on securing electronic systems, submitting a report to Congress, and issuing joint safety principles with automakers. NHTSA was instrumental in supporting the automobile industry's formation of the Auto-ISAC, a global information-sharing organization on cybersecurity risks for industry members to share cybersecurity vulnerabilities anonymously. In October 2016, NHTSA released "Cybersecurity Best Practices for Modern Vehicles," in which NHTSA made cybersecurity recommendations to the automotive industry.[54]

NHTSA's recall power also gives NHTSA significant authority over cybersecurity of automated vehicles. Without a Safety Standard, or even with a Safety Standard, the primary focus will be on whether a cybersecurity risk or vulnerability is a defect related to motor vehicle safety. A manufacturer is under a good-faith obligation to report safety defects in its vehicles. An important question will be whether all cybersecurity vulnerabilities in a motor vehicle are safety defects or whether a certain threshold, such as an actual breach, is needed to trigger the recall power.

NHTSA has already used its recall power to demand a recall due to cybersecurity vulnerabilities. In 2015, two white-hat hackers,[55] in connection with a writer from *Wired*, demonstrated that they could remotely take control of a 2014 Jeep Cherokee traveling on the highway and turn on the air conditioner, blast music over the radio, activate the windshield wipers, burst windshield wiper fluid, and stop the vehicle.[56] The hackers gained access to the Jeep Cherokee through the cellular connection to the entertainment system in the vehicle. NHTSA required the recall before cyberattackers exploited the vulnerabilities. The recall applied to 1,416,903 vehicles, and Chrysler mailed vehicle owners a

---

54. Nat'l Highway Traffic Safety Admin., *Cybersecurity Best Practices for Modern Vehicles*, U.S. Dep't of Transp. (Oct. 2016), https://www.nhtsa.gov/sites/nhtsa.dot.gov/files/documents/812333_cybersecurityformodernvehicles.pdf.

55. A white-hat hacker is someone who tests systems and vulnerabilities to improve security before malicious hackers can exploit those vulnerabilities.

56. Andy Greenberg, *Hackers Remotely Kill a Jeep on the Highway—With Me in It*, Wired (July 21, 2015), https://www.wired.com/2015/07/hackers-remotely-kill-jeep-highway/.

USB drive that included the software update, created a downloadable software patch, and allowed customers to have the vulnerabilities fixed at a dealership.[57]

D. FTC

In the absence of a federal act regulating and governing data and privacy, the FTC will continue to exercise its powers under section 5 to fill that void. Beginning in the mid-1990s, the FTC shifted part of its focus to online privacy.[58] The FTC's initial approach to online privacy was to foster industry self-regulation, and the FTC explicitly recommended against congressional regulation of online privacy.[59] The FTC has also used its powers to investigate data breaches and data security practices. The FTC will be the primary regulator of automated vehicle data privacy and security. Because the FTC is expected to have a significant role in the automated vehicle industry, this section begins by describing the FTC's powers and its procedures in investigating claims and violations. This section then considers how the FTC has exercised its powers for data privacy and data breaches.

1. Overview of the FTC and the FTC Act

Section 5 declares unlawful "unfair or deceptive acts or practices in or affecting commerce."[60] The FTC uses both prongs—the unfairness prong and the deceptive prong—to pursue privacy protection and data security. The FTC's power in this regard is independent of any other laws, and the FTC's enforcement is not dependent upon a particular act or practice being unlawful in some other manner.[61]

---

57. Fiat Chrysler Autos., Safety Recall R40/NHTSA 15V-461 Radio Security Vulnerability (July 2015), https://static.nhtsa.gov/odi/rcl/2015/RCRIT-15V461-7681.pdf.

58. Fed. Trade Comm'n, *Privacy Online: A Report to Congress* 2 (June 1998), https://www.ftc.gov/sites/default/files/documents/reports/privacy-online-report-congress/priv-23a.pdf.

59. *Id.*; Fed. Trade Comm'n, *Self-Regulation and Privacy Online: A Report to Congress* 7 (July 1999), https://www.ftc.gov/sites/default/files/documents/public_statements/prepared-statement-federal-trade-commission-self-regulation-and-privacy-online/privacyonlinetestimony.pdf (stating that "the Commission believes that legislation to address online privacy is not appropriate at this time"); *but see* Fed. Trade Comm'n, *Privacy Online: Fair Information Practices in the Electronic Marketplace: A Report to Congress* at ii (May 2000), https://www.ftc.gov/sites/default/files/documents/reports/privacy-online-fair-information-practices-electronic-marketplace-federal-trade-commission-report/privacy2000.pdf (recommending legislation in conjunction with self-regulation).

60. 15 U.S.C. § 45(a).

61. *See, e.g.*, F.T.C. v. Accusearch Inc., 570 F.3d 1187, 1194 (10th Cir. 2009) (rejecting an argument because "[i]ts premise appears to be that a practice cannot be an unfair one unless it violates some law independent of the FTCA. But the FTCA imposes no such constraint. On the contrary, the FTCA enables the FTC to take action against unfair practices that have not yet been contemplated by more specific laws." (internal citation omitted)); Spiegel, Inc. v. F.T.C., 540 F.2d 287, 292 (7th Cir. 1976) ("In *F.T.C. v. Sperry & Hutchinson Co.*, 405 U.S. 233, 92 S. Ct. 898, 31 L. Ed. 2d 170 (1972), the Supreme Court

The unfairness prong is purposely vague to give the FTC flexibility to declare new acts or practices unfair.[62] Historically, the FTC believed there were three separate components of "unfairness": "(1) whether the practice injures consumers; (2) whether it violates established public policy; [or] (3) whether it is unethical or unscrupulous."[63] In a 1980 Policy Statement, the FTC clarified these three prongs. The consumer injury prong, the most important prong, has three parts: "It must be substantial; it must not be outweighed by any countervailing benefits to consumers or competition that the practice produces; and it must be an injury that consumers themselves could not reasonably have avoided."[64] The public policy prong looks at "whether the conduct violates public policy as it has been established by statute, common law, industry practice, or otherwise."[65] The FTC acknowledged that it typically uses public policy to support the prevention of consumer injury but that it can serve and has served as an independent justification. The FTC abandoned the "unethical or unscrupulous" prong. In 1994, Congress codified the consumer injury test:

> The Commission shall have no authority . . . to declare unlawful an act or practice on the grounds that such act or practice is unfair unless the act or practice causes or is likely to cause substantial injury to consumers which is not reasonably avoidable by consumers themselves and not outweighed by countervailing benefits to consumers or to competition. In determining whether an act or practice is unfair, the Commission may consider established public policies as evidence to be considered with all other evidence. Such public policy considerations may not serve as a primary basis for such determination.[66]

---

left no doubt that the FTC had the authority to prohibit conduct that, although legally proper, was unfair to the public.").

62. *See* H.R. Rep. No. 1142, 63d Cong., 2d Sess. 19 (1914) ("It is impossible to frame definitions which embrace all unfair practices. There is no limit to human inventiveness in this field. Even if all known unfair practices were specifically defined and prohibited, it would be at once necessary to begin over again. If Congress were to adopt the method of definition, it would undertake an endless task."). Although this House Report quote is about the unfairness prong for competition, courts have considered the House Report quote relevant to the unfairness prong for consumer protection. *See, e.g., In re* Pfizer Inc., 81 F.T.C. 23 (1972) ("Unfairness is potentially a dynamic analytical tool capable of a progressive, evolving application which can keep pace with a rapidly changing economy. Thus as consumer products and marketing practices change in number, complexity, variety, and function, standards of fairness to the consumer may also change.").

63. Fed. Trade Comm'n, *FTC Policy Statement on Unfairness* (Dec. 17, 1980), https://www.ftc.gov/public-statements/1980/12/ftc-policy-statement-unfairness.

64. *Id.*

65. *Id.*

66. 15 U.S.C. § 45(n).

The FTC has issued a policy statement on its enforcement policy against deceptive acts or practices.[67] The FTC stated that it "will find deception if there is a representation, omission or practice that is likely to mislead the consumer acting reasonably in the circumstances, to the consumer's detriment."[68] The FTC has stated that "in determining whether a representation is deceptive, we conduct a three-step inquiry, determining: (1) what claims are conveyed; (2) whether those claims are false, misleading, or unsubstantiated; and (3) whether the claims are material."[69] "A representation is considered material if it involves information that is important to consumers and, hence, likely to affect their choice of, or conduct regarding a product."[70] The FTC presumes that express representations are material.[71]

The FTC Act gives the FTC authority to investigate practices and acts that it believes violate the FTC Act.[72] In conducting investigations, the FTC can subpoena production of documents and information, and it has the right to request testimony of witnesses. These precomplaint investigations are usually not public.

If the FTC has reason to believe that section 5 is being violated, it can initiate an administrative process or judicial process. In the administrative process, the FTC issues a proposed complaint setting forth the violations and the remedy. The respondent can either settle or contest the items in the complaint. If the respondent contests the complaint, then the complaint is adjudicated by an administrative law judge (ALJ). The ALJ issues an initial decision that is appealable to the FTC. The FTC will then issue its final order. The Supreme Court has

---

67. Fed. Trade Comm'n, *FTC Policy Statement on Deception* (Oct. 14, 1983), https://www.ftc.gov/system/files/documents/public_statements/410531/831014deceptionstmt.pdf.

68. *Id.* at 2; *see also F.T.C. v. World Travel Vacation Brokers, Inc.*, 861 F.2d 1020, 1029 (7th Cir. 1988) (stating that an act is deceptive if "the representations, omissions, or practices likely would mislead consumers, acting reasonably, to their detriment").

69. *In re* Cambridge Analytica, LLC, No. 9383, at 11 (F.T.C. Nov. 25, 2019), *available at* https://www.ftc.gov/system/files/documents/cases/d09389_comm_final_opinionpublic.pdf.

70. *Id.* (internal quotation marks and citations omitted).

71. *Id.*

72. 15 U.S.C. § 43 ("The Commission may . . . prosecute any inquiry necessary to its duties in any part of the United States."); 15 U.S.C. § 46(a) ("The Commission shall also have power . . . to gather and compile information concerning, and to investigate from time to time the organization, business, conduct, practices, and management of any person, partnership, or corporation engaged in or whose business affects commerce.").

given the FTC wide latitude in crafting its orders.[73] A final order of the FTC is appealable to an appropriate United States Circuit Court of Appeals.[74]

The FTC is not required to commence an administrative process and can proceed with a judicial process in certain circumstances.[75] Section 13(b) of the FTC Act also empowers the FTC to seek a temporary, preliminary, or permanent injunction when it has "reason to believe" that section 5 is being violated and such injunction is in "the interest of the public."[76] In seeking to enjoin a trade practice or act, courts have overwhelmingly found, except for a recent decision of the Seventh Circuit, that the injunction power includes the power to seek restitution or other remedies such as freezing assets.[77]

Once the cease and desist order and the injunction become final, the FTC is given additional powers to enforce the order. The FTC can file a civil action against violators of that cease and desist order.[78] If the violator is the original party to whom the FTC issued the cease and desist order, the FTC only needs to show the violation occurred. If the violator is not the original party, then the FTC can seek civil remedies if it can show that "the act or practice to which the cease and desist order relates is one which a reasonable man would have known under the circumstances was dishonest or fraudulent."[79]

---

73. F.T.C. v. Henry Broch & Co., 368 U.S. 360, 364, 82 S. Ct. 431, 433–34, 7 L. Ed. 2d 353 (1962) ("The Commission has a wide discretion to formulate a remedy adequate to prevent Broch's repetition of the violation he was found to have committed."); Jacob Siegel Co. v. F.T.C., 327 U.S. 608, 611–12, 66 S. Ct. 758, 760, 90 L. Ed. 888 (1946) ("The Commission has wide discretion in its choice of a remedy deemed adequate to cope with the unlawful practices in this area of trade and commerce. Here, as in the case of orders of other administrative agencies under comparable statutes, judicial review is limited.").

74. 15 U.S.C. § 45(c).

75. *E.g.*, United States v. JS & A Grp., Inc., 716 F.2d 451, 457 (7th Cir. 1983) ("[W]e hold that section 13(b) of the FTC Act authorizes the Commission to seek, in a proper case, and the court to grant, after proper proof, permanent injunctive relief, irrespective of whether a Commission proceeding regarding the alleged violations is pending or contemplated."); F.T.C. v. H.N. Singer, Inc., 668 F.2d 1107, 1111 (9th Cir. 1982) ("We hold that [section] 13(b) gives the Commission the authority to seek, and gives the district court the authority to grant, permanent injunctions in proper cases even though the Commission does not contemplate any administrative proceedings.").

76. 15 U.S.C. § 53(b).

77. *Compare H.N. Singer, Inc.*, 668 F.2d at 1113 ("We hold that Congress, when it gave the district court authority to grant a permanent injunction against violations of any provisions of law enforced by the Commission, also gave the district court authority to grant any ancillary relief necessary to accomplish complete justice because it did not limit that traditional equitable power explicitly or by necessary and inescapable inference."); F.T.C. v. Sec. Rare Coin & Bullion Corp., 931 F.2d 1312, 1315 (8th Cir. 1991) (holding that the district court had the authority to award equitable monetary relief); F.T.C. v. U.S. Oil & Gas Corp., 748 F.2d 1431, 1434 (11th Cir. 1984) (holding that a "Court's inherent equitable powers may be employed to issue a preliminary injunction, including a freeze of assets, during the pendency of an action for permanent injunctive relief"), *with* F.T.C. v. Credit Bureau Ctr., LLC, 937 F.3d 764, 786 (7th Cir. 2019) (overturning prior cases and holding "that section 13(b)'s permanent-injunction provision does not authorize monetary relief").

78. 15 U.S.C. § 57b.

79. *Id.* § 57b(a)(2).

The FTC can also commence a civil action to recover civil penalties once a cease and desist order becomes final if any person engages in the act or practice "with actual knowledge that such act or practice is unfair or deceptive and is unlawful."[80] The FTC typically establishes actual knowledge by showing that it "provided the violator with a copy of the Commission determination about the act or practice in question, or a 'synopsis' of that determination."[81] The current penalty for each violation as adjusted for inflation is $42,530.[82] Each day of continued violation is treated as a separate violation. Courts have held that each contact with a consumer constitutes a separate violation.[83] In determining the penalty amount, courts are supposed to consider "the degree of culpability, any history of prior such conduct, ability to pay, effect on ability to continue to do business, and such other matters as justice may require."[84]

Very few FTC privacy or data actions end up in court, and a consent decree resolves most disputes between the FTC and the potential offender.[85] These consent orders are in effect contracts between the FTC and the persons under investigation. Many of these consent decrees remain in effect for 20 years, but the duration varies.[86] In crafting its remedies in consent decrees, the FTC has used a variety of the following remedies: (1) prohibitions on wrongful activities; (2) fines, remuneration, or freezing of assets; (3) notification to consumers and remediation; (4) the deletion of data or the agreement not to use data; (5) changes to privacy policies; (6) creation of comprehensive programs on data, privacy, or security; (7) independent assessments by third-party professionals; (8) recordkeeping and reporting to the FTC; and (9) notification of material changes.[87]

Because there are few published court decisions, practitioners use the complaints and orders published by the FTC as guidance. These complaints shed light on what acts and practices the FTC considers unlawful and where the FTC is allocating its resources. The complaints and orders issued by the FTC are published online at https://www.ftc.gov/enforcement/cases-proceedings.

---

80. *Id.* § 45(m)(1)(B).

81. Fed. Trade Comm'n, *A Brief Overview of the Federal Trade Commission's Investigative, Law Enforcement, and Rulemaking Authority* (Oct. 2019), https://www.ftc.gov/about-ftc/what-we-do/enforcement-authority.

82. 16 C.F.R. § 1.98.

83. *E.g.*, United States v. Reader's Digest Ass'n, Inc., 662 F.2d 955, 966 (3d Cir. 1981) (holding that "each letter included as part of a mass mailing constitutes a separate violation"); United States v. Floersheim, No. CV 74-484-RF, 1980 WL 1852, at *9 (C.D. Cal. May 1, 1980) (holding that each form sent to consumers constitutes a separate violation).

84. 15 U.S.C. § 45(m)(1)(C).

85. The consent order procedures are set forth in 16 C.F.R. § 2.31 *et seq.*

86. Daniel J. Solove & Woodrow Hartzog, *The FTC and the New Common Law of Privacy*, 114 Colum. L. Rev. 583, 614 (2014).

87. *Id.* at 614–19.

There have been two significant challenges to the FTC's authority in data breach cases that are relevant to both privacy and data security cases. *F.T.C. v. Wyndham Worldwide Corporation*[88] (*Wyndham*) presented the first court challenge to the authority of the FTC over data breaches. In *Wyndham*, the FTC issued a complaint against Wyndham following three data breaches. The FTC alleged that Wyndham's conduct constituted an unfair trade practice and that Wyndham's privacy policy was deceptive. Wyndham did not challenge that privacy policy claim in the Third Circuit but did test two aspects of the FTC's authority over cybersecurity. Wyndham contended that the FTC lacked authority under section 5 to regulate cybersecurity and, in the alternative, that it did not receive fair notice that its conduct violated section 5.

The Third Circuit held that Wyndham's data security conduct falls within the meaning of "unfair" under section 5 and that the FTC has the authority to regulate data security under section 5. The court then considered whether Wyndham had fair notice that its conduct would violate section 5. The court ruled that Wyndham is not entitled to "know with ascertainable certainty the FTC's interpretation of what cybersecurity practices are required by [section 5]. Instead, the relevant question . . . is whether Wyndham had fair notice that its conduct could fall within the meaning of the statute." The court had little trouble finding that Wyndham had notice that its conduct could fall within section 5. The court reviewed the soft law publications and consent decrees as evidence that Wyndham could have figured out that weak cybersecurity practices could be prosecuted under section 5. *Wyndham* gave strong support of the FTC's power in data security, and its reasoning may prevent similar challenges in other circuits or challenges to the FTC's authority to regulate privacy.

In *LabMD, Inc. v. F.T.C.*,[89] the Eleventh Circuit ruled that a cease and desist order issued by the FTC was unenforceable for lack of specificity. LabMD was a medical laboratory. Its billing manager downloaded Limewire onto his work computer in violation of LabMD policy, and he inadvertently shared a file with personal information of 9,300 customers. A company specializing in data security, Tiversa Holding Corporation (Tiversa), accessed the file and contacted LabMD to inquire about LabMD retaining it to provide data security services. LabMD removed Limewire from the employee's computer. LabMD refused Tiversa's services, and at a later date, Tiversa gave the file to the FTC. The FTC filed a complaint against LabMD, claiming that LabMD failed to perform several data-security measures. LabMD sought to dismiss the claim, which was denied by the FTC. After hearing evidence, the ALJ dismissed the complaint against LabMD, finding that the FTC failed to prove an unfair or deceptive

88. 799 F.3d 236 (3d Cir. 2015).
89. 894 F.3d 1221 (11th Cir. 2018).

trade practice. The FTC appealed it to the full FTC, which reversed the ALJ. LabMD appealed the case to the Eleventh Circuit.

The Eleventh Circuit reversed the FTC. The Eleventh Circuit focused on how the FTC had used the Limewire security breach to broadly contend that LabMD's entire data security practices were deficient. The FTC then used the broad attack to contend that LabMD was required "to implement and maintain a data-security program 'reasonably designed' to the Commission's satisfaction." The Eleventh Circuit held that requiring a "reasonably designed" data-security program without specifying the components of such a program lacked specificity: "[T]he cease and desist order contains no prohibitions. It does not instruct LabMD to stop committing a specific act or practice. Rather, it commands LabMD to overhaul and replace its data-security program to meet an indeterminable standard of reasonableness. This command is unenforceable."[90]

Under *LabMD*, the FTC will need to specifically tailor the conditions and obligations of companies that are accused of failing to protect consumer privacy or data security. Footnote 24 in the case could have even greater implications for future FTC developments; it provides:

> Section 5(n) now states, with regard to public policy, "In determining whether an act or practice is unfair, the Commission may consider established public policies as evidence to be considered with all other evidence. Such public policy considerations may not serve as a primary basis for such determination." We do not take this ambiguous statement to mean that the Commission may bring suit purely on the basis of substantial consumer injury. The act or practice alleged to have caused the injury must still be unfair under a well-established legal standard, whether grounded in statute, the common law, or the Constitution.[91]

As described above, Congress, in enacting section 45(n), had appeared to signal that the consumer injury test was the only test to establish unfairness. The Eleventh Circuit appears to reject that analysis and will require that the act or practice violate the traditional public policy test in addition to the consumer injury test. Thus, in the Eleventh Circuit, the FTC may have a heightened burden to prove its case. Given that venue can be found in the Eleventh Circuit if the "act or practice was used" in the Eleventh Circuit,[92] litigants may be able to forum shop when it is unclear if the FTC can establish a violation under the policy prong. Litigants should exercise caution before fully

---

90. *Id.* at 1236.

91. *Id.* at n.24. The Eleventh Circuit reiterated this point later on in the case: "The Commission must find the standards of unfairness it enforces in 'clear and well-established' policies that are expressed in the Constitution, statutes, or the common law." *Id.* at 1231.

92. 15 U.S.C. § 45(c). The Eleventh Circuit consists of Florida, Georgia, and Alabama.

relying on this portion of the *LabMD* opinion because the Supreme Court had previously stated, before the enactment of section 45(n), that the consumer injury prong is an independent prong that does not require proof of the other two historic prongs.[93]

2. The FTC "Common Law" for Privacy and Data Security

The FTC has used the deceptive and unfairness prongs of section 5 to pursue companies for privacy violations and data security. In reviewing the FTC's activities for privacy, Daniel Solove and Woodrow Hartzog described the FTC's activities as establishing a "common law of privacy."[94] Unlike traditional common law, which is deduced by reading case law, the FTC's common law is deduced by reviewing the publications and the enforcement activities of the FTC. The FTC, in this regard, is establishing, in the words of Solove and Hartzog, two common law regimes that can help shed light on what data privacy and data security practices or acts are considered unfair or deceptive. Practitioners can review these authorities as they would case law.

The FTC has provided guidance through different avenues. The FTC publishes guidance and hosts workshops, conferences, forums, and roundtables.[95] The FTC advises in speeches, videos, and blogs, as well as through social media such as Twitter. The FTC has issued complaints and entered into consent decrees that detail specific, identifiable acts that the FTC considers grave enough to file a complaint.

a. FTC's Privacy Common Law

The FTC's role in privacy began in the 1990s. The FTC identified five fair information principles that establish core privacy protection: (1) notice and awareness; (2) choice and consent; (3) access and participation; (4) integrity and security; and (5) enforcement and redress.[96] Consumer notice and awareness, the "most fundamental principle," concerns disclosing data collection and use practices.[97] The choice and consent principle is about allowing consumers to

---

93. F.T.C. v. Sperry & Hutchinson Co., 405 U.S. 233, 244 n.5 (1972); *see also Wyndham Worldwide Corp.*, 799 F.3d at 243 (citing *Sperry* and stating that "[t]he Court also held that, under the policy statement, the FTC could deem a practice unfair based on the third prong—substantial consumer injury—without finding that at least one of the other two prongs was also satisfied").

94. Solove & Hartzog, *supra* note 86.

95. The FTC makes its events available online. One of the most important vehicle-related workshops was the FTC and NHTSA's joint workshop on June 28, 2017, entitled "Connected Cars: Privacy, Security Issues Related to Connected, Automated Vehicles," which is available at https://www.ftc.gov/news-events/events-calendar/2017/06/connected-cars-privacy-security-issues-related-connected. The FTC also hosts an annual Privacy Conference entitled "PrivacyCon."

96. Fed. Trade Comm'n, *Privacy Online: A Report to Congress* at 7 (June 1998), https://www.ftc.gov/sites/default/files/documents/reports/privacy-online-report-congress/priv-23a.pdf.

97. *Id.*

have control over how a company uses the personal information collected. The access and participation principle means that consumers have access to their data and the ability to correct inaccuracies and fix incomplete data. The integrity and security principle concerns company use of reputable data and the security of data. The final principle, enforcement and redress, establishes the credibility of the other core principles by making them enforceable.

The FTC has used its section 5 powers in the privacy realm mostly to enforce the notice/awareness and choice/consent core privacy principles. The early actions by the FTC focused on deception in privacy policies—a privacy policy says that the company does not do X, but, in reality, the company does do X or something that could be considered X.[98] Over time, the FTC has expanded its approach to tackle deception throughout all communications by a company to consumers concerning privacy policies,[99] including user manuals.[100] For example, in its complaint against Uber Technologies, Inc., the FTC claimed that Uber failed to fulfill promises made in a statement released on its website about employee access to personal data of drivers and riders.[101] The FTC has

98. *See, e.g.*, *In re* Snapchat, Inc., No. C-4501 (F.T.C. Dec. 23, 2014), *available at* https://www.ftc.gov/system/files/documents/cases/140508snapchatcmpt.pdf; United States v. Path, Inc., No. 3:13-cv-0448 (N.D. Cal. Jan. 31, 2013), *available at* https://www.ftc.gov/enforcement/cases-proceedings/122-3158/path-inc. In its complaint against Snapchat, one of the FTC's allegations was that Snapchat misrepresented that videos would only last as long as the user set and then disappear forever. The FTC found that the video messages were stored in an unrestricted area that a person could access using certain tools and thereby save the video message. The FTC stated that this was publicly known in December 2012, but Snapchat did not remedy the flaw until October 2013. The FTC also found that third-party applications allow people to access video messages without starting the video timer.

99. *See, e.g.*, *In re* Uber Techs., Inc., No. C-4662 (F.T.C. Aug. 15, 2017), *available at* https://www.ftc.gov/enforcement/cases-proceedings/152-3054/uber-technologies-inc; *see In re* HTC Am., Inc., No. 122-3049 (F.T.C. Feb. 22, 2013), *available at* https://www.ftc.gov/enforcement/cases-proceedings/122-3049/htc-america-inc-matter.

100. In its complaint against HTC America, Inc., an allegation against HTC concerned false or misleading statements made by HTC in its user manual. *In re* HTC Am., Inc., No. 122-3049 (F.T.C. Feb. 22, 2013), *available at* https://www.ftc.gov/enforcement/cases-proceedings/122-3049/htc-america-inc-matter. The FTC stated that HTC expressly or impliedly represented in its user manual "that, through the Android permission-based security model, a user of an HTC Android-based mobile device would be notified when a third-party application required access to the user's personal information or to certain functions or settings of the user's device before the user completes installation of the third-party application." The FTC found that the statement was not true because a user "would not be notified when a third-party application required access to the user's personal information or to certain functions or settings of the user's device before the user completes installation of the third-party application."

101. *In re* Uber Techs., Inc., No. C-4662 (F.T.C. Aug. 15, 2017), *available at* https://www.ftc.gov/enforcement/cases-proceedings/152-3054/uber-technologies-inc. The statement provided as follows:

> Uber has a strict policy prohibiting all employees at every level from accessing a rider or driver's data. The only exception to this policy is for a limited set of legitimate business purposes. Our policy has been communicated to all employees and contractors. . . .
>
> The policy is also clear that access to rider and driver accounts is being closely monitored and audited by data security specialists on an ongoing basis, and any violations of the policy will result in disciplinary action, including the possibility of termination and legal action.

also examined whether companies have properly apprised consumers of how they collect and use data—deception by omission when nothing about data collection is disclosed to consumers.[102] Importantly for the automated vehicle industry, a remedy in *In re Aspen Way* required Aspen Way to obtain express consent before tracking geolocation information of consumers.[103]

In March 2012, the FTC issued its best practices for consumer privacy entitled "Protecting Consumer Privacy in an Era of Rapid Change: Recommendations for Businesses and Policymakers" (the Privacy Framework).[104] The Privacy Framework applies to "all commercial entities that collect or use consumer data that can be reasonably linked to a specific consumer, computer, or other device, unless the entity collects only non-sensitive data from fewer than 5,000 consumers per year and does not share the data with third parties."[105]

The FTC claimed that Uber did not in fact closely monitor or audit its employees' access to data and that Uber failed to timely follow up on alerts of potential misuse of consumer personal information provided by its automated monitoring system. The FTC challenged the publicly released statement as being false and misleading.

102. *See, e.g., In re* Aspen Way Enters., Inc., No. 112-3151 (F.T.C. Sept. 25, 2012), *available at* https://www.ftc.gov/enforcement/cases-proceedings/112-3151/aspen-way-enterprises-inc-matter; *In re* Sony BMG Music Enm't, No. 062-3019 (F.T.C. Jan. 30, 2007), *available at* https://www.ftc.gov/enforcement/cases-proceedings/062-3019/sony-bmg-music-entertainment-matter; F.T.C. v. Vizio, Inc., No. 2:17-cv-00758 (N.J. filed Feb. 6, 2017).

Aspen Way rents computers to users. In one of its allegations against Aspen Way, the FTC claimed that Aspen Way failed to disclose to consumers that it could remotely—and surreptitiously—install software that tracked keystrokes, took screenshots, and photographed users with the computer's webcam. Aspen Way used the software and obtained confidential and sensitive information from users to assist with collecting debts or money under the rental agreements. The FTC found that the gathering of personal information constituted an unfair trade practice or act.

In *Sony BMG*, the FTC alleged, among other things, that Sony BMG failed to disclose its collection and use of personal information. The FTC alleged that Sony BMG collected personal information about CD users and transmitted the personal information to Sony BMG, and then Sony BMG used that data to display images and promotional messages on the computers. The FTC claimed that this was an unfair act or practice.

103. *In re* Aspen Way Enters., Inc., No. C-4392 (F.T.C. Apr. 15, 2013), *available at* https://www.ftc.gov/enforcement/cases-proceedings/112-3151/aspen-way-enterprises-inc-matter (ruling that Aspen Way is "permanently restrained and enjoined from: A. Gathering any information or data from any computer via any geophysical location tracking technology without providing clear and prominent notice to the computer user at the time the computer is rented and immediately prior to each use of the geophysical location tracking technology, and also obtaining affirmative express consent from the computer's renter at the time the computer is rented").

104. Fed. Trade Comm'n., *Protecting Consumer Privacy in an Era of Rapid Change: Recommendations for Businesses and Policymakers*, at 22 (Mar. 2012), https://www.ftc.gov/sites/default/files/documents/reports/federal-trade-commission-report-protecting-consumer-privacy-era-rapid-change-recommendations/120326privacyreport.pdf. The March 2012 guide was the final report and was issued after a preliminary FTC staff report released in December 2010. *See* Fed. Trade Comm'n., *Protecting Consumer Privacy in an Era of Rapid Change: A Proposed Framework for Businesses and Policymakers*, Preliminary FTC Staff Report (Dec. 2010), https://www.ftc.gov/sites/default/files/documents/reports/federal-trade-commission-bureau-consumer-protection-preliminary-ftc-staff-report-protecting-consumer/101201privacyreport.pdf.

105. Fed. Trade Comm'n, *Protecting Consumer Privacy in an Era of Rapid Change: Recommendations for Businesses and Policymakers*, at 22 (Mar. 2012), https://www.ftc.gov/sites/default/files/documents

The Privacy Framework calls on companies to, among other things, implement privacy by design to promote consumer privacy throughout the entire design of a product, provide simplified consumer choice, and be transparent about data practices. Companies developing their privacy policies and data programs need to consult with the Privacy Framework and structure their policies and programs following its terms.

The FTC will be active in the automated vehicle industry about data collection and use. Those who collect data from automated vehicles will need to provide explicit notice of the information collected by an automated vehicle and what the company will do with that data. Companies will want their notifications to be clear and readable—the FTC may examine the means of describing data collection activities. Automated vehicles also will have the ability to collect significant personal or highly sensitive information about the habits and associations of individuals. The FTC's enforcement activities have shown that the collection of this type of data may necessitate better notification than for information that is not personal or highly sensitive.

The FTC can enforce voluntary principles adopted by the automobile industry. One such set of voluntary principles is the Consumer Privacy Protection Principles: Privacy Principles for Vehicle Technologies and Services (the Principles) adopted by the Alliance of Automobile Manufacturers, Inc. and the Association of Global Automakers, Inc. on November 12, 2014.[106] Twenty automakers have pledged to meet or exceed the commitments of the Principles.[107] The Principles are about "covered information"—defined as "1) Identifiable Information that vehicles collect, generate, record, or store in an electronic form that is retrieved from the vehicles by or on behalf of a Participating Member in connection with Vehicle Technologies and Services; or 2) Personal Subscription Information provided by individuals subscribing or registering for Vehicle Technologies and Services." Data that can no longer be linked to an individual, such as through anonymization, is not covered information. "Vehicle technologies and services" are those "technologies and services provided

/reports/federal-trade-commission-report-protecting-consumer-privacy-era-rapid-change-recommendations/120326privacyreport.pdf.

106. Alliance of Auto. Mfrs., Inc. & Ass'n of Global Automakers, Inc., *Consumer Privacy Protection Principles: Privacy Principles for Vehicle Technologies and Services* (Nov. 12, 2014; reviewed May 2018), https://autoalliance.org/wp-content/uploads/2017/01/Consumer_Privacy_Principlesfor_VehicleTechnologies_Services-03-21-19.pdf.

107. These automakers are American Honda Motor Co., Inc.; Aston Martin Lagonda of North America, Inc.; BMW of North America, LLC; FCA US LLC; Ferrari North America, LLC; Ford Motor Company; General Motors; Hyundai Motor America; Jaguar Land Rover North America LLC; Kia Motors America; Maserati North America, Inc.; Mazda North American Operations; Mercedes-Benz USA, LLC; Mitsubishi Motors North America, Inc.; Nissan North America, Inc.; Porsche Cars North America; Subaru of America, Inc.; Toyota Motor Sales, USA; Volkswagen Group of America, Inc.; and Volvo Car Group.

by, made available through, or offered on behalf of Participating Members that involve the collection, use, or sharing of information that is collected, generated, recorded, or stored by a vehicle."

The Principles establish seven principles for privacy and data collection in automobiles. These seven principles are summarized as follows:

- Transparency: Participating Members commit to providing Owners and Registered Users with ready access to clear, meaningful notices about the Participating Member's collection, use, and sharing of Covered Information.
- Choice: Participating Members commit to offering Owners and Registered Users with certain choices regarding the collection, use, and sharing of Covered Information.
- Respect for Context: Participating Members commit to using and sharing Covered Information in ways that are consistent with the context in which the Covered Information was collected, taking account of the likely impact on Owners and Registered Users.
- Data Minimization, De-Identification & Retention: Participating Members commit to collecting Covered Information only as needed for legitimate business purposes. Participating Members commit to retaining Covered Information no longer than they determine necessary for legitimate business purposes.
- Data Security: Participating Members commit to implementing reasonable measures to protect Covered Information against loss and unauthorized access or use.
- Integrity & Access: Participating Members commit to implementing reasonable measures to maintain the accuracy of Covered Information and commit to giving Owners and Registered Users reasonable means to review and correct Personal Subscription Information.
- Accountability: Participating Members commit to taking reasonable steps to ensure that they and other entities that receive Covered Information adhere to the Principles.[108]

The members committed to begin using the Principles for vehicle model years 2017 or later and vehicle technologies and services starting on January 2, 2016. In a joint letter, the Auto Alliance and Global Automakers submitted the Principles to the FTC and publicly committed to abide by them.[109] Because the Principles are public representations by the automakers that they will comply with the provisions, the failure to do so may lead to prosecution by the FTC.

---

108. *Id.*

109. Letter to the Honorable Edith Ramirez, Chairwoman of the U.S. F.T.C. (Nov. 12, 2014), https://autoalliance.org/wp-content/uploads/2017/01/Auto_Alliance_Global_Automakers_Letter_to_FTCRE_Privacy.pdf.

b. FTC's Data Breach Common Law

The FTC's data breach common law comes from numerous sources. The FTC has used the unfairness and the deceptive prongs against companies following data breaches or for data security. These enforcements helped shape the FTC's early activities with data breach cases and began to establish a baseline for data security. The FTC has also given speeches, provided workshops, and issued staff reports and closing letters. These soft law guidances help give clarity on the expectations of the FTC and what acts and practices it considers unfair or deceptive. The FTC has also effectively used guides and blog posts.

Two of these guides have been particularly important in establishing the FTC's soft law policy on data security. The FTC has issued a guide on securing personal information entitled "Protecting Personal Information: A Guide for Business."[110] In this guide, the FTC outlined five principles that it believes make up a sound data security plan: (1) "take stock," (2) "scale down," (3) "lock it," (4) "pitch it," and (5) "plan ahead." "Take stock" means to know what personal information is collected and stored. "Scale down" means to limit the collection and storage of personal information unless there is a good business justification for storing it. "Lock it" means to ensure that data is secure in digital and physical form in transmission and storage, appropriate data protection mechanisms are used, access is controlled, and breaches are monitored. "Pitch it" is about implementing appropriate disposal policies to delete personal information once it is no longer needed. "Plan ahead" is about developing a plan to address security incidents in advance.

In 2015, the FTC released a guide on data security, entitled "Start with Security: A Guide for Business," that outlined ten lessons from its prosecutions in data security cases.[111] In 2017, the FTC penned a series of blog posts entitled "Stick with Security: A Business Blog Series" that further elucidated these ten lessons.[112] Members of the automated vehicle industry that collect or use data will need to familiarize themselves with these principles. These principles will help prevent cybersecurity breaches from occurring in the first place. Even when the breaches occur, compliance (or noncompliance) will be relevant to whether the FTC issues a complaint[113] and whether litigants can establish that

110. Fed. Trade Comm'n, *Protecting Personal Information: A Guide for Business* (Oct. 2016), https://www.ftc.gov/system/files/documents/plain-language/pdf-0136_proteting-personal-information.pdf.

111. Fed. Trade Comm'n, *Start with Security: A Guide for Business* (June 2015), https://www.ftc.gov/system/files/documents/plain-language/pdf0205-startwithsecurity.pdf.

112. Links to all of the blog posts can be found at https://www.ftc.gov/tips-advice/business-center/guidance/stick-security-business-blog-series.

113. *See* Thomas B. Pahl, *Stick with Security: Insights into FTC Investigations* (July 21, 2017), https://www.ftc.gov/news-events/blogs/business-blog/2017/07/stick-security-insights-ftc-investigations ("A preliminary question we often get from businesses is if there are recurring themes that run through the investigations that are ultimately closed without law enforcement. One thing we've noticed is that those companies' practices often lined up with the common-sense security funda-

the data collector breached a duty in a civil lawsuit or a class action. This section describes the ten security principles.

Principle 1: Start with security. The basic premise of starting with security is that companies need to incorporate security in all aspects of their business. Companies need to provide staff training on the 2015 Guide and cybersecurity principles and empower employees to make cybersecurity recommendations. Companies should not collect personal information that they do not need, and companies should only retain personal information for as long as they have a legitimate business need for such data and only use personal information when it is necessary.[114] Companies need to train staff about their cybersecurity standards and make sure employees follow them.[115] The FTC would like companies to offer consumers more secure choices when feasible. Companies in the automated vehicle industry that collect, use, or store data will need to consider data security comprehensively, what types of data they collect, why they collect data, and how long business necessities justify keeping such data. The collection and retention of data, especially sensitive information, with no apparent business justification may subject the company to scrutiny from the FTC. Automated vehicle industry members need to develop cybersecurity practices and train their staff accordingly.

Principle 2: Control access to data sensibly. This principle requires companies to provide access to data on a need-to-know basis and to treat data like a bank treats its vault. This means that companies need to restrict access to sensitive data by not allowing employees who do not use personal information to have access to it.[116] Companies also need to limit those who have administrative access to the network settings and administrative rights over data.[117] For automated vehicles, this principle will mean that data collectors need to ensure that

---

mentals in Start with Security. For example, the companies typically had effective procedures in place to train their staff, keep sensitive information secure, address vulnerabilities, and respond quickly to new threats.").

114. *See, e.g., In re* foru Int'l Corp., No. C-4457 (F.T.C. Jan. 7, 2014), *available at* https://www.ftc.gov/enforcement/cases-proceedings/112-3095/forutm-international-corporation-matter. In this matter, an allegation was that foru should have provided service providers with fictitious data sets to develop new applications rather than providing service providers with consumers' complete personal information.

115. *See, e.g., In re* Franklin's Budget Car Sales, Inc., No. C-4371 (F.T.C. June 7, 2012). The FTC alleged, among other things, that Franklin's Budget Car Sales failed to "[a]dequately train employees about information security to prevent unauthorized disclosures of personal information."

116. *See, e.g., In re* Uber Techs., Inc., No. C-4662 (F.T.C. Aug. 15, 2017), *available at* https://www.ftc.gov/enforcement/cases-proceedings/152-3054/uber-technologies-inc. Allegations against Uber included claims that Uber "i. failed to require programs and engineers that access the Amazon S3 Datastore to use distinct access keys, instead permitting all programs and engineers to use a single AWS access key that provided full administrative privileges over all data in the Amazon S3 Datastore; ii. failed to restrict access to systems based on employees' job functions; and iii. failed to require multi-factor authentication for access to the Amazon S3 Datastore."

117. *See id.*

employees have access to data on a need-to-know basis. For example, locational data about a vehicle should not be available to all employees. Access to data collected by automated vehicles needs to be restricted.

Principle 3: Require secure passwords and authentication. This principle is about password protection for employees and consumers. Companies need to insist that employees use long, complex, and unique passwords.[118] A better practice is for the system to reject obvious choices, such as the use of "payroll" as the password to access the payroll database. Companies also need to conduct training on password security.[119] Companies must prevent brute force attacks by suspending or disabling login attempts after several unsuccessful login attempts.[120] Sensitive information should be protected through multiple authentication techniques and should not be accessible with a single password. The FTC also wants companies to test authentication bypass vulnerability to make sure that the need for a password or authentication cannot be bypassed.[121] The automated vehicle industry will need to comply with this principle for their employees and customers. Companies that allow access to and control over automated vehicles via a phone application will want to ensure that passwords and authentication on the phone are robust.

Principle 4: Store sensitive personal information securely and protect it during transmission. This principle is about securing sensitive data through transmission and storage, such as by using Transport Layer Security/Secure Sockets Layer (TLS/SSL) encryption, data-at-rest encryption, or an iterative cryptographic hash.[122] To comply with this principle, a company needs to know where and how it obtains sensitive data so that it can secure data upon receiving it. The

---

118. *See, e.g.*, *In re* Twitter, Inc., No. C-4316 (F.T.C. June 24, 2010), *available at* https://www.ftc.gov/enforcement/cases-proceedings/092-3093/twitter-inc-corporation. In one of its allegations against Twitter, the FTC claimed that Twitter failed to "establish or enforce policies sufficient to make administrative passwords hard to guess, including policies that: (1) prohibit the use of common dictionary words as administrative passwords; and (2) require that such passwords be unique—i.e., different from any password that the employee uses to access third-party programs, websites, and network."

119. *See, e.g.*, *In re* Uber Techs., Inc., No. C-4662 (F.T.C. Aug. 15, 2017), *available at* https://www.ftc.gov/enforcement/cases-proceedings/152-3054/uber-technologies-inc. One of the allegations against Uber included the claim that it failed to "implement reasonable security training and guidance."

120. *See, e.g.*, *In re* Twitter, Inc., No. C-4316 (F.T.C. June 24, 2010), *available at* https://www.ftc.gov/enforcement/cases-proceedings/092-3093/twitter-inc-corporation. An allegation against Twitter was that it failed to "suspend or disable administrative passwords after a reasonable number of unsuccessful login attempts."

121. *See, e.g.*, *In re* Lookout Servs., Inc., No. C-4326 (F.T.C. filed May 3, 2011), *available at* https://www.ftc.gov/enforcement/cases-proceedings/102-3076/lookout-services-inc-matter. One of the allegations against Lookout Services was that it "allowed users to bypass the authentication procedures on Lookout's website when they typed in a specific URL."

122. *See, e.g.*, *In re* foru Int'l Corp., No. C-4457 (F.T.C. Jan. 7, 2014), *available at* https://www.ftc.gov/enforcement/cases-proceedings/112-3095/forutm-international-corporation-matter; *In re* Guidance Software, Inc., No. C-4187 (Nov. 16, 2006), *available at* https://www.ftc.gov/enforcement/cases-proceedings/062-3057/guidance-software-inc-matter. In *foru*, the FTC alleged, among other things, that foru "created unnecessary risks to personal information by maintaining consumers' personal

FTC also cautions against using nonstandard or proprietary forms of encryption instead of tried-and-true industry-accepted methods; companies that use their own encryption methods need to ensure that they do not have vulnerabilities that could be eliminated by using an industry method. The FTC also wants companies to configure their encryption methods properly so that they do not create security vulnerabilities. Automated vehicle industry members will need to determine what type of sensitive data they obtain and how they obtain it and ensure that data is encrypted through transmission and in storage.

Principle 5: Segment your network and monitor who's trying to get in and out. Principle 5 is about securing the network itself. One way to protect a network is to segment the network by having confidential or sensitive information segmented into a separate network. The FTC equates segmenting a network to having watertight compartments on a ship that minimize flooding if damage happens to one part of the ship. The other component of this principle is to monitor activity on a network.[123] This component is typically accomplished by using tools that monitor intrusion into the network and activities of network users to look for data transfers and other anomalies. Automated vehicle companies will want to consider segmenting automated vehicle data from other parts of their network or from other information collected on consumers.

Principle 6: Secure remote access to your network. Principle 6 concerns securing the network by limiting who can access the network and how. This includes ensuring endpoint security for employees, contractors, and consumers who can remotely access the network. Securing remote access also means limiting those who have access according to their needs—in other words, people should not have an "all-access, backstage pass" to the network. Automated vehicle industry members will need to secure their networks and connections to automated vehicles.

Principle 7: Apply sound security practices when developing new products. Sound security practices during the development of a new product or feature require a comprehensive approach to coding and designing the product. They begin with training engineers on secure coding practices and prioritizing coding

information, including consumers' names, addresses, email addresses, telephone numbers, dates of birth, Social Security numbers, and bank account numbers, in clear text."

123. *See, e.g., In re* Dave & Buster's, Inc., No. C-4291 (F.T.C. filed Mar. 25, 2010), *available at* https://www.ftc.gov/enforcement/cases-proceedings/082-3153/dave-busters-incin-matter. In its complaint, the FTC alleged that Dave & Buster's "(a) failed to employ sufficient measures to detect and prevent unauthorized access to computer networks or to conduct security investigations, such as by employing an intrusion detection system and monitoring system logs; (b) failed to adequately restrict third-party access to its networks, such as by restricting connections to specified IP addresses or granting temporary, limited access; (c) failed to monitor and filter outbound traffic from its networks to identify and block export of sensitive personal information without authorization; (d) failed to use readily available security measures to limit access between in-store networks, such as by employing firewalls or isolating the payment card system from the rest of the corporate network; and (e) failed to use readily available security measures to limit access to its computer networks through wireless access points on the networks."

security during development. The FTC recommends that companies follow platform guidelines when designing features or products. Another common problem that the FTC has encountered and prosecuted is situations where companies advertise or represent that they have a specific security feature, but the security feature did not work as advertised or represented.[124] An example was the FTC's prosecution of Snapchat for representations that Snapchat made that messages would "disappear forever," even though the messages were saved outside of the app's sandbox and could be accessed.[125] Finally, the FTC says that companies need to test their systems for common vulnerabilities.[126] Automated vehicle companies making representations about system security need to ensure the veracity of those statements. Automated vehicle companies need to test automated vehicles for vulnerabilities throughout their lifecycles.

Principle 8: Make sure your service providers implement reasonable security measures. It is rare for a company not to utilize service providers or contractors in their development process, in data collection or retention, or in their company more generally. This principle is about verifying that those service providers and contractors are committed to data security, that contracts include commitments to data security, and that companies verify that contractual requirements for data security are upheld.[127] An automated vehicle manufacturer needs to examine the security practices of suppliers and service providers.

---

124. *See, e.g., In re* Snapchat, Inc., No. C-4501 (F.T.C. Dec. 23, 2014), *available at* https://www.ftc.gov/system/files/documents/cases/140508snapchatcmpt.pdf.

125. *Id.*

126. *See, e.g., In re* Guidance Software, Inc., No. C-4187 (Nov. 16, 2006), *available at* https://www.ftc.gov/enforcement/cases-proceedings/062-3057/guidance-software-inc-matter; F.T.C. v. D-Link Corp., No. 3:17-cv-00039 (N.D. Cal. filed Jan. 5, 2017), *available at* https://www.ftc.gov/enforcement/cases-proceedings/132-3157/d-link. The FTC alleged, among other things, that Guidance Software did "not adequately assess the vulnerability of its web application and network to commonly known or reasonably foreseeable attacks, such as 'Structured Query Language' (or 'SQL') injection attacks." In its complaint against D-Link, the FTC stated, "Defendants repeatedly have failed to take reasonable software testing and remediation measures to protect their routers and IP cameras against wellknown and easily preventable software security flaws, such as 'hard-coded' user credentials and other backdoors, and command injection flaws, which would allow remote attackers to gain control of consumers' devices."

127. *See, e.g., In re* foru Int'l Corp., No. C-4457 (F.T.C. Jan. 7, 2014), *available at* https://www.ftc.gov/enforcement/cases-proceedings/112-3095/forutm-international-corporation-matter; *In re* Blu Prods. & Samuel Ohev-Zion, No. C-4657 (F.T.C. Apr. 30, 2018), *available at* https://www.ftc.gov/enforcement/cases-proceedings/172-3025/blu-products-samuel-ohev-zion-matter. Two allegations against foru were that it "[f]ailed to require by contract that service providers implement and maintain appropriate safeguards for consumers' personal information; [and] [f]ailed to provide reasonable oversight of service providers, for instance by requiring that service providers implement simple, low-cost, and readily available defenses to protect consumers' personal information." The allegations against Blu Products included that Blu Products

> failed to implement appropriate security procedures to oversee the security practices of their service providers, such as by:

Principle 9: Put procedures in place to keep your security current and address vulnerabilities that may arise. This principle reflects that data security is a process over the lifetime of the product. The FTC has prosecuted companies for failing to install updates and security patches on third-party software.[128] Companies must fix known security vulnerabilities and plan how they will deliver security updates. The FTC also gives a nod to white-hat hackers (i.e., security researchers) by providing that companies need to have a direct line for these researchers to inform companies of known vulnerabilities. For automated vehicle companies, this principle may mean that for personally owned automated vehicles, companies could have a duty to test for vulnerabilities and update software for the lifetime of the vehicle. Given how long some people own vehicles, this may mean that companies could have these duties for decades. The other concept here is providing access to third parties to report vulnerabilities. A good option is a bounty program that rewards white-hat hackers for revealing vulnerabilities in the system. Tesla Motors currently runs a bounty program.

Principle 10: Secure paper, physical media, and devices. The FTC's final principle shows that the FTC is concerned with more than digital and electronic security. Security also includes securing sensitive information in physical form. This obligation also includes limiting the physical transportation of data unless needed and then using proper security when data is transported physically, such as in a laptop. If devices collect data, then companies need to ensure that devices themselves are secure and protected from tampering or theft. Finally, companies are required to securely dispose of sensitive data, whether in paper or digital form.[129]

E. The Interplay of the FTC's Authority and NHTSA's Authority

The FTC's and NHTSA's oversight of the automobile industry and cybersecurity overlap, but the two regulatory agencies focus on cybersecurity from two

---

a. failing to perform adequate due diligence in the selection and retention of service providers; for example, [Blu Products] failed to assess or evaluate the privacy or security practices of [the software supplier] prior to entering into an agreement with that company; b. failing to adopt and implement written data security standards, policies, procedures or practices that apply to the oversight of their service providers, including [the software supplier]; c. failing to contractually require their service providers to adopt and implement data security standards, policies, procedures or practices; and d. failing to adequately assess the privacy and security risks of third-party software, such as [the software supplier].

128. *See, e.g., In re* The TJX Cos., Inc., No. C-4227 (F.T.C. Mar. 27, 2008), *available at* https://www.ftc.gov/enforcement/cases-proceedings/072-3055/tjx-companies-inc-matter. The FTC alleged, among other things, that the TJX Companies "failed . . . to conduct security investigations, such as by patching or updating anti-virus software or following up on security warnings and intrusion alerts."

129. *See, e.g., In re* Rite Aid Corp., No. C-4308 (F.T.C. filed July 27, 2010), *available at* https://www.ftc.gov/enforcement/cases-proceedings/072-3121/rite-aid-corporation-matter. The allegations against Rite Aid concerned its disposal procedures and failures to adequately train employees on how to dispose of sensitive information.

separate angles. The FTC's oversight is, in a sense, broader than that of NHTSA. The FTC monitors cybersecurity risks of any electronic system, including electronic systems in a motor vehicle and an automaker's general computer network, while NHTSA's current authority focuses on vulnerabilities that create risks to motor vehicle safety. A hack of an automaker's computer network that is unrelated to its fleet of vehicles would be analyzed as an unfair or deceptive trade practice under the FTC Act and not under the Motor Vehicle Safety Act.

In another sense, NHTSA's authority is broader for vulnerabilities that exist within vehicles or their equipment. NHTSA does not care whether a security vulnerability in a vehicle is due to an unfair or deceptive trade practice; if the vulnerability creates a risk to motor vehicle safety, then NHTSA will exercise its recall powers. NHTSA's recall power is strict or absolute and does not depend upon the care the automaker exercised. The FTC lacks a recall power, and not all cyberattacks on automated vehicles would be unfair or deceptive trade practices. The FTC has never considered a breach, in and of itself, evidence of an unfair or deceptive trade practice. Instead, the FTC will review the security practices of the company to determine whether the practices created an unfair or deceptive trade practice or act. For these vulnerabilities in vehicles and equipment, both NHTSA and the FTC may have the authority to investigate and potentially seek recourse.

Chapter 17

State Data Privacy and Security Laws

Of the automated vehicle laws that have been enacted, none, other than California's, has specifically addressed data privacy and security. Instead, ownership, use, privacy, and security of automated vehicle data will be determined under existing data privacy and security law. This chapter has three sections. The first section describes state privacy laws. As will be seen, approximately a quarter of the states have constitutional provisions that protect a right to privacy. Still, only a couple of states have enacted a statute resembling an omnibus privacy law for private parties. The most principled privacy statute in the United States is the California Consumer Privacy Act of 2018 (CCPA). The first section also describes the invasion of privacy torts and their limited application to automated vehicles, as well as state laws on data brokers and biometrics. The second section explores state data security laws and data breach laws. Because many states have narrowly defined their data security laws, many of those laws seemingly do not apply to the collection of most data by automated vehicles. However, as new and broader privacy laws pass in various states, the way states define terms like "personal information" will almost certainly pull in the automated vehicle industry. The third section will describe two statutory regimes that apply to data privacy and security. The first is state event data recorder (EDR) laws, which, like their federal counterpart, will apply to automated vehicles but leave unanswered who or what owns most data generated by automated vehicles. The second statutory regime is state consumer protection laws that resemble or replicate the Federal Trade Commission (FTC) Act.

I. STATE PRIVACY LAWS

State constitutions in 12 states include a right to privacy, and other states have recognized a right of privacy under their state laws.[1] Although states have these broad constitutional principles of privacy in or derived from their state constitutions, states have not widely adopted omnibus privacy laws that protect consumer privacy of data collected in the modern economy. Most states have a privacy tort that may have limited applicability to the privacy of automated vehicles. Most state data laws concern notice requirements by companies that have suffered data breaches, which are discussed later in this chapter. The only state with an omnibus privacy law currently is California (the CCPA).[2] Georgia has a specific law that regulates automobile industry data. Vermont has a specific law that addresses privacy relating to data brokers. Utah has a Notice of Intent to Sell Nonpublic Personal Information Act that requires companies that collect nonpublic data with the intent to sell that data to third parties to provide notice of such practice to consumers.[3] Other states have biometric privacy laws that may apply to the automated vehicle industry to the extent that automated vehicles use biometrics to identify passengers. This section also details the CCPA and California's automated vehicle privacy law.

A. Privacy Torts

With the significant data collection activities of automated vehicles, people's private lives could be revealed by someone obtaining access to that data and possibly publicizing such data. For example, a hacker or an employee of an automaker who supports a certain political candidate may hack or gain access to data about disfavored political candidates' travel habits and publicly reveal any unfavorable data. The common law recognizes four torts involving an invasion of privacy;[4] the tort of intrusion is likely the most relevant tort when

1. The states with explicit constitutional provisions are as follows: ALASKA CONST. art. I, § 22; ARIZ. CONST. art. II, § 8; CAL. CONST. art. I, § 1; FLA. CONST. art. I, § 23; HAW. CONST. art. I, § 6; ILL. CONST. art. I, § 6; LA. CONST. ANN. art. I, § 5; MONT. CONST. art. II, § 10; S.C. CONST. art. I, § 10; WASH. CONST. art. I, § 7.

2. Nevada and Maine both have recently enacted laws that provide certain privacy protections for Internet use. Other states, like New York, are currently considering privacy laws similar to or more stringent than the CCPA.

3. UTAH CODE ANN. § 13-37-101 (West) *et seq.*

4. The four invasion of privacy torts are "1. Intrusion upon the plaintiff's seclusion or solitude, or into his private affairs. 2. Public disclosure of embarrassing private facts about the plaintiff. 3. Publicity which places the plaintiff in a false light in the public eye. 4. Appropriation, for the defendant's advantage, of the plaintiff's name or likeness." William L. Prosser, *Privacy*, 48 CALIF. L. REV. 383, 389 (1960).

data from an automated vehicle is improperly accessed or revealed. Some states have statutory enactments that provide for invasion of privacy torts.[5]

The tort of intrusion requires proof of an "(1) intrusion into a private place, conversation or matter, (2) in a manner highly offensive to a reasonable person."[6] The California Supreme Court has described this in the realm of data as requiring proof that "the defendant penetrated some zone of physical or sensory privacy surrounding, or obtained unwanted access to data about, the plaintiff. The tort is proven only if the plaintiff had an objectively reasonable expectation of seclusion or solitude in the place, conversation or data source."[7] One court has found that hacking into a private computer to steal personal correspondence is sufficient proof to establish the tort of intrusion.[8] It seems likely that hacking into, or otherwise gaining unlawful access to, vehicle data could be actionable under the tort of intrusion. Reasonable people would expect that vehicle data, which could show "their political and religious beliefs, sexual habits,"[9] and more, would be kept private and that the computer systems recording that data are private places. Therefore, victims may be able to seek recourse under the tort of intrusion.

B. Georgia Senate Bill 122

Georgia Senate Bill 122 provides specific rules with respect to consumer data and nonpublic personal information collected by dealers and provided to manufacturers or designees of manufacturers.[10] The bill requires, among other things, "a franchisor, manufacturer, distributor, or a third party" acting on behalf of one of them to (1) comply with applicable "restrictions on reuse or disclosure of consumer data established by federal or state law" and not to cause dealers to violate such laws; (2) "provide a written statement to the dealer upon request describing [their] established procedures which meet or exceed federal or state requirements to safeguard the consumer data"; (3) provide, upon written request of a dealer, a "written list of the consumer data obtained from the dealer and all persons to whom any consumer data has been provided [by them] during the preceding six months"; and (4) not require a dealer to give

5. *See, e.g.*, Wis. Stat. § 995.50.

6. Shulman v. Grp. W Prods., Inc., 18 Cal. 4th 200, 231, 955 P.2d 469, 490, *modified on denial of reh'g* (July 29, 1998).

7. *Id. Shulman* itself is not a data case.

8. *See, e.g.*, Coal. for an Airline Passengers' Bill of Rights v. Delta Air Lines, Inc., 693 F. Supp. 2d 667, 675 (S.D. Tex. 2010) ("The court concludes that hacking into a person's private computer and stealing personal correspondence would represent an intentional intrusion on the victim's private affairs and that such an intrusion would be highly offensive to a reasonable person.").

9. United States v. Jones, 565 U.S. 400, 416, 132 S. Ct. 945, 956, 181 L. Ed. 2d 911 (2012) (Sotomayor, J., concurring).

10. *Consumer Protection—Motor Vehicles*, 2019 Ga. Laws Act 157 (S.B. 122).

them "direct or indirect access to such dealer's data management system to obtain consumer data."[11]

Georgia Senate Bill 122 applies to the interaction between the manufacturer and members of the supply chain, on the one hand, and dealers, on the other hand. To the extent that automated vehicles are delivered and sold to consumers through traditional distribution channels, the law will restrict supply chain members' access to dealer data and empower dealers to obtain certain records of supply chain members.

C. Vermont's Data Broker Law

Vermont's law may limit an automated vehicle company's ability to utilize data or become a data broker without providing notice to consumers. Industry members that desire to monetize consumer data may find it beneficial to become a data broker and register under certain laws, like the one in Vermont. The CCPA, for example, has relaxed notice requirements for registered data brokers in certain instances.

A data broker is a "business, or unit or units of a business, separately or together, that knowingly collects and sells or licenses to third parties the brokered personal information of a consumer with whom the business does not have a direct relationship."[12] Brokered personal information is data elements that are "categorized or organized for dissemination to third parties," and such data elements include biometric information, basic information about a consumer, Social Security number, and any "other information that, alone or in combination with the other information sold or licensed, would allow a reasonable person to identify the consumer with reasonable certainty."[13] A violation of Vermont's law is considered an "unfair or deceptive act." Vermont grants its attorney general the authority to promulgate rules and to enforce the law.

Vermont prohibits the acquisition of brokered personal information through fraudulent means, as well as the acquisition of brokered personal information to stalk or harass someone, to commit fraud, or to engage in unlawful discrimination.[14] Vermont requires data brokers to register with the Vermont secretary of state and provide specific information in its registration, such as whether the data broker allows a consumer to opt out or suffered a data breach in the

---

11. GA. CODE ANN. § 10-1-632 (West).

12. VT. STAT. ANN. tit. 9, § 2430(4)(A) (West). The Vermont statute carves out of the definition of "business" certain business activities—such as maintaining or developing third-party applications, providing directory information services, and providing publicly available information about a person's business or a profession—to which the collection, sale, or licensing of brokered personal information is incidental. *Id.* § 2430(4)(C).

13. *Id.* § 2430(1)(A).

14. *Id.* § 2433.

prior year.[15] Data brokers are required to develop, implement, and maintain a written comprehensive information security program tailored to the scope of the data broker, the type of information collected, and its operations.[16] Vermont law provides specific minimal requirements for any comprehensive information security program.

D. Utah's Notice of Intent to Sell Nonpublic Personal Information Act

Utah enacted its Notice of Intent to Sell Nonpublic Personal Information Act in 2003 (the Utah Act), to be effective as of January 1, 2004.[17] The Utah Act applies to commercial entities[18] that seek to sell to third parties nonpublic personal information obtained through commercial transactions with consumers.[19] Nonpublic personal information is described broadly to "mean[] information that: (i) is not public information; and (ii) either alone or in conjunction with public information, identifies a person in distinction from other persons."[20]

The Utah Act prohibits disclosure of such information to third parties unless the commercial entity provides notice to consumers.[21] The notice must "be given before the earlier of: (a) the point at which the person is requested to provide the nonpublic personal information; or (b) the commercial entity otherwise obtains the nonpublic personal information as a result of the consumer transaction."[22] The notice "(a) shall read substantially as follows: 'We may choose to disclose nonpublic personal information about you, the consumer, to a third party for compensation'; (b) may be made: (i) orally, if the consumer transaction itself is entirely conducted orally; or (ii) in writing, if the notice is written in dark bold; and (c) shall be sufficiently conspicuous so that a reasonable person would perceive the notice before providing the nonpublic personal information."[23] A private cause of action is created for breaches of

---

15. *Id.* § 2446.
16. *Id.* § 2447.
17. Utah Code Ann. § 13-37-101 *et seq.* (West).
18. A "'commercial entity' means a person that: (i) has an office or other place of business located in the state; and (ii) in the ordinary course of business transacts a consumer transaction in this state. (b) 'Commercial entity' does not include: (i) a governmental entity; or (ii) an entity providing services on behalf of a governmental entity." *Id.* § 13-37-102(2).
19. *Id.* § 13-37-201.
20. *Id.* § 13-37-102(5). It explicitly includes "(i) a person's Social Security number; (ii) information used to determine a person's credit worthiness including a person's: (A) income; or (B) employment history; (iii) the purchasing patterns of a person; or (iv) the personal preferences of a person." *Id.* Public information is defined narrowly to mean "a person's: (a) name; (b) telephone number; or (c) street address." *Id.*
21. *Id.* § 13-37-202.
22. *Id.* § 13-37-201(2).
23. *Id.* § 13-37-201(3).

the Utah Act.[24] Because the Utah Act only provides for limited damages, $500 per violation and court costs, and prohibits class action lawsuits,[25] practically speaking, the Utah Act may be difficult for plaintiffs to enforce.

The Utah Act would apply when members of the automated vehicle industry seek to monetize consumer data. Industry members that seek to monetize such data need to comply with the Utah Act and provide notice to Utah residents about such monetization practices.

E. Biometric Laws

Automated vehicles may utilize biometrics to verify occupants or users. Biometrics is the use of physical or behavior characteristics, such as facial recognition or voiceprint or fingerprint authentication, to identify or authenticate an individual. Biometrics are commonly used in cell phones today. Biometrics could be used in automated vehicles to verify or authenticate who is using the automated vehicle or in a phone application for an automated vehicle service.

Biometrics that become compromised create difficulties unlike any other type of passcode or means of authentication. Passwords, passcodes, e-mail addresses, and cell phone numbers can be changed if any of them becomes compromised in data security breaches. A person's iris scan, fingerprint, or voiceprint cannot be changed; if an unauthorized person obtains access to a person's biometrics, the victim could effectively be unable to use biometrics as a means of identification or authentication.

Currently, only four states have regulated biometric privacy: California, Illinois, Texas, and Washington.[26] The CCPA covers biometrics and is described at length in the next part. This part focuses on the laws enacted in Illinois, Texas, and Washington. Of these three acts, the Illinois act, the Illinois Biometric Information Privacy Act (the BIPA), is the broadest and most protective of consumers. The BIPA allows for private enforcement of its act: "Any person

---

24. *Id.* § 13-37-203.

25. *Id.*

26. 740 Ill. Comp. Stat. Ann. 14/1 *et seq.*; Tex. Bus. & Com. Code Ann. § 503.001; Wash. Rev. Code § 19.375.010 *et seq.* Other states include biometrics within their data breach notification statutes. *See* Emily Matta, *Kansans at Risk: Strengthened Data Breach Notification Laws as a Deterrent to Reckless Data Storage*, 67 U. Kan. L. Rev. 823, 835 (2019) (collecting statutes); Ala. Code § 8-38-5(b); Ariz. Rev. Stat. Ann. § 18-552(B); Md. Code Ann., Com. Law § 14-3504(b)(3) (West); N.M. Stat. § 57-12C-6(A); Ohio Rev. Code Ann. §§ 1347.12(B)(2), 1349.19(B)(2) (West); Or. Rev. Stat. § 646A.604(3)(a); 11 R.I. Gen. Laws § 11-49.3-4(a)(2) (2018); Tenn. Code Ann. § 47-18-2107(b) (West); Vt. Stat. Ann. tit. 9, § 2435(b)(1) (West); Wash. Rev. Code §§ 19.255.010(16), 42.56.590(15); Wis. Stat. § 134.98(3)(a). The FTC has issued a best practices manual on the use of facial recognition. Fed. Trade Comm'n, *Facing Facts: Best Practices for Common Uses of Facial Recognition Technologies* (Oct. 2012), https://www.ftc.gov/sites/default/files/documents/reports/facing-facts-best-practices-common-uses-facial-recognition-technologies/121022facialtechrpt.pdf.

aggrieved by a violation of this Act shall have a right of action."[27] The BIPA also includes a liquidated damages provision that allows for a base recovery, if greater than the actual damages, of $1,000 for negligent violations and $5,000 for intentional or reckless violations, in addition to attorneys' fees and costs and other relief ordered by the court.[28] Texas and Washington prohibit private actions and only allow their respective attorney generals to enforce their acts.[29] Generally, the statutes regulate the security, use, sale, and sharing of biometrics. Because a business that complies with the Illinois Act can comply with Texas's and Washington's biometric privacy laws, but the converse is not true, businesses that operate nationally need to develop their biometric policies to comply with the BIPA and the CCPA.

The BIPA uses two concepts: biometric identifier and biometric information. A biometric identifier is a "retina or iris scan, fingerprint, voiceprint, or record of hand or face geometry."[30] The biometric identifier does not include certain types of information, such as photographs, written signatures, tattoo descriptions, physical descriptions like hair or eye color or height, and certain types of health information, such as an X-ray used for treatment. Biometric information is "any information, regardless of how it is captured, converted, stored, or shared, based on an individual's biometric identifier used to identify an individual."[31]

Under the BIPA, a business that possesses biometric identifiers or information is required to develop and make available publicly a written policy about the retention schedule and guidelines for permanent destruction.[32] The written policy must provide for the destruction of biometric identifiers and information when the initial need for the identifier or information has been satisfied or within three years following the individual's last interaction with the company. A private entity is not allowed to collect, capture, receive, or obtain a biometric identifier or biometric information without providing notice of the collection

---

27. 740 Ill. Comp. Stat. Ann. 14/20.

28. *Id.*

29. Tex. Bus. & Com. Code Ann. § 503.001(d) (West); Wash. Rev. Code § 19.375.030.

30. 740 Ill. Comp. Stat. Ann. 14/10; *see also* Tex. Bus. & Com. Code Ann. § 503.001(a) (West). Washington uses a slightly different base definition:

> "Biometric identifier" means data generated by automatic measurements of an individual's biological characteristics, such as a fingerprint, voiceprint, eye retinas, irises, or other unique biological patterns or characteristics that is used to identify a specific individual. "Biometric identifier" does not include a physical or digital photograph, video or audio recording or data generated therefrom, or information collected, used, or stored for health care treatment, payment, or operations under the federal health insurance portability and accountability act of 1996.

Wash. Rev. Code § 19.375.010.

31. 740 Ill. Comp. Stat. Ann. 14/10.

32. *Id.* at 14/15.

and storage; providing the specific purpose for the collection, storage, and use of the biometric identifier or biometric information; and obtaining a written release from the individual or the individual's authorized representative. Entities are prohibited from selling, trading, leasing, or profiting from an individual's biometric identifier or information. Entities can disclose, redisclose, or disseminate biometric identifiers or information only under four circumstances: with the written consent of the individual or the individual's representative, when necessary to complete a financial transaction authorized or requested by the individual, when required by applicable state or local law, or pursuant to a valid warrant or subpoena. The BIPA requires companies to exercise the same care that they do to protect other confidential information in their possession.[33]

Courts have found that plaintiffs are not required to prove actual damages in addition to a statutory violation to assert claims under the BIPA. The Illinois Supreme Court recently held that a BIPA violation is actionable even if a plaintiff has not suffered any actual harm other than the statutory violation.[34] In one of the more important cases under the BIPA, the Ninth Circuit Court of Appeals held that a putative class had standing to sue Facebook for violating the BIPA without requiring the putative class to show an actual injury in addition to the statutory violation.[35] The Ninth Circuit ruled that the BIPA created substantive privacy rights, the violation of which satisfies the injury-in-fact requirement of standing. Facebook ended up settling the pending cases against it for $550 million.[36] These cases will make it so that technical noncompliance with the BIPA can lead to a class-action lawsuit and make it so that no automated vehicle company can use biometrics in Illinois without first ensuring absolute compliance with the BIPA.

F. The California Consumer Privacy Act of 2018

In June 2018, California enacted the CCPA. The CCPA became effective on January 1, 2020. The California attorney general, the regulatory body charged with promulgating regulations and enforcing the CCPA, issued proposed regulations in October 2019, modifications to proposed regulations

---

33. *Id.* at 14/15(c); *see also* Tex. Bus. & Com. Code Ann. § 503.001(c)(2) (West); Wash. Rev. Code § 19.375.010(4)(a).

34. Rosenbach v. Six Flags Entm't Corp., 2019 IL 123186, 129 N.E.3d 1197 (2019).

35. Patel v. Facebook, Inc., 932 F.3d 1264 (9th Cir. 2019), *cert. denied*, No. 19-706, 2020 WL 283288 (U.S. Jan. 21, 2020).

36. Jeff Horwitz, *Facebook Reaches $550 Million Settlement in Facial-Recognition Lawsuit*, Wall St. J. (Jan. 29, 2020), https://www.wsj.com/articles/facebook-reaches-550-million-settlement-in-facial-recognition-lawsuit-11580347594.

on February 10, 2020, a second set of modifications on March 11, 2020, and the final proposed regulations on June 1, 2020. Broadly speaking, the CCPA takes a dual track by granting certain rights to California residents concerning personal information collected about them and obligating companies to provide certain notices about their collection and use of data.

The CCPA applies to "personal information" of a "consumer" collected by a "business." Personal information is defined broadly to mean "information that identifies, relates to, describes, is reasonably capable of being associated with, or could reasonably be linked, directly or indirectly, with a particular consumer or household."[37] The CCPA provides a list of 11 categories of Personal information (PI or personal information) that includes nonexhaustive examples of PI within each respective category—in short, if something "identifies, relates to, describes, is reasonably capable of being associated with, or could be reasonably linked, directly or indirectly, with a particular consumer or household," it is PI.[38] These categories include identifiers such as names, addresses, e-mail addresses, Internet protocol addresses, and Social Security numbers; commercial information; biometric information; "internet or other electronic network activity information"; geolocation data; and "audio, electronic, visual, thermal, olfactory, or similar information."[39] Because the meaning of PI is based on the ability to link the information to a consumer, "consumer information that is deidentified or aggregate consumer information" is excluded from the meaning of "personal information."[40] A historic form of PI in the automobile industry has been the vehicle identification number (VIN).

Based on that definition, data collected by an automated vehicle that can be linked to a specific individual or household would be PI. For example, geolocation data linked to a specific vehicle or user of ridesharing would be PI. In contrast, data that is aggregated or in which the identities of the individual consumers are permanently removed would not be considered PI. The ability

---

37. Cal. Civ. Code § 1798.140(o)(1) (West).

38. *Id.*

39. *Id.*

40. *Id.* § 1798.140(o)(3). Deidentified information is "information that cannot reasonably identify, relate to, describe, be capable of being associated with, or be linked, directly or indirectly, to a particular consumer, provided that a business that uses deidentified information: (1) Has implemented technical safeguards that prohibit reidentification of the consumer to whom the information may pertain. (2) Has implemented business processes that specifically prohibit reidentification of the information. (3) Has implemented business processes to prevent inadvertent release of deidentified information. (4) Makes no attempt to reidentify the information." *Id.* § 1798.140(h). "Aggregate consumer information" means "information that relates to a group or category of consumers, from which individual consumer identities have been removed, that is not linked or reasonably linkable to any consumer or household, including via a device. 'Aggregate consumer information' does not mean one or more individual consumer records that have been deidentified." *Id.* § 1798.140(a).

to aggregate data or anonymize data will present the automated vehicle industry the option to use data within their machine learning algorithm software without worrying about consumer deletion rights. However, since the VIN of sold or leased vehicles can be used as an identifier by automakers, the question of feasibility is a glaring one in many circumstances when it comes to anonymization or aggregation.

The CCPA does not define "consumer" colloquially to mean a person who purchases or uses a service or product. Instead, a "consumer" is a "natural person who is a California resident."[41] This broad definition of "consumer" effectively means that the CCPA could apply to a California resident who has never directly interacted with the business, so long as the business has personal information about the consumer. For example, any California resident captured on a camera by an automated vehicle is a consumer who is granted certain rights under the CCPA—the image of a consumer would certainly be deemed PI in, at least, the audio, thermal, or visual categories. The definition can also capture service providers. For example, a car dealership mechanic may have his or her name associated with a vehicle, which would be consumer personal information. If a business does not collect any data on California residents, then it is not required to comply with the CCPA.

Not all businesses that collect data on California residents are required to comply with the CCPA. The CCPA does not cover nonprofits. A for-profit business needs to comply with the CCPA if it (i) has annual gross revenues greater than $25 million, (ii) "alone or in combination, annually buys, receives for the business's commercial purposes, sells, or shares for commercial purposes, alone or in combination, the personal information of 50,000 or more California residents, households, or devices," or (iii) "derives 50 percent or more of its annual revenues from selling California residents' personal information."[42] Under this definition, all automakers that collect PI on California residents would be businesses. Other automated vehicle industry members that have gross revenues of at least $25 million and collect PI on California residents would also be businesses.

California residents are granted four broad types of rights with respect to PI: the right to know what categories of PI are collected about the consumer, the right to access the consumer's PI, the right to delete PI, and the right to opt out of the sale of PI. The CCPA grants California residents the following information rights:

41. *Id.* § 1798.140(g).
42. *Id.* § 1798.140(c).

1. California residents can request a business that collects[43] personal information about the consumer to disclose to the consumer the following information:
    a. The categories of personal information it has collected about that consumer.
    b. The categories of sources from which the personal information is collected.
    c. The business or commercial purpose for collecting or selling personal information.
    d. The categories of third parties with whom the business shares personal information.
    e. The specific pieces of personal information it has collected about that consumer.[44]
2. A consumer shall have the right to request that a business that sells the consumer's personal information, or that discloses it for a business purpose, disclose to that consumer:
    a. The categories of personal information that the business collected about the consumer.
    b. The categories of personal information that the business sold about the consumer and the categories of third parties to whom the personal information was sold, by category or categories of personal information for each category of third parties to whom the personal information was sold.
    c. The categories of personal information that the business disclosed about the consumer for a business purpose.[45]

A consumer exercises his or her right to this information by making a verifiable consumer request.[46] A "verifiable consumer request" is "a request that is made by a consumer, by a consumer on behalf of the consumer's minor child, or by a natural person or a person registered with the Secretary of State, authorized by the consumer to act on the consumer's behalf, and that the business can reasonably verify . . . to be the consumer about whom the business has collected personal information."[47] Upon receiving a request, the business must "promptly take steps to disclose and deliver, free of charge to the con-

43. The CCPA defines "collects" as "buying, renting, gathering, obtaining, receiving, or accessing any personal information pertaining to a consumer by any means. This includes receiving information from the consumer, either actively or passively, or by observing the consumer's behavior." *Id.* § 1798.140(e). Note that, with respect to automated vehicles, "collect" includes observing consumers.

44. *Id.* § 1798.110(a).

45. *Id.* § 1798.115(a).

46. *Id.* §§ 1798.110(b), 1798.115(b).

47. *Id.* § 1798.140(y).

sumer, the personal information."[48] A consumer is limited to two requests per 12-month period.[49]

Automated vehicle industry members—and all other companies—need to be very cautious in implementing a verification system. To the extent that data points cannot be verified, those data points should not be disclosed. A paradox of these data privacy laws is that by giving consumers protection and control over their PI, these laws make it easier for unauthorized access to a person's PI to occur. Before the CCPA and the European Union's General Data Protection Regulation, personal information about an individual may have been disjointed and not systematically organized or mapped. Additionally, these laws create opportunities for persons to impersonate someone to obtain data. Automated vehicles could be particularly pernicious in this regard. Spouses, who have most verifiable information about their significant other, could have a strong incentive to obtain location data or other automated vehicle data. Companies will want to consider how to authenticate and verify that a request is actually from the person whom the requestor purports to be.

The deletion right allows a consumer to request the deletion of PI that the business has collected about the consumer.[50] A business that receives a verifiable consumer request must comply with the request and cause service providers to delete the consumer's PI, unless the business determines that declining to delete an item of PI is "necessary for the business or service provider to maintain the consumer's personal information" for one of nine enumerated reasons.[51] These enumerated reasons include to "complete the transaction for which the personal information was collected, fulfill the terms of a written warranty or product recall conducted in accordance with federal law, provide a good or service requested by the consumer, or reasonably anticipated within the context of a business' ongoing business relationship with the consumer, or otherwise perform a contract between the business and the consumer"; "detect security incidents, protect against malicious, deceptive, fraudulent, or illegal activity; or prosecute those responsible for that activity"; "debug to identify and repair errors that impair existing intended functionality"; and "otherwise use the consumer's personal information, internally, in a lawful manner that is compatible with the context in which the consumer provided the information."[52] Automated vehicle companies will need to determine which personal information is not subject to the deletion rights. Automakers will want to build the right to require deletion into contracts with component manufacturers and

48. *Id.* § 1798.100(d).
49. *Id.*
50. *Id.* § 1798.105(a).
51. *Id.* § 1798.105(c), (d).
52. *Id.* § 1798.105(d).

service providers. Automakers will also want to revisit existing relationships to ensure agreements account for CCPA requirements.

The opt-out right does not allow California residents to opt out of the collection of PI about the consumer. Instead, the opt-out right enables a consumer to "direct a business that sells personal information about the consumer to third parties not to sell the consumer's personal information."[53] The opt-out rule is the general rule for the sale of personal information of California residents over the age of 16. For California residents that the business has actual knowledge of being under the age of 16, the business cannot sell personal information without affirmative authorization from the consumer, if the consumer is at least 13, or the consumer's parent or guardian if the consumer is younger than 13.[54] The opt-out right is not applicable to vehicle information or ownership information shared between a new motor vehicle dealer and a vehicle manufacturer, "if the vehicle or ownership information is shared for the purpose of effectuating, or in anticipation of effectuating, a vehicle repair covered by a vehicle warranty or a recall" conducted pursuant to the Motor Vehicle Safety Act.[55] Businesses are required to make the opt-out right reasonably accessible to California residents by, among other things, having a link on their home pages that is entitled "Do Not Sell My Personal Information" and providing a description of the consumer's rights to opt out on the page that is linked.[56] Members of the automated vehicle industry may interact with this rule differently. This rule is also somewhat ambiguous. For example, what responsibilities does an automated ridesharing company have to determine whether its passengers include a minor? What happens if an automated ridesharing company provides a ride to an adult and a minor—is the geolocation data of the adult permitted to be sold without an opt-out, even though such data may be revealing of the minor within the automated vehicle?

To help California residents be able to exercise their rights and require businesses to be more transparent about their data collection, use, and sharing activities, the CCPA requires businesses to provide specific information in their online privacy policies and any California-specific description of privacy rights of:

(1) Their specific rights under the CCPA;
(2) A list of the following information about the personal information collected by the business, organized by category of personal information listed in the enumerated categories:

53. *Id.* § 1798.120(a).
54. *Id.* § 1798.120(c).
55. *Id.* § 1798.145(g)(1).
56. *Id.* § 1798.135.

a. The categories of personal information it has collected about California residents.
b. The categories of sources from which the personal information is collected.
c. The business or commercial purpose for collecting or selling personal information.
d. The categories of third parties with whom the business shares personal information.
e. That a consumer has the right to request the specific pieces of personal information the business has collected about that consumer.

(3)[T]he following information about the sale or disclosure of personal information, organized by category of personal information listed in the enumerated categories:

a. The category or categories of California residents' personal information it has sold, or if the business has not sold California residents' personal information, it shall disclose that fact.
b. The category or categories of California residents' personal information it has disclosed for a business purpose, or if the business has not disclosed the California residents' personal information for a business purpose, it shall disclose that fact.[57]

Businesses are obligated to make sure employees responsible for handling requests are trained and aware of consumer rights and their obligations under the CCPA.

The CCPA is, with a limited exception, only enforceable by the California attorney general. The CCPA allows for private lawsuit by a consumer only if "nonencrypted and nonredacted personal information . . . is subject to an unauthorized access and exfiltration, theft, or disclosure as a result of the business's violation of the duty to implement and maintain reasonable security procedures and practices appropriate to the nature of the information to protect the personal information."[58] A consumer is required to provide notice and an opportunity to cure to the business before instituting a lawsuit or filing a class action unless the consumer suffered "actual pecuniary damages."[59] California residents are otherwise prohibited from enforcing the CCPA.[60] The attorney

57. *Id.* § 1798.130(b)(5).
58. *Id.* § 1798.150(a)(1).
59. *Id.* § 1798.150(b).
60. *Id.* § 1798.150(c). One way plaintiffs are attempting to circumvent this requirement is by alleging that businesses that fail to comply with the CCPA are de facto breaching the California constitutional right to privacy.

general enforces the CCPA by providing notice to an alleged violator. If the alleged violator fails to cure the noncompliance, then the attorney general can seek an injunction and fine unintentional violators $2,500 for each violation and fine intentional violators $7,500 for each violation.[61]

By granting rights to California residents and, at the same time, obligating businesses to provide certain notices, the CCPA means two things for non-California residents. First, companies are not obligated to grant these rights to nonresidents of California. Second, companies are providing the notices about their collection and use of PI to everyone, not just California residents. This latter point is likely due to the near impossibility or the significant costs of geofencing website privacy notices to California residents. Nonresidents of California can receive the secondary benefits of increased disclosure in privacy statements about the information collected by a company. This is important for automated and connected vehicles. California residents may have more control over data collected by automated and connected vehicles than nonresidents. Still, nonresidents will receive the benefits of the CCPA's increased disclosures of data collection practices of the automated vehicle industry.

The CCPA will require automated vehicle companies that do business in California to review their data practices. Companies will need to inventory the personal information they collect, the category of each item of PI (under the CCPA definition), how they maintain or use PI, whether PI is sold or shared with others, and whether PI is encrypted, deidentified, or aggregated. The comprehensive review and inventory will apprise automated vehicle companies of their disclosure obligations.

G. California's Automated Vehicle Law

California's automated vehicle law addresses data collection by automated vehicles. California has specific requirements for manufacturers that collect personal information that is not necessary for the safe operation of automated vehicles.[62] The manufacturer must provide a written disclosure to the driver or, if driverless, the passengers of the automated vehicle that describes the personal information collected and how the manufacturer intends to use the information. Alternatively, the manufacturer can anonymize personal information. For automated vehicles sold or leased to consumers, to the extent that the personal information is not anonymized, the manufacturer must obtain written approval to collect the personal information. A manufacturer cannot condition sale on the written approval.

61. *Id.* § 1798.155.

62. CAL. CODE REGS. tit. 13, § 228.24.

II. STATE DATA SECURITY LAWS

States have enacted two types of laws that can apply in the data security and breach context for automated vehicles. The first type is data security laws that apply when a company possesses personal information of a state resident. The second type of laws is data breach notification laws that require companies to notify consumers when they suffer a data breach.

A. Data Security Laws

Because companies in the automated vehicle industry will collect personal information, these companies will need to review state laws that govern security of their residents' personal information. These data security laws require security measures to be taken by entities that collect, own, or maintain personal information about a resident of the state. Approximately 25 states have enacted these types of laws.[63] Because automated vehicles will likely collect data on residents of all states, data collectors will need to review all of these statutes to determine their applicability.

These statutes differ in their complexity and requirements. A great many of these laws simply require that a company take reasonable measures to protect personal information.[64] Some statutes are more complex and require companies to take specific action or do more than take reasonable measures.[65] At the heart of any of these statutes is the statute's definition of personal information (or sensitive personal information). Most statutes define these terms narrowly.[66] Many states focus on healthcare information or information that, when coupled with a person's name, such as a Social Security number, could

63. 2018 Alabama S.B. 318; Ark. Code Ann. § 4-110-104(b) (West); Cal. Civ. Code § 1798.81.5 (West); Colo. Rev. Stat. § 6-1-713.5; Conn. Gen. Stat. § 38a-999b; Del. Code § 12B-100; Fla. Stat. § 501.171(2); 815 Ill. Comp. Stat. Ann. 530/45; Ind. Code § 24-4.9-3-3.5; Kan. Stat. Ann. § 50-6,139b; La. Rev. Stat. § 3074; Md. Code Ann., Com. Law § 14-3501 to -3503 (West); Mass. Gen. Laws ch. 93H, § 2(a); Minn. Stat. § 325M.05; Neb. Rev. Stat. Ann. § 87-801 to -807 (West); Nev. Rev. Stat. § 603A.210, 603A.215(2); N.M. Stat. § 57-12C-4, 57-12C-5; Ohio Rev. Code § 1354.01 to 1354.05; Or. Rev. Stat. § 646A.622; 11 R.I. Gen. Laws § 11-49.3-2; S.C. Code Ann. § 38-99-10 to -100; Tex. Bus. & Com. Code Ann. § 521.052 (West); Utah Code Ann. § 13-44-101, -201, -301 (West); Vt. Stat. Ann. tit. 9, §§ 2446–2447 (West).

64. For example, Texas law provides: "A business shall implement and maintain reasonable procedures, including taking any appropriate corrective action, to protect from unlawful use or disclosure any sensitive personal information collected or maintained by the business in the regular course of business." Tex. Bus. & Com. Code Ann. § 521.052 (West); *see also* Ark. Code Ann. § 4-110-104(b) (West); Cal. Civ. Code § 1798.81.5 (West); Colo. Rev. Stat. § 6-1-713.5; Conn. Gen. Stat. § 42-471(a); Fla. Stat. § 501.171(2); Ind. Code § 24-4.9-3-3.5(c); Kan. Stat. Ann. § 50-6, 139b; La. Stat. Ann. § 51:3074(B); Md. Code Ann., Com. Law § 14-3503(a) (West); Neb. Rev. Stat. Ann. § 87-808 (West); N.M. Stat. § 57-12C-4; Utah Code Ann. § 13-44-201(1) (West).

65. Or. Rev. Stat. § 646A.622.

66. For example, Texas law uses the defined term "sensitive personal information," which is defined to mean

result in financial harm to the individual. Most data collected by an automated vehicle or the industry would fall outside the narrow scope of these statutes, but certain data points, like VINs, may identify individuals and be viewed as personal information.

A new important data security law is California's data security law that applies to connected devices that comprise the Internet of things.[67] A connected device is defined broadly and could include a connected automated vehicle.[68] A manufacturer of a connected device is required to "equip the device with a reasonable security feature or features" tailored to the nature and function of the device and the information collected, stored, or transmitted and designed to protect against unauthorized access, use, disclosure, modification, or destruction.[69] A connected device authenticated outside of a local area network is deemed to have a reasonable security feature if it has a preprogrammed password unique to each device or the device contains a security feature that requires a user to generate a new authentication means prior to first use. No private cause of action is permitted under the act.[70]

B. Data Breach Notification Laws

All 50 states and the District of Columbia have data breach laws that require companies that suffer a data breach to notify persons affected.[71] These statutes

---

> (A) an individual's first name or first initial and last name in combination with any one or more of the following items, if the name and the items are not encrypted: (i) social security number; (ii) driver's license number or government-issued identification number; or (iii) account number or credit or debit card number in combination with any required security code, access code, or password that would permit access to an individual's financial account; or (B) information that identifies an individual and relates to: (i) the physical or mental health or condition of the individual; (ii) the provision of health care to the individual; or (iii) payment for the provision of health care to the individual.

TEX. BUS. & COM. CODE ANN. § 521.002(a)(2) (West). Other states narrowly limit their definition of personal information. *See, e.g.*, ARK. CODE ANN. § 4-110-103(7); IND. CODE § 24-4.9-2-10; FLA. STAT. § 501.171(1)(g); LA. STAT. ANN. § 51:3073(4); MASS. GEN. LAWS ch. 93H, § 1; OR. REV. STAT. § 646A.602(12)(a).

67. CAL. CIV. CODE § 1798.91.04 *et seq.* (West).

68. A "connected device" is "any device, or other physical object that is capable of connecting to the Internet, directly or indirectly, and that is assigned an Internet Protocol address or Bluetooth address." *Id.* § 1798.91.05(b).

69. *Id.* § 1798.91.04.

70. *Id.* § 1798.91.06(e).

71. ALA. CODE § 8-38-1 *et seq.*; ALASKA STAT. § 45.48.010 *et seq.*; ARIZ. REV. STAT. ANN. § 18-551 *et seq.*; ARK. CODE ANN. § 4-110-101 *et seq.* (West); CAL. CIV. CODE § 1798.80 *et seq.* (West); COLO. REV. STAT. § 6-1-716; CONN. GEN. STAT. § 36a-701b; DEL. CODE ANN. tit. 6, § 12B-101 *et seq.* (West); D.C. CODE ANN. § 28-3851 *et seq.* (West); FLA. STAT. § 501.171; GA. CODE ANN. § 10-1-910 *et seq.* (West); HAW. REV. STAT. § 487N-1 *et seq.*; IDAHO CODE § 28-51-104 *et seq.*; 815 ILL. COMP. STAT. ANN. 530/5 *et seq.*; IND. CODE § 24-4.9-1-1 *et seq.*; IOWA CODE § 715C.1 *et seq.*; KAN. STAT. ANN. § 50-7a01 *et seq.* (West); KY. REV. STAT. ANN. § 365.732 (West); LA. STAT. ANN. § 51:3071 *et seq.*; ME. STAT. tit. 10, § 1346 *et seq.*; MD. CODE ANN., COM. LAW § 14-3501 *et seq.* (West); MASS. GEN. LAWS ch. 93H, § 1 *et seq.*; MICH. COMP. LAWS § 445.63, .72; MINN. STAT. § 325E.61; MISS. CODE. ANN. § 75-24-29 (West); MO. REV. STAT. § 407.1500; MONT. CODE

generally apply when there is a data security breach that reveals personal information of a state's residents.[72] Upon such a breach, the breached company must notify persons reasonably believed to have been affected by the breach, as well as credit agencies and state regulators.[73] Data breaches often involve residents of almost all 50 states and the District of Columbia, so a company could be under the obligation to comply with 51 different data breach regimes (that are not always consistent). These data breach notification statutes will find their most obvious application in the automated vehicle industry when industry members are hacked and data they collect is compromised. The hacking of individual vehicles could also implicate these statutes when the hacker obtains personal information contained within the vehicle.

III. OTHER LAWS

States have enacted general laws that may implicate or complicate data privacy and security. This section describes laws that protect data collected by an event data recorder (EDR). This section then considers state unfair and deceptive trade practices acts. Chapter 16 detailed the FTC's use of section 5 of the FTC Act to enforce data privacy and security. States have what are known by many as baby FTC Acts or unfair and deceptive acts and practices acts that states and individuals could use to enforce data privacy or security.

ANN. § 30-14-1701 to -1702, -1704 (West); NEB. REV. STAT. ANN. § 87-801 *et seq.* (West); NEV. REV. STAT. § 603A.010 *et seq.*; N.H. REV. STAT. ANN. § 359-C:19–C:21, § 332-I:5; N.J. STAT. ANN. § 56:8-161, -163, -165 to -166 (West); N.M. STAT. § 57-12C-1 to 57-12C-12; N.Y. GEN. BUS. LAW § 899-aa (McKinney); N.C. GEN. STAT. § 75-61, 75-65; N.D. CENT. CODE § 51-30-01 to -07; OHIO REV. CODE ANN. § 1349.19 to .192 (West); OKLA. STAT. tit. 24, §§ 161–166; OR. REV. STAT. § 646A.600 to .628; 73 PA. CONS. STAT. ANN. § 2301 *et seq.*; 11 R.I. GEN. LAWS § 11-49.3-1 to 11-49.3-6; S.C. CODE ANN. § 39-1-90; S.D. CODIFIED LAWS § 22-40-19 to -26; TENN. CODE ANN. § 47-18-2105 to -2107; TEX. BUS. & COM. CODE ANN. § 521.002, 521.053, 521.151 to .152 (West); UTAH CODE ANN. § 13-44-101 *et seq.* (West); VT. STAT. ANN. tit. 9, §§ 2430, 2435 (West); VA. CODE ANN. § 18.2-186.6, § 32.1-127.1:05, § 58.1-341.2; WASH. REV. CODE § 19.255.010 *et seq.*; W. VA. CODE § 46A-2A-101 *et seq.*; WIS. STAT. § 134.98; WYO. STAT. ANN. § 40-12-501, -502 (West).

72. *See, e.g.*, CAL. CIV. CODE § 1798.82 (West) ("A person or business that conducts business in California, and that owns or licenses computerized data that includes personal information, shall disclose a breach of the security of the system following discovery or notification of the breach in the security of the data to a resident of California (1) whose unencrypted personal information was, or is reasonably believed to have been, acquired by an unauthorized person, or, (2) whose encrypted personal information was, or is reasonably believed to have been, acquired by an unauthorized person and the encryption key or security credential was, or is reasonably believed to have been, acquired by an unauthorized person and the person or business that owns or licenses the encrypted information has a reasonable belief that the encryption key or security credential could render that personal information readable or usable.").

73. *See, e.g.*, *id.*

A. Event Data Recorder Laws

In 2015, Congress enacted the Driver Privacy Protection Act of 2015 (the DPPA) to identify the owner of EDR data.[74] States have also enacted laws that provide rules and regulations on the collection of data by EDRs.[75] The DPPA did not include an express preemption clause, but it seems broad enough to override the application of a state law that is inconsistent with the DPPA. Likewise, NHTSA has enacted regulations that mandate certain information be collected by EDRs and certain disclosures be made to consumers.[76] To the extent that a state law conflicts with NHTSA's requirements, it, too, may be preempted.

Under the DPPA, the vehicle owner is the owner of data recorded by an EDR; other people cannot access EDR data unless they fall within an exception. All of the states, like the DPPA, provide that the vehicle owner owns EDR data. States also clarify that after a crash, EDR data remains the property of the vehicle owner, even if the title to the vehicle passes from the vehicle owner to a lienholder or an insurer.[77] Although spoliation of evidence would arise, some states explicitly prohibit the deleting of EDR data for a reasonable time following a crash resulting in death or serious physical injury.[78]

Like the DPPA, states also provide enumerated exceptions to allow other persons to access EDR data, including pursuant to a court order and for vehicle safety research, legal discovery, or vehicle maintenance. Because states do not have the same restrictions as the DPPA, it is important to consider whether the DPPA would preempt an enumerated exception by a state. Under the DPPA, non–vehicle owners are authorized to access EDR data in limited circumstances: (1) a court or other judicial or administrative authority can authorize the retrieval of data from an EDR, so long as EDR data is subject to standards of review of evidence; (2) written consent is given for any purpose, including

---

74. Driver Privacy Act of 2015, S. 766, 114th Congress. The Driver Privacy Act of 2015 is described in Chapter 16.

75. Ark. Code Ann. § 23-112-107 (West); Cal. Veh. Code § 9951 (West); Colo. Rev. Stat. § 12-6-401 *et seq.*; Conn. Gen. Stat. § 14-164aa; Del. Code Ann. tit. 18, § 3918 (West); Me. Stat. tit. 29-A, § 1971 *et seq.*; Mont. Code Ann. § 61-12-1001 *et seq.* (West); Nev. Rev. Stat. § 484D.485; N.H. Rev. Stat. Ann. § 357-G:1; N.J. Stat. Ann. § 39:10B-7 to -9 (West); N.Y. Veh. & Traf. Code § 416-b; N.D. Cent. Code § 51-07-28; Or. Rev. Stat. § 105.925 *et seq.*; Tex. Transp. Code § 547.615; Utah Code Ann. § 41-1a-1501 *et seq.* (West); Va. Code Ann. § 46.2-1088.6 (West); Wash. Rev. Code § 46.35.010 *et seq.*

76. 49 C.F.R. § 563.1 *et seq.*

77. Ark. Code Ann. § 23-112-107(e)(1) (West); Mont. Code Ann. § 61-12-1003(1) (West); Or. Rev. Stat. § 105.932; Utah Code Ann. § 41-1a-1504(1) (West); Va. Code Ann. § 46.2-1088.6.E.

78. Conn. Gen. Stat. § 14-164aa(d) ("No person may knowingly alter or delete data on an event data recorder, or knowingly destroy an event data recorder, after a crash event that resulted in a death or a serious physical injury, as defined in section 53a-3, within a reasonable amount of time sufficient for a peace officer to obtain a search warrant."); N.J. Stat. Ann. § 39:10B-9 (West) ("No person shall knowingly alter or delete data on a recording device, or knowingly destroy a recording device with the intent to prevent access to or destroy the recorded data, within two years after a crash event that resulted in bodily injury or death.").

a disclosure in a subscription agreement; (3) EDR data can be retrieved pursuant to an NHTSA investigation if the personally identifiable information is removed; (4) EDR data is used to determine the need for or facilitate an emergency medical response to a crash; or (5) EDR data is retrieved for traffic safety research if the personally identifiable information is removed. State laws that authorize the removal of EDR data for other purposes may very well be preempted. Because an exception to the DPPA's ownership rule is a court order, state laws that authorize courts or other judicial or administrative authorities (such as an administrative law judge) to authorize retrieval of EDR data will continue to be relevant and apply.

A slight difference between the DPPA and state laws is the definition of EDR (and thus the scope of their statutes). Under the DPPA, an EDR means "a device or function in a vehicle that records the vehicle's dynamic time-series data during the time period just prior to a crash event (e.g., vehicle speed vs. time) or during a crash event (e.g., delta–V vs. time), intended for retrieval after the crash event. For the purposes of this definition, the event data do not include audio and video data."[79] Except for Montana and Utah,[80] states do not incorporate the federal definition of EDR into their statutes. States use a variety of definitions, with some seemingly broader than others.[81] However, most of the EDR statutes define an EDR to mean a device that records limited information precrash. No state defines an EDR broadly enough to capture all data captured by an automated vehicle, but Virginia and Washington law come the closest.[82] Accordingly, the EDR laws are very limited in scope and leave unanswered who or what owns the most important data of an automated vehicle.

---

79. 49 C.F.R. § 563.5.

80. MONT. CODE ANN. § 61-12-1001(1) (West); UTAH CODE ANN. § 41-1a-1502(2) (West).

81. *Compare* ARK. CODE ANN. § 23-112-107(a)(2) (West) ("'Motor vehicle event data recorder' means a factory-installed feature in a motor vehicle that does one (1) or more of the following: (A) Records, stores, transmits, or dispenses any of the following information for the purpose of retrieval after a crash: (i) Vehicle speed; (ii) Vehicle direction; (iii) Vehicle location; (iv) Steering performance; or (v) Seat belt restraint status; (B) Has the capacity to transmit information concerning a crash in which the motor vehicle has been involved to a central communications system when a crash occurs; or (C) Includes a sensing and diagnostic module, restraint control module, electronic throttle control, or other similar components"), *with* WASH. REV. CODE § 46.35.010 ("'Recording device' includes event data recorders, sensing and diagnostic modules, electronic control modules, automatic crash notification systems, geographic information systems, and any other device that records and preserves data that can be accessed related to that motor vehicle. 'Recording device' does not include onboard diagnostic systems whose exclusive function is to capture fault codes used to diagnose or service the motor vehicle.").

82. Virginia's statute defines recorded data as follows:

> "Recorded data" means the data stored or preserved electronically in a recording device identifying performance or operation information about the motor vehicle including, but not limited to: 1. Speed of the motor vehicle or the direction in which the vehicle is traveling, or both; 2. Vehicle location data; 3. Vehicle steering performance; 4. Vehicle brake performance including, but not limited to, whether brakes were applied before a crash;

All states require disclosure to the vehicle owner that the vehicle has an EDR and the type of data recorded.[83] Some states require that this notice be provided in the owner's manual, while other states allow a dealer to give the notice.[84] NHTSA requires manufacturers to provide a specific statement about EDRs in the owner's manual.[85]

Another area where states have specifically regulated access to EDR data is that they prohibit insurers insisting on access to EDR data as a condition of a claim or policy renewal or to cause increased premiums due to a refusal.[86] For example, Virginia prohibits insurance companies from denying or refusing to renew an insurance policy if the motor vehicle owner refuses to provide access

---

> 5. The driver's seatbelt status; and 6. Information concerning a crash in which the motor vehicle has been involved, including the ability to transmit such information to a central communications system.

VA. CODE ANN. § 46.2-1088.6 (West). A recording device is also defined broadly as "an electronic system, and the physical device or mechanism containing the electronic system, that primarily, or incidental to its primary function, preserves or records, in electronic form, data collected by sensors or provided by other systems within the vehicle. 'Recording device' includes event data recorders (EDRs), sensing and diagnostic modules (SDMs), electronic control modules (ECMs), automatic crash notification (ACN) systems, geographic information systems (GIS), and any other device that records and preserves data that can be accessed related to that vehicle." *Id.*

Washington does not use the concept of recorded data and defines a recording device as "an electronic system, and the physical device or mechanism containing the electronic system, that primarily, or incidental to its primary function, preserves or records, in electronic form, data collected by sensors or provided by other systems within a motor vehicle. 'Recording device' includes event data recorders, sensing and diagnostic modules, electronic control modules, automatic crash notification systems, geographic information systems, and any other device that records and preserves data that can be accessed related to that motor vehicle. 'Recording device' does not include onboard diagnostic systems whose exclusive function is to capture fault codes used to diagnose or service the motor vehicle." WASH. REV. CODE § 46.35.010. By not using a "data recorded" concept and using "information recorded or transmitted" by a recording device, Washington's law is the broadest and could conceivably apply to a lot more data recorded by an automated vehicle than merely traditional EDR data.

83. *See, e.g.*, ARK. CODE ANN. § 23-112-107(b) (West).

84. *Compare* ARK. CODE ANN. § 23-112-107(b) (West) ("At the time of a new vehicle purchase by a consumer from a dealership, an owner of a motor vehicle shall be given written notice by the seller or manufacturer that includes the following: (1) The presence of the motor vehicle event data recorder in the motor vehicle; (2) The type of motor vehicle event data recorder in the motor vehicle; and (3) The type of data that is recorded, stored, or transmitted on the motor vehicle event data recorder."), *with* CAL. VEH. CODE § 9951(a) (West) ("A manufacturer of a new motor vehicle sold or leased in this state that is equipped with one or more recording devices commonly referred to as 'event data recorders (EDR)' or 'sensing and diagnostic modules (SDM),' shall disclose that fact in the owner's manual for the vehicle."); COLO. REV. STAT. § 42-4-2402 ("A manufacturer of a motor vehicle that is sold or leased in Colorado with an event data recorder shall in bold-faced type disclose, in the owner's manual, that the vehicle is so equipped and, if so, the type of data recorded. A disclosure made by means of an insert into the owner's manual shall be deemed a disclosure in the owner's manual.").

85. 49 C.F.R. § 563.11.

86. ARK. CODE ANN. § 23-112-107(e) (West); MONT. CODE ANN. § 61-12-1003 (West); N.D. STAT. § 51-07-28.6; OR. REV. STAT. § 105.932; UTAH CODE ANN. § 41-1a-1504(2) (West); VA. CODE ANN. § 38.2-2212.C; *but see* DEL. CODE ANN. tit. 18, § 3918 (West) (authorizing insurers to install data-reporting devices upon written consent of the policyholder).

to the vehicle's recorded data.[87] Virginia also prohibits increasing insurance premiums based on the denial.[88] Other states, like Maine, explicitly say that their EDR law does not impact an insured's duty to cooperate with an insurer.[89]

B. State Consumer Protection Laws

State attorneys general and private parties may use their state's unfair and deceptive acts and practices (UDAP) act for privacy and data security. UDAP acts could create unique opportunities for state attorneys general and private plaintiffs to pursue claims against members of the automated or connected vehicle industry, even in the absence of a state law governing data privacy or security in an automated vehicle. States have taken different approaches with their UDAP acts.[90] Many states replicate the federal FTC Act by using the "unfair or deceptive" trade practices or acts language without explicitly enumerating specific trade practices or acts that are unfair or deceptive.[91] Other states list specific prohibited acts and have a catchall of "unfair or deceptive" trade practices or acts.[92] UDAP acts are enforced by state governments, typically the attorney general, and consumers. In contrast to the FTC Act, which does not allow for private enforcement, UDAP acts allow consumers to enforce the acts and seek civil remedies against violators.

An initial limitation in many of the data privacy and breach cases has been whether the plaintiff (and usually putative class) has standing to assert a UDAP claim. A plaintiff pursuing a claim in federal court must prove that he or she has Article III standing. A major hurdle in privacy and data breach cases has been the injury-in-fact element of standing. A plaintiff proves an injury in fact by showing "an invasion of a legally protected interest which is (a) concrete and particularized and (b) actual or imminent, not conjectural or hypothetical."[93] Standing, and its limitations in these cases, is detailed in Chapter 18. There is currently a circuit split between the Seventh and Ninth Circuits in automaker unexploited vulnerability cases about whether plaintiffs

87. Va. Code Ann. § 38.2-2212.C.s ("No insurer shall refuse to renew a motor vehicle insurance policy solely because of any one or more of the following factors. . . . The refusal of a motor vehicle owner as defined in § 46.2-1088.6 to provide access to recorded data from a recording device as defined in § 46.2-1088.6.").

88. Va. Code Ann. § 38.2-2213.1 (West).

89. Me. Stat. tit. 29-A, § 1972.

90. Jolina C. Cuaresma, *§ 2:10. Model Acts for State UDAP Statutes*, Consumer Protection and the Law (Dec. 2019).

91. *See, e.g.*, 815 Ill. Comp. Stat. Ann. 505/2; Mass. Gen. Laws Ann. ch. 93A, § 2 (West); N.C. Gen. Stat. Ann. § 75-1.1.

92. *See, e.g.*, Alaska Stat. Ann. § 45.50.471 (West); Ala. Code § 8-19-1 *et seq.*; Ga. Code Ann. § 10-1-393 (West); Tex. Bus. & Com. Code Ann. § 17.46 (West).

93. Lujan v. Defs. of Wildlife, 504 U.S. 555, 560, 112 S. Ct. 2130, 119 L. Ed. 2d 351 (1992) (internal citations omitted).

can establish standing in the absence of the vulnerability being exploited. This development will be important to follow as the Seventh Circuit's finding of standing for a putative class against Fiat Chrysler for an unexploited vulnerability could have devastating effects for the automated vehicle industry (and the motor vehicle industry more broadly) as automakers update and patch new security vulnerabilities.

Even if plaintiffs can get over the standing hurdle, plaintiffs may also have to get over the substantive hurdle within a UDAP act that they have suffered an actual property or monetary loss to institute a suit under the UDAP act.[94] This limitation in many states' UDAP laws may effectively curtail privacy claims. In *Deacon v. Pandora Media, Inc.*,[95] a putative class brought a number of privacy claims against Pandora Media, Inc. (Pandora) pursuant, in relevant part, to the Michigan UDAP act. The putative class alleged that Pandora automatically created a personal page for each subscriber identifying his or her listening habits to anyone who located the personal page. The class further alleged that Pandora unilaterally integrated Pandora with Facebook accounts, allowing a subscriber's Facebook friends to see the subscriber's listening habits. The District Court for the Northern District of California noted that the Michigan UDAP act requires a putative class to show an actual loss to bring a class action. Because the class could not prove any actual loss, the court found that it could not pursue a class action against Pandora.[96] In many automated vehicle privacy claims, plaintiffs may struggle to establish actual monetary or property loss.

These hurdles for private plaintiffs will limit their ability to enforce these laws against the automated vehicle industry. In contrast to private plaintiffs, state attorneys general do not need to establish an actual monetary or property loss to enforce their UDAP acts. However, due to political pressures and resource constraints, state attorneys general rarely enforce their UDAP acts, at least in comparison to private actions.

---

94. The following statutes require some ascertainable loss of money or property: ALA. CODE § 8-19-10(a); ALASKA STAT. § 45.50.531(a); ARK. CODE ANN. § 4-88-113(f) (West); CAL. BUS. & PROF. CODE § 17204 (West); CONN. GEN. STAT. § 42-110g(a); IDAHO CODE § 48-608(1); IOWA CODE § 714H.5; KY. REV. STAT. ANN. § 367.220(1) (West); LA. STAT. ANN. § 51:1409(A); ME. STAT. tit. 5, § 213(1); MISS. CODE ANN. § 75-24-15(1) (West); MO. REV. STAT. § 407.025(1); MONT. CODE ANN. § 30-14-133(1) (West); N.J. STAT. ANN. § 56:8-19 (West); OR. REV. STAT. § 646.638(1); 73 PA. CONS. STAT. ANN. § 201-9.2(a); 6 R.I. GEN. LAWS § 6-13.1-5.2(a); S.C. CODE ANN. § 39-5-140(a); TENN. CODE ANN. § 47-18-109(a)(1) (West); WASH. REV. CODE § 19.86.090; W. VA. CODE § 46A-6-106; WIS. STAT. § 100.18(11)(b)(2).

95. 901 F. Supp. 2d 1166, 1168 (N.D. Cal. 2012).

96. The court stated that the Michigan UDAP act did allow an individual, but not a class, to seek injunctive relief.

Chapter 18

Liability for Cybersecurity Breaches and Cyberattacks

Motor vehicles today are potentially vulnerable to cyberattacks. In 2015, two white-hat hackers,[1] in connection with a writer from *Wired*, demonstrated that they could remotely take control of a 2014 Jeep Cherokee traveling on the highway and turn on the air conditioner, blast music over the radio, activate the windshield wipers, burst windshield wiper fluid, and stop the vehicle.[2] The hackers gained access to the Jeep Cherokee through the cellular connection to the entertainment system in the vehicle. This article led Fiat Chrysler Automobiles to conduct a safety recall.[3] Researchers from the University of Washington and the University of San Diego have shown that vehicles can be hacked through the electronic control unit that monitors tire pressure.[4] Tesla Motors currently has a "bug bounty program" that pays rewards to groups that provide details on vulnerabilities of Tesla's software.[5] The automotive industry has created the Automotive Information Sharing and Analysis Center to address vehicle cybersecurity risks and facilitate the sharing of information

1. A white-hat hacker is someone who tests systems and vulnerabilities to improve security before malicious hackers can exploit those vulnerabilities.

2. Andy Greenberg, *Hackers Remotely Kill a Jeep on the Highway—With Me in It*, Wired (July 21, 2015), https://www.wired.com/2015/07/hackers-remotely-kill-jeep-highway/.

3. Fiat Chrysler Autos., Safety Recall R40/NHTSA 15V-461 Radio Security Vulnerability (July 2015), https://static.nhtsa.gov/odi/rcl/2015/RCRIT-15V461-7681.pdf.

4. Peter Bright, *Cars Hacked through Wireless Tire Sensors*, Ars Technica (Aug. 10, 2010), https://arstechnica.com/information-technology/2010/08/cars-hacked-through-wireless-tyre-sensors/.

5. *See Product Security*, Tesla, https://www.tesla.com/about/security (last visited Nov. 12, 2019).

concerning cybersecurity threats, vulnerabilities, and incidents. The opportunities to cyberattack vehicles increase as they become more automated and connected.[6]

Cyberattacks of automated vehicles may be used to steal automated vehicles or their data, seek ransomware, or cause significant physical harm or damage.[7] Cyberattacks could be isolated—limited to a single, targeted vehicle—or systematic—carried out on groups of vehicles. An isolated attack may lead to a systematic attack once the vulnerabilities of a single vehicle are duplicated for other vehicles.[8] There are two reasons why isolated attacks of a single automated vehicle may be unlikely. First, there may be simpler means than hacking into a vehicle for a criminal to carry out his or her ultimate goals. Second, for cyberattackers who seek ransoms, obvious targets would be the automated vehicle manufacturers or entity-owners of automated vehicles, such as the owner of an automated ridesharing fleet. These companies would be more willing to pay a ransom if a cyberattack affected a fleet or many vehicles and not just a single vehicle.

Cyberattackers can gain access to an automated vehicle using devices or objects that connect with the automated vehicle, during the production of the vehicle or components, or while automated vehicles are repaired or updated.[9] A cyberattacker gains access to vehicles through vehicle communication buses or electronic control units.

In many cyberattack cases, a significant limitation is standing. This chapter initially explores standing and the injury-in-fact element that plaintiffs have struggled to prove in cybersecurity cases. This chapter then introduces the types of causes of action that plaintiffs have brought in cybersecurity cases and may bring when automated vehicles and automated vehicle industry members are subject to cyberattacks. This chapter then explores two broad types of cybersecurity cases. The first is unexploited vulnerability cases where class actions are filed before a cyberattacker actually exploits a vulnerability (outside of a controlled environment) or where the plaintiffs' vehicles had not been exploited. The automobile industry has confronted two of these unexploited vulnerability cases with mixed results. The second is where the plaintiffs were subject to hacking or directly harmed as a result of the cyberattack.

---

6. The focus of this section is on cyberattacks and not situations where the owner or user of the automated vehicle modifies the code by jailbreaking. For one of the more comprehensive analyses of civil liability for cyberattacks, see Zev Winkelman et al., *When Autonomous Vehicles Are Hacked, Who Is Liable?*, Rand Corp. (2019) [hereinafter Rand Report].

7. *See* Melissa L. Griffin, *Steering (or Not) through the Social and Legal Implications of Autonomous Vehicles*, 11 J. Bus. Entrepreneurship & L. 81, 98–99 (2018).

8. *See* Rand Report, *supra* note 6, at 13.

9. *See id.* at 5–11.

I. STANDING

Putative classes have filed class actions against companies whose computer systems or products suffered from security vulnerabilities or breaches. In many of these cases, the defendants have challenged the right of the putative class to establish standing. Standing is a constitutional requirement for a plaintiff invoking federal jurisdiction. A plaintiff must establish standing by showing the following requirements:

> First, the plaintiff must have suffered an "injury in fact"—an invasion of a legally protected interest which is (a) concrete and particularized, and (b) actual or imminent, not "conjectural" or "hypothetical." Second, there must be a causal connection between the injury and the conduct complained of—the injury has to be fairly traceable to the challenged action of the defendant, and not the result of the independent action of some third party not before the court. Third, it must be "likely," as opposed to merely "speculative," that the injury will be redressed by a favorable decision.[10]

Typically, plaintiffs find the most difficulty establishing an injury in fact, which has three components: the injury must be concrete, particularized, and imminent. A concrete injury is one that actually exists.[11] A breach of a statute, in and of itself, is not sufficient to establish concreteness.[12] A particular injury is one that "affect[s] the plaintiff in a personal and individualized way."[13] The Supreme Court has clarified "imminence" to mean that the injury must be "certainly impending" or there must be a "substantial risk of harm."[14] Three recent cases have clarified the injury-in-fact element and have had an impact on data breach and cybersecurity cases: *Clapper v. Amnesty International USA*,[15] *Susan B. Anthony List v. Driehaus*,[16] and *Spokeo, Inc. v. Robins*.[17]

In *Clapper*, attorneys and human rights, labor, and media organizations filed suit challenging the constitutionality of an amendment to the Federal Intelligence Surveillance Act (FISA) that allows surveillance under specific circumstances of non–United States persons reasonably believed to be located outside of the United States. The plaintiffs alleged that the amendments compromise their abilities to communicate with clients, find sources, and obtain

---

10. Lujan v. Defs. of Wildlife, 504 U.S. 555, 560–61, 112 S. Ct. 2130, 119 L. Ed. 2d 351 (1992) (internal quotation marks and citations omitted).
11. Spokeo, Inc. v. Robins, 136 S. Ct. 1540, 1548, 194 L. Ed. 2d 635, *revised* (May 24, 2016).
12. *Id.* at 1549.
13. *Id.* (quoting *Lujan*, 504 U.S. at 560 n.1).
14. Susan B. Anthony List v. Driehaus, 573 U.S. 149, 134 S. Ct. 2334, 189 L. Ed. 2d 246 (2014).
15. Clapper v. Amnesty Int'l USA, 568 U.S. 398, 133 S. Ct. 1138, 185 L. Ed. 2d 264 (2013).
16. 573 U.S. 149.
17. 136 S. Ct. 1540.

information, and the threat of surveillance would require them to engage in costly and burdensome measures to maintain confidentiality, like traveling overseas. The plaintiffs sought to enjoin the amendments, claiming that it was reasonably likely that their communications would be acquired. The Supreme Court held that the plaintiffs lacked standing because they could not prove an injury that was "certainly impending."[18] The Supreme Court also rejected the argument that the additional measures the plaintiffs undertook to protect confidentiality could establish standing, stating that they "cannot manufacture standing merely by inflicting harm on themselves based on their fears of hypothetical future harm that is not certainly impending."

In *Susan B. Anthony List v. Driehaus*, a unanimous Supreme Court found that Susan B. Anthony List (SBA), together with the Coalition Opposed to Additional Spending and Taxes (COAST), had standing to challenge an Ohio statute that criminalized the dissemination of false statements during a political campaign. SBA had been subject to inquiry over statements that it sought to publish against an Ohio congressman for his vote on the Affordable Care Act. The inquiry settled after the congressman lost his bid for reelection. COAST was not subject to an inquiry for violating the Ohio statute at the time of the lawsuit. The Supreme Court considered whether SBA and COAST had standing, which turned on whether they had suffered an injury in fact. The Supreme Court clarified what injury in fact requires for future injuries: "An allegation of future injury may suffice if the threatened injury is 'certainly impending,'

---

18. The Supreme Court elaborated:

> Respondents assert that they can establish injury in fact that is fairly traceable to § 1881a because there is an objectively reasonable likelihood that their communications with their foreign contacts will be intercepted under § 1881a at some point in the future. This argument fails. As an initial matter, the Second Circuit's "objectively reasonable likelihood" standard is inconsistent with our requirement that "threatened injury must be certainly impending to constitute injury in fact." . . . Furthermore, respondents' argument rests on their highly speculative fear that: (1) the Government will decide to target the communications of non-U.S. persons with whom they communicate; (2) in doing so, the Government will choose to invoke its authority under § 1881a rather than utilizing another method of surveillance; (3) the Article III judges who serve on the Foreign Intelligence Surveillance Court will conclude that the Government's proposed surveillance procedures satisfy § 1881a's many safeguards and are consistent with the Fourth Amendment; (4) the Government will succeed in intercepting the communications of respondents' contacts; and (5) respondents will be parties to the particular communications that the Government intercepts. As discussed below, respondents' theory of standing, which relies on a highly attenuated chain of possibilities, does not satisfy the requirement that threatened injury must be certainly impending. . . . Moreover, even if respondents could demonstrate injury in fact, the second link in the above-described chain of contingencies—which amounts to mere speculation about whether surveillance would be under § 1881a or some other authority—shows that respondents cannot satisfy the requirement that any injury in fact must be fairly traceable to § 1881a.

Clapper, 568 U.S. at 410–11 (internal citations omitted).

or there is a 'substantial risk' that the harm will occur."[19] The Court found that SBA and COAST sufficiently alleged a substantial risk that they would be subject to a threat, arrest, or prosecution under the Ohio statute. This substantial risk prong has made it easier for plaintiffs to establish standing for data breaches.[20]

Standing was again at issue before the Supreme Court in *Spokeo.* Spokeo operated a "search engine" that allowed a person to search for information about a specific individual, and upon such a request, Spokeo provided information that it could gather about that individual. The plaintiff had examined the information provided about himself and determined it was incorrect. The plaintiff brought a claim under the Fair Credit Reporting Act of 1970 (the FCRA), claiming that Spokeo violated the FCRA. Spokeo challenged the plaintiff's standing because he did not introduce proof that he suffered an actual injury. The Ninth Circuit held that the plaintiff had standing because he claimed a statutory violation of his rights. The Supreme Court reversed and remanded the case back to the Ninth Circuit because the Ninth Circuit only examined the particularized requirement of injury in fact and did not consider whether the plaintiff suffered a concrete injury. According to the Supreme Court, the statutory violation, in and of itself, could not establish a concrete injury. On remand, the Ninth Circuit found that the plaintiff established a concrete injury.[21]

Standing will be particularly relevant to many cases that involve unexploited vulnerabilities or cases where the manufacturer updates, patches, or fixes vulnerabilities. When automated vehicles themselves are hacked, standing may be a hurdle for all other vehicle owners and users who sue as a result of the hacking. For example, if an automated vehicle manufactured by ABC Manufacturing is hacked, a putative class of owners of automated vehicles manufactured by ABC Manufacturing may file suit against ABC Manufacturing for the security vulnerability. The person whose vehicle was hacked should easily establish standing, but the other persons whose vehicles were not hacked may have more difficulty establishing standing. This topic is described below in regard to unexploited vulnerabilities and the effect of software patches.

---

19. *Susan B. Anthony List*, 573 U.S. at 158 (internal quotation marks and citation omitted).

20. *See, e.g.*, Attias v. Carefirst, Inc., 865 F.3d 620, 627 (D.C. Cir. 2017); Galaria v. Nationwide Mut. Ins. Co., 663 F. App'x 384, 388 (6th Cir. 2016); Remijas v. Neiman Marcus Grp., LLC, 794 F.3d 688, 693 (7th Cir. 2015).

21. Robins v. Spokeo, Inc., 867 F.3d 1108 (9th Cir. 2017).

II. CAUSES OF ACTION

Plaintiffs in cybersecurity cases assert numerous causes of action against the companies who suffer breaches. Frequently, plaintiffs' claims include negligence, misrepresentation, breach of warranties, and state consumer protection laws. In cases where physical injury or damage occurs in an automated vehicle crash, traditional products liability claims may be asserted against the automaker and possibly component manufacturers that are responsible for the vulnerability. Courts have had little difficulty in finding that plaintiffs have a duty, whether statutorily or in the common law, to provide reasonable security from cyberattacks.[22]

A significant hindrance to plaintiffs in cybersecurity cases is proving a cognizable injury under the substantive law. The economic loss doctrine may limit claims that are based in negligence, and consumer protection laws and other state law claims may require actual harm or actual loss. The economic loss doctrine bars negligence claims unless the plaintiffs can allege certain noneconomic injuries. For example, under California law, "[i]n the absence of (1) personal injury, (2) physical damage to property, (3) a 'special relationship' existing between the parties, or (4) some other common law exception to the rule, recovery of purely economic loss is foreclosed."[23] Each state defines the economic loss doctrine differently, and some state negligence claims may be able to proceed if a state has narrowly interpreted the economic loss doctrine.[24] The actual injury requirement of consumer protection laws in some states

---

22. *See, e.g., In re* Sony Gaming Networks & Customer Data Sec. Breach Litig., 996 F. Supp. 2d 942, 966, *order corrected*, No. 11MD2258 AJB (MDD), 2014 WL 12603117 (S.D. Cal. Feb. 10, 2014) ("Although neither party provided the Court with case law to support or reject the existence of a legal duty to safeguard a consumer's confidential information entrusted to a commercial entity, the Court finds the legal duty well supported by both common sense and California and Massachusetts law."); Daly v. Metro. Life Ins. Co., 4 Misc. 3d 887, 893, 782 N.Y.S.2d 530, 535 (Sup. Ct. 2004) (holding "that Met Life had a duty to protect the confidential personal information provided by the plaintiff"); *In re* Target Corp. Customer Data Sec. Breach Litig., 64 F. Supp. 3d 1304, 1310 (D. Minn. 2014).

23. Kalitta Air, L.L.C. v. Cent. Tex. Airborne Sys., Inc., 315 F. App'x 603, 605 (9th Cir. 2008); *see also* Aardema v. U.S. Dairy Sys., Inc., 147 Idaho 785, 792, 215 P.3d 505, 512 (2009) ("There are two exceptions to the general rule which prevents a party from recovering purely economic loss in a tort claim; those two exceptions are, (1) where a special relationship exists between the parties, or (2) where unique circumstances require a reallocation of the risk."); *In re* Michaels Stores Pin Pad Litig., 830 F. Supp. 2d 518, 528 (N.D. Ill. 2011) ("Illinois law recognizes three exceptions to the economic loss rule: (1) where plaintiff sustains personal injury or property damage resulting from a sudden or dangerous occurrence; (2) where plaintiff's damages were proximately caused by defendant's intentional, false representation; and (3) where plaintiff's damages were proximately caused by the negligent misrepresentation of a defendant in the business of supplying information for the guidance of others in business transactions." (citing Moorman Mfg. Co. v. Nat'l Tank Co., 91 Ill. 2d 69, 435 N.E.2d 443, 450–52 (1982))).

24. The court in *In re Target Corp. Data Security Breach Litigation* surveyed state interpretations of the economic loss doctrine and dismissed some but not all of the claims. 66 F. Supp. 3d 1154, 1171–76 (D. Minn. 2014).

has foreclosed these claims following cybersecurity breaches.[25] The hurdle of proving a cognizable injury will be relevant when members of the automated vehicle industry suffer cybersecurity breaches or when other automated vehicle owners (other than the victim of the hacking) sue over a hack into an automated vehicle brand they own or use.

III. VULNERABILITIES OR PATCHES BEFORE EXPLOITATION

In recent years, putative classes have filed claims against companies for vulnerabilities before their exploitation.[26] These cases are based on the opportunities that third-party hackers could have exploited. Automated vehicle manufacturers may face these types of claims when they identify and fix unexploited vulnerabilities or when white-hat hackers identify vulnerabilities. Automated vehicle industry members will likely patch vulnerabilities through over-the-air updates or other software updates to their driving automation systems. These claims are particularly pernicious because, by identifying and fixing security vulnerabilities, manufacturers create the cause of action for which plaintiffs bring consumer protection claims, typically under an economic loss benefit-of-the-bargain-type claim.[27] Under the benefit-of-the-bargain test, plaintiffs claim that they suffer economic harm through overpayment, loss in value, or loss of usefulness by purchasing a vehicle or a feature with the alleged defect.[28]

25. *See, e.g.*, Shafran v. Harley-Davidson, Inc., No. 07 CIV. 01365 (GBD), 2008 WL 763177, at *3 (S.D.N.Y. Mar. 20, 2008) ("Courts have uniformly ruled that the time and expense of credit monitoring to combat an increased risk of future identity theft is not, in itself, an injury that the law is prepared to remedy. Plaintiff has not presented any case law or statute, from any jurisdiction, indicating otherwise. Plaintiff's alleged injuries are solely the result of a perceived and speculative risk of future injury that may never occur. Plaintiff has failed to show an actual resulting injury that might support a claim for damages. As damages are an essential element of each of plaintiff's claims, plaintiff's claims fail as a matter of law."); Hendricks v. DSW Shoe Warehouse, Inc., 444 F. Supp. 2d 775, 782 (W.D. Mich. 2006) ("Here, plaintiff has alleged not that she suffered a 'loss' as a result of DSW's alleged violation of the statute. Instead, she has alleged only that she purchased a credit monitoring product to protect herself from identity theft. However, the court is aware of no Michigan authority which recognizes, as injury in an action under the MCPA, recovery for a potential future loss which has not actually occurred."). See Chapter 17 for a list of statutes requiring actual loss.

26. *See, e.g.*, *In re* Apple Processor Litig., No. 5:18-CV-00147-EJD, 2019 WL 3533876 (N.D. Cal. Aug. 2, 2019); Beyer v. Symantec Corp., No. 18-CV-02006-EMC, 2019 WL 935135 (N.D. Cal. Feb. 26, 2019); Cahen v. Toyota Motor Corp., 147 F. Supp. 3d 955 (N.D. Cal. 2015), *aff'd*, 717 F. App'x 720 (9th Cir. 2017).

27. See Chapter 8 for a discussion on these benefit-of-the-bargain claims.

28. *See, e.g.*, *In re* Toyota Motor Corp. Unintended Acceleration Mktg., Sales Practices, & Prod. Liab. Litig., 754 F. Supp. 2d 1145, 1162 (C.D. Cal. 2010); *see also* Mazza v. Am. Honda Motor Co., 666 F.3d 581, 595 (9th Cir. 2012) (holding that plaintiffs satisfy the injury-in-fact element when they "contend that class members paid more for the [collision mitigation braking system] than they otherwise would have paid, or bought it when they otherwise would not have done so, because Honda made deceptive claims and failed to disclose the system's limitations"); Cole v. Gen. Motors Corp., 484 F.3d

In other words, plaintiffs claim that they overpaid for a product that suffered from vulnerabilities or that the existence of vulnerabilities devalued their products. Courts are split on whether plaintiffs have standing based on the speculative nature of the unexploited vulnerability claims or the impact of security patches.[29]

Automakers have already encountered two of these cases with mixed results. The first case was *Cahen v. Toyota Motor Corp.*,[30] which the court dismissed for lack of standing, and the second case was *Flynn v. FCA US LLC*,[31] in which the court found that the plaintiffs had standing. In *Cahen*, a putative class filed suit against Ford Motor Company; General Motors, LLC; Toyota Motor Corporation; and Toyota Motor Sales (collectively, the Defendant Automakers), alleging that they had equipped "their vehicles with computer technologies that were susceptible to being hacked by third parties" and had collected and transmitted data in violation of their right to privacy. The putative class did "not allege that any of their vehicles have actually been hacked, or that they are aware of any vehicles that have been hacked outside of controlled environments, but instead alleged that hacking is an 'imminent eventuality.'" The main causes of action were state consumer protection acts, state warranty acts, contractual claims, and fraudulent concealment claims. The Defendant Automakers moved to dismiss the claims for, among other reasons, the failure of the plaintiffs to establish Article III standing. The Defendant Automakers claimed that risk of future hacking is too speculative to constitute a "certainly impending" injury. The plaintiffs attempted to use the benefit-of-the-bargain test to show that they had suffered economic harm by having vehicles subject to vulnerabilities. The court found that the plaintiffs lacked standing because the risk of harm was too speculative and that they had failed to introduce evidence that they incurred out-of-pocket costs and a demonstrable and documented effect of the vulnerability on the vehicle market. The Ninth Circuit affirmed this ruling.[32]

The *Wired* article and the subsequent recall discussed in the introduction led to a class action against FCA.[33] In this class, owners of Chrysler vehicles asserted warranty and consumer protection claims about the vulnerabilities

717, 723 (5th Cir. 2007) (finding that the putative class has suffered an injury in fact where they based their "recovery [on] their actual economic harm (e.g., overpayment, loss in value, or loss of usefulness) emanating from the loss of their benefit of the bargain").

29. Compare these cases finding lack of standing—*Cahen*, 147 F. Supp. 3d 955; *Katz v. Pershing, LLC*, 672 F.3d 64, 79–80 (1st Cir. 2012); and *Reilly v. Ceridian Corp.*, 664 F.3d 38, 42 (3d Cir. 2011)—with *Flynn v. FCA US LLC*, No. 15-CV-0855-MJR-DGW, 2016 WL 5341749 (S.D. Ill. Sept. 23, 2016), finding standing.

30. 147 F. Supp. 3d 955.

31. No. 15-CV-0855-MJR-DGW, 2016 WL 5341749, at *1 (S.D. Ill. Sept. 23, 2016).

32. Cahen v. Toyota Motor Corp., 717 F. App'x 720, 724 (9th Cir. 2017).

33. *Flynn*, 2016 WL 5341749, at *1.

exposed in the *Wired* article. No plaintiff's vehicle had suffered any hack, nor had there been any verified hacks outside of the *Wired* article incident[34]—and future hacks were prevented due to the recall. FCA claimed that the plaintiffs lacked standing. The court found that the plaintiffs lacked standing to present claims based on the serious risk of harm to the plaintiffs. However, the court found that they had standing under an overpayment and diminished value theory; the plaintiffs claimed they overpaid for vehicles with vulnerabilities and that the vulnerabilities caused an appreciable drop in their vehicles' values. FCA filed an interlocutory appeal to the Seventh Circuit, which denied the appeal, and then appealed the decision to the United States Supreme Court, which denied the petition on January 7, 2019.

Another recent case was *In re Apple Processor Litigation*,[35] where a putative class, purchasers of certain Apple products, claimed they suffered harm because their Apple products were defective by having a vulnerability that would have, if exploited, allowed unauthorized third parties access to their sensitive data. None of the named plaintiffs or any consumer had their devices' vulnerabilities exploited. The plaintiffs, relying on the benefit-of-the-bargain test, claimed that they overpaid for their Apple products, that the value of their products diminished, and that Apple's software patches that fixed the vulnerabilities slowed the performance of their devices. The putative class introduced a regression analysis to establish diminution of value. Apple argued, among other things, that the putative class lacked standing for failing to demonstrate an injury in fact. The court found that the putative class lacked standing. As to the performance slowdown, the court ruled that the named plaintiffs did not suffer from performance degradation after the updates, and, thus, they could not establish a particularized injury. As to the diminution-in-value claims, the court held that the plaintiffs needed more proof than the regression analysis—"[t]he measurement for the value of electronics in these circumstances must be more well-known or widespread." The court also focused on whether, given the widespread nature of the vulnerabilities, plaintiffs could establish that they would have purchased alternative devices.

In *Beyer v. Symantec Corp.*,[36] the court found security patches issued to fix software vulnerabilities relevant in determining whether a putative class could establish an injury in fact. A putative class filed suit claiming that they had overpaid for network security software because there were security vulnerabilities in Symantec's AntiVirus Decomposer Engine. A cybersecurity team at Google had notified Symantec of the vulnerabilities. Symantec notified the public of the

34. Thirty people had reported possible hacking during the recall, but most were theft-related hacks unrelated to risk of injury or death. *Id.* at *2.

35. No. 5:18-CV-00147-EJD, 2019 WL 3533876 (N.D. Cal. Aug. 2, 2019).

36. No. 18-CV-02006-EMC, 2019 WL 935135 (N.D. Cal. Feb. 26, 2019).

vulnerabilities and issued security patches on the same date. As to the overpayment theory, the court found that the plaintiffs failed to establish an injury in fact for that claim. The court focused on the speculative nature of the vulnerabilities, stating that the claim was based "on a purported *past* risk that has never been alleged to manifest and presumably never will" after the patches.

The overpayment theory approved by the *Flynn* court but rejected by other courts could have significant repercussions for the automotive industry and the automated vehicle industry. In most cases where an automated vehicle has been subject to vulnerabilities, plaintiffs may claim that it resulted in a diminution of value. To protect against this risk, manufacturers will want to take cybersecurity seriously throughout the design process and implement security by design. Manufacturers will also want to manage public relations surrounding a cyberattack to minimize the risk of financial repercussions for automated vehicle owners, lessees, or subscribers. One way to manage public relations and minimize diminution-of-value cases is to patch security vulnerabilities immediately. The *Semantec Corp.* court gave significant weight to an immediate patch as defeating the putative class's injury in fact. Companies will also want to examine their obligations and responsibilities to indemnify for cybersecurity risks in their supply contracts.

IV. CYBERATTACKS THAT CAUSE DAMAGE OR HARM

When cyberattacks cause crashes, persons harmed may seek to recover for their harm. Cyberattacks create problems for our sense of justice and fairness. With cyberattacks, there is a clear bad actor who should be held to account by both our civil and criminal justice systems.[37] However, the aggrieved parties (and government authorities) may be unable to identify the bad actor, or, even if identifiable, the bad actor may be judgment proof. The inability to identify the cyberattacker or recover against an identified cyberattacker will place undue pressure on others who could conceivably be responsible for the harm caused by the cyberattack.

Victims of these crashes may initially seek recourse against other persons involved in the crash. This colloquial claim may be the first instinct of victims and police officers who may view claims of cyberattacks with skepticism. Defendants, the vehicle users or owners, will, depending upon the level of automated vehicle, have their normal defenses and arguments for why they should not be liable, including the principal argument that they are not at fault for the

37. *See generally* Jeffrey K. Gurney, *Driving into the Unknown: Examining the Crossroads of Criminal Law and Autonomous Vehicles*, 5 Wake Forest J.L. & Pol'y 393 (2015); Rand Report, *supra* note 6, at 60–62.

crash. Assuming that they could prove that a cyberattack caused the vehicle's action, then the victims will probably need to look elsewhere for recovery.

The next choice for recovery may be under the victim's own insurance policy. In states that have no-fault insurance, victims of the crash may be able to seek recourse under their own insurance policies (assuming they have one). In fault states, for people harmed by a cyberattacked vehicle, perhaps this type of claim may be covered under the uninsured or personal injury protection coverage of the victims. In some cases, victims may be able to recover under a homeowner's insurance policy. Whether this type of crash is covered under an insurance policy would require a review of the policy's terms.

If neither of these routes for recovery exists, the aggrieved parties would then need to pursue recovery against those who could conceivably be responsible for allowing or creating the circumstances of the crash. Because cyberattacks are criminal acts, an initial inquiry is whether other parties can be liable for the harm caused by the criminal actors. The answer to who could conceivably be responsible depends upon the duties, if any, of certain persons to prevent cyberattacks or harm caused by cyberattacks. Generally, "[w]hether an individual must protect another from the unreasonable risk of foreseeable harm depends on the nature of the risk and the particular level of knowledge of the alleged tortfeasor."[38] Given the frequency of cyberattacks generally and the risk of harm to occupants of vehicles and others, courts will likely impose duties on certain persons to guard against cyberattacks on a vehicle,[39] and victims of cyberattacks will be foreseeable.[40] The question then becomes whose acts or omissions created the opportunity for the criminal act.[41]

---

38. Stanford v. Kuwait Airways Corp., 89 F.3d 117, 123 (2d Cir. 1996) (New York law); *see also* Restatement (Second) of Torts § 302B cmt f. (1965) ("Factors to be considered are the known character, past conduct, and tendencies of the person whose intentional conduct causes the harm, the temptation or opportunity which the situation may afford him for such misconduct, the gravity of the harm which may result, and the possibility that some other person will assume the responsibility for preventing the conduct or the harm, together with the burden of the precautions which the actor would be required to take. Where the risk is relatively slight in comparison with the utility of the actor's conduct, he may be under no obligation to protect the other against it.").

39. Ryan J. Duplechin, *The Emerging Intersection of Products Liability, Cybersecurity, and Autonomous Vehicles*, 85 Tenn. L. Rev. 803, 834–37 (2018); *see also* Mark A. Geistfeld, *A Roadmap for Autonomous Vehicles: State Tort Liability, Automobile Insurance, and Federal Safety Regulation*, 105 Cal. L. Rev. 1611, 1661 (2017) (stating that as a matter of tort law, "the manufacturer's tort obligations encompass the cybersecurity of the vehicle"); Scott L. Wenzel, *Not Even Remotely Liable: Smart Car Hacking Liability*, 2017 U. Ill. J.L. Tech. & Pol'y 49, 56 (2017) ("[D]ue to the recent highly publicized controlled hacking events, car manufacturers should foresee cyberattacks against their connected cars.").

40. *See, e.g.*, *In re* Sept. 11 Litig., 280 F. Supp. 2d 279, 295–96 (S.D.N.Y. 2003) (describing how a reasonable jury could find that ground victims are foreseeable victims of a plane hijacking).

41. *In re* Michaels Stores Pin Pad Litig., 830 F. Supp. 2d 518, 528 (N.D. Ill. 2011) ("Generally, a defendant will not be held liable for negligence if an intervening criminal act causes the plaintiff's injury. However, an exception exists where the defendant's acts or omissions create a condition conducive to a foreseeable intervening criminal act." (internal citation omitted)).

Victims of cyberattacks may sue persons who interact with the automated vehicle in any meaningful way. These persons include the vehicle owner, current and prior vehicle users, repair shops that may have repaired the vehicle, component manufacturers, the vehicle manufacturer, or an infrastructure owner operator (IOO). Each of these groups may have duties to prevent cyberattacks: automated vehicle owners and users may have duties to update vehicle software and install the latest security patches; repair shops may have duties to ensure that employees do not carry out or enable digital interception or create vulnerabilities; component manufacturers will have duties to ensure that their components are reasonably secure in the event of a cyberattack and to minimize vulnerabilities; vehicle manufacturers will have duties to ensure that their vehicles are reasonably safe and secure from cyberattacks; and IOOs will need to use reasonably secure infrastructure that is not vulnerable to cyberattacks.

In filing these causes of action, those harmed in a crash resulting from a cyberattack will need to identify the cause of the cyberattack to be able to properly identify the parties who caused the cyberattack and who may have breached their duties to protect against the cyberattack. The manufacturer will be a likely target in all lawsuits because the hacker would have exploited some vulnerability in the vehicle. The potential risks for a manufacturer of a vehicle or a component could be significant. Those harmed may seek to hold the manufacturer of vehicles that were susceptible to the cyberattack liable for the crash. For systematic cyberattacks that involve multiple manufacturers, each transmitter of the cyberattack may be sued. In these crashes, plaintiffs will likely contend that the driving automation system was defectively designed by being susceptible to hacking or by identifying a security patch that could have prevented the cyberattack. These cases may seem like absolute liability cases for manufacturers because once a vulnerability is identified, the costs to patch the vulnerability may be insignificant. The struggle for courts will be deciding how to define the risk-utility test and whether to permit plaintiffs to introduce a driving automation system that could have prevented the cyberattack, with the post-hoc knowledge of the vulnerability, when the costs of detecting and identifying the vulnerability prior to the cyberattack may have been significant. This could turn on whether the court admits the patch, a subsequent remedial action.[42]

---

42. See Chapter 8 for a discussion on the admissibility of subsequent remedial actions.

Part 5

Criminal Law

INTRODUCTION

The effects of automated vehicles on criminal behavior, police enforcement, and criminal law will be significant. Automated vehicles will lower transaction costs and make certain types of crimes easier to commit. Drug dealers will be able to deploy automated vehicles carrying contraband that comply with traffic codes, making it more difficult for police to catch them. Terrorists could use driverless vehicles to cause destruction without any personal risk.

Police will also lose one of their most effective tools to combat drug and firearm crimes: the traffic stop. In 2015, police in the United States stopped approximately 19,204,500 vehicles for potential traffic violations.[1] Given the abundance of traffic laws and the impossibility that drivers will always comply with them, police have historically used traffic stops to stop whomever they desire.[2] These often-pretextual stops target, in the best of cases, known

1. Elizabeth Davis, Anthony Whyde & Lynn Langton, *Special Report: Contacts between Police and the Public, 2015*, U.S. Dep't of Justice (Oct. 2018), https://www.bjs.gov/content/pub/pdf/cpp15.pdf.

2. David A. Sklansky, *Traffic Stops, Minority Motorists, and the Future of the Fourth Amendment*, 1997 Sup. Ct. Rev. 271, 273 (1997) ("Since virtually everyone violates traffic laws at least occasionally, the upshot of these decisions is that police officers, if they are patient, can eventually pull over almost

criminals and, in the worst of cases, racial minorities.[3] Automated vehicles will be programmed to comply with traffic laws, resulting in reduced traffic stops, which, in turn, may reduce police opportunities "to detect criminal activity, investigate crime, and apprehend criminal suspects"[4] but also may result in greater general trust of the police as pretextual stops become a relic of a bygone era.[5]

Automated vehicles will impact criminal law by placing significant pressure on historical assumptions underlying criminal law's application to vehicles and drivers. Most traffic laws and automobile crimes assume that a vehicle cannot operate itself and that any action or inaction taken by an automobile is the result of human control over the vehicle. Courts have broadly interpreted the acts of driving or operating to fully enforce the fundamental assumption that the human driver is necessary and responsible for the safe operation of a motor vehicle. As will be seen in Chapter 19, these broad interpretations may complicate enforcing traffic laws and prosecuting automobile crimes for highly automated vehicles.

Automated vehicles will create new opportunities for criminal mischief. People may physically interfere with automated vehicles by obstructing the vehicle or creating false blockades or traffic signs. People may cyberattack automated vehicles to cause terrorism or seek ransoms. Chapter 20 describes physical interference and digital interference with automated vehicles and how police can prosecute those who interfere with automated vehicles.

Automated vehicles will both hamper opportunities for police to collect information about occupants and create new opportunities for police. Different deployments of automated vehicles create different rights and obligations under the Fourth Amendment. As society potentially begins to embark on a world where vehicles are no longer personally owned, courts will need to consider how the Fourth Amendment applies in an automated ridesharing world. Chapter 21 describes physical searches and seizures of automated vehicles and their occupants. Chapter 22 explores searches of automated vehicle data.

---

anyone they choose, order the driver and all passengers out of the car, and then ask for permission to search the vehicle without first making clear the detention is over.").

3. For a discussion on the latter, see, for example, Devon W. Carbado, *From Stopping Black People to Killing Black People: The Fourth Amendment Pathways to Police Violence*, 105 Cal. L. Rev. 125 (2017).

4. Jordan Blair Woods, *Autonomous Vehicles and Police De-Escalation*, 114 Nw. U.L. Rev. Online 74, 84 (2019).

5. *See generally* Chapters 11 and 12 in Malcolm Gladwell, Talking to Strangers: What We Should Know about the People We Don't Know (2019) (describing, among other things, how excessive traffic stops by police create mistrust in communities).

Chapter 19

Automobile Crimes

This chapter explores the intersection of automated vehicles with traffic laws and more substantive automobile crimes. The inquiries here are similar to the ones that occupied Part 3 in the discussion of civil liability: what are the impacts of the shift in responsibility and control over driving to automated vehicles? Does criminal responsibility shift from the traditional driver to some other party? Are acts that are criminal for human drivers no longer criminal for human users of automated vehicles? What new crimes are created by automated vehicles?

Unlike in the context of civil law, which is mostly governed by common law decisions, the problems created by using historical assumptions about driving and responsibility are more significant and pronounced with criminal law. Motor vehicle codes, as that name denotes, are significant statutory regimes that have specified statutory definitions. Courts and legislatures have broadly interpreted many definitions to allow for maximum enforcement of penalties against bad acts, such as driving or operating a vehicle while under the influence of drugs or alcohol. The broad interpretations by courts and legislatures may make any shift in criminal responsibility away from the user of a highly automated vehicle more complicated or impossible without legislative action or rulings from appellate courts. This problem becomes more pronounced when one considers that crimes carry with them greater societal reprehension and consequences than civil liability for automobile crashes. People charged and convicted could face jail or prison time and lose their jobs, career opportunities, and community status based on antiquated concepts within motor vehicle codes and for automobile crimes.

This chapter begins where the root of the statutory problem exists in motor vehicle codes: the definitions of "person," "driver," and "operator" and the concepts of "driving" and "operating." These terms drive most of the regulation of motor vehicles, from basic traffic offenses to the substantive crimes. This chapter then uses that discussion to explore traffic offenses specifically and whether automation in vehicles could require a change in the enforcement of traffic laws. Following that discussion, this chapter describes three substantive automobile crimes: driving under the influence, reckless driving, and vehicular manslaughter. This chapter concludes by examining the use of data in trials to either rebut eyewitness testimony or help reconstruct the vehicle's movements to assist the jury (or court, on a motion to suppress).

I. "PERSON," "OPERATOR," "DRIVER," "OPERATING," AND "DRIVING"

The starting point for considering criminal responsibility is the definition of the actor whose conduct is proscribed or required under the applicable statute and what that actor is doing. Most statutes use one of three terms to define the actor: "person," "operator," or "driver." In comparing these terms, "person" is the broadest, then "operator," and then "driver"—though, in some states, "driver" and "operator" are defined interchangeably.[1] Statutes also say what the actor is doing—most times, an actor is either "operating" or "driving" a motor vehicle. Operating is broader than driving and includes driving.

"Person" is defined to include humans and entities.[2] Except in Tennessee, which has included an engaged automated driving system within its definition of "person,"[3] an automated driving system would not be a person. Effectively, the options for a "person" are the automated vehicle owner, whether human or an entity; the users of the automated vehicle; the companies operating automated vehicles; the automated vehicle manufacturer; or a component manufacturer. Deciding who or what the person is in a statutory provision for automated vehicles may depend upon the level of automation, the deployment method, the statutory prohibition, and the underlying facts. There could also

---

1. *Compare* N.Y. Veh. & Traf. Law § 113 (McKinney) (defining "driver" as "[e]very person who operates or drives or is in actual physical control of a vehicle. Whenever the terms 'chauffeur' or 'operator' or 'chauffeur's license' or 'operator's license' are used in this chapter, such terms shall be deemed to mean driver and driver's license respectively."), *with* Mich. Comp. Laws § 257.13 (defining driver), § 257.36 (defining operator).

2. *See, e.g.*, N.Y. Veh. & Traf. Law § 131 (McKinney) ("Every natural person, firm, copartnership, association, or corporation."); Mich. Comp. Laws § 257.40; Okla. Stat. tit. 47, § 1-144.

3. Tenn. Code Ann. § 55-8-101 (West) ("'Person' means a natural person, firm, copartnership, association, corporation, or an engaged [automated driving system].").

be multiple people that could be responsible for a particular traffic violation or other automobile crime.

The terms "operator" and "driver" have received far more attention from legislatures in regulating automated vehicles; however, not every state that has enacted an automated vehicle law has addressed who or what is the operator or driver of automated vehicles. Additionally, in the states that have legislatively addressed this issue, they have (except for Nevada) done so only in the context of automated driving systems (levels 3, 4, and 5) and not for lower levels of automation. Thus, for those levels of automation, the default state law definitions of "operator" and "driver" would continue to apply.

States that have redefined these terms for highly automated vehicles have taken different approaches. Many of these states have made the automated driving system the operator or driver of the vehicle.[4] Florida makes the automated driving system the operator "unless the context otherwise requires," which provides less certainty as to who the operator could be at a given time but more flexibility in applying its motor vehicle code.[5]

Texas makes the automated driving system owner (i.e., the automated vehicle owner) the operator: "When an automated driving system installed on a motor vehicle is engaged . . . the owner of the automated driving system is considered the operator of the automated motor vehicle solely for the purpose of assessing compliance with applicable traffic or motor vehicle laws, regardless of whether the person is physically present in the vehicle while the vehicle is operating."[6]

Michigan defines "operator" as a person who "operates an automated motor vehicle upon a highway or street."[7] Michigan's statute further defines an "automated motor vehicle" as a "motor vehicle on which an automated driving system has been installed, either by a manufacturer of automated driving systems or an upfitter that enables the motor vehicle to be operated without

---

4. Fla. Stat. § 316.85(3)(a) ("For purposes of this chapter, unless the context otherwise requires, the automated driving system, when engaged, shall be deemed to be the operator of an autonomous vehicle, regardless of whether a person is physically present in the vehicle while the vehicle is operating with the automated driving system engaged."); Nev. Rev. Stat. § 484A.080 (for level 4 and level 5 automated vehicles); Tenn. Code Ann. § 55-30-106(b)(2) (West) ("When the [automated driving system] is fully engaged, operated reasonably and in compliance with manufacturer instructions and warnings, the [automated driving system] shall be considered the driver or operator of the motor vehicle for purposes of determining . . . [l]iability for non-conformance to applicable traffic or motor vehicle laws."); Utah Code Ann. § 41-26-104(1)(a) (West) ("For the purpose of assessing compliance with applicable traffic or motor vehicle laws . . . [w]hen an [automated driving system] is operating a motor vehicle, the [automated driving system] is the operator, and shall satisfy electronically all physical acts required by a conventional driver in operation of the vehicle.").

5. Fla. Stat. § 316.85(3)(a).

6. Tex. Transp. Code Ann. § 545.453 (West).

7. Mich. Comp. Laws § 257.36.

any control or monitoring by a human operator."[8] The use of "human operator" means that there are nonhuman operators; arguably, under this definition, any provision under Michigan law that applies to control or monitoring by a human driver should not apply to a human operator.

Nevada uses a three-tier approach to clarify who or what is the driver of different types of automated vehicles: for levels 1 and 2, the driver is the "person who drives or is in actual physical control of" the vehicle; for level 3, the driver is the person who engages the automated driving system; for levels 4 and 5, the person who engages the automated driving system is not the driver unless he or she is the owner of the automated vehicle.[9]

North Dakota takes an entirely different approach by not defining "operator" or "driver" in the context of highly automated vehicles and instead providing that "[a]n individual using [a level 4 or level 5] autonomous vehicle is not driving or in actual physical control of the autonomous vehicle" while the automated driving system is engaged.[10] Because North Dakota defines "driver" or "operator" as someone who drives or is in actual physical control of the vehicle, North Dakota effectively treats the human user as not being an operator or driver without providing clarity on who or what is the operator or driver.[11]

California takes a convoluted approach to defining "operator" by providing that the operator "of an autonomous vehicle is the person who is seated in the driver's seat, or, if there is no person in the driver's seat, causes the autonomous technology to engage."[12] California's statute penalizes the person who sits in a driver's seat of a level 4 or level 5 automated vehicle that does not

8. *Id.* § 257.2b(2).

9. Nev. Rev. Stat. § 484A.080. Curiously, for level 4 and level 5 automated vehicles, the statute uses the term "natural person" rather than "person" for determining whether the owner of the vehicle is the driver: "If a vehicle is a fully autonomous vehicle, as defined in NRS 482A.036, and the automated driving system, as defined in NRS 482A.025, of the fully autonomous vehicle is engaged, 'driver' does not include a natural person who causes the automated driving system of the fully autonomous vehicle to engage unless the natural person is the owner of the fully autonomous vehicle." *Id.* Seemingly, this would exclude automated ridesharing companies, car rental companies, trucking companies, delivery companies, and, potentially more arbitrarily, vehicles owned by companies.

10. N.D. Cent. Code Ann. § 39-01-01.2.

11. North Dakota law defines a driver as "every person who drives or is in actual physical control of a vehicle." N.D. Cent. Code Ann. § 39-01-01.18. This language mirrors the language above. "Operator" is defined more broadly to include "every person who drives or is in actual physical control of a motor vehicle upon a highway or who is exercising control over or steering a vehicle being towed by a motor vehicle." *Id.* § 39-01-01.53. This definition is broad enough to include three types of activities: (1) a person who drives, (2) a person who is in actual physical control, and (3) a person who has control over a vehicle or is steering a vehicle being towed. The automated vehicle provision does not apply to the portion of the definition for towing. This is a potential problem in not being clear on what the statute is trying to say—a more straightforward approach for North Dakota would have been to state that "[a]n individual using [a level 4 or level 5] autonomous vehicle is not" an operator or driver while the automated driving system is engaged.

12. Cal. Veh. Code § 38750(a)(4) (West).

require or rely on a driver. Under the California definition, it is not hard to imagine people avoiding summoning automated vehicles or racing for every seat except the driver's seat. The loser of the race ought to consider walking or riding in a (more dangerous) human-driven vehicle.

The Uniform Law Commission released its Uniform Automated Operation of Vehicles Act in December 2019 (the Uniform Act), which makes the automated-driving provider the driver and the operator.[13] The automated-driving provider is an entity that makes a specified declaration to the relevant state agency. The Uniform Act provides that the "automated-driving provider is responsible for a violation of the state's rules of the road during automated operation of an associated automated vehicle." The Reporter's Comments expound upon this view:

> [A]n automated driving provider is responsible for a violation of the rules of the road by an associated automated vehicle under automated operation. This retrospective principle merely recognizes that the automated driving provider is the legal driver in these circumstances and is therefore subject to corresponding sanctions under the state's vehicle code. In other words, the automated driving provider should receive the speeding ticket when an associated automated vehicle under automated operation is caught speeding. At the same time, this section does not address the appropriate level of enforcement. It is expected that federal, state, and local authorities will continue to evaluate the role of various forms of automated enforcement (including self-reporting obligations) in improving road traffic safety.[14]

These different approaches highlight the difficulties that manufacturers, vehicle owners, and users of highly automated vehicles may encounter when users travel through different states using highly automated vehicles. A highly automated vehicle operated across state lines could encounter situations where the automated-driving provider, the automated driving system, the vehicle owner, and the user could all, at one point during a trip, be responsible for a traffic violation. The remainder of this section explores the meaning of "operator" or "driver" where the state law does not explicitly identify the driver or operator of automated vehicles.

In the absence of a specific automated vehicle law, the inquiry into who or what could be considered the operator or driver of the vehicle depends upon how legislatures and courts have defined these terms. However, because the definitions of "driver" and "operator" typically depend on a "person" acting, the driving automation system could not be considered a person without

---

13. The Uniform Act is detailed in Chapter 7.

14. Unif. Automated Operation of Vehicles Act Reporter's Comments to Section 9.

legislative action redefining "person" (or judicial action in the absence of a legislative definition)—however, the designer of the code and software behind a driving automation system could be considered a person.

"Driver" and "operator" can be defined interchangeably, as typified by the definition of "driver" in New York: "Every person who operates or drives or is in actual physical control of a vehicle. Whenever the term[] . . . 'operator' [is] used in this chapter, such term[] shall be deemed to mean driver."[15] The definition of "driver" is not limited to a single person, and more than one person could be considered a driver. For example, one person may be driving the automated vehicle, and another person may be in actual physical control of the automated vehicle. Even in those states that define the term "operator" more broadly than "driver" or differently than "driver," the most important context is the definition of the action that the person is alleged to have performed.

Courts have generally defined "operate" more broadly than "drive."[16] For example, in New York, "a person operates a motor vehicle within the meaning of the statute when, in the vehicle, he intentionally does any act or makes use of any mechanical or electrical agency which alone or in sequence will set in motion the motive power of the vehicle."[17] Other courts also broadly interpret "operate."[18] Based on the definition of "operating," a user of an automated

---

15. N.Y. VEH. & TRAF. LAW § 113 (McKinney); NEV. REV. STAT. § 484A.080; *see also* Bryant Walker Smith, *Automated Vehicles Are Probably Legal in the United States*, 1 TEX. A&M L. REV. 411, 464 & n.307 (2014) (collecting statutes and stating that "'[d]river' and 'operator' are broad terms that, in general, refer to anyone who 'drives,' 'operates,' or 'is in actual physical control of' a vehicle").

16. *See, e.g.*, Jacobson v. State, 551 P.2d 935, 937 (Alaska 1976) ("As a general proposition, it appears that 'to operate' includes a larger class of activities than 'to drive.' While one who drives a vehicle must necessarily in that process operate it, the reverse is not necessarily so."); McDuell v. State, 231 A.2d 265, 267 (Del. 1967) ("[T]he term 'driving' is encompassed within the term 'operating'; but the reverse is not necessarily so. One may not drive a vehicle without operating it; but one may operate the engine or devices of a vehicle without driving it."); Thomas v. State, 277 Md. 314, 353 A.2d 256, 258–59 (1976) (stating that "driving" has a "more strict and limited meaning" than "operating").

17. People v. Prescott, 95 N.Y.2d 655, 745 N.E.2d 1000, 1004 (2001) (quoting People v. Alamo, 34 N.Y.2d 453, 315 N.E.2d 446, 459 (1974)).

18. *See, e.g.*, Barton v. State, 882 S.W.2d 456, 459 (Tex. App. 1994) (holding that "the evidence is sufficient to show operation of a motor vehicle where the totality of the circumstances demonstrates that the defendant took action to affect the functioning of his vehicle in a manner that would enable the vehicle's use"); Williams v. City of Petersburg, 216 Va. 297, 300, 217 S.E.2d 893, 896 (1975) ("We pointed out that 'driving' an automobile means putting it in motion but that 'operating' a vehicle has a broader meaning. 'Operating' not only includes the process of moving the vehicle from one place to another, but also includes starting the engine, or manipulating the mechanical or electrical equipment of the vehicle without actually putting the car in motion. It means engaging the machinery of the vehicle which alone, or in sequence, will activate the motive power of the vehicle." (citation omitted)); Commw. v. Plowman, 28 Mass. App. Ct. 230, 548 N.E.2d 1278, 1280 (1990) (citing Massachusetts v. Uski, 160 N.E. 305, 306 (Mass. 1928)); State v. Ducatt, 22 Conn. App. 88, 575 A.2d 708, 710 (1990) (citing State v. Swift, 125 Conn. 399, 6 A.2d 359, 361 (Conn. 1939)); Flournoy v. Georgia, 128 S.E.2d 528, 530 (Ga. Ct. App. 1962) (citations omitted); *but see* Corl v. State, 544 N.E.2d 211, 213 (Ind. Ct. App. 1989) (the mere starting of a vehicle, in and of itself, is not sufficient to find that the defendant operated the motor vehicle). The all-encompassing nature of the meaning of "operate" has been applied to new

vehicle who, through a key, button in the vehicle, or phone application, starts an automated vehicle could be considered an "operator." This seemingly broad definition of "operating" may make sense in the context of enforcing driving-under-the-influence laws for human-driven vehicles but may prevent certain persons, such as those who cannot currently obtain a license, like a minor or someone who suffers from disabilities, from using automated vehicles.

The term "drive" has been defined both legislatively and by courts. Legislatures broadly and sometimes circularly define "drive."[19] Courts typically require motion or manipulation of the vehicle while in motion for an act to constitute driving.[20] For example, in California, "[a] person drives a vehicle when he or she intentionally causes it to move by exercising actual physical control over it. The person must cause the vehicle to move, but the movement may be slight."[21] Under these definitions of "drive," a user of a moving automated vehicle could be considered a driver.

Actual physical control, a concept used in defining "operating" or "driving," has typically been used in situations where the person has control over a vehicle incapable of being operated or driven, such as due to it being in a snowbank[22] or after an engine failure.[23] As one scholar has noted, "'actual physical control' can be broader than operation but probably does involve physical presence."[24] A user who occupies an automated vehicle almost certainly could be considered in actual physical control of the automated vehicle, and, depending upon the breadth of the definition, remote users who have control over the vehicle may be considered in actual physical control.

These definitions help narrow who could be an operator or driver or who could be said to have operated or driven an automated vehicle. Although occupants of a highly automated vehicle resemble passengers more than drivers or

---

technology. For example, in *State v. Cyr*, 291 Conn. 49, 967 A.2d 32 (2009), the Connecticut Supreme Court held that a defendant operated his motor vehicle when he, while under the influence of alcohol, used a remote start outside of his vehicle and then entered the vehicle.

19. *See, e.g.*, CONN. GEN. STAT. § 14-1(28) ("'Drive' means to drive, operate or be in physical control of a motor vehicle, including a motor vehicle being towed by another."); HAW. REV. STAT. § 286-2 ("'Drive' means to drive, operate, or be in physical control of a motor vehicle in any place open to the general public for purposes of vehicular traffic.").

20. James O. Pearson, Jr., *What Constitutes Driving, Operating, or Being in Control of Motor Vehicle for Purposes of Driving While Intoxicated Statute or Ordinance*, 93 A.L.R.3D 7 (1979) ("The courts which have defined 'driving,' as used in statutes prohibiting driving while intoxicated, have ordinarily held that the term requires motion. Thus, many courts have stated that driving requires that the vehicle be in motion in order for the offense of drunk driving to be committed. Similarly, other courts have stated that driving means steering and controlling a vehicle while in motion. In view of this requirement of motion, some courts have specifically recognized that driving of necessity includes operation.").

21. Judicial Council of California Criminal Jury Instruction 2241 (2017 ed.).

22. *See, e.g.*, Commw. v. Kallus, 212 Pa. Super. 504, 507, 243 A.2d 483, 485 (1968).

23. *See, e.g.*, Chamberlain v. State, 163 Tex. Crim. 529, 529, 294 S.W.2d 719, 720 (1956).

24. Smith, *supra* note 15, at 473.

operators,[25] the broad definitions may mean that one or more of the occupants and the user of the highly automated vehicle could be considered its operator or driver. The user activates the highly automated vehicle and sets it in motion, which would be sufficient to constitute operation or driving under most statutes. A user inside the highly automated vehicle would be in actual physical possession of the vehicle. Nonuser occupants could also be in actual physical control over the vehicle, especially in those instances where the user is not actually in the vehicle. Ownership of a highly automated vehicle, in and of itself, is insufficient to constitute driving and operating a motor vehicle—ownership coupled with something more, like occupying the vehicle, summoning the vehicle, or using the vehicle, could be sufficient.

In contrast to the user of highly automated vehicles, police would have more difficulty in finding that the manufacturers of these vehicles are operating or driving their automated vehicles. Manufacturers are not in actual physical control of their vehicles or the one that activates or sets the vehicles in motion. Manufacturers are not driving the vehicles—arguably, the automated driving systems are the drivers. To impose criminal liability in these instances would require courts to consider the manufacturer in an agency relationship with its vehicles or to impose vicarious liability. No such agency could be imputed to a driver assist feature like a level 1 or level 2 driving automation system because the responsibility for driving the vehicle would be on the human driver. Only for those systems where the human user is excused from driving and supervising should courts even consider whether a manufacturer should be an operator or driver.

Courts have entertained and found entities criminally responsible, even for such crimes as manslaughter.[26] New York provides a typical formation of corporate criminal responsibility:

---

25. Courts have found that mere presence in the vehicle of a passenger does not mean that person is driving or operating the vehicle, but courts have not hesitated to find the passenger in actual physical control of the vehicle or operating the vehicle when the passenger grabs the steering wheel. *See, e.g.*, Commw., Dep't of Transp., Bureau of Driver Licensing v. Hoover, 161 Pa. Commw. 517, 521, 637 A.2d 721, 723 (1994); Dugger v. Commw., 40 Va. App. 586, 593, 580 S.E.2d 477, 481 (2003); *In re* F.H., 192 Cal. App. 4th 1465, 1472, 122 Cal. Rptr. 3d 43, 49 (2011); People v. Yamat, 475 Mich. 49, 57, 714 N.W.2d 335, 339 (2006).

26. *See, e.g.*, People v. Gen. Dynamics Land Sys., Inc., 175 Mich. App. 701, 703, 438 N.W.2d 359, 361 (1989) (stating that "a corporation is sufficiently a 'person' to be the perpetrator of a manslaughter"); Sea Horse Ranch, Inc. v. Super. Ct., 24 Cal. App. 4th 446, 456, 30 Cal. Rptr. 2d 681, 687 (1994), *as modified*; Granite Constr. Co. v. Super. Ct., 149 Cal. App. 3d 465, 468, 197 Cal. Rptr. 3, 5 (Ct. App. 1983); New York Cent. & H.R.R. Co. v. United States, 212 U.S. 481, 495, 29 S. Ct. 304, 307, 53 L. Ed. 613 (1909); *see also* Colo. Rev. Stat. § 18-1-606(1); Del. Code Ann. tit. 11, § 281 (West); Me. Stat. tit. 17-A, § 60(1)(A); N.Y. Penal Law § 20.20(2)(a) (McKinney); Utah Code Ann. § 76-2-204(1) (West); Wash. Rev. Code § 9A.08.030(2)(a); § 2.07. *Liability of Corporations, Unincorporated Associations and Persons Acting, or Under a Duty to Act, in Their Behalf*, Model Penal Code.

> A corporation is guilty of an offense when:
>
> (a) The conduct constituting the offense consists of an omission to discharge a specific duty of affirmative performance imposed on corporations by law; or
>
> (b) The conduct constituting the offense is engaged in, authorized, solicited, requested, commanded, or recklessly tolerated by the board of directors or by a high managerial agent acting within the scope of his employment and in behalf of the corporation; or
>
> (c) The conduct constituting the offense is engaged in by an agent of the corporation while acting within the scope of his employment and in behalf of the corporation, and the offense is (i) a misdemeanor or a violation, (ii) one defined by a statute which clearly indicates a legislative intent to impose such criminal liability on a corporation, or (iii) any offense set forth in title twenty-seven of article seventy-one of the environmental conservation law.[27]

An agent is defined as "any director, officer or employee of a corporation, or any other person who is authorized to act in behalf of the corporation."[28]

Subsection (a) requires a duty imposed on corporations by law, which may, in due course, become applicable but is probably inapplicable now. Subsection (b) allows for the imposition of criminal liability when the "conduct constituting the offense" is "authorized" or "recklessly tolerated" by high-ranking officials of the company. A statute like this could be applied to companies that release their automated driving systems with significant problems or without adequately testing their automated driving systems, but it would hardly apply to companies that make an informed decision on the testing and deployment of their automated driving systems. Those systems, though adequately tested and deployed, will violate traffic laws either intentionally or unintentionally.[29] Under subsection (c), no employee is acting within his or her scope of employment when an automated driving system operates an automated vehicle. Thus, a corporate criminal responsibility statute like the one in New York may not fit well with treating a manufacturer as the operator or driver.

Vicarious criminal liability can be imposed in certain circumstances. However, the common law has not typically imposed vicarious criminal liability

---

27. N.Y. PENAL LAW § 20.20 (McKinney).

28. *Id.*

29. Automated vehicles programmed to always comply with traffic codes would not be desirable. There are many reasons why violating traffic laws may make sense, such as swerving into the other lane to avoid a child.

in the absence of statutory language or legislative intent.[30] In the absence of legislative action, manufacturers probably would not be vicariously liable for the actions of their automated driving systems.

As some states have done with their automated vehicle laws, states should provide clarity and specifically define the concepts of "operator," "driver," "operating," and "driving" for automated vehicles. Conceivably, in all automated vehicles, the driving automation system "operates" some function of the dynamic driving task. The Taxonomy defines the dynamic driving task as "[a]ll of the real-time operational and tactical functions required to operate a vehicle in on-road traffic, excluding the strategic functions such as trip scheduling and selection of destinations and waypoints."[31] Level 1 driving automation systems perform the lateral or longitudinal vehicle motion. Level 2 driving automation systems combine both of these elements. Level 3 automated driving systems perform the entire dynamic driving task but still utilize a human driver as the dynamic driving task fallback-ready user. Level 4 and level 5 automated driving systems remove the human driver from the loop while the automated driving system is engaged.

Driver-operator concepts can be viewed on a spectrum. A human driver-operator can be on one side and an automated driving system driver-operator on the other side. The levels of automation can help define the location on the spectrum. Level 1 and level 2 systems, dubbed by the Taxonomy as driver support systems, appear closer to the human driver-operator side of the spectrum. Level 4 and level 5 systems, with the human wholly removed from the loop, appear to fall on the automated driving system driver-operator side. Level 3 automated driving systems are more challenging because the human driver is partially removed from the loop until summoned by the automated driving system. However, because the human user remains an integral component of the safe operation of level 3 automated driving systems, there is seemingly a good reason to treat that person as an operator or driver for some automobile crimes, such as driving or operating under the influence.

Additional data points can be added to the spectrum to help define where a driving automation system falls. The configuration of the vehicle could be considered. A vehicle that lacks human-driving components would be closer to the automated driving system driver-operator side than a vehicle with such equipment.[32] That an automated vehicle lacks human occupants is another data point

30. Wayne R. LaFave, Substantive Criminal Law § 13.4(b) *Statutory Crimes—Interpretation* (3d ed. 2019) (stating that "vicarious liability cannot be created by the courts but can only be delineated by statute" (internal quotation marks and citation omitted)).

31. *Taxonomy and Definitions for Terms Related to On-Road Motor Vehicle Automated Driving Systems*, Document J3016, SAE Int'l (June 18, 2018).

32. Though, at the time of any traffic violation, the existence or nonexistence of human-driving controls seems not to make any difference.

in determining where a vehicle falls on the spectrum. Another determinant could be the amount of control that an automated vehicle gives to its users. For example, if an automated vehicle allows occupants to adjust the vehicle's speed above the posted limit (perhaps, within reason), then this control may impact whether the human user is operating the vehicle or in actual physical control of the vehicle for purposes of speed-related offenses.

Legislatures could also consider the purposes of punishment in determining who or what should bear responsibility for automobile crimes.[33] Inordinate pressure may be placed on the purposes of punishment when human users or owners of automated driving systems are responsible for automobile crimes. The five general purposes of punishment—specific deterrence, general deterrence, incapacitation, rehabilitation, and retribution[34]—seem off-kilter when applied to users of highly automated vehicles. For highly automated vehicles, ticketing the human user or owner cannot deter future traffic

33. Jeffrey K. Gurney, *Driving into the Unknown: Examining the Crossroads of Criminal Law and Autonomous Vehicles*, 5 Wake Forest J.L. & Pol'y 393, 416 (2015).

34. As I have described elsewhere, these purposes of punishment can be described as follows:

> Scholars have put forth five traditional theories of punishment to support this objective. The first theory of punishment is called "prevention" or "specific deterrence." Under this theory, punishment serves the function of "dissuad[ing] potential criminals from offending." Prevention is concerned with the specific individual who is being punished and "deterring him from committing similar crimes." The second theory, which is related to prevention or specific deterrence, is called "deterrence" or "general deterrence." This theory is concerned with deterring other people from "committing future crimes, lest they suffer the same unfortunate fate." Thus, while prevention or specific deterrence is concerned with dissuading the particular offender from committing future offenses, general deterrence focuses on discouraging society from committing offenses. The third theory of punishment is "restraint" or "incapacitation," which is the belief that society can protect itself by isolating persons deemed dangerous to society. The fourth theory of punishment is "rehabilitation," in which punishment is used to improve criminals so that they do not reoffend. The fifth theory of punishment is "retribution." Retribution "is not a single concept," and it is usually "justified on the grounds that wrongdoing merits punishment." Retribution takes many forms: it can be based on vengeance, just desserts, or proportionality.
>
> These theories provide a moral basis to the legitimacy of a punishment. A punishment cannot be "conceptually legitimate" where:
>
>> [T]here is nothing that deserves denunciation or blame (retribution); no action we care to and/or can deter (general deterrence); no individual we want to and/or can deter (specific deterrence); nothing that can be gained by isolation and/or no individual we can isolate for more than a de minimis time (incapacitation); and we are not going to try to change the individual (rehabilitation).
>
> At the heart of the theories is a sense of moral blameworthiness on the part of the offender. The offender committed some act that needs punishment (retribution), that others engage in (general deterrence), that the offender may do again (specific deterrence), that can be prevented by removing the individual (incapacitation), and that can be corrected (rehabilitation).

Id. at 404–06 (citations omitted but internal quotation marks remain).

violations or other automobile crimes. The only deterrent effect is that the user may not use highly automated vehicles. Incapacitation is not furthered either—the proper party to incapacitate in these circumstances would seem to be the vehicle itself. The human user also does not need to be rehabilitated, nor is the human user a wrongdoer meriting punishment. Human users or owners of highly automated vehicles would have strong arguments that punishing them for traffic violations or automobile crimes does not further the purposes of punishment.

The purposes of punishment may be better served by ticketing or punishing the manufacturer for traffic violations or other automobile crimes or by incapacitating the vehicle itself. The manufacturer could be incentivized to program its vehicles to follow traffic laws and better test and maintain software if it bore responsibility for these violations and crimes. General deterrence could occur too because other manufacturers may worry about being subjected to criminal responsibility. Incapacitation could be served by impounding automated vehicles, pending an update to fix the software.

In determining responsibility for traffic violations and other automobile crimes, legislatures will want to be cognizant of the effects that their selection has on civil liability. As described in Chapter 8, traffic regulations are prominent examples of where negligence per se is applicable.[35]

II. DEFENDING CASES WITH HIGHLY AUTOMATED VEHICLES

Because "operator," "driver," "operate," and "drive" are defined broadly, prosecutors will be able to seek recourse against persons who may not be blameworthy. Based on the prior discussion of these defined terms, three situations may be currently applicable. First, a state may not have a law that addresses criminal responsibility or defines "operator," "driver," "operate," or "drive." Second, a state law may make the automated driving system the driver or operator. Third, a state law could impose responsibility on the automated vehicle owner, like the law in Texas. This section describes different problems under these three regimes that states may want to consider before prosecuting people for violations and that defense attorneys may consider in defending these claims.

A. Basic State Law Definition

In these cases, counsel will want to determine whether the offense is a strict liability offense or whether it requires culpability on the part of the defendant.

35. See Chapter 8 for a discussion of negligence per se.

As described in the context of the more substantive automobile crimes, prosecutors may have difficulty establishing a mens rea against human users and occupants removed from the loop of level 4 and level 5 automated vehicles. Even if the statute is a strict liability offense, the party who commits the actus reus of the crime is important to identify. Although many rules of the road are strict liability offenses, they are not absolute liability offenses and still require the defendant to have committed the actus reus.

The actus reus requirement means that the defendant must have voluntarily committed the act or omission that constitutes the offense.[36] With level 4 and level 5 automated vehicles, courts may need to grapple with whether the user or occupants of the automated vehicle commit a voluntary act when a traffic law is violated or another automobile crime is committed. For example, does a user voluntarily act if his or her level 5 automated vehicle violates a traffic law while he or she is asleep?

In these highly automated vehicles, the occupants resemble passengers who have no actual control over the driving or operating of the vehicle. Courts have considered whether a passenger commits the actus reus of the offense.[37] For example, in *New York v. Shaughnessy*,[38] a passenger of a vehicle was charged with trespassing because the driver drove the vehicle on private property. The court found that the state had failed to prove that a passenger committed the actus reus of trespass:

> She merely was a passenger in a vehicle. Any action taken by the vehicle was caused and guided by the driver thereof and not by the Defendant. If the Defendant were to be held guilty under these circumstances, it would dictate that she would be guilty if she had been unconscious or asleep at the time or even if she had been a prisoner in the automobile. There are many situations which can be envisioned and in which the trespass statute in question would be improperly applied to an involuntary act. One might conceive of a driver losing control of a vehicle through mechanical failure and the vehicle proceeding onto private property which is the subject of a trespass.[39]

---

36. *See, e.g.*, N.H. Rev. Stat. Ann. § 626:1 ("A person is not guilty of an offense unless his criminal liability is based on conduct that includes a voluntary act or the voluntary omission to perform an act of which he is physically capable."); N.Y. Penal Law § 15.10 (McKinney) ("The minimal requirement for criminal liability is the performance by a person of conduct which includes a voluntary act or the omission to perform an act which he is physically capable of performing.").

37. *See supra* note 25.

38. 66 Misc. 2d 19, 19, 319 N.Y.S.2d 626, 627 (Dist. Ct. 1971).

39. *Id.* at 20–21.

The passenger argument would be particularly effective for traffic offenses and automobile crimes where the user of the automated vehicle is, by definition, not involved in the functioning of the automated vehicle.

Alternatively, courts could analogize these events to situations where there have been mechanical failures in vehicles.[40] In *Minnesota v. Kremer*,[41] the defendant was charged with running a red light after he ran a red light when his brakes failed. The defendant had never experienced any brake problems, and he lacked knowledge of a brake problem. The court ruled that the defendant could not be convicted in these circumstances because he did not commit a voluntary act.

Other courts have reached a different conclusion.[42] For example, in *State v. Packin*,[43] the defendant was ticketed for speeding, but he contended that his cruise control caused the excessive speed. The court held that the defendant was responsible for the excessive speed: "A motorist who entrusts his car to the control of an automatic device is 'driving' the vehicle and is no less responsible for its operation if the device fails to perform a function which under the law he is required to perform. The safety of the public requires that the obligation of a motorist to operate his vehicle in accordance with the Traffic Act may not be avoided by delegating a task he normally would perform to a mechanical device."

The broad language in *Packin* could, of course, apply to all levels of automation—a person using the system cannot delegate driving responsibility to a mechanical device. A deeper dive, however, shows that the court decision can be limited. With a cruise control device, a driver merely delegates one task over which the driver maintains complete control and is otherwise driving.

Kremer and *Packin* are also not necessarily irreconcilable. In *Kremer*, the defendant violated a traffic law when a mechanical failure, outside of his control, caused his vehicle to commit the traffic violation. In *Packin*, the defendant could have easily prevented the excessive speed by applying the brakes or by lowering the cruise control speed. In this regard, the *Packin* defendant could be said to have voluntarily sped by continuing to drive at the excess speeds without deactivating cruise control or reducing the vehicle's speed. Thus, the

---

40. *See, e.g.*, State v. Kremer, 262 Minn. 190, 192, 114 N.W.2d 88 (1962); City of Kettering v. Greene, 9 Ohio St. 2d 26, 222 N.E.2d 638 (1966); *see also* People v. Jones, 132 Mich. App. 368, 371, 347 N.W.2d 235, 237 (1984) (noting, in dicta, that "[t]he facts here do not present a situation in which there was no actus reus (e.g., a vehicle behind the motorist's car failed to stop and pushed him past the stop sign into the intersection) or in which the motorist failed to stop due to a mechanical failure previously unknown to him"). In *City of Kettering v. Greene*, the Ohio Supreme Court ruled that a bus driver could receive a citation for running a stop sign, even when the violation was caused by brake failure.

41. 262 Minn. 190, 192 (1962).

42. *See, e.g.*, State v. Baker, 1 Kan. App. 2d 568, 570, 571 P.2d 65, 67 (1977); *City of Kettering*, 9 Ohio St. 2d 26; State v. Packin, 107 N.J. Super. 93, 96, 257 A.2d 120, 121 (App. Div. 1969).

43. 107 N.J. Super. 93, 96, 257 A.2d 120, 121 (App. Div. 1969).

distinction between the two cases is that allegations of a mechanical failure causing a traffic violation can be a defense when the consequences of the alleged mechanical failure are not preventable by the driver properly operating or using the vehicle.

For vehicles operated by level 2 driving automation systems, the driver must supervise the drive. The mechanical failure defense should be viewed under the *Kremer-Packin* distinction and provide a limited escape hatch of liability when some mechanical failure makes it impossible for the driver to avoid the traffic offense. Unless the circumstances are such that the malfunction or error makes it so even an attentive driver would have been incapable of preventing the traffic violation, criminal liability could be imposed.[44] Users of automated driving systems, however, are no longer "driving" the vehicle and have different responsibilities for failures or mistakes caused by the automated driving system. They have been sold a product that allows them to become distracted or engage in other activities. Level 4 and level 5 systems completely remove the human from the loop. Users of automated driving systems more closely resemble the *Kremer* driver or passengers who encounter a vehicle malfunction that is not their responsibility to prevent.

B. Automated Driving System as the Driver or Operator

Some states provide that the automated driving system is the driver or operator for purposes of their motor vehicle codes. These statutes help provide clarity on whether a user or an occupant needs a license to use a level 4 or level 5 automated vehicle. Still, they create some ambiguity as to who or what bears criminal responsibility for traffic violations. "Although there is a body to kick, there is no soul to damn."[45] The automated driving system is incapable of paying the fine or penalty incurred for a traffic violation. This raises an important question in these jurisdictions about which party would bear the fine, penalty, or conceivably arrest for a traffic violation committed by the automated driving system operator. A significant hurdle to imposing criminal liability—as opposed to a civil penalty that is not considered a criminal penalty—is that vicarious liability is a statutory creature and generally not known to the common law.[46] Without defining which party bears responsibility for traffic violations, merely designating the automated driving system as the operator of the

---

44. *See* Tracy Hresko Pearl, *Fast & Furious: The Misregulation of Driverless Cars*, 73 N.Y.U. Ann. Surv. Am. L. 19, 51–52 (2017) (stating that criminal liability would not be appropriate for a user of a level 2 driving automation system when a sudden malfunction caused a crash).

45. Jeffrey K. Gurney, *Crashing into the Unknown: An Examination of Crash-Optimization Algorithms through the Two Lanes of Ethics and Law*, 79 Alb. L. Rev. 183, 241 (2016) (internal quotation marks and citations omitted).

46. *See supra* note 30 and accompanying text.

vehicle fails to resolve the question and creates ambiguity. In these states, it is unclear whether any party would bear responsibility for the actions of the automated driving system, and these states should further clarify their statutes.

C. Automated Vehicle Owner Responsibility

Texas has enacted a statute that makes the owner of the automated driving system the "operator" of the vehicle: "When an automated driving system installed on a motor vehicle is engaged . . . the owner of the automated driving system is considered the operator of the automated motor vehicle solely for the purpose of assessing compliance with applicable traffic or motor vehicle laws, regardless of whether the person is physically present in the vehicle while the vehicle is operating."[47] A statute like this is particularly pernicious and creates numerous problems.

The first problem is that the statute lacks the nuance needed for automated driving systems. Does this statute preclude other persons from being considered the operator of the vehicle, such as the user of the automated driving system or the manufacturer? If an automated vehicle allows users to enable the vehicle to speed, is the automated vehicle owner responsible for the user causing the vehicle to travel at excessive speed? Texas law provides that "[a]n operator may not drive at a speed greater than is reasonable and prudent under the circumstances then existing."[48] Driving in excess of the posted speed limit is prima facie evidence of driving at a speed that is unreasonable or imprudent.[49] If the user of the automated driving system is not the operator, then conceivably, that person could not be responsible for causing the vehicle to speed. The liability imposed on the vehicle owner in this context raises an important limitation on strict liability offenses—the need for the defendant to have committed the actus reus of the crime.

Vehicle owner criminal responsibility in this context is not strict liability—an offense devoid of a mens rea requirement—because the vehicle owner would not have personally committed the actus reus of the crime.[50] Instead,

47. TEX. TRANSP. CODE ANN. § 545.453 (West).

48. *Id.* § 545.351.

49. *Id.* § 545.352.

50. *See* LAFAVE, *supra* note 30, § 13.4 ("Vicarious liability must also be distinguished from strict liability, for there is a clear and important difference between these two types of liability without fault, although courts have not always distinguished carefully between the two. With strict liability, there must be a showing that the defendant personally engaged in the necessary acts or omissions; only the requirement of mental fault is dispensed with altogether. By contrast, with vicarious liability it is the need for a personal actus reus that is dispensed with, and there remains the need for whatever mental fault the law requires on the part of the employee. It is common, however, for a vicarious liability statute to also impose strict liability; in such an instance, there is no need to prove an act or omission by the defendant-employer (one by his employee will do), and there is no need to prove mental fault by anyone.").

this type of liability is vicarious criminal liability. Criminal liability is being imposed whenever an automated vehicle commits a traffic violation or other automobile crime. Courts have considered vicarious liability in the context of motor vehicles.

In *Iowa City v. Nolan*,[51] the Iowa Supreme Court examined a statute that, if the operator of the vehicle cannot be determined, imposed prima facie responsibility on the vehicle owner for an illegally parked vehicle. The punishment was a civil penalty. The court upheld the civil penalty and ruled that it establishes prima facie responsibility "unless the evidence indicated defendant was not in fact responsible for the violation. This permits defendant to come forward with evidence that someone was operating the vehicle without his consent or with other facts which would rebut the prima facie inference that the registered owner of a vehicle is responsible for its operation." Other courts have also upheld vicarious liability on a vehicle owner for civil penalties.[52] These cases, typically involving parking violations, focus on the inability of the police to locate the operator, the minimal penalty, the lack of incarceration, and the lack of public stigma involved in the penalty.[53]

Courts have considered whether vicarious liability for more significant traffic offenses, such as speeding or red-light violations, is constitutional. In *Tupper v. City of St. Louis*,[54] a city enacted an ordinance that presumed the vehicle owner was responsible for red-light violation tickets issued in conjunction with an automated traffic control system that records red lights and has employees identify those who commit possible red-light violations. Tupper received two fines of $100 each. The Missouri Supreme Court found that the ordinance was unconstitutional because it shifted the burden of persuasion to the vehicle owner to prove that he or she was not the operator of the vehicle. In doing so, it distinguished a prior decision that had upheld vicarious liability in the parking violation context because, to the court, a $100 fine is not insignificant and running a red light results in the assessment of two points on the owner's driver's license.

---

51. 239 N.W.2d 102 (Iowa 1976).

52. *See, e.g.*, Commw. v. Minicost Car Rental, Inc., 354 Mass. 746, 749, 242 N.E.2d 411, 413 (1968) (parking violation imposed on the vehicle owner—a rental company); Commw. v. Ober, 286 Mass. 25, 32, 189 N.E. 601, 603–04 (1934) (parking violations); City of Kansas City v. Hertz Corp., 499 S.W.2d 449, 453 (Mo. 1973) (parking violations); Kinney Car Corp. v. City of New York, 28 N.Y.2d 741, 742, 269 N.E.2d 829 (1971) (parking violations).

53. *See Hertz Corp.*, 499 S.W.2d at 453.

54. 468 S.W.3d 360 (Mo. 2015) (en banc).

Other courts have overwhelmingly upheld these camera tickets.[55] In *Idris v. City of Chicago, Illinois,*[56] the Seventh Circuit upheld an ordinance that ticketed a vehicle owner for red-light violations captured on a camera unless the vehicle owner could prove that the vehicle or the license plates were stolen at the time of the violation—proof that the vehicle owner was not driving the vehicle was irrelevant. The penalty was $90. The Seventh Circuit called a substantive due process right argument a "dud" because the interest at stake—a $90 civil penalty—was so modest and because no one has a fundamental right to run a red light or avoid being recorded on camera. The court noted that the law was not arbitrary, finding that vicarious liability has numerous justifications: "A camera can show reliably which cars and trucks go through red lights but is less likely to show who was driving. That would make it easy for owners to point the finger at friends or children—and essentially impossible for the City to prove otherwise. A system of photographic evidence reduces the costs of law enforcement and increases the proportion of all traffic offenses that are detected; these benefits can be achieved only if the owner is held responsible." The court also identified that vehicle owner liability may achieve deterrence by making owners more vigilant, and owners may pass the penalty along, indirectly, to the driver.

Under the line of authority that upholds these infractions, courts have focused primarily on the civil penalty itself and how the vehicle owner is typically not subject to a criminal penalty.[57] The Texas statute—which seemingly imposes vicarious liability in all circumstances—may cause defendants to

---

55. *See, e.g.,* Idris v. City of Chicago, Ill., 552 F.3d 564 (7th Cir. 2009); Fischetti v. Vill. of Schaumburg, 2012 IL App (1st) 111008, ¶ 16, 967 N.E.2d 950, 960 (2012); Morales v. Parish of Jefferson, 13-486 (La. App. 5 Cir. 4/30/14), 140 So. 3d 375, 395, *writ denied,* 2014-1293 (La. 10/10/14), 151 So. 3d 582, *writ denied,* 2014-1296 (La. 10/10/14), 151 So. 3d 582, *writ denied,* 2014-1299 (La. 10/10/14), 151 So. 3d 583; City of Knoxville v. Brown, 284 S.W.3d 330, 338 (Tenn. Ct. App. 2008).

56. 552 F.3d 564 (7th Cir. 2009).

57. Matthew S. Maisel, *Slave to the Traffic Light: A Road Map to Red Light Camera Legal Issues,* 10 Rutgers J.L. & Pub. Pol'y 401, 415 (2013) ("[M]ost federal challenges to red light cameras have claimed violations of the U.S. Constitution but have proved fruitless. As a threshold issue, courts have uniformly concluded that the penalties imposed for a violation detected by a red light camera are civil rather than criminal in nature."). In the double jeopardy context, the United States Supreme Court has provided guideposts in *Kennedy v. Mendoza-Martinez,* 372 U.S. 144, 83 S. Ct. 554, 9 L. Ed. 2d 644 (1963), for determining whether a penalty or punishment is criminal in nature:

> (1) whether the sanction involves an affirmative disability or restraint; (2) whether it has historically been regarded as a punishment; (3) whether it comes into play only on a finding of scienter; (4) whether its operation will promote the traditional aims of punishment-retribution and deterrence; (5) whether the behavior to which it applies is already a crime; (6) whether an alternative purpose to which it may rationally be connected is assignable for it; and (7) whether it appears excessive in relation to the alternative purpose assigned. It is important to note, however, that these factors must be considered in relation to the statute on its face, and only the clearest proof will suffice to override legislative intent and transform what has been denominated a civil remedy into a criminal penalty.

challenge whether the law violates their substantive due process rights. For example, a vehicle owner would have a strong due process argument if Texas tried to charge him or her with vehicular manslaughter for nothing more than being an automated vehicle owner.[58] To the extent that a court narrowly construes the statute to avoid this result, a whole host of crimes may be excused because other parties are statutorily excluded as operators.

Another problem with this type of statute is that it draws arbitrary lines between the vehicle owner and other occupants of an automated vehicle. For example, under this distinction, an owner could not use his personal level 5 automated vehicle to take him home drunk from a pub because he would be operating the vehicle under the influence, but his friends could use his vehicle. Or he and his friends could engage in an unnecessary vehicle swap whereby each friend takes another friend's automated vehicle home. Indeed, one has to wonder whether both friends would be considered operating under the influence for the automated vehicle they own but do not occupy. More offensive would be a situation where an automated vehicle owner, while at home intoxicated, receives a charge of operating under the influence if his daughter was using the vehicle. These are all seemingly possible under Texas's statute.

For more substantive automobile crimes that have a mens rea, the treatment of the highly automated vehicle owner as the operator may make these crimes unchargeable. The highly automated vehicle owner would lack the mens rea in almost all cases. If someone else has a mens rea, they may be excused by not being the operator.

In considering whether to challenge a statute that imposes vicarious liability on an automated vehicle owner, the owner would need to consider whether to challenge the statute facially or to challenge it as applied to the particular defendant.[59] In jurisdictions that have upheld the application of vicarious liability in specific contexts for human-driven vehicles, a facial challenge may not be successful. There seems to be more justification, at least for some tickets and vehicle configurations, for imposing vicarious liability on owners of level 4 and level 5 automated vehicles than possibly for human-driven vehicles. These automated vehicles, with their ability to operate without any occupants, may create a greater need for police to have someone, such as the vehicle owner,

---

Hudson v. United States, 522 U.S. 93, 99–100, 118 S. Ct. 488, 493, 139 L. Ed. 2d 450 (1997) (internal citations and quotation marks omitted).

58. This due process argument could be in addition to challenging the conviction under Tex. Penal Code Ann. § 6.01 (West) for lack of a voluntary act.

59. "[A] plaintiff can only succeed in a facial challenge by 'establish[ing] that no set of circumstances exists under which the Act would be valid,' i.e., that the law is unconstitutional in all of its applications." Wash. State Grange v. Wash. State Republican Party, 552 U.S. 442, 449, 128 S. Ct. 1184, 170 L. Ed. 2d 151 (2008) (quoting United States v. Salerno, 481 U.S. 739, 107 S. Ct. 2095, 95 L. Ed. 2d 697 (1987)).

to cite for a traffic violation. Because courts have upheld the validity of these statutes as applied to human-driven vehicles, it seems likely that those courts would reject a facial attack of a statute like the one in Texas.

Instead, a defendant will want to challenge the statute as applied to the particular offense. In looking at the justifications where courts have upheld vicarious liability on vehicle owners, courts have focused on the lack of stigma and the civil nature of the penalty. Thus, an ideal case would be one where the penalty could involve imprisonment and carries a stigma if the vehicle owner were to be convicted of it. A perfect example would be charging an owner of a level 5 automated vehicle for operating under the influence when the owner was at home and someone else was using the vehicle. A charge of operating under the influence can result in imprisonment, immense stigma in the community, and potentially other significant indirect costs, like the inability to travel to certain countries or to hold certain jobs. Thus, to the extent a police officer sought to impose vicarious liability on an automated vehicle owner for operating under the influence, the automated vehicle owner could have a strong case that vicarious liability is unconstitutional as applied.

III. CRIMINAL RESPONSIBILITY FOR VIOLATING RULES OF THE ROAD

Rules of the road are the most pervasive laws that people interact with on a day-to-day basis. There is a Uniform Vehicle Code and there has been one since 1924, but states have not widely adopted the Uniform Vehicle Code, and it has merely influenced the traffic codes in all 50 states.[60] Each state has its own rules of the road. One of the better automated vehicle laws (but also one of the least likely to occur) would be national uniformity of the rules of the road. At a recent talk I attended, a legislator's response to this suggestion was for manufacturers to geofence their vehicles for each state's and each locality's laws.

Programming an automated vehicle to comply with traffic codes can be complicated by the ambiguous and often overlapping nature of traffic codes. As one scholar stated:

> The rules of the road as codified assume human judgment, and the rules of the road as observed reflect that judgment. These dependencies may complicate the lawful operation of automated vehicles.
>
> Consider speeding. "It is unlawful for a driver of a vehicle to fail to obey" a posted speed limit, and "[n]o person shall drive a vehicle upon

60. *See* Smith, *supra* note 15, at 417 n.11.

> a highway at a speed greater than is reasonable or prudent." These rules are not coextensive: A driver can violate both, one, or neither. Because the first rule is nominally quantitative, determining ex ante the requirements for compliance is generally straightforward. In contrast, because the second rule is nominally qualitative, determining those requirements ex ante may be more difficult.[61]

The qualitative nature of rules-of-the-road violations may complicate criminal responsibility. For example, it may seem fairer to impose a rule like the one in the Uniform Highly Automated Vehicle Act for quantitative rules than it is to impose a similar rule on context-specific qualitative rules.

The mens rea or lack thereof of rules-of-the-road violations further complicates the analysis of criminal responsibility for violations. Most traffic violations are strict liability offenses, requiring only proof of a violation but not any mental state. Thus, for most of these violations, criminal responsibility can be imposed on the party that commits the actus reus of the violation, and the focus will be on which party commits the voluntary act.

The punishment for a traffic violation can be a civil infraction, a misdemeanor, or a felony.[62] A civil infraction is generally a noncriminal offense that results in only a civil penalty.[63] These offenses can include speeding violations, registration requirements, parking violations, and seat belt requirements. Misdemeanors and felonies typically can involve a fine or imprisonment or both.[64] A misdemeanor is punished by a fine and, in some cases, imprisonment (typically not to exceed a year, but some states limit imprisonment to shorter periods).[65] A felony allows for imprisonment in excess of one year. An important distinction between civil infractions, on the one hand, and misdemeanors and felonies, on the other hand, is that police can arrest the driver for the latter. The arrest of the driver can allow the police to possibly search the vehicle incident to arrest or to conduct an inventory search if the vehicle is impounded.[66]

---

61. *Id.* at 498.

62. Clint W. Westbrook, *The Google Made Me Do It: The Complexity of Criminal Liability in the Age of Autonomous Vehicles*, 2017 Mich. St. L. Rev. 97, 102–03 (2017).

63. *See, e.g.*, Mich. Comp. Laws § 257.6a ("'Civil infraction' means an act or omission prohibited by law which is not a crime . . . and for which civil sanctions may be ordered.").

64. *See, e.g.*, S.C. Code Ann. § 56-5-6190 ("It is a misdemeanor for any person to violate any of the provisions of [the motor vehicle code] unless such violation is by this chapter or other law of this State declared to be a felony."); Ohio Rev. Code Ann. § 4511.99 (West) (providing that the default punishment is a minor misdemeanor).

65. *See, e.g.*, S.C. Code Ann. § 56-5-6190 ("Every person convicted of a misdemeanor for a violation of any of the provisions of this chapter for which another penalty is not provided shall be punished by a fine of not more than one hundred dollars or by imprisonment for not more than thirty days.").

66. See Chapter 21 for a discussion of searches and seizures.

Generally, traffic codes provide that a person commits a violation by doing any act prohibited by the traffic code or by failing to perform an act required.[67] Which party bears criminal responsibility requires an analysis of the statute that criminalizes the actions and the above-described definitions of "operator," "driver," "drive," and "operate." These types of offenses would be analyzed based on the discussion in the prior two sections.

IV. SUBSTANTIVE AUTOMOBILE CRIMES

In addition to the basic traffic violations, states have substantive automobile crimes that could create unique problems for highly automated vehicles. This section describes issues relating to three types of offenses: driving under the influence, reckless driving, and vehicular homicide. These crimes share some commonality with the traffic violations. They all use the "driver," "operator," "drive," and "operate" concepts and will depend upon how states have defined those terms. These crimes, except for driving under the influence, require proof of a mens rea. The mens rea requirement may help shield an otherwise blameless person who falls within the broad definitions of "operating" and "driving."

The need for the government to prove a mens rea, whether purposefully, knowingly, recklessly, or negligently, adds (perhaps) much-needed color to the definitions of "driver," "operator," "drive," and "operate." In a state like Texas that treats the owner as the operator or in a Uniform Act state that treats the automated-driving provider as the operator, the mens rea requirement may protect these persons from prosecution where they lack culpability due to a potentially arbitrary definition of the specific person as the operator.

Most of the automobile offenses use a reckless or negligent mens rea or a form thereof.[68] Recklessness is generally defined as follows: "A person acts recklessly when he consciously disregards a substantial and unjustifiable risk

67. *See, e.g.*, S.C. Code Ann. § 56-5-730 ("It is unlawful and, unless otherwise declared in this chapter with respect to particular offenses, it is a misdemeanor for any person to do any act forbidden or to fail to perform any act required in this chapter."); N.Y. Veh. & Traf. Law § 1101 (McKinney) ("It is unlawful and, unless otherwise declared in this title with respect to particular offenses, it is a traffic infraction for any person to do any act forbidden or fail to perform any act required in this title.").

68. Another mens rea seen in automobile crime statutes is "wanton disregard," a mens rea that resembles or is synonymous with recklessness. *E.g.*, Commw. v. Welansky, 316 Mass. 383, 398, 55 N.E.2d 902, 910 (1944) ("The words 'wanton' and 'reckless' are practically synonymous in this connection, although the word 'wanton' may contain a suggestion of arrogance or insolence or heartlessness that is lacking in the word 'reckless.'"); Dahl v. Sittner, 474 N.W.2d 897, 900 (S.D. 1991) (stating that "'reckless' is synonymous with 'wanton'"); Lipscomb v. News Star World Pub. Corp., 5 So. 2d 41, 45 (La. Ct. App. 1941) ("Wanton and reckless are virtually synonymous in meaning.").

that a result will occur or that a circumstance exists."[69] A negligent mens rea typically requires the defendant to "fail to become aware of a substantial and unjustifiable risk that the material element exists or will result from his conduct. The risk must be of such a nature and degree that his failure to become aware of it constitutes a gross deviation from the conduct that a reasonable person would observe in the situation."[70] The difference between recklessness and negligence is the awareness of the defendant—in the former, the defendant must be aware of a risk and disregard it, while in the latter, the defendant is not aware of a risk that he or she should have been aware of.[71]

When faced with an automobile crime that requires a particular mens rea, defendants charged while using highly automated vehicles will likely challenge their culpability in committing an offense.[72] For example, level 4 and level 5 automated vehicles remove humans from the loop; human users, though perhaps technically operators, likely lack a mens rea for crimes based on the decisions or lack thereof of the automated driving system. In these circumstances, courts may pay particularly close attention to acts or decisions by that person that could establish culpability, such as whether the automated vehicle was being operated outside of its operational design domain, alterations made to the automated vehicle, and use of the automated vehicle without installing

69. Colo. Rev. Stat. § 18-1-501(8); *see also* Conn. Gen. Stat. § 53a-3(13) ("A person acts 'recklessly' with respect to a result or to a circumstance described by a statute defining an offense when he is aware of and consciously disregards a substantial and unjustifiable risk that such result will occur or that such circumstance exists. The risk must be of such nature and degree that disregarding it constitutes a gross deviation from the standard of conduct that a reasonable person would observe in the situation."); N.Y. Penal Law § 15.05 (McKinney) ("A person acts recklessly with respect to a result or to a circumstance described by a statute defining an offense when he is aware of and consciously disregards a substantial and unjustifiable risk that such result will occur or that such circumstance exists. The risk must be of such nature and degree that disregard thereof constitutes a gross deviation from the standard of conduct that a reasonable person would observe in the situation. A person who creates such a risk but is unaware thereof solely by reason of voluntary intoxication also acts recklessly with respect thereto."); N.H. Rev. Stat. Ann. § 626:2 (same as New York); § 2.02. *General Requirements of Culpability,* Model Penal Code ("A person acts recklessly with respect to a material element of an offense when he consciously disregards a substantial and unjustifiable risk that the material element exists or will result from his conduct. The risk must be of such a nature and degree that, considering the nature and purpose of the actor's conduct and the circumstances known to him, its disregard involves a gross deviation from the standard of conduct that a law-abiding person would observe in the actor's situation.").

70. § 2.02. *General Requirements of Culpability,* Model Penal Code; *see* Conn. Gen. Stat. § 53a-3 (same); N.Y. Penal Law § 15.05 (McKinney) (same); *see also* Colo. Rev. Stat. § 18-1-501(3) ("A person acts with criminal negligence when, through a gross deviation from the standard of care that a reasonable person would exercise, he fails to perceive a substantial and unjustifiable risk that a result will occur or that a circumstance exists.").

71. Charles E. Torcia, Wharton's Criminal Law, § 27. *The Mental State* (15th ed. 2019).

72. Another potential challenge by defendants will be the concurrence doctrine. Under the concurrence doctrine, the mens rea and the actus rea must concur. LeFave, *supra* note 30, at § 6.3(a); *see also* Cal. Penal Code § 20 (West) ("In every crime or public offense there must exist a union, or joint operation of act and intent, or criminal negligence.").

a necessary update or with maintenance or mechanical problems. These concepts are described below in the context of reckless driving.

A. Driving/Operating under the Influence

Statutes on driving or operating a motor vehicle under the influence, for simplicity referred to as DUIs, could, as presently written, be a significant obstacle to the use of highly automated vehicles by intoxicated persons. These restrictions are remnants of the assumption that human drivers are necessary for the safe operation of motor vehicles. By driving, intoxicated drivers place themselves, occupants, and innocent bystanders at risk. This assumption is still valid for users of level 1, level 2, and level 3 automated vehicles, which all rely on a human driver. With level 4 and level 5 automated vehicles, those fundamental assumptions are no longer accurate because the human driver is completely removed from the loop and is unnecessary for their safe operation. These automated vehicles may usher in a world where people no longer drive while intoxicated.

Whether a defendant can be convicted of a DUI while using an automated vehicle depends upon the terms of the statute, the meanings ascribed to the terms (like "driver," "operator," "drive," and "operate"), and the facts (such as the type of automated vehicle). State statutes define the offense of DUI differently. Some states base their statutes solely on whether a "person" has driven a motor vehicle while intoxicated,[73] while others base their statutes on whether a "person" has driven a vehicle or is in physical control of a vehicle[74] and others utilize the term "operate" within their statutory prohibition.[75] Almost all

73. *See, e.g.*, N.M. Stat. § 66-8-102 ("It is unlawful for a person who is under the influence of intoxicating liquor to drive a vehicle within this state."); Or. Rev. Stat. § 813.010(1) ("A person commits the offense of driving while under the influence of intoxicants if the person drives a vehicle while the person [is intoxicated.]"); S.C. Code Ann. § 56-5-2933(A) ("It is unlawful for a person to drive a motor vehicle within this State while his alcohol concentration is eight one-hundredths of one percent or more.").

74. *See, e.g.*, Ala. Code § 32-5A-191(a) ("A person shall not drive or be in actual physical control of any vehicle while [intoxicated.]"); Tenn. Code Ann. § 55-10-401 (West) ("It is unlawful for any person to drive or to be in physical control of any automobile or other motor driven vehicle on any of the public roads and highways of the state, or on any streets or alleys, or while on the premises of any shopping center, trailer park, or apartment house complex, or any other premises that is generally frequented by the public at large, while [intoxicated.]").

75. *See* Mo. Rev. Stat. § 577.010 ("A person commits the offense of driving while intoxicated if he or she operates a vehicle while in an intoxicated condition."); 75 Pa. Cons. Stat. Ann. § 3802(a)(1) ("An individual may not drive, operate or be in actual physical control of the movement of a vehicle after imbibing a sufficient amount of alcohol such that the individual is rendered incapable of safely driving, operating or being in actual physical control of the movement of the vehicle."); 31 R.I. Gen. Laws § 31-27-2 ("Whoever drives or otherwise operates any vehicle in the state while under the influence" is guilty of a DUI.).

DUI statutes are strict liability offenses,[76] though additional charges for death or property damage may require a mens rea or additional conduct. States have enacted "per se" statutes providing that a blood alcohol content at or above a certain limit (typically 0.08 percent for noncommercial drivers and 0.04 percent for commercial drivers) creates a per se finding that the person is intoxicated. These per se statutes treat alcohol differently than other intoxicants (or even proof of driving or operating a vehicle under the per se limit) in a meaningful way for highly automated vehicles. For example, Alabama's statute provides five ways to violate its prohibition:

> A person shall not drive or be in actual physical control of any vehicle while: (1) There is 0.08 percent or more by weight of alcohol in his or her blood; (2) Under the influence of alcohol; (3) Under the influence of a controlled substance to a degree which renders him or her incapable of safely driving; (4) Under the combined influence of alcohol and a controlled substance to a degree which renders him or her incapable of safely driving; or (5) Under the influence of any substance which impairs the mental or physical faculties of such person to a degree which renders him or her incapable of safely driving.[77]

Under a statute like this, it matters whether the defendant consumed alcohol or drugs when using a level 4 or level 5 automated vehicle.[78] Regardless of whether the defendant is drunk or under the influence of a drug, the defendant's intoxication would not render him or her "incapable of safely driving" a level 4 or level 5 automated vehicle that completely removed him or her from the loop. However, in the case of the drunk, he or she could be convicted because the statute does not require proof of the driver being incapable of safely driving if his or her blood alcohol content exceeds 0.08 percent. The drug user would have a strong argument that the state is unable to satisfy its burden because intoxication has no impact on his or her ability to drive safely.

States should revisit these statutes for level 4 and level 5 automated vehicles that negate the fundamental assumptions of DUI laws. Allowing the use of level 4 and level 5 automated vehicles by intoxicated persons may be safer than

---

76. *See, e.g.*, Bodoh v. D.C. Bureau of Motor Vehicle Servs., 377 A.2d 1135, 1137 (D.C. 1977) ("The public safety requires strict liability under the regulation now before us, and a motorist who consumes alcohol while using medication acts at his own peril."); State v. Goding, 126 N.H. 50, 51, 489 A.2d 579, 580 (1985) ("In interpreting the DWI statutes, this court has stated that it is the commission of the prohibited conduct alone which constitutes the offense."); State v. Hammond, 118 N.J. 306, 314, 571 A.2d 942, 946 (1990) ("[D]riving under the influence has generally been considered an absolute liability offense requiring no culpable mental state, including knowledge of one's intoxication.").

77. Ala. Code § 32-5A-191(a).

78. *See* Smith v. State, 470 So. 2d 1365, 1368 (Ala. Crim. App. 1985) (holding that the state does not need to prove that the defendant is "incapable of safely driving his vehicle" when he is prosecuted under subsection (1) for having a blood alcohol content in excess of 0.08 percent).

requiring them to ride with human drivers, such as in taxi cabs, via ridesharing, or with designated drivers. These human drivers suffer from the human tendencies that make driving so dangerous—distraction, fatigue, poor decision-making, and, perhaps, having themselves consumed a little bit of alcohol or other intoxicants. Human drivers may also complicate an otherwise connected and automated world by unnecessarily creating variables—i.e., human drivers—who may not drive rationally or predictably.

Another potential complication from the broad definition of "operator" or "operates" to a user of a level 4 or level 5 automated vehicle is that some DUI laws impose vicarious liability on a vehicle owner that lets an intoxicated person operate the vehicle. For example, New Jersey's statute provides, in pertinent part, that a person commits the offense of driving while intoxicated when the person "permits another person who is under the influence of intoxicating liquor, narcotic, hallucinogenic or habit-producing drug to operate a motor vehicle the person owns."[79] The "allowing statute" requires proof that the vehicle owner had actual or constructive knowledge of the person's intoxication.[80] People may allow intoxicated people to use their level 4 and level 5 automated vehicles to take them home, rather than hailing a cab, or an automated ridesharing company may allow an intoxicated person to use its services. In either case, but more so for the private owner than the ridesharing owner due to the requirement of knowing the intoxicated state of the operator, if "operator" were defined to include a user of a level 4 or level 5 automated vehicle, then conceivably the private owner and the ridesharing company may have technically violated an allowing statute.

In addition to the typical defenses of a DUI prosecution, a new attack will be to challenge the meanings of "operator" and "user" as applied to the user or occupant of a highly automated vehicle. In these circumstances, defendants will want to argue that DUI laws do not prohibit them, as passengers, from riding intoxicated in highly automated vehicles.[81] To the extent that they are

---

79. N.J. STAT. ANN. § 39:4-50(a) (West).

80. State v. Stas, 212 N.J. 37, 50, 50 A.3d 632, 639 (2012). "The knowledge element of the offense is assessed by an objective standard: what a permittor in the exercise of reasonable care reasonably should have known about the permittee's condition to drive." State v. Zanger, 370 N.J. Super. 360, 368–69, 851 A.2d 134, 138–39 (Law. Div. 2004).

81. State v. Votava, 149 Wash. 2d 178, 184, 66 P.3d 1050, 1053 (2003) ("The actual physical control statute was enacted to protect the public by (1) deterring anyone who is intoxicated from getting into a car except as a passenger, and (2) enabling law enforcement to arrest an intoxicated person before that person strikes."); People v. Marzulli, 76 Misc. 2d 971, 972, 351 N.Y.S.2d 775, 776 (App. Term 1973) (reversing a DUI conviction and finding that a passenger was not an "operator" when the passenger unintentionally caused movement of the motor vehicle); Jones v. Dir. of Revenue, 189 S.W.3d 187, 190 (Mo. Ct. App. 2006) (declining to find a passenger, the vehicle owner, in violation of a DUI statute where there was no evidence that he drove the vehicle).

successful in analogizing themselves to passengers, they may be successful in convincing courts not to apply DUI laws against them.

B. Reckless Driving

States prohibit reckless driving.[82] These statutes typically carry a punishment of a fine or a short imprisonment or both. An example of a reckless driving statute is Oregon's prohibition on recklessly endangering another person: "A person commits the offense of reckless driving if the person recklessly drives a vehicle upon a highway or other premises described in this section in a manner that endangers the safety of persons or property."[83] States use different terms to define the concept of reckless driving, such as "recklessly," "in willful or wanton disregard of the rights or safety of others," or "gross negligence."[84] Mere negligence by the driver is insufficient to convict him or her of reckless driving.

Recklessness based on human conduct within an automated vehicle depends upon the type of automated vehicle. Occupant conduct constituting reckless driving should only be a problem with lower levels of automation. These lower levels of automation, such as level 2, create a false sense of security for their users that can create misuse.[85] By engaging in activities other than supervising the drive while the driving automation system operates, the driver may be engaging in behavior that is merely negligent or reckless. The distinc-

---

82. *E.g.*, FLA. STAT. § 316.192; 625 ILL. COMP. STAT. ANN. 5/11-503; N.C. GEN. STAT. § 20-140; N.Y. VEH. & TRAF. LAW § 1212 (McKinney); OR. REV. STAT. § 811.140; TENN. CODE ANN. § 55-10-205 (West); TEX. TRANSP. CODE ANN. § 545.401 (West); VA. CODE ANN. § 46.2-852 (West).

83. OR. REV. STAT. § 811.140. This offense is a Class A misdemeanor. Oregon defines "recklessly" as follows: "'Recklessly,' when used with respect to a result or to a circumstance described by a statute defining an offense, means that a person is aware of and consciously disregards a substantial and unjustifiable risk that the result will occur or that the circumstance exists. The risk must be of such nature and degree that disregard thereof constitutes a gross deviation from the standard of care that a reasonable person would observe in the situation." *Id.* § 161.085(9).

84. *Compare* N.C. GEN. STAT. § 20-140(a) ("Any person who drives any vehicle upon a highway or any public vehicular area carelessly and heedlessly in willful or wanton disregard of the rights or safety of others shall be guilty of reckless driving."), *with* People v. Garo, 208 Misc. 496, 498, 144 N.Y.S.2d 107, 108 (Cnty. Ct. 1955) ("It has been frequently held by the courts of this State that mere negligence is not sufficient to justify a conviction under this statute, but that the conduct of the defendant must be so reckless as to constitute culpable or gross negligence."). States also have used a variety of terminology in describing reckless driving, using such terms as "recklessly," "reckless manner," or "reckless disregard," and similarly for the terms "willful or wanton disregard"; "negligent manner" or "negligent disregard"; "endangered," "endangering," or the like; "careless" or "carelessly and heedlessly"; "without due caution"; "with disregard for the safety of others"; "material deviation"; and "reasonable and proper control." Barbara J. Van Arsdale et al., § 355 *Validity of Reckless Driving Statutes*, 7A AM. JUR. 2D AUTOMOBILES (Feb. 2020).

85. One of the significant complications with criminal convictions for reckless driving in level 2 driving automation system cases is that the drivers may not fully perceive the system's limitations or the system's effects on them. See Chapter 9 for a discussion of driver misuse of level 2 driving automation systems.

tion between negligent and reckless behavior may very well depend upon the activities of the driver.[86]

For level 3 automated driving systems, where the driver can engage in other activities and remain merely receptive to a request to intervene, the fact that the dynamic driving task fallback-ready user is engaged in other activities would not, in and of itself, be recklessness. The user of a level 3 automated driving system has purchased a system that is capable of driving without human driver assistance except as requested. Thus, these users are not aware of a substantial or unjustifiable risk that they disregard, and their behavior does not endanger the safety of other people or property.[87]

For level 4 or level 5 automated vehicles, engaging in other activities plainly and more clearly than level 3 could not constitute reckless driving. Level 4 and level 5 automated driving systems remove users from the loop; thus, activities of users do not create risks to others.

There are four areas where recklessness may be applicable or become an issue for a user of almost all driving automation systems: use of the driving automation system outside of its operational design domain, the failure to install safety-critical updates, the physical or digital alteration of an automated vehicle, and the use of a driving automation system with mechanical or maintenance problems. Whether the activity constitutes recklessness or mere negligence will depend upon the facts and the knowledge of the user of the system. A decision not to clean a sensor and use an automated driving system is different than taking an automated vehicle mudding and activating the automated driving system on the mud-covered automated vehicle. If automated vehicles are not geofenced to their operational design domains, then there is a risk that users may knowingly or unknowingly use the systems outside of their operational design domains. A decision to use a driving automation system that has an operational design domain of limited access highways on a non–limited access highway seems different than using that system within a city. As to software updates, if automated vehicles can operate without necessary software updates or when there are mechanical or maintenance problems, then

---

86. Compare, for example, a driver who is sleeping or watching a movie while driving with a driver who briefly checks his or her phone. Drivers have fallen asleep using Tesla Autopilot. *See* Dalvin Brown, *Video Appears to Show Tesla Driver "Literally Asleep at the Wheel"*, USA Today (Sept. 9, 2019), https://www.usatoday.com/story/money/cars/2019/09/09/tesla-driver-recorded-asleep-car-drives-down-massachusetts-highway/2262113001/.

87. On this latter point, courts have found that even when the defendant engages in conduct that constitutes recklessness, the defendant cannot be convicted of reckless driving when there is no proof that the defendant endangered the safety of other people or property. *See* Jackson v. State, 576 N.E.2d 607, 609 (Ind. Ct. App. 1991) ("[T]he record is utterly bereft of any indication, either from direct or circumstantial evidence, that [the defendant] endangered the safety or property of another. The common denominator of convictions upheld under the reckless driving statute is the endangerment presented by the operator of a motor vehicle to the safety or property of another.").

users may knowingly or unknowingly use their automated vehicles without such updates or repairs or may not appreciate the significance of the updates or repairs.[88]

Some of these uses of driving automation systems may be preventable by engineering solutions. Manufacturers could geofence their driving automation systems to their operational design domains; prevent the use of the systems when safety-critical updates are needed; or prevent their use while maintenance or mechanical problems alter their operation. For the most part, then, these four areas may be engineering problems that could be prevented through design. Manufacturer attempts at solving these problems may complicate a factual analysis about whether the driver or occupants of the vehicles were reckless. A driver may assume (rightfully or wrongfully) that if the driving automation system engages, the system can safely operate.

C. Vehicular Homicide

States have different categories of vehicular homicide statutes. Depending upon a state's definition of "operator" or "driver," these statutes can create pernicious problems for highly automated vehicles. The first inquiry in these cases would be how the term "operator" or "driver" has been defined. In states where the automated driving system is considered the operator or driver, the human user would not be an operator or driver for purposes of these statutes. This section focuses on prosecutions where "operator" or "driver" has not been redefined or has been defined to include the human user.

There are two broad categories of vehicular homicide cases, though there certainly can be overlap. The first type is vehicular homicide based on reckless driving or criminal negligence. The above analysis of reckless driving and the conclusions reached therein should help shape whether a person could be convicted of vehicular homicide while using a driving automation system. For these statutes, the offense that constitutes reckless driving is typically a lesser included offense. The second type of vehicular homicide cases are those that are based on a death that results when the driver or operator is intoxicated.

1. Reckless or Negligent Vehicular Homicide

Vehicular homicide statutes allow states to criminally punish a driver or operator of a vehicle for a crime relating to the death of another person.[89] For exam-

---

88. "Knowingly or unknowingly" here may depend upon the type of notification that the vehicle gives to the user about the need for repair or update. Other circumstances, such as being away from home or actually using the vehicle when a safety-critical update is released, may negate the voluntariness or culpability that a user faces by not immediately updating the vehicle.

89. *See, e.g.*, Fla. Stat. § 782.071; Mass. Gen. Laws ch. 90, § 24G(c); N.J. Stat. Ann. § 2C:11-5 (West); S.C. Code Ann. § 56-5-2910.

ple, Florida's vehicular homicide statute provides: "'Vehicular homicide' is the killing of a human being, or the killing of an unborn child by any injury to the mother, caused by the operation of a motor vehicle by another in a reckless manner likely to cause the death of, or great bodily harm to, another."[90] Under the Florida statute, vehicular homicide is essentially reckless driving that results in the death of a human being or an unborn child.[91]

A person can also be convicted of vehicular homicide by operating a motor vehicle in a grossly[92] or simply[93] negligent manner. Gross negligence is a substantially higher level of culpability than simple negligence.[94] Gross

---

90. FLA. STAT. § 782.071.

91. Santisteban v. State, 72 So. 3d 187, 195 (Fla. Dist. Ct. App. 2011) ("Vehicular homicide cannot be proven without also proving the elements of reckless driving."). Florida's reckless driving statute provides, in pertinent part: "Any person who drives any vehicle in willful or wanton disregard for the safety of persons or property is guilty of reckless driving." FLA. STAT. § 316.192. Other states have similar elements. *See, e.g.*, State v. Watson, 349 S.C. 372, 376, 563 S.E.2d 336, 338 (2002) ("Reckless homicide requires proof that the defendant (1) operated an automobile (2) in reckless disregard for the safety of others; (3) the defendant's conduct proximately caused injury to the victim, and (4) within one year, the victim died as a result of these injuries.").

92. *See, e.g.*, MINN. STAT. § 609.2112 ("[A] person is guilty of criminal vehicular homicide and may be sentenced to imprisonment for not more than ten years or to payment of a fine of not more than $20,000, or both, if the person causes the death of a human being not constituting murder or manslaughter as a result of operating a motor vehicle: (1) in a grossly negligent manner"); CAL. PENAL CODE § 192(c)(1) (West) ("Manslaughter is the unlawful killing of a human being without malice. It is of three kinds: . . . (c) Vehicular— . . . driving a vehicle in the commission of an unlawful act, not amounting to a felony, and with gross negligence; or driving a vehicle in the commission of a lawful act which might produce death, in an unlawful manner, and with gross negligence.").

93. CAL. PENAL CODE § 192(c)(2) (West) ("Manslaughter is the unlawful killing of a human being without malice. It is of three kinds: . . . (c) Vehicular— . . . Driving a vehicle in the commission of an unlawful act, not amounting to a felony, but without gross negligence; or driving a vehicle in the commission of a lawful act which might produce death, in an unlawful manner, but without gross negligence."); *see also* People v. Bussel, 97 Cal. App. 4th Supp. 1, 9, 118 Cal. Rptr. 2d 159, 165 (Cal. App. Dep't Super. Ct. 2002) ("Most jurisdictions that have addressed this issue have held that a vehicular manslaughter conviction may be premised on ordinary negligence."); State v. Russo, 38 Conn. Supp. 426, 433, 450 A.2d 857, 863 (Super. Ct. 1982) ("The crime of negligent homicide has three elements: (1) the death of a person; (2) by the instrumentality of a motor vehicle; (3) which is operated in a negligent manner.").

94. As the Minnesota Supreme Court has described,

> Gross negligence is substantially and appreciably higher in magnitude than ordinary negligence. It is materially more want of care than constitutes simple inadvertence. It is an act or omission respecting legal duty of an aggravated character as distinguished from a mere failure to exercise ordinary care. It is very great negligence, or the absence of slight diligence, or the want of even scant care. It amounts to indifference to present legal duty, and to utter forgetfulness of legal obligations so far as other persons may be affected. It is a heedless and palpable violation of legal duty respecting the rights of others. The element of culpability which characterizes all negligence is in gross negligence magnified to a high degree as compared with that present in ordinary negligence. Gross negligence is a manifestly smaller amount of watchfulness and circumspection than the circumstances require of a person of ordinary prudence. Ordinary and gross negligence differ in degree of inattention.

State v. Al-Naseer, 690 N.W.2d 744, 752 (Minn. 2005) (quoting Altman v. Aronson, 231 Mass. 588, 121 N.E. 505, 506 (1919)).

negligence will help delineate the appropriateness of allowing a state to convict a human user of a highly automated vehicle for vehicular homicide when the automated vehicle causes the death of a person. In these instances, it seems that courts should examine the same factors described above for recklessness: did the user use the automated driving system outside of its operational design domain, fail to install safety-critical updates, physically or digitally alter the automated vehicle, or use a driving automation system when there were mechanical or maintenance problems? For lower levels of automation, the focus, in addition to those factors, can also shift to the driver's (or, for level 3 automated vehicles, the dynamic driving task fallback-ready user's) conduct. In states that only require simple negligence or that have statutes based on violations of traffic laws, the question becomes whether the user of the automated vehicle has a duty or responsibility to prevent the violation of traffic laws. This inquiry, again, turns on the level of automation. In level 1 and level 2 automated vehicles, the driver has the duty and responsibility to drive his or her vehicle reasonably and to comply with traffic laws. In level 3, level 4, and level 5 automated vehicles, the user of the system does not have a duty to drive the vehicle. Because the user is not responsible for driving the vehicle, the user should also not be responsible for the negligent behavior or death.

2. *Vehicular Homicide While Intoxicated*

States have criminal laws that apply when a driver, under the influence of drugs or alcohol, causes a crash that kills someone. Of all the potential operators or drivers of an automated vehicle, these types of statutes would only apply to human users or owners of automated vehicles. There are variations in these statutes that can provide opportunities for defending these cases for higher levels of automated vehicles.

States are split on whether their statute includes a separate element of negligence. For example, Colorado law provides as follows:

> If a person operates or drives a motor vehicle while under the influence of alcohol or one or more drugs, or a combination of both alcohol and one or more drugs, and such conduct is the proximate cause of the death of another, such person commits vehicular homicide. This is a strict liability crime.[95]

In comparison, South Carolina's statute provides as follows:

95. COLO. REV. STAT. § 18-3-106(b)(I); People v. Garner, 781 P.2d 87, 89 (Colo. 1989) (en banc) ("There is no requirement under the current statute that the prosecution also prove that the defendant's driving was negligent.").

> A person who, while under the influence of alcohol, drugs, or the combination of alcohol and drugs, drives a motor vehicle and when driving a motor vehicle does any act forbidden by law or neglects any duty imposed by law in the driving of the motor vehicle, which act or neglect proximately causes great bodily injury or death to another person, is guilty of the offense of felony driving under the influence.[96]

In those states where the prosecution needs to prove a separate element of negligence, the prosecution may struggle to do so in the context of highly automated vehicles. In a state like South Carolina, a user of a level 4 or level 5 automated vehicle charged under this statute would be able to argue that he or she did not commit "any act forbidden by law or neglect[] any duty imposed by law."[97] Users of level 3 automated driving systems may also be able to contend that they did not do any act forbidden by law or neglect a duty so long as the system had not requested intervention. Users of level 1 and level 2 driving automation systems lack that argument because they are responsible for driving their vehicles.

Under either type of statute, the defendant would not lose his or her right to argue that the prosecutor fails to establish proximate cause. Proximate cause would be the primary argument that defendants could use to challenge a conviction. Under the Colorado statute, the proximate cause required is that operating or driving a motor vehicle under the influence proximately caused the death of another human. In a level 4 or level 5 automated vehicle, the user's intoxication while using the system could not cause the death.

An example of a case examining proximate cause is *State v. Benoit*,[98] where the defendant, who had a blood alcohol content of 0.208 at the time of the crash, was traveling with a passenger when a vehicle crossed into their lane of traffic and hit them. The passenger died and the defendant was significantly injured. The police charged the defendant under a vehicular homicide statute that provided: "When the death of any person other than the operator ensues as a proximate result of an injury received by the operation of any vehicle, the operator of which is under the influence of, any intoxicating liquor . . . the person so operating such vehicle shall be guilty of 'driving under the influence of liquor or drugs, resulting in death.'" The defendant was also charged under a similar

96. S.C. Code Ann. § 56-5-2945; State v. Grampus, 288 S.C. 395, 396–97, 343 S.E.2d 26, 27 (1986), *abrogated on other grounds by* State v. Easler, 327 S.C. 121, 489 S.E.2d 617 (1997) ("Felony D.U.I. requires proof of three elements: (1) the actor drives a vehicle while under the influence of alcohol and/or drugs; (2) the actor does an act forbidden by law or neglects a duty imposed by law; and (3) the act or neglect proximately causes great bodily injury or death to another person." (citation omitted)).

97. *See* S.C. Code Ann. § 56-5-2945.

98. 650 A.2d 1230 (R.I. 1994).

statute with causing serious bodily injury. The defendant moved to dismiss the case for lack of proximate cause, arguing that the state failed to prove that his operation of the motor vehicle caused the death or serious bodily injury. The Rhode Island Supreme Court ruled that "the state must produce sufficient evidence for a jury to conclude that the defendant's manner of operating his or her motor vehicle was a proximate cause of the victim's death and that the collision occurred while the defendant was legally intoxicated." The court ruled that the statute requires proof that the operation of the vehicle, and not the intoxication, was the proximate cause.

V. USE OF DATA TO DEFEND AUTOMOBILE CRIMES

Similar to the way that body cameras have called testimony into question and shed new light on the circumstances of a police encounter, driving automation systems and data recorded by automated vehicles may shed new light on traffic stops. In these circumstances, there may or may not be a police recording of the traffic violation, and officer testimony about a witnessed traffic violation may be the significant piece of evidence needed to find a traffic violation. Traffic violations and the fines that accompany them may not be the main impetus to challenge a traffic stop. In many of those cases, the costs of obtaining vehicle data and employing an expert to reconstruct the vehicle's actions will far exceed the cost of the fine. However, if the traffic stop eventually leads to the discovery of other evidence, such as drugs or other contraband, the criminal defendant may be facing significant imprisonment. Depending upon the facts, vehicle data may be an effective way to establish that the officer lacked probable cause or reasonable suspicion for the stop and allow defendants to challenge the initial stop and exclude evidence that is discovered.

The United States Supreme Court has given significant weight to hard and objective evidence over subjective eyewitness testimony. In *Scott v. Harris*,[99] the Court noted that a court should not adopt a factual story that a "videotape quite clearly contradicts." The Supreme Court noted that "[w]hen opposing parties tell two different stories, one of which is blatantly contradicted by the record, so that no reasonable jury could believe it, a court should not adopt that version of the facts for purposes of ruling on a motion for summary judgment."[100]

Courts have considered how a party seeking to introduce an event data recorder (i.e., the black box of a vehicle) or data from a vehicle computer system

99. 550 U.S. 372, 380, 127 S. Ct. 1769, 167 L. Ed. 2d 686 (2007).
100. *Id.* at 380.

into evidence can establish its reliability.[101] Generally, the admission of such systems into evidence requires an expert witness to establish the system's reliability and the reliability of data. An expert will need to have expertise in using data to reconstruct the movements of the vehicle.

---

101. Admissibility has been considered in both criminal and civil cases. *See, e.g.*, Bachman v. Gen. Motors Corp., 332 Ill. App. 3d 760, 776 N.E.2d 262 (2002); People v. Christmann, 3 Misc. 3d 309, 315–16, 776 N.Y.S.2d 437, 442 (Just. Ct. 2004); Easter v. State, 223 Md. App. 65, 115 A.3d 239, 248 (2015); Commw. v. Zimmermann, 70 Mass. App. Ct. 357, 360, 873 N.E.2d 1215, 1218 (2007); Commw. v. Safka, 2014 Pa. Super. 131, 95 A.3d 304 (2014), *aff'd*, 636 Pa. 169, 141 A.3d 1239 (2016); State v. Shabazz, 400 N.J. Super. 203, 946 A.2d 626 (2005); State v. Byard, 2018 WL 2077324 (Del. Super. Ct. May 1, 2018), *as corrected* (May 3, 2018); State v. Diaz, No. 35,563, 2017 WL 1017356, at *3 (N.M. Ct. App. Feb. 27, 2017); *see also* Marjorie A. Shields, *Admissibility of Evidence Taken from Vehicle Event Data Recorders (EDR), Sensing Diagnostic Modules (SDM), or "Black Boxes"*, 40 A.L.R.6TH 595 (2008).

Chapter 20

Physical and Digital Interference with Automated Vehicles

Automated vehicles will create new opportunities for criminal mischief. People could physically interfere with automated vehicles, such as by blocking a road or by manipulating infrastructure. These acts may become more pronounced with level 4 and level 5 automated vehicles that may be more obedient than human drivers who may meet such acts with a foot on the accelerator. Digital interference is hacking into or cyberattacking an automated vehicle. Not all hacks into an automated vehicle are bad, and some hacking, such as white-hat hacking,[1] is beneficial. Other hacking that can be beneficial—or at least that requires a different analysis than cyberattacks—is jailbreaking a device.[2] The real evil is cyberattacks, the unwanted third-party attacks. Cyberattacks will present new opportunities for criminals to terrorize occupants of automated vehicles by taking control of automated vehicles or by preventing automated vehicles from being operated until a ransom is paid.

The existing automated vehicle laws have not created new crimes for physical or digital interference with an automated vehicle. Prosecutors will need to look to existing laws to determine criminal responsibility. For the most part, there are not "unique" automated vehicle defenses to these crimes, so defendants will need to use traditional defenses. Current law creates a

1. A white-hat hacker is someone who tests systems and vulnerabilities to improve security before malicious hackers can exploit those vulnerabilities.

2. Jailbreaking is "the act of lifting manufacturer-imposed restrictions on software operating systems, unlocking a host of latent, unauthorized capabilities." Michael Sinanian, *Jailbreak!: What Happens When Autonomous Vehicle Owners Hack into Their Own Cars*, 23 MICH. TELECOMM. & TECH. L. REV. 357, 358 (2017). Jailbreaks are user-initiated hacks.

square-peg-into-a-round-hole problem. Many of these laws were enacted at a time when physical interference with vehicles meant stealing, damaging, or altering parts that could not, unbeknownst to the user, cause the vehicle to malfunction. Cyberattacks were historically financial crimes that did not create the same risks and terror as hacking automated vehicles.

I. PHYSICAL INTERFERENCE WITH AUTOMATED VEHICLES

Physical interference with an automated vehicle could occur in numerous ways. A person could physically interfere with an automated vehicle by removing, damaging, or altering a physical component. A person could stand in a roadway or place objects on the road to interfere with a driving automation system. The placement of cones in a roadway could impede or confuse an automated vehicle. A person could also manipulate traffic signs, such as by creating traffic lines or placing unauthorized traffic signs. This section describes state and federal laws that could be used to hold those who physically interfere with automated vehicles criminally responsible.

A. State Laws

States, in contrast to the federal government, have comprehensive motor vehicle codes and criminal codes that apply when someone physically interferes with an automated vehicle. Depending upon the facts, other crimes, such as kidnapping, larceny, or murder, may be committed. This section introduces three offenses that may be applicable to physical interferences: tampering with a motor vehicle, malicious mischief, and the use of traffic control devices.

The most prominent state statutes are prohibitions on tampering with motor vehicles. These antitampering statutes may be applicable when a person alters, uses, or possesses someone else's automated vehicle without their permission.[3] Tampering statutes would be applicable whenever a person physically alters

3. *See, e.g.*, CAL. VEH. CODE § 10852 (West) ("No person shall either individually or in association with one or more other persons, wilfully injure or tamper with any vehicle or the contents thereof or break or remove any part of a vehicle without the consent of the owner."); MO. REV. STAT. § 569.080 ("A person commits the offense of tampering in the first degree if he or she . . . [k]nowingly receives, possesses, sells, or unlawfully operates an automobile, airplane, motorcycle, motorboat or other motor-propelled vehicle without the consent of the owner thereof."); N.C. GEN. STAT. § 20-107 ("Any person who either individually or in association with one or more other persons willfully injures or tampers with any vehicles or breaks or removes any part or parts of or from a vehicle without the consent of the owner is guilty of a Class 2 misdemeanor."); R.I. GEN. LAWS § 31-9-3 ("Any person who either individually or in association with one or more other persons willfully injures or tampers with any vehicle or breaks or removes any part or parts of or from a vehicle without the consent of the owner is guilty of a misdemeanor.").

or damages an automated vehicle or interferes with a person's use of an automated vehicle.

Malicious mischief is the criminal offense generally characterized by intentionally damaging the property of another person.[4] Malicious mischief will be an offense when someone destroys part of an automated vehicle. Statutes that require "physical damage" may mean that their malicious mischief statutes would not apply to digital alterations or digital damage to an automated vehicle's computer system.

Depending upon the method of interference, states may be able to pursue defendants for violating statutes on possession of traffic control devices. An example of this type of offense would be a defendant holding up a stop sign to trick an automated vehicle into stopping. Ohio's statute, for example, defines traffic control devices as "any sign, traffic control signal, or other device conforming to and placed or erected in accordance with" the state's manual for uniform traffic control devices.[5] These traffic device statutes generally prohibit the possession or use of traffic control devices except for specified statutory exceptions.[6] For example, Texas law provides:

> A person may not place, maintain, or display on or in view of a highway an unauthorized sign, signal, marking, or device that:
>
> (1) imitates or resembles an official traffic-control device or railroad sign or signal;
>
> (2) attempts to direct the movement of traffic; or
>
> (3) hides from view or hinders the effectiveness of an official traffic-control device or railroad sign or signal.[7]

These statutes could be used to prosecute those who use traffic devices or signs to interfere with automated vehicles.

B. Federal Law

There are limited federal laws that could apply when someone physically interferes with an automated vehicle. This section will describe three such statutes

---

4. *See, e.g.*, MASS. GEN. LAWS ch. 266, § 127 ("Whoever destroys or injures the personal property, dwelling house or building of another in any manner or by any means not particularly described or mentioned in this chapter shall, if such destruction or injury is willful and malicious, be punished."); N.Y. PENAL LAW § 145.00 (McKinney) ("A person is guilty of criminal mischief in the fourth degree when, having no right to do so nor any reasonable ground to believe that he or she has such right, he or she intentionally damages property of another person"); WASH. REV. CODE § 9A.48.090 ("A person is guilty of malicious mischief in the third degree if he or she: (a) Knowingly and maliciously causes physical damage to the property of another, under circumstances not amounting to malicious mischief in the first or second degree. . . .").

5. OHIO REV. CODE ANN. § 4511.18(A) (West).

6. *See id.* § 4511.18; GA. CODE ANN. § 40-6-25 (West); TEX. TRANSP. CODE ANN. § 544.006 (West).

7. TEX. TRANSP. CODE ANN. § 544.006 (West).

and their limited application. The first statute that may apply would be the federal "destruction of motor vehicles or motor vehicle facilities" statute codified at 18 U.S.C. § 33:

> Whoever willfully, with intent to endanger the safety of any person on board or anyone who he believes will board the same, or with a reckless disregard for the safety of human life, damages, disables, destroys, tampers with, or places or causes to be placed any explosive or other destructive substance in, upon, or in proximity to, any motor vehicle which is used, operated, or employed in interstate or foreign commerce, or its cargo or material used or intended to be used in connection with its operation . . . shall be fined under this title or imprisoned not more than twenty years, or both.

Under this statute, "motor vehicle" is defined narrowly only to include a motor vehicle used for commercial purposes and would exclude privately owned automated vehicles.[8] The interstate commerce element can be satisfied, as one court has found, if the motor vehicles are "used in connection with or in furtherance of the interstate market activities of the entities operating or employing the vehicles."[9] One court has interpreted willfulness to require the government to show two types of intent: (1) that the defendant willfully damages, disables, destroys, tampers with, or places an explosive or destructive device in or in the proximity of a motor vehicle (2) with the intent of endangering the safety of persons on board or with reckless disregard to their safety.[10] Accordingly, this statute could be applicable when someone willfully seeks to damage or tamper with an automated vehicle used for commercial purposes to endanger the safety of occupants or with reckless disregard for their safety. However, this statute would not apply to unoccupied automated vehicles or noncommercial automated vehicles.

A statute that would require unique facts is the federal prohibition on nonqualifying uses of traffic signal preemption transmitters. A traffic signal preemption transmitter is a "mechanism that can change or alter a traffic signal's

---

8. "'[M]otor vehicle' means every description of carriage or other contrivance propelled or drawn by mechanical power and used for commercial purposes on the highways in the transportation of passengers, passengers and property, or property or cargo." 18 U.S.C. § 31(a)(6). "Used for commercial purposes" is defined as "the carriage of persons or property for any fare, fee, rate, charge or other consideration, or directly or indirectly in connection with any business, or other undertaking intended for profit." *Id.* § 31(a)(10).

9. United States v. Lowe, 65 F.3d 1137, 1147 (4th Cir. 1995); *see also* United States v. Heightland, 678 F. Supp. 159, 163 (E.D. Ky. 1987), *aff'd*, 865 F.2d 94 (6th Cir. 1989) (holding that "[t]he fact that the trucks themselves may have been engaged in intrastate travel does not alter the fact that the coal hauled by the trucks entered the stream of interstate commerce").

10. United States v. Kurka, 818 F.2d 1427 (9th Cir. 1987).

phase time or sequence."[11] Thus, to the extent someone uses a mechanism to alter traffic signals to trick an automated vehicle, they could be prosecuted under this statute.

Automated vehicles, especially those that lack a human override, may be more susceptible to theft or carjacking. For example, a person may stand in front of an automated vehicle and insist that the occupants exit the vehicle by brandishing a weapon or making other threats. With human-driven vehicles, such threats may have been met by a hard push on the accelerator by the human driver. Automated vehicles almost certainly will not be able to determine the intentions of a person standing in front of the vehicle and may not be able to evade such tactics. The federal carjacking statute provides that "[w]hoever, with the intent to cause death or serious bodily harm takes a motor vehicle that has been transported, shipped, or received in interstate or foreign commerce from the person or presence of another by force and violence or by intimidation, or attempts to do so" is guilty of carjacking.[12] Essentially, the offense of carjacking has four elements: "(1) that [the defendant] took a motor vehicle from the person or presence of another; (2) that he did so by force, violence or intimidation; (3) that [the defendant] intended to cause death or serious bodily harm; and (4) that the motor vehicle had been transported, shipped, or received in interstate or foreign commerce."[13] The third element—that the defendant intended to cause death or serious bodily harm—requires proof that "the defendant would have at least attempted to seriously harm or kill the driver if that action had been necessary to complete the taking of the car."[14]

The carjacking statute would not criminalize the stealing or apprehension of unoccupied automated vehicles. Carjacking requires a threat of bodily harm to the occupants. Thus, so long as defendants apprehend and take unoccupied automated vehicles, carjacking would not be applicable. Carjacking may also have limited application or no application to cyberattacks because the cyberattacker would arguably not be taking a vehicle from someone with force, violence, or intimidation.

II. DIGITAL INTERFERENCE

Digital interference with automated vehicles will be a significant problem that manufacturers need to confront in developing and designing their vehicles and that governments need to consider in regulating automated vehicles. Modern,

11. 18 U.S.C. § 39.
12. 18 U.S.C. § 2119.
13. United States v. Gurule, 461 F.3d 1238, 1243 (10th Cir. 2006).
14. Holloway v. United States, 526 U.S. 1, 12, 119 S. Ct. 966, 972, 143 L. Ed. 2d 1 (1999).

nonautomated vehicles have already been shown to suffer vulnerabilities that could be exploited for criminal reasons.[15] Cyberattacks may be used for theft, to extort economic control, to introduce ransomware, or to cause significant physical harm or damage. Cyberattackers may gain access to an automated vehicle through devices or objects that connect with the automated vehicle, such as vehicle-to-vehicle, vehicle-to-infrastructure, and vehicle-to-everything communications; during the production of the vehicle or components; or while automated vehicles are repaired or updated.[16] A cyberattacker can gain access to vehicles through the vehicle communication buses or electronic control units.

Cyberattacks are criminally punished at the state and federal level, and this section describes these criminal prohibitions.[17] The federal government has historically been better positioned to prosecute interstate cyberattacks than an individual state whose residents are affected by the cyberattack. States have shown their effectiveness at regulating the consequences of a cyberattack, such as by requiring notification, a topic covered in Chapter 17.

A. State Regulation

States have two main avenues for prosecuting cyberattackers. The first avenue is to use the state laws that address hacking or digital trespass. These laws directly address the crime that the state is prosecuting. However, states may find, at least at first, that the crime of cyberattacks on an automated vehicle does not fit the punishment. Underlying most of these statutes is the fundamental assumption that most computer crimes are financial or personal privacy crimes, not crimes that can involve death or significant bodily harm or damage. Because of that, state prosecutors may consider alternative and more creative ways to prosecute hackers of vehicles under their noncomputer crime statutes, the second avenue of prosecution.

---

15. See Chapter 18 for a discussion of cyberattacks on vehicles.

16. Zev Winkelman et al., *When Autonomous Vehicles Are Hacked, Who Is Liable?*, RAND CORP. 5–11 (2019), https://www.rand.org/pubs/research_reports/RR2654.html.

17. In any cyberattack, there may be two separate criminal acts that need to be analyzed. The first is the cyberattack itself, the substance of this section. The second is any criminal violations that occur as a result of the cyberattack. For example, a cyberattacker may cause a vehicle to speed or run a red light—either of which are themselves traffic violations. Criminal violations that occur as a result of the cyberattack may, initially, give rise to fault being placed on the customary party under that jurisdiction's laws—whether the vehicle owner, the user, or the manufacturer. Once the cyberattack has made itself known, however, those customary parties should not bear criminal fault because they have not committed a voluntary act, a requirement of almost every criminal prosecution. Thus, these criminal violations would also be the responsibility of the cyberattacker.

1. Computer Crimes

All 50 states criminalize hacking.[18] Many of these statutes explicitly create civil causes of action that allow victims to seek recourse against the culprit.[19] These laws differ in their breadth and punishments. Punishable conduct includes unauthorized access,[20] unauthorized use of a computer,[21] computer

---

18. *See Computer Crime Statutes*, NAT'L CONF. OF STATE LEGS. (June 14, 2018), http://www.ncsl.org/research/telecommunications-and-information-technology/computer-hacking-and-unauthorized-access-laws.aspx (listing statutes); Fernando M. Pinguelo & Bradford W. Muller, *Virtual Crimes, Real Damages: A Primer on Cybercrimes in the United States and Efforts to Combat Cybercriminals*, 16 VA. J.L. & TECH. 116, 150–88 (2011) (collecting statutes); *see also* ALA. CODE §§ 13A-8-112, 13A-8-113; ALASKA STAT. § 11.46.740; ARIZ. REV. STAT. ANN. §§ 13-2316, 13-2316.01, 13-2316.02; ARK. CODE ANN. §§ 5-41-101 to -206; CAL. PENAL CODE § 502 (West); COLO. REV. STAT. §§ 18-5.5-101 to -102; CONN. GEN. STAT. §§ 53a-250 to -261; DEL. CODE ANN. tit. 11, §§ 931–941 (West); FLA. STAT. §§ 815.01 to .07, 668.801 to .805; GA. CODE ANN. §§ 16-9-90 to -94, 16-9-150 to -157 (West); HAW. REV. STAT. §§ 708-890 to -895.7; IDAHO CODE §§ 18-2201, -2202; 720 ILL. COMP. STAT. ANN. 5/17-50 to -55; IND. CODE §§ 35-43-1-8, 35-43-2-3; IOWA CODE § 716.6B; KAN. STAT. ANN. § 21-5839 (West); KY. REV. STAT. ANN. §§ 434.840, .845, .850, .851, .853, .855, .860 (West); LA. STAT. ANN. §§ 14:73.1 to .8; ME. STAT. tit. 17-A, §§ 431–435; MD. CODE ANN., CRIM. LAW § 7-302 (West); MASS. GEN. LAWS ch. 266, § 33A; MICH. COMP. LAWS §§ 752.791, .792, .793, .794, .795, .796, .797; MINN. STAT. §§ 609.87 to .893; MISS. CODE. ANN. §§ 97-45-1 to -33 (West); MO. REV. STAT. §§ 537.525, 569.095, 569.097, 569.099; MONT. CODE ANN. § 45-6-311 (West); NEB. REV. STAT. ANN. §§ 28-1341 to -1348 (West); NEV. REV. STAT. §§ 205.473 to .513; N.H. REV. STAT. ANN. §§ 638:16, :17, :18, :19; N.J. STAT. ANN. §§ 2A:38A-1 to -3, 2C:20-2, 2C:20-23 to -34 (West); N.M. STAT. §§ 30-45-1 to -7; N.Y. PENAL LAW §§ 156.00 to .50 (McKinney); N.C. GEN. STAT. §§ 14-453 to -458; N.D. CENT. CODE § 12.1-06.1-08; OHIO REV. CODE ANN. §§ 2913.01 to .04 (West); OKLA. STAT. tit. 21, § 1951 *et seq.*; OR. REV. STAT. § 164.377; 18 PA. CONS. STAT. ANN. § 5741; 11 R.I. GEN. LAWS § 11-52-1 to -8; S.C. CODE ANN. § 16-16-10 *et seq.*; S.D. CODIFIED LAWS § 43-43B-1 *et seq.*; TENN. CODE ANN. § 39-14-601 *et seq.* (West); TEX. PENAL CODE ANN. § 33.02 (West); UTAH CODE ANN. § 76-6-702 to -705 (West); VT. STAT. ANN. tit. 13, § 4101 *et seq.* (West); VA. CODE ANN. § 18.2-152.1 *et seq.* (West); WASH. REV. CODE § 9A.90.010 *et seq.*; W. VA. CODE § 61-3C-3 *et seq.*; WIS. STAT. § 943.70; WYO. STAT. ANN. § 6-3-501 *et seq.*, § 40-25-101 (West).

19. *See, e.g.*, ARK. CODE ANN. § 5-41-106(a)(1) (West) ("Any person whose property or person is injured by reason of a violation of any provision of this subchapter may sue for the injury and recover for any damages sustained and the costs of suit."); CAL. PENAL CODE § 502(e)(1) (West) ("In addition to any other civil remedy available, the owner or lessee of the computer, computer system, computer network, computer program, or data who suffers damage or loss by reason of a violation of any of the provisions of subdivision (c) may bring a civil action against the violator for compensatory damages and injunctive relief or other equitable relief."); GA. CODE ANN. § 16-9-93(g) (West).

20. *See, e.g.*, IND. CODE § 35-43-2-3(b) ("A person who knowingly or intentionally accesses: (1) a computer system; (2) a computer network; or (3) any part of a computer system or computer network; without the consent of the owner of the computer system or computer network, or the consent of the owner's licensee, commits computer trespass, a Class A misdemeanor."); MD. CODE ANN., CRIM. LAW § 7-302(c)(1) (West) ("A person may not intentionally, willfully, and without authorization: (i) access, attempt to access, cause to be accessed, or exceed the person's authorized access to all or part of a computer network, computer control language, computer, computer software, computer system, computer service, or computer database."); N.H. REV. STAT. ANN. § 638:17 ("A person is guilty of the computer crime of unauthorized access to a computer or computer network when, knowing that the person is not authorized to do so, he or she knowingly accesses or causes to be accessed any computer or computer network without authorization."); TEX. PENAL CODE ANN. § 33.02(a) (West) ("A person commits an offense if the person knowingly accesses a computer, computer network, or computer system without the effective consent of the owner.").

21. *See, e.g.*, MONT. CODE ANN. § 45-6-311 (West) ("A person commits the offense of unlawful use of a computer if the person knowingly or purposely: (a) obtains the use of any computer, computer

crime to defraud,[22] theft of computer services,[23] or damage or alteration of computers.[24]

The punishment of these offenses depends upon the type of crime, with some being misdemeanors, such as unauthorized access,[25] and others being felony offenses, such as damage or alteration of computers.[26] Some of these statutes have more significant punishments where there is monetary damage, such as damages exceeding $10,000, than where there is "a risk of serious physical injury to another person."[27] An inherent problem with many of these statutes is that the punishment reflects an assumption of a computer crime and not significant bodily harm or death. States may want to consider tailoring their computer crime statutes to the risk of harm that could be caused by cyberattacking an automated vehicle.

As an example of the types of crimes that are covered by these statutes, this section examines the California Computer Data Access and Fraud Act (the

---

system, or computer network without consent of the owner. . . ."). This offense is referred to differently in different states, with some referring to it as computer tampering. *See, e.g.*, 720 ILL. COMP. STAT. ANN. 5/17-51; LA. STAT. ANN. § 14:73.7.

22. *See, e.g.*, KY. REV. STAT. ANN. § 434.845(1) (West) ("A person is guilty of unlawful access to a computer in the first degree when he or she, without the effective consent of the owner, knowingly and willfully, directly or indirectly accesses, causes to be accessed, or attempts to access any computer software, computer program, data, computer, computer system, computer network, or any part thereof, for the purpose of: (a) Devising or executing any scheme or artifice to defraud; or (b) Obtaining money, property, or services for themselves or another by means of false or fraudulent pretenses, representations, or promises.").

23. *See, e.g.*, DEL. CODE ANN. tit. 11, § 933 (West) ("A person is guilty of the computer crime of theft of computer services when the person accesses or causes to be accessed or otherwise uses or causes to be used a computer system with the intent to obtain unauthorized computer services, computer software or data."); N.H. REV. STAT. ANN. § 638:17 ("A person is guilty of the computer crime of theft of computer services when he or she knowingly accesses or causes to be accessed or otherwise uses or causes to be used a computer or computer network with the purpose of obtaining unauthorized computer services.").

24. *See, e.g.*, CAL. PENAL CODE § 502(c) (West) (including numerous crimes relating to damaging computers).

25. *See, e.g.*, TENN. CODE ANN. § 39-14-602 (West) ("Whoever intentionally and without authorization, directly or indirectly: (1) Accesses any computer, computer system, or computer network commits a Class C misdemeanor."); *see also* IOWA CODE § 716.6B (the most significant penalty is an aggravated misdemeanor); *but see* MONT. CODE ANN. § 45-6-311(2) (West) (providing a penalty of up to ten years of imprisonment if the offense involves "property exceeding $1,500 in value"); GA. CODE ANN. § 16-9-93 (West) ("Any person convicted of the crime of computer theft, computer trespass, computer invasion of privacy, or computer forgery shall be fined not more than $50,000.00 or imprisoned not more than 15 years, or both.").

26. *See, e.g.*, DEL. CODE ANN. tit. 11, § 939 (West).

27. Delaware's statute provides that a person who commits the computer crimes is guilty of a Class D felony if the property damage exceeds $10,000, but a person is only guilty of a Class G felony if the computer crime "creates a risk of serious physical injury to another person." *Id.* Under this statute, a cyberattacker of an automated vehicle who terrorizes occupants of the vehicle but does not damage the vehicle would face a lower penalty than a cyberattacker who crashes an unoccupied automated vehicle.

CDAFA).[28] The CDAFA provides 14 different crimes that can be committed based on the actor knowingly and without permission committing some act—some of these prohibitions are based on gaining access to a computer or computer system, while others are not based on access.[29] Importantly, the CDAFA explicitly includes computer systems and devices that interact with a motor vehicle within the definition of "computer system."[30] Devices and systems that comprise the automated vehicle and devices and systems that the automated vehicle interacts with using vehicle-to-vehicle, vehicle-to-infrastructure, and vehicle-to-everything communications are computer systems under the CDAFA. The following list includes statutory language that may be relevant to automated vehicles and examples and context for how a given prohibition may be violated by cyberattacking an automated vehicle:

(1) "Knowingly accesses and without permission alters, damages, deletes, destroys, or otherwise uses any data, computer, computer system, or computer network in order to either (A) devise or execute any scheme or artifice to defraud, deceive, or extort, or (B) wrongfully control or obtain money, property, or data." This type of prohibition would apply to ransomware attacks whereby the cyberattacker hacks into an automated vehicle to demand money.

(2) "Knowingly accesses and without permission takes, copies, or makes use of any data from a computer, computer system, or computer network, or takes or copies any supporting documentation, whether existing or residing internal or external to a computer, computer system, or computer network." A cyberattacker could copy data or otherwise make use of data to effectuate the cyberattack on an automated vehicle, all in violation of this prohibition.

(3) "Knowingly and without permission uses or causes to be used computer services." The CDAFA defines "computer services" as follows: "'Computer services' includes, but is not limited to, computer time, data processing, or storage functions, internet services, electronic mail services, electronic message services, or other uses of a computer, computer system, or computer network."[31] This type of statute could be used where

---

28. Cal. Penal Code § 502 (West).

29. Although conduct may be prohibited under the CDAFA and the federal Computer Fraud and Abuse Act described below, these acts are not identical, and even the similar provisions, such as the unlawful access, have different elements. *See* Facebook, Inc. v. Power Ventures, Inc., 844 F.3d 1058, 1069 (9th Cir. 2016) (stating that the CDAFA and the federal Computer Fraud and Abuse Act are different).

30. "A 'computer system' includes, without limitation, any such device or system that is located within, connected to, or otherwise integrated with, any motor vehicle as defined in Section 415 of the Vehicle Code." Cal. Penal Code § 502(a)(5) (West).

31. *Id.* § 502(a)(4).

a cyberattacker uses an automated vehicle's computer system for his or her own benefit. For example, the cyberattacker breaks into the system to get a "free" ride.

(4) "Knowingly accesses and without permission adds, alters, damages, deletes, or destroys any data, computer software, or computer programs which reside or exist internal or external to a computer, computer system, or computer network." A cyberattacker can commit this offense by affecting automated vehicle data.

(5) "Knowingly and without permission disrupts or causes the disruption of computer services or denies or causes the denial of computer services to an authorized user of a computer, computer system, or computer network." A cyberattacker could be charged with this offense when the cyberattacker interferes with a person's use of an automated vehicle. A cyberattacker could also violate this provision by using ransomware and denying a user access to the automated vehicle.

(7) "Knowingly and without permission accesses or causes to be accessed any computer, computer system, or computer network." This is the basic computer trespass offense. A cyberattacker who knowingly accesses an automated vehicle or connected infrastructure would violate this statute.

(8) "Knowingly introduces any computer contaminant into any computer, computer system, or computer network." The CDAFA defines "computer contaminant" as "any set of computer instructions that are designed to modify, damage, destroy, record, or transmit information within a computer, computer system, or computer network without the intent or permission of the owner of the information. They include, but are not limited to, a group of computer instructions commonly called viruses or worms, that are self-replicating or self-propagating and are designed to contaminate other computer programs or computer data, consume computer resources, modify, destroy, record, or transmit data, or in some other fashion usurp the normal operation of the computer, computer system, or computer network."[32] This type of offense would focus on whether the cyberattacker unleashed a virus, worm, or other contaminant to affect an automated vehicle or connected infrastructure.

(10) "Knowingly and without permission disrupts or causes the disruption of government computer services or denies or causes the denial of government computer services to an authorized user of a

32. *Id.* § 502(a)(12).

government computer, computer system, or computer network." This would prohibit cyberattacks on government-owned automated vehicles or connected infrastructure.

(11) "Knowingly accesses and without permission adds, alters, damages, deletes, or destroys any data, computer software, or computer programs which reside or exist internal or external to a public safety infrastructure computer system computer, computer system, or computer network." "Public safety infrastructure computer system" is defined as "any computer system, or part thereof, that is necessary for the health and safety of the public including computer systems owned, operated, or used by drinking water and wastewater treatment facilities, hospitals, emergency service providers, telecommunication companies, and gas and electric utility companies."[33] It seems like a connected infrastructure system, which should be considered necessary for the safety of the public, would be a public safety infrastructure computer system. This offense would be most relevant to any attacks on infrastructure that automated vehicles rely upon and would not be limited to infrastructure that is connected with automated vehicles.

(12) "Knowingly and without permission disrupts or causes the disruption of public safety infrastructure computer system computer services or denies or causes the denial of computer services to an authorized user of a public safety infrastructure computer system computer, computer system, or computer network." This would occur whenever a cyberattack interferes with connected infrastructure. A cyberattacker could also violate this provision by using ransomware and denying use of connected infrastructure until that ransom was paid.

(13) "Knowingly and without permission provides or assists in providing a means of accessing a computer, computer system, or public safety infrastructure computer system computer, computer system, or computer network in violation of this section." A person violates this prohibition by creating a means for the cyberattack or assisting with a cyberattack on connected infrastructure.

(14) "Knowingly introduces any computer contaminant into any public safety infrastructure computer system computer, computer system, or computer network." This could apply whenever someone uses worms or spyware or otherwise affects connected infrastructure.

---

33. *Id.* § 502(a)(7).

The first-time offender punishments for some of these offenses are not significant, at least in the context of automated vehicle cyberattacks. The paragraphs (1), (2), (4), (5), (10), (11), and (12) offenses are felonies or misdemeanors that can be punished with up to, as a felony, three years' imprisonment and a fine not to exceed $10,000. A paragraph (3) violation is a misdemeanor unless the victim suffered $5,000 in damages, injury results,[34] or the value of the computer services exceeds $950, in which case it is a felony carrying up to three years' imprisonment and a fine not to exceed $5,000. A person who violates paragraph (7) or (13) is subject to a fine up to $1,000. None of the punishments are increased by risking serious bodily harm or death, and none of the punishments are increased if such bodily injury or death results. Given that California has included hacking into motor vehicles in the CDAFA, California presumably knew of this risk and elected not to address it in its computer crime statute.

2. Noncomputer Crimes

Noncomputer crimes, with their more robust punishments, can be used to prosecute cyberattackers when certain harm or bad acts occur. The ability to prosecute a defendant for noncomputer crimes will depend upon the facts. A cyberattacker that moves an occupied automated vehicle against the will of its occupants could be charged with kidnapping.[35] A cyberattacker who uses or steals an automated vehicle by hacking into the vehicle can be charged with larceny or joyriding.[36] A cyberattacker who causes an occupied vehicle to crash, killing its occupants, could be charged with manslaughter or murder. Prosecutors should always consider noncomputer crimes in any prosecution for cyberattacks on automated vehicles.

B. Federal Regulation

Bills have been introduced into Congress to address cyberattacks on automated vehicles but not to create new crimes. The federal government has two

---

34. Injury "means any alteration, deletion, damage, or destruction of a computer system, computer network, computer program, or data caused by the access, or the denial of access to legitimate users of a computer system, network, or program." *Id.* § 502(a)(10).

35. For example, South Carolina's prohibition on kidnapping provides as follows: "Whoever shall unlawfully seize, confine, inveigle, decoy, kidnap, abduct or carry away any other person by any means whatsoever without authority of law, except when a minor is seized or taken by his parent, is guilty of a felony and, upon conviction, must be imprisoned for a period not to exceed thirty years unless sentenced for murder as provided in Section 16-3-20." S.C. Code Ann. § 16-3-910.

36. Joyriding offenses became popular alternatives to larceny when the state struggled to prove or could not prove an intent to permanently deprive the vehicle owner, as common law larceny had required. Joyriding offenses are defined differently among states, but the gist of the offense is that the defendant has taken or used an automobile without authority. *See, e.g.*, Mich. Comp. Laws § 750.414; Ohio Rev. Code Ann. § 2913.03 (West); S.C. Code Ann. § 16-21-60(A); Tenn. Code Ann. § 39-14-106 (West); W. Va. Code § 17A-8-4.

statutory regimes that may be applicable to cyberattacks: the Computer Fraud and Abuse Act (the CFAA) and the Stored Communications Act (the SCA).[37]

1. The Computer Fraud and Abuse Act

The CFAA is the federal government's principal regulation to criminalize computer crimes.[38] The CFAA criminalizes unauthorized access to computer systems and creates a private cause of action.[39] The CFAA, codified at 18 U.S.C. § 1030, creates seven different violations listed in sections (a)(1)–(a)(7).[40] One scholar has categorized these subsections as follows: section (a)(1): hacking to commit espionage; section (a)(2): hacking to obtain information; section (a)(3): hacking a federal government computer; section (a)(4): hacking to commit fraud; section (a)(5): hacking to commit damage; section (a)(6): trafficking in passwords; and section (a)(7): threats of hacking.[41] Most cyberattacks on automated vehicles will fall within one or more of sections (a)(2), (a)(5), and (a)(7). Generally, these sections are based on the defendant obtaining unauthorized access or exceeding authorized access to a computer system. This section begins by considering the meaning of "unauthorized access" and "exceeding authorized access." This section then briefly discusses the definitions of "computer" and "protected computer," terms frequently used in the CFAA.

a. "Unauthorized Access" and "Exceeds Authorized Access"

The CFAA's principal terms, "access" and "unauthorized access," are undefined. The CFAA defines "exceeds authorized access" as "access[ing] a computer with authorization and [using] such access to obtain or alter information in the computer that the accessor is not entitled so to obtain or alter."[42] One court has used Merriam-Webster's definition of "access," stating that the term

---

37. The federal government also has authority to apprehend and prosecute international cybercriminals to the extent that a cyberattack on automated vehicles or intelligent transportation systems crosses international borders. *See* 6 U.S.C. § 1531.

38. The United States Department of Justice publishes a helpful guide on the CFAA entitled "Prosecuting Computer Crimes." *Prosecuting Computer Crimes*, U.S. DEP'T OF JUSTICE, https://www.justice.gov/sites/default/files/criminal-ccips/legacy/2015/01/14/ccmanual.pdf (last visited May 19, 2020) [hereinafter DOJ Manual].

39. A private cause of action is created under 18 U.S.C. § 1030(g), which provides, in part, that "[a]ny person who suffers damage or loss by reason of a violation of this section may maintain a civil action against the violator to obtain compensatory damages and injunctive relief or other equitable relief." The creation of a private remedy extends to anyone "proximately harmed by unauthorized access" even if they were not the owner of the computer. Theofel v. Farey-Jones, 359 F.3d 1066, 1078 (9th Cir. 2004).

40. P.C. Yonkers, Inc. v. Celebrations the Party & Seasonal Superstore, LLC, 428 F.3d 504, 510 (3d Cir. 2005) ("18 U.S.C. § 1030 lists seven different types of conduct punishable by fines or imprisonment.").

41. JEFF KOSSEFF, CYBER SECURITY LAW 185–86 (2nd ed. 2020).

42. 18 U.S.C. § 1030(e)(6).

can mean, as a verb, "to gain access to" or, as a noun, "to exercise the freedom or ability to make use of something."[43] Unauthorized access, at least in the case of cyberattacks on automated vehicles, will be of less importance than in other contexts.[44]

b. "Computer" and "Protected Computer"

A computer is "an electronic, magnetic, optical, electrochemical, or other high speed data processing device performing logical, arithmetic, or storage functions, and includes any data storage facility or communications facility directly related to or operating in conjunction with such device, but such term does not include an automated typewriter or typesetter, a portable hand held calculator, or other similar device."[45] The computer systems that comprise a driving automation system would be computers for purposes of the CFAA.

A protected computer is defined, in pertinent part, as a computer "which is used in or affecting interstate or foreign commerce or communication, including a computer located outside the United States that is used in a manner that affects interstate or foreign commerce or communication of the United States."[46] A computer used in an automated vehicle would qualify itself as either being used in interstate commerce, due to being connected to other computer systems or the Internet,[47] or being part of and used in an automobile and on highways. Thus, computer systems used in automated vehicles would be "protected computers."

---

43. Am. Online, Inc. v. Nat'l Health Care Disc., Inc., 121 F. Supp. 2d 1255, 1273 (N.D. Iowa 2000) (internal quotation marks and citation omitted).

44. There is currently a circuit split about whether "unauthorized access" or "exceeds authorized access" includes violating company or website use restrictions. Compare these cases narrowly construing the CFAA, *United States v. Nosal*, 676 F.3d 854 (9th Cir. 2012); *WEC Carolina Energy Sols. LLC v. Miller*, 687 F.3d 199 (4th Cir. 2012); and *United States v. Valle*, 807 F.3d 508, 528 (2d Cir. 2015), with these cases broadly construing the CFAA, *United States v. John*, 597 F.3d 263 (5th Cir. 2010); *United States v. Rodriguez*, 628 F.3d 1258 (11th Cir. 2010); *Int'l Airport Ctrs., L.L.C. v. Citrin*, 440 F.3d 418 (7th Cir. 2006); and *EF Cultural Travel BV v. Explorica, Inc.*, 274 F.3d 577 (1st Cir. 2001). See generally KOSSEFF, *supra* note 41, at 173–84, for a detailed discussion of this circuit split.

45. 18 U.S.C. § 1030(e)(1).

46. *Id.* § 1030(e)(2).

47. *See, e.g.*, Patrick Patterson Custom Homes, Inc. v. Bach, 586 F. Supp. 2d 1026, 1032 (N.D. Ill. 2008) ("[A] computer that provides access to worldwide communications through applications accessible through the internet qualifies as a protected computer."); United States v. Drew, 259 F.R.D. 449, 458 (C.D. Cal. 2009) (holding that the interstate commerce requirement is satisfied "where contact is made between an individual's computer and an Internet website"); Paradigm Alliance, Inc. v. Celeritas Techs., LLC, 248 F.R.D. 598, 602 (D. Kan. 2008) ("[A] computer providing a 'web-based' application accessible through the internet would satisfy the 'interstate communication' requirement."); *see also* Reno v. Am. Civil Liberties Union, 521 U.S. 844, 849, 117 S. Ct. 2329, 2334, 138 L. Ed. 2d 874 (1997) ("The Internet is an international network of interconnected computers.").

c. Section (a)(2) Hacking to Obtain Information

Section (a)(2) criminalizes, in pertinent part, "[w]hoever . . . intentionally accesses a computer without authorization or exceeds authorized access, and thereby obtains . . . information from any protected computer."[48] There are four elements to this offense: (1) intentional access of a computer, (2) without or in excess of authorization, (3) to obtain information, and (4) from a protected computer. The section (a)(2) crime is a misdemeanor unless the offense was committed "for purposes of commercial advantage or private financial gain" or "in furtherance of any criminal or tortious act" in violation of federal or state laws, or if "the value of the information obtained exceeds $5,000."[49] The felony offense carries a fine and imprisonment up to five years.

The use of "intentional" is to limit the offense to only persons "whose conduct evinces a clear intent" to hack. The only intent required under the statute is the "intent to obtain unauthorized access [to] a protected computer."[50] Thus, a cyberattacker commits this offense even if he or she lacked "the intent to defraud in obtaining the information or that the information was used to any particular ends."[51] Congress used "intentional" to avoid prosecution of "mistaken, inadvertent, or careless" acts and "to focus Federal criminal prosecutions on those whose conduct evinces a clear intent to enter, without proper authorization, computer files or data belonging to another."[52]

The obtaining-of-information element can be satisfied without obtaining data in the normal sense of that term. One court examined a Senate Committee Report and concluded that the Senate Committee Report intended the statute to include the "mere observation of the data."[53] The Senate Committee Report

---

48. 18 U.S.C. § 1030(a)(2).

49. *Id.* § 1030(c)(2)(B). The furtherance of a federal or state criminal or tort law is very broad and has been held to include common law torts. *See* United States v. Powers, No. 8:09CR361, 2010 WL 1418172, at *2 (D. Neb. Mar. 4, 2010), *report and recommendation adopted*, No. 8:09CR361, 2010 WL 1418168 (D. Neb. Apr. 6, 2010) (stating that the tortious acts were invasion of privacy and intentional infliction of emotional distress). The DOJ Manual states that "the legislative history of section 1030 reveals that Congress intended the phrase to have the same meaning as identical language under the Wiretap Act, and cases construing that language hold the phrase encompasses state common law torts." DOJ Manual, *supra* note 38, at 20.

50. United States v. Willis, 476 F.3d 1121, 1125 (10th Cir. 2007).

51. *Id.*; *see also* KOSSEFF, *supra* note 41, at 185–86 (reviewing cases and concluding that "to satisfy the 'intentional' requirement of Section (a)(2), the government or civil plaintiff must establish that the defendant intentionally obtained the information through unauthorized hacking. However, they need not establish that the defendant intended to cause harm, defraud, or support the commission of another crime.").

52. United States v. Morris, 928 F.2d 504, 507–08 (2d Cir. 1991) (quoting S. REP. No. 99-432).

53. *Id.*

elaborates that "the term 'obtaining information' includes merely reading it. There is no requirement that the information be copied or transported."[54]

Section (a)(2) will be relevant whenever cyberattackers access automated vehicles and obtain or view their data. Data may reveal significant information about the target automated vehicle, its owners and users, or the coding or other intellectual property of the manufacturer. In any of these circumstances, the federal government could prosecute the hackers under section (a)(2) and the victims could bring private causes of action against the violators.

d. Section (a)(5) Hacking to Commit Damage

Section (a)(5) creates three different criminal acts:

> Whoever . . . (A) knowingly causes the transmission of a program, information, code, or command, and as a result of such conduct, intentionally causes damage without authorization, to a protected computer; (B) intentionally accesses a protected computer without authorization, and as a result of such conduct, recklessly causes damage; or (C) intentionally accesses a protected computer without authorization, and as a result of such conduct, causes damage and loss [commits a criminal offense.]

Because the section (a)(5) offenses relate to damage to the computer system and have aggravating penalties for bodily harm to individuals, these offenses will be particularly relevant for cyberattacks on automated vehicles. These offenses are broader than mere damage to an automated vehicle and include harm to the computer system itself. A single cyberattack could be a violation of more than one of these offenses.

(i) Knowingly Transmits and Intentionally Damages

Section (a)(5)(A) provides a distinct offense that requires proof of five elements: the defendant (1) knowingly caused (2) the transmission of a program, information, code, or command, and (3) such transmission intentionally caused (4) damage to a protected computer (5) without authorization. This offense has two specific mens rea requirements: the defendant must knowingly make a transmission, and the defendant must have the intent to cause damage to a protected computer. Another unique aspect of this offense is that it lacks an "unauthorized access" or "exceeds authorized access" requirement, and the "without authorization" provision relates to the damages.[55]

---

54. Senate Judiciary Committee Report, S. Rep. No. 104-357—The National Information Infrastructure Protection Act of 1995, at 7 (1996).

55. Kosseff, *supra* note 41, at 198; Charles Doyle, Cong. Research Serv., RS20830, Cybercrime: An Overview of the Federal Computer Fraud and Abuse Statute and Related Federal Criminal Laws 33 (2014), *available at* https://fas.org/sgp/crs/misc/97-1025.pdf.

The United States Department of Justice broadly interprets the "transmission of a program, information, code, or command" element to cover "all transmissions that are capable of having an effect on a computer's operation," including worms, software commands, and network packets.[56] The Seventh Circuit has called into question a broad interpretation of transmission in a case where one of the alleged transmissions was erasing a computer file: "Pressing a delete or erase key in fact transmits a command, but it might be stretching the statute too far (especially since it provides criminal as well as civil sanctions for its violation) to consider any typing on a computer keyboard to be a form of 'transmission' just because it transmits a command to the computer."[57] Courts have held that installing a program, whether electronically or physically through the insertion of a disk, is a transmission.[58]

The defendant need not have intended to cause the ultimate harm, whatever that harm may be. The defendant merely needed to intend that the transmission would cause damage to a protected computer. A defendant acts with intent when he or she acts with the "conscious purpose of causing damage" to a protected computer.[59]

Damage means "any impairment to the integrity or availability of data, a program, a system, or information."[60] In a civil suit, the Sixth Circuit used a dictionary definition to determine what "impairment," "integrity," and "availability" meant in the statute and concluded that "a transmission that weakens a sound computer system—or, similarly, one that diminishes a plaintiff's ability to use data or a system—causes damage."[61] Accordingly, damage to a protected computer does not require physical damage to the computer.

---

56. DOJ Manual, *supra* note 38, at 37.

57. Int'l Airport Ctrs., L.L.C. v. Citrin, 440 F.3d 418, 419 (7th Cir. 2006).

58. *Id.* (installing a secure-erasure program on a computer is a transmission); United States v. Sullivan, 40 F. App'x 740, 741 (4th Cir. 2002) (installing a computer code (dubbed a logic bomb) to disable communication functions in handheld computers was a transmission).

59. Pulte Homes, Inc. v. Laborers' Int'l Union of N. Am., 648 F.3d 295, 303 (6th Cir. 2011) (holding that "to satisfy its pleading burden, Pulte must allege that LIUNA acted with the conscious purpose of causing damage (in a statutory sense) to Pulte's computer system—a standard that does not require perfect knowledge"); United States v. Carlson, 209 F. App'x 181, 184 (3d Cir. 2006) ("Although the statute itself does not define 'intentionally,' this Court has defined it in the criminal context as performing an act deliberately and not by accident. Accordingly, the Government was required to prove at trial that Carlson deliberately caused an impairment to the integrity or availability of data, a program, a system, or information." (internal citation omitted)).

60. 18 U.S.C. § 1030(e)(8).

61. *Pulte Homes, Inc.*, 648 F.3d at 301; *see also* Condux Int'l, Inc. v. Haugum, No. CIV 08-4824 ADM/JSM, 2008 WL 5244818, at *8 (D. Minn. Dec. 15, 2008) ("The 'damage' contemplated by subsection (a)(5)(A)(i) requires some diminution in the completeness or useability of data or information on a computer system." (internal quotation marks and citation omitted)); Becker v. Toca, No. CIV. A. 07-7202, 2008 WL 4443050, at *5 (E.D. La. Sept. 26, 2008) ("Error messages and slow processing constitute impairments to the integrity or availability of data.").

A section (a)(5)(A) offense could occur with automated vehicles whenever a cyberattacker transmits any program into an automated vehicle to affect the system or the driving behavior of the vehicle. This offense would also be applicable when a cyberattacker uses malware or worms to affect automated vehicles.

(ii) Intentionally Accesses and Recklessly Damages

In contrast to the section (a)(5)(A) offense, the section (a)(5)(B) offense focuses on intentional access and reckless damage. There are four components to this offense: (1) intentional access of (2) a protected computer (3) without authorization, and (4) the intentional access recklessly causes damage. Intentional access, defined and described above, will almost certainly be satisfied in automated vehicle cyberattacks. Cyberattackers who hack into automated vehicles will have the intention of accessing the vehicle's computer system. The major element in the section (a)(5)(B) offense is that the intentional access recklessly causes damage—the hacker did not directly intend to cause damage, but his or her conduct was such that damage recklessly occurs. "Recklessly" is undefined under the CFAA, but courts generally define "recklessness" to mean a conscious disregard of a substantial and unjustifiable risk.[62]

A section (a)(5)(B) offense can occur with automated vehicles whenever a cyberattacker intentionally accesses the driving automation system and recklessly affects data. A situation where this could occur is where a person hacks into a vehicle to transmit ransomware—which is also a separate offense under the CFAA—and, while not intending to affect data unless the person does not pay, affects data of the automated vehicle. When the damage to data is negligent but not reckless, the appropriate offense would be under section (a)(5)(C).

(iii) Intentionally Accesses and Causes Damage and Loss

Section (a)(5)(C) requires proof that the defendant (1) intentionally accessed (2) a protected computer (3) without authorization, and (4) the access caused damage and loss. There are two differences between this offense and the section (a)(5)(B) offense. The section (a)(5)(B) offense requires recklessness to cause the damage, while this offense only requires negligence to cause the damage. Additionally, a section (a)(5)(B) offense only requires damage, while this offense requires damage and loss. Loss means "any reasonable cost to any victim, including the cost of responding to an offense, conducting a damage assessment, and restoring the data, program, system, or information to its condition prior to the offense, and any revenue lost, cost incurred, or other consequential

62. See Chapter 19 for a discussion of recklessness.

damages incurred because of interruption of service."[63] Thus, mere impairment of data or a protected system is insufficient if there is not a reasonable cost to the victim.

(iv) Penalties for Section (a)(5) Offenses

The penalty for committing any of the three offenses, for a first time, is a misdemeanor. Section (a)(5)(A) and (a)(5)(B) offenses can become felonies if certain aggravating factors exist. A section (a)(5)(C) offense is always a misdemeanor for first-time offenders. The aggravating factors to increase the section (a)(5)(A) and (a)(5)(B) offenses to felonies are (1) loss to any one person in excess of $5,000; (2) modification or impairments, or potential thereof, of medical examination, treatment, or care; (3) physical injury to any person; (4) threat to public health or safety; (5) damage to certain U.S. government computers; and (6) damage to more than ten protected computers within one year.[64] The penalty for a felony section (a)(5)(A) offense can be imprisonment up to ten years and a fine, and the penalty for a felony section (a)(5)(B) offense can be imprisonment up to five years and a fine. These penalties increase substantially if the defendant attempts to cause or knowingly or recklessly causes serious bodily injury (imprisonment up to 20 years and a fine) or death (imprisonment up to life and a fine) in a section (a)(5)(A) offense.[65]

Cyberattacks on automated vehicles, connected infrastructure, or other components of the transportation will almost certainly cause a threat to public safety or physical injury. Thus, it seems likely that almost any section (a)(5)(A) or (a)(5)(B) offense involving automated vehicles will be a felony. If the cyberattacker attempts to cause an occupied automated vehicle to crash and the elements of a section (a)(5)(A) offense are met, the defendant could receive a sentence of up to 20 years or life imprisonment.

e. Section (a)(7) Threats of Hacking

Section (a)(7) criminalizes the following:

> Whoever . . . with intent to extort from any person any money or other thing of value, transmits in interstate or foreign commerce any communication containing any— (A) threat to cause damage to a protected computer; (B) threat to obtain information from a protected computer without authorization or in excess of authorization or to impair the confidentiality of information obtained from a protected computer without authorization or by exceeding authorized access; or (C) demand or request for money or

63. 18 U.S.C. § 1030(e)(11).
64. *Id.* § 1030(c)(4)(A).
65. *Id.* § 1030(c)(4)(E), (F).

> other thing of value in relation to damage to a protected computer, where such damage was caused to facilitate the extortion [commits a criminal offense.]

This offense has four elements: (1) intent to extort from any person (2) money or anything of value (3) via transmission in interstate commerce of a communication (4) containing a threat to cause damage to a protected computer. Section (a)(7) is a cyberextortion statute that can apply when cyberattackers use ransomware attacks on automated vehicles.

The key to determining whether section (a)(7) is applicable will be based on whether the cyberattacker has a specific intent to extort the person for money. The obvious examples would be a cyberattacker locking down a parked automated vehicle until the owner paid a ransom or threatening to take control of a moving automated vehicle unless a ransom is paid. In either situation, whether the defendant received money or intended to carry out the threats is irrelevant.[66]

The communication could be "by mail, a telephone call, electronic mail, or through a computerized messaging service."[67] With automated vehicles, the likely communication would be by telephone call, an electronic message on the infotainment center or an application used to summon the ride, or voice-over control using the vehicle's speakers. These communications would be considered in interstate commerce because the cyberattacker uses the Internet or telecommunications to communicate the threat.

Damage, with its broad definition, would include any impairment or unavailability of data. Given that the statute only requires a threat of impairment, any threat to alter or affect the driving automation system would satisfy that requirement, as would a threat to crash an automated vehicle.

The penalty for a section (a)(7) offense is a fine and imprisonment up to five years for a first offense. In many cases, a violation of section (a)(7) could also be a violation of section (a)(5) with its heightened penalties. This certainly would be true when, under section (a)(7)(A), the defendant follows through with the damage to the protected computer.

2. Section 2701 of the Stored Communications Act

Section 2701 of the SCA is an antihacking statute that can apply in certain circumstances but is generally more limited in its scope and difficult to establish

---

66. This statute is similar to the federal extortion statute in 18 U.S.C. § 1951, which has a long historical interpretation. *See* DOJ Manual, *supra* note 38, at 53.

67. S. Rep. No. 104-357, *supra* note 54.

a violation under the CFAA.[68] Like the CFAA, section 2701 is both civil and criminal in scope.[69] Section 2701 creates two separate offenses:

> [W]hoever—(1) intentionally accesses without authorization a facility through which an electronic communication service is provided; or (2) intentionally exceeds an authorization to access that facility; and thereby obtains, alters, or prevents authorized access to a wire or electronic communication while it is in electronic storage in such system shall be punished as provided in subsection (b) of this section.[70]

A person guilty of this offense can be fined and imprisoned up to five years if the offense is "in furtherance of any criminal or tortious act" in violation of federal or state law. The significant question is whether a cyberattack on an automated vehicle would fall within the scope of section 2701.

The key term of section 2701 is "facility," an undefined term. Courts have considered whether a personal device, such as a cell phone or a computer, is a facility for purposes of section 2701. These courts have overwhelmingly concluded that these personal devices are not facilities.[71] These courts would likely find that automated vehicles are not facilities. Courts have ruled that section 2701 applies to electronic communication services. An electronic communication service is "any service which provides to users thereof the ability to send or receive wire or electronic communications."[72] Based on this interpretation, section 2701 applies only to a limited circumstance: "[A]ccessing a computer without authorization or exceeding authorization is an offense when the computer is acting as an [electronic communication service] and the person 'obtains, alters, or prevents authorized access' to a file in 'electronic storage.'"[73] Accordingly, section 2701 would not apply to a cyberattack on an automated vehicle.

---

68. *See* Sherman & Co. v. Salton Maxim Housewares, Inc., 94 F. Supp. 2d 817, 820 (E.D. Mich. 2000) ("The general purpose of the ECPA was to create a cause of action against computer hackers (e.g., electronic trespassers)." (internal quotation marks and citation omitted)).

69. *See* 18 U.S.C. § 2707.

70. *Id.* § 2701(a).

71. *See, e.g.*, Garcia v. City of Laredo, Tex., 702 F.3d 788, 793 (5th Cir. 2012) (holding that text messages and photos stored on a phone are outside the scope of the SCA); United States v. Steiger, 318 F.3d 1039, 1049 (11th Cir. 2003) ("The SCA . . . does not appear to apply to the source's hacking into Steiger's computer to download images and identifying information stored on his hard-drive."); *In re* Google Inc. Cookie Placement Consumer Privacy Litig., 806 F.3d 125, 148 (3d Cir. 2015) (an individual personal computing device is not a facility); Morgan v. Preston, No. 3:13-00403, 2013 WL 5963563, at *6 (M.D. Tenn. Nov. 7, 2013) (holding that "plaintiff's personal computer is not covered by the SCA"); *In re* iPhone Application Litig., 844 F. Supp. 2d 1040, 1057 (N.D. Cal. 2012) (an "individual's computer, laptop, or mobile device" is not a facility); Bailey v. Bailey, No. 07-11672, 2008 WL 324156, at *6 (E.D. Mich. Feb. 6, 2008) ("Stored Communications Act protection does not extend to emails and messages stored only on Plaintiff's personal computer.").

72. 18 U.S.C. § 2510(15).

73. Orin S. Kerr, *A User's Guide to the Stored Communications Act, and a Legislator's Guide to Amending It*, 72 GEO. WASH. L. REV. 1208, 1239 (2004) (quoting 18 U.S.C. § 2701).

Chapter 21

The Right to Seize and Search Highly Automated Vehicles and Their Occupants

Three factors combined in a perfect storm to make traffic stops one of the police's most effective ways to prevent drug and firearm crimes. The first factor is rather obvious. Drug trafficking at any scale requires vehicles to transport drugs quickly, and those who illegally possess firearms typically keep those firearms with them. The second factor is how heavily driving is regulated, making it nearly impossible for any driver to comply with every rule of the road at all times. The multitude of rules of the road is magnified when many of these rules are subjective, and police only need probable cause to stop a vehicle for violating an objective or subjective traffic offense. The near impossibility of a human driver complying with all rules of the road at all times may not have created such a significant opportunity had the United States Supreme Court limited traffic stops in some meaningful way. The third factor is Supreme Court precedent that has often failed to place effective limitations on traffic stops, even those alleged to have been pretextual.[1] A unanimous Supreme Court stated that "[s]ubjective intentions play no role in ordinary, probable-cause Fourth Amendment analysis."[2] The Supreme Court held that if an officer has probable cause to believe a driver violated a traffic law, then the traffic stop is reasonable under the Fourth Amendment.

1. But see *Rodriguez v. United States*, 575 U.S. 348, 135 S. Ct. 1609, 191 L. Ed. 2d 492 (2015), and *Arizona v. Gant*, 556 U.S. 332, 129 S. Ct. 1710, 173 L. Ed. 2d 485 (2009), for two recent cases where the Supreme Court constrained police authority to prolong traffic stops or to search incident to arrest for traffic violations.

2. Whren v. United States, 517 U.S. 806, 813, 116 S. Ct. 1769, 135 L. Ed. 2d 89 (1996).

Automated vehicles will do nothing to impact the first factor. Criminals will use automated vehicles to transport drugs, and users and occupants will still carry their firearms. The third factor—existing Supreme Court jurisprudence on traffic stops—is unlikely to change. Automated vehicles will, however, significantly alter the second factor. Automated vehicles will have the ability to know and remember every traffic law and the ability to obey those laws by being able to control the mechanics of the automated vehicle perfectly. Because of their ability to comply with traffic laws, traffic stops should become rare. Indeed, automated vehicles' near-perfect compliance with traffic codes will lower the transaction costs of trafficking drugs and may lead to more drug trafficking.

Automated vehicles will have significantly more impact on Fourth Amendment jurisprudence than merely removing the main arrow from the quiver of the police. The Fourth Amendment issues surrounding level 1 and level 2 automated vehicles are almost identical to those involving nonautomated vehicles, and those levels of automation will not create new ripples to existing jurisprudence. This chapter will focus on the Fourth Amendment issues surrounding vehicles operated by automated driving systems and specifically focus on level 4 and level 5 automated vehicles.

The Fourth Amendment contains two clauses. The first guarantees "[t]he right of the people to be secure in their persons, houses, papers, and effects, against unreasonable searches and seizures."[3] The second includes the requirements—probable cause, a particular description of the target of the search or seizure, and oath or affirmation—for a warrant. These two clauses, when read together, effectively prohibit warrantless searches and seizures unless they fall within an exception to the warrant requirement.[4] Exceptions to the warrant

---

3. U.S. Const. amend. IV. A vehicle is an "effect." United States v. Jones, 565 U.S. 400, 404, 132 S. Ct. 945, 181 L. Ed. 2d 911 (2012) ("It is beyond dispute that a vehicle is an 'effect' as that term is used in the Amendment." (citation omitted)).

4. *See* Katz v. United States, 389 U.S. 347, 357, 88 S. Ct. 507, 19 L. Ed. 2d 576 (1967) ("[S]earches conducted outside the judicial process, without prior approval by judge or magistrate, are per se unreasonable under the Fourth Amendment—subject only to a few specifically established and well-delineated exceptions."); Riley v. California, 573 U.S. 373, 382, 134 S. Ct. 2473, 189 L. Ed. 2d 430 (2014) ("In the absence of a warrant, a search is reasonable only if it falls within a specific exception to the warrant requirement."); Brigham City, Utah v. Stuart, 547 U.S. 398, 403, 126 S. Ct. 1943, 164 L. Ed. 2d 650 (2006) ("[B]ecause the ultimate touchstone of the Fourth Amendment is 'reasonableness,' the warrant requirement is subject to certain exceptions." (citations omitted)); Vernonia Sch. Dist. 47J v. Acton, 515 U.S. 646, 653, 115 S. Ct. 2386, 132 L. Ed. 2d 564 (1995) ("Where a search is undertaken by law enforcement officials to discover evidence of criminal wrongdoing, this Court has said that reasonableness generally requires the obtaining of a judicial warrant. . . . Warrants cannot be issued, of course, without the showing of probable cause required by the Warrant Clause. But a warrant is not required to establish the reasonableness of all government searches; and when a warrant is not required (and the Warrant Clause therefore not applicable), probable cause is not invariably required either.").

requirement include the automobile exception, exigent circumstances, search incident to arrest, plain view, and consent.

There are four steps to analyzing a Fourth Amendment search or seizure. The first step is to determine whether a search or seizure occurred for which the defendant has "standing"—the short (and perhaps erroneous) name for determining whether a defendant has a legitimate interest protected by the Fourth Amendment. The second step is to determine whether the search or seizure was committed by a state actor or an actor acting on behalf of a state actor. An automated vehicle manufacturer may search data of an automated vehicle, but that is not a search protected by the Fourth Amendment, unless the search was at the request of law enforcement. The third step is to determine whether the search or seizure happened pursuant to a warrant. If police had obtained a warrant, then the focus is on whether there was probable cause for a judge to issue the warrant and whether the search or seizure was within the scope of the warrant. If there was not a warrant, then the fourth step is to determine whether an exception to the warrant requirement applied.

This chapter focuses on the Fourth Amendment clause in the United States Constitution. State constitutions or laws can be more restrictive than the Fourth Amendment, and attorneys need to review whether state law provides further protections. State restrictions on searches and seizures are beyond the scope of this book.

I. THE RIGHT TO SEIZE A HIGHLY AUTOMATED VEHICLE AND ITS OCCUPANTS

This section introduces seizures of vehicles and then identifies specific issues regarding seizures of highly automated vehicles. "A person is seized by the police and thus entitled to challenge the government's action under the Fourth Amendment when the officer, by means of physical force or show of authority, terminates or restrains his freedom of movement, through means intentionally applied."[5] "A 'seizure' of property occurs when there is some meaningful interference with an individual's possessory interests in that property."[6] Traffic stops are seizures under the Fourth Amendment.[7]

5. Brendlin v. California, 551 U.S. 249, 254, 127 S. Ct. 2400, 168 L. Ed. 2d 132 (2007) (internal quotation marks and citations omitted).

6. United States v. Jacobsen, 466 U.S. 109, 113, 104 S. Ct. 1652, 80 L. Ed. 2d 85 (1984).

7. *Brendlin*, 551 U.S. at 255. Typically, questions concerning whether a seizure has occurred with respect to vehicles involve police questioning on public buses. *See, e.g.*, Florida v. Bostick, 501 U.S. 429, 111 S. Ct. 2382, 115 L. Ed. 2d 389 (1991); United States v. Drayton, 536 U.S. 194, 122 S. Ct. 2105, 153 L. Ed. 2d 242 (2002). Neither of these cases involved a traffic stop of a bus.

Generally, seizure cases can be broken down into two types of inquiries: the initial stop and the duration of the stop.[8] The first inquiry is about the suspicion required before stopping a vehicle. Stops to arrest or cite the driver for a criminal infraction require probable cause.[9] Investigative stops to determine whether a crime has been committed require police to have an "articulable and reasonable suspicion" before making the stop.[10] The reasonable suspicion standard requires police to have "specific and articulable facts which, taken together with rational inferences from those facts, reasonably warrant" the traffic stop.[11] As to the second inquiry, "the tolerable duration of police inquiries in the traffic-stop context is determined by the seizure's 'mission'—to address the traffic violation that warranted the stop and attend to related safety concerns."[12] "Authority for the seizure thus ends when tasks tied to the traffic infraction are—or reasonably should have been—completed."[13] Police typically run into problems with the duration of traffic stops when they use stops to investigate other crimes.[14]

This section focuses on six issues that can arise with seizing automated vehicles: (1) standing, (2) noncrime seizures, (3) checkpoints, (4) impoundments of automated vehicles, (5) remote stops of automated vehicles by police, and (6) duration of traffic stops for automated vehicles.

A. Standing

Automated vehicle occupants will have standing to challenge a seizure. In *Brendlin v. California*, the Supreme Court held that police seize passengers of a vehicle when they conduct a traffic stop.[15] Thus, even if the automated driving system is considered the driver, occupants will have standing to challenge the seizure.

A new standing arises with unoccupied automated vehicles. Who can challenge the seizure of an unoccupied automated vehicle? When police stop an

---

8. United States v. Sharpe, 470 U.S. 675, 682, 105 S. Ct. 1568, 84 L. Ed. 2d 605 (1985) ("In Terry v. Ohio, 392 U.S. 1, 88 S.Ct. 1868, 20 L.Ed.2d 889 (1968), we adopted a dual inquiry for evaluating the reasonableness of an investigative stop. Under this approach, we examine 'whether the officer's action was justified at its inception, and whether it was reasonably related in scope to the circumstances which justified the interference in the first place.'").

9. Whren v. United States, 517 U.S. 806, 810 (1996) ("As a general matter, the decision to stop an automobile is reasonable where the police have probable cause to believe that a traffic violation has occurred.").

10. Delaware v. Prouse, 440 U.S. 648, 663, 99 S. Ct. 1391, 59 L. Ed. 2d 660 (1979).

11. Terry v. Ohio, 392 U.S. 1, 21, 88 S. Ct. 1868, 20 L. Ed. 2d 889 (1968).

12. Rodriguez v. United States, 575 U.S. 348, 354, 135 S. Ct. 1609, 191 L. Ed. 2d 492 (2015) (internal citations omitted).

13. *Id.*

14. *See, e.g., Rodriguez*, 575 U.S. 348; Illinois v. Caballes, 543 U.S. 405, 125 S. Ct. 834, 160 L. Ed. 2d 842 (2005).

15. 551 U.S. 249, 257–58, 127 S. Ct. 2400, 168 L. Ed. 2d 132 (2007).

unoccupied automated vehicle, two different principles may be at play. The first principle is that no person is being detained, though the traffic stop may be depriving someone of his or her ability to proceed with an itinerary.[16] The second principle is that the traffic stop is depriving someone of his or her possessory interest in the vehicle.[17] Under these principles, those who have a current possessory interest in the vehicle should have standing to challenge an unlawful seizure of an automated vehicle. Courts will need to wrestle with whether a future occupant has a possessory interest in an automated vehicle such that the future occupant could challenge the traffic stop. What if a user summons an automated vehicle for a ride? In these circumstances, presumably, the stop would deprive this person of a possessory interest and may prevent the person from proceeding with his or her itinerary. What if, instead, a user did not summon the vehicle but someone else did or a friend sent his or her automated vehicle to the person?[18] At what point does this person's possessory interest vest for Fourth Amendment purposes? Does the delay or deprivation of the automated vehicle give this person standing? The analogy here would likely be the same as the standing rights or lack thereof of future vehicle users.

B. Noncrime Seizures

Police may believe in good faith that a traffic violation or a crime has occurred, but the vehicle's data may reveal that there was not a traffic violation or that the police's belief may be erroneous on further inspection. An example of the latter may be a police officer who mistakenly believes a crime is being committed because the officer is unaware that the vehicle is highly automated . For example, a police officer could stop a level 5 automated vehicle with a person sleeping in the traditional driver seat. In these situations, where there is a disconnect between officer perceptions and reality, defendants may claim that the officer violated the Fourth Amendment. The reason why noncrime seizures matter is that they may be the first (or perhaps only) way to challenge the admission of evidence uncovered after that stop. Because neither reasonable suspicion nor

16. *See* United States v. Place, 462 U.S. 696, 708, 103 S. Ct. 2637, 77 L. Ed. 2d 110 (1983) (stating that "in the case of detention of luggage within the traveler's immediate possession, the police conduct intrudes on both the suspect's possessory interest in his luggage as well as his liberty interest in proceeding with his itinerary"). Although *Place* dealt with a direct interference with a traveler's "liberty interest in proceeding with this itinerary," given that the traveler accompanied his luggage, a traffic stop of an automated vehicle will also interfere with a traveler's or multiple travelers' itinerary—though in a less direct way than in *Place*.

17. *See, e.g.*, Soldal v. Cook Cnty., Ill., 506 U.S. 56, 61, 113 S. Ct. 538, 121 L. Ed. 2d 450 (1992); United States v. Jacobsen, 466 U.S. 109, 124 (1984); *Place*, 462 U.S. at 708.

18. A drug deal could involve one person deploying an unoccupied automated vehicle with the drugs to another person who then deploys the unoccupied automated vehicle to the ultimate recipient. If the vehicle is seized before reaching the drug recipient, who could challenge the seizure of the vehicle?

probable cause requires absolute truth, conceivably police may contend that an officer's stop, if supported by probable cause or reasonable suspicion, can be reasonable under the Fourth Amendment even though no crime was committed.[19] Probable cause or reasonable suspicion to stop an automated vehicle may be different than that for a human-driven vehicle. For example, an officer seeing a person asleep at the steering wheel today may be sufficient evidence to stop the vehicle. With the advent of highly automated vehicles, the officer may need to couple that observation with the knowledge that the vehicle model is nonautomated or with other observations, such as swerving.

Practically speaking, at least for traffic stops, it may be difficult for an officer to establish that he or she had probable cause to stop a vehicle for a traffic violation if vehicle data shows that the vehicle did not commit the traffic offense.[20] In cases where the defendant believes data of the vehicle would establish that a traffic violation did not occur, the defendant would need to hire an expert to obtain and read the vehicle's data, establish the reliability of the vehicle's data, and recreate the vehicle's action to establish that the police officer's beliefs were erroneous such that he or she did not have probable cause to stop the vehicle.

C. Checkpoints

If highly automated vehicles succeed in effectively ending traffic stops, police will need to find new ways to fight drug and firearm crimes. The use of checkpoints for the specific purpose of investigating drug and firearm crimes may very well be foreclosed by *City of Indianapolis v. Edmond*,[21] where the Supreme Court found unconstitutional a "highway checkpoint program whose primary purpose is the discovery and interdiction of illegal narcotics."[22] The Supreme Court declined to forbid the use of drug detection dogs at checkpoints or checkpoints that have a secondary purpose of detecting narcotics.[23] However, this raises an important question for police stopping unoccupied automated vehicles: what would the legitimate purpose be for a checkpoint that stops an empty automated vehicle? Presumably, police would be unable to check for a driver's license or registration (other than to look at the license plate, which

19. *See* Illinois v. Rodriguez, 497 U.S. 177, 185, 110 S. Ct. 2793, 111 L. Ed. 2d 148 (1990) ("It is apparent that in order to satisfy the 'reasonableness' requirement of the Fourth Amendment, what is generally demanded of the many factual determinations that must regularly be made by agents of the government—whether the magistrate issuing a warrant, the police officer executing a warrant, or the police officer conducting a search or seizure under one of the exceptions to the warrant requirement—is not that they always be correct, but that they always be reasonable.").

20. *See* Scott v. Harris, 550 U.S. 372, 380–81, 127 S. Ct. 1769, 167 L. Ed. 2d 686 (2007) (stating that eyewitness testimony should be disregarded when a videotape shows that the eyewitness testimony is a "visible fiction").

21. 531 U.S. 32 (2000).

22. *Id.* at 34.

23. *Id.* at 47 n.2.

may not take much time).[24] The problem with these checkpoints may not be the initial stop, which may be permissible, but rather the prolonged duration of the stop once the impossibility of finding information from the automated vehicle becomes evident.

D. Impounding an Automated Vehicle

The ability to conduct an inventory search before impounding a vehicle has been an effective means for police officers to search vehicles following an arrest of the driver.[25] These two acts—the inventory search and impounding vehicles—are distinct events for Fourth Amendment purposes. Following an arrest of the driver or for other reasons, such as following an accident or in police's "community caretaking" role,[26] police can exercise discretion in deciding whether to impound a vehicle "so long as that discretion is exercised according to standard criteria and on the basis of something other than suspicion of evidence of criminal activity."[27] Inventory searches conducted pursuant to police procedures serve specific functions, such as protecting the owner's property, guarding police against dangers, and protecting police from claims of lost, stolen, or vandalized property.[28]

Impoundment procedures and inventory searches may gain prominence with highly automated vehicles and could be a way for police to regain opportunities to prevent drug trafficking. When an unoccupied highly automated vehicle commits a traffic violation, police policies may allow them to impound that vehicle (and presumably conduct an inventory search). These policies make sense, given that the alternative would be to allow the automated vehicle to continue driving in its rule-breaking form or to leave it alongside the road. Impounding the vehicle could also allow the manufacturer to update the system to prevent future infractions and, depending upon who or what is responsible for the traffic violation, allow the police to interact with (and possibly ticket) the owner. A future pastime for automated vehicle owners may be getting their vehicles out of police timeout.

With occupied highly automated vehicles, police policies will need to determine whether the occupants would have the right to drive (if possible) or tow the vehicle away themselves. The fact that the presumption may very well be that no occupant is responsible (at least morally) for a traffic offense may mean

---

24. Delaware v. Prouse, 440 U.S. 648, 663 (1979) (suggesting that "[q]uestioning of all oncoming traffic at roadblock-type stops is one possible alternative" to check for driver's licenses).

25. The right to conduct an inventory search is not dependent upon the vehicle's impoundment at a police facility.

26. S. Dakota v. Opperman, 428 U.S. 364, 368–69, 96 S. Ct. 3092, 49 L. Ed. 2d 1000 (1976) (describing police's "community caretaking functions").

27. Colorado v. Bertine, 479 U.S. 367, 375, 107 S. Ct. 738, 93 L. Ed. 2d 739 (1987).

28. *Id.* at 372.

that the Fourth Amendment could prohibit arbitrary decisions to impound these automated vehicles.[29] If the police were to impound the occupied automated vehicle, it seems as though the Fourth Amendment may require police to let the occupants take their belongings—the impounding of the belongings itself would be a separate seizure in these instances.

E. Remote Control by Police

Police officer control over an automated vehicle will occur.[30] This control could include control of an automated vehicle by a police officer through towing the vehicle or inputting a destination for the vehicle.[31] Police currently tow vehicles under their community caretaking role or following traffic stops. Another method and more intrusive and controversial means of control would be if police could stop an automated vehicle through a vehicle-to-vehicle communication or a remote "stop button." Regardless of the merits of such a system or whether a state could even compel such a vehicle design,[32] remote access to automated vehicles would need to be assessed under the Fourth Amendment.

The government would argue that because police currently have broad authority to stop a vehicle that police have probable cause to believe committed a traffic offense, police have broad authority to stop automated vehicles and that authority could extend to a remote stop. The government would argue that there is nothing inherently unreasonable about this form of seizure. The Supreme Court has stated that "[i]n determining the reasonableness of the manner in which a seizure is effected, we must balance the nature and quality of the intrusion on the individual's Fourth Amendment interests against the importance of the governmental interests alleged to justify the intrusion."[33] It seems that the government could put forth legitimate justifications for a remote stop, such as the need to stop unoccupied automated vehicles or the need to stop occupied automated vehicles when occupants may be preoccupied (for example, asleep). Thus, the government would argue that as long as the remote

---

29. Arbitrary impounding of vehicles may also be a taking under the Fifth Amendment.

30. See Elizabeth E. Joh, *Automated Seizures: Police Stops of Self-Driving Cars*, 94 N.Y.U. L. Rev. Online 113 (2019), https://nyulawreview.org/wp-content/uploads/2019/10/Joh-fin.pdf, for a discussion of this issue.

31. *See Opperman*, 428 U.S. at 368–69 ("Police will also frequently remove and impound automobiles which violate parking ordinances and which thereby jeopardize both the public safety and the efficient movement of vehicular traffic. The authority of police to seize and remove from the streets vehicles impeding traffic or threatening public safety and convenience is beyond challenge.").

32. See generally Part 2 for a discussion of the limitations of states to regulate the design of motor vehicles. A mandated design by a state that requires such remote access would impact the design of motor vehicles and possibly create additional security vulnerabilities by creating a new access point to the vehicle. Because of the impact on vehicle design, an avenue to attack such a system may be to claim that the law is preempted.

33. Scott v. Harris, 550 U.S. 372, 383 (2007) (internal quotation marks and citation omitted).

stop does not create a significant risk of harm—and even if it does[34]—the use of a remote stop would be a reasonable way to seize an automated vehicle.

Defendants would counter with the intrusiveness of the police effectively having access to their automated vehicles through the remote access and challenge the mechanism of the remote access. A vehicle stop where the driver can voluntarily stop on seeing blue lights is far less coercive than a vehicle stop where the police (perhaps with no warning) stop an automated vehicle. A person may believe the automated vehicle is being hacked. This remote stop ability may turn upon what information about vehicles and the roadway police need to have to utilize the remote stop. There is a significant difference between a remote access system that requires the government to know an automated vehicle's whereabouts at all times and one that can be effected by a vehicle-to-vehicle communication between the police officer and the automated vehicle.

F. Duration of Stop

Automated vehicles may change the calculus for the duration of a stop, depending upon how police decide to ticket violators. A typical traffic stop with a human-driven vehicle includes the officer blue-lighting the vehicle; the vehicle pulling to the side of the road; an interaction between the officer and the driver and passengers of the vehicle; the officer obtaining the driver's license, registration, and proof of insurance; a record check on the driver and possibly passengers; and the writing of a warning or citation. Based on the interactions with the driver and passengers, the officer could obtain additional clues that create a reasonable suspicion of other criminal activity that could extend the duration of the stop. Regardless of whether reasonable suspicion is found, the duration of the stop may create opportunities for police to investigate other crimes, for example, by using a drug dog or general questioning of the occupants.[35]

Automated vehicles may prolong or shorten the duration of the stop depending upon police procedures. For occupied privately owned highly automated vehicles, the traffic stop could resemble a traditional traffic stop. For automated ridesharing vehicles, the occupants may not know the location of the registration or insurance information, nor would they be responsible for the failure to register or insure the vehicle. Police will need to establish policies that provide general rules for traffic stops involving automated ridesharing vehicles. To the extent those procedures create unnecessary delays or unnecessarily prolong stops, such as to provide time for a drug detection dog to sniff vehicles,

34. *See id.* at 386 ("A police officer's attempt to terminate a dangerous high-speed car chase that threatens the lives of innocent bystanders does not violate the Fourth Amendment, even when it places the fleeing motorist at risk of serious injury or death.").

35. *See, e.g.,* Illinois v. Caballes, 543 U.S. 405 (2005) (drug dog); Arizona v. Johnson, 555 U.S. 323, 129 S. Ct. 781, 172 L. Ed. 2d 694 (2009) (general questioning).

defendants may challenge the duration. In unoccupied automated vehicles, the timing of the stop may shorten if police ticket the responsible party through a computer system or lengthen if police policy establishes that the unoccupied automated vehicle needs to be retrieved by a human driver or be impounded.

II. THE RIGHT TO SEARCH HIGHLY AUTOMATED VEHICLES OR OCCUPANTS

Once an automated vehicle is seized, the other prominent issue will surround police searches of the vehicle or its occupants. By stopping a vehicle, police will have the ability to order the occupants out of the vehicle and conduct *Terry* patdowns on them upon reasonable suspicion that they may be armed and dangerous.[36] There are two tests to determine whether a search violates the Fourth Amendment. The Fourth Amendment protects an individual's expectation of privacy "[w]hen an individual 'seeks to preserve something as private,' and his expectation of privacy is 'one that society is prepared to recognize as reasonable.'"[37] This test, known as the *Katz*[38] test, requires a defendant to show a reasonable expectation of privacy. The *Katz* test is not the sole test and is supplemented by the historical trespass test that was resurrected in *United States v. Jones*.[39] Under the *Jones* test, police conduct a search when they commit a trespass.

This section focuses on five topics involving automated vehicle searches: (1) the automobile exception, (2) the search-incident-to-arrest exception, (3) expectation of privacy and standing in automated ridesharing vehicles, (4) the scope of searches of shared automated ridesharing vehicles, and (5) mixed consents in shared automated ridesharing vehicles.

---

36. *See Johnson*, 555 U.S. at 326–27 ("[T]o proceed from a stop to a frisk, the police officer must reasonably suspect that the person stopped is armed and dangerous."); Knowles v. Iowa, 525 U.S. 113, 117–18, 119 S. Ct. 484, 142 L. Ed. 2d 492 (1998) (describing that, based on the concern for officer safety, police "may order out of a vehicle both the driver and any passengers; perform a 'patdown' of a driver and any passengers upon reasonable suspicion that they may be armed and dangerous; conduct a '*Terry* patdown' of the passenger compartment of a vehicle upon reasonable suspicion that an occupant is dangerous and may gain immediate control of a weapon; and even conduct a full search of the passenger compartment, including any containers therein, pursuant to a custodial arrest" (internal citations omitted)).

37. Carpenter v. United States, 138 S. Ct. 2206, 2213, 201 L. Ed. 2d 507 (2018).

38. Katz v. United States, 389 U.S. 347, 357, 88 S. Ct. 507, 19 L. Ed. 2d 576 (1967).

39. 565 U.S. 400, 132 S. Ct. 945, 181 L. Ed. 2d 911 (2012).

A. Automobile Exception

The "automobile exception" is an exception to the warrant requirement that allows police to search a vehicle upon probable cause to believe that the vehicle contains contraband.[40] The automobile exception is based on the ready mobility of an automobile and the lowered expectations of privacy that a person has when in a vehicle in contrast to a home or office.[41] The ready mobility of an automobile has been held to depend upon the inherent mobility of the vehicle and not necessarily if the vehicle searched was readily mobile.[42] "[T]he justification to conduct such a warrantless search does not vanish once the car has been immobilized; nor does it depend upon a reviewing court's assessment of the likelihood in each particular case that the car would have been driven away, or that its contents would have been tampered with, during the period required for the police to obtain a warrant."[43] Police are also not required to search vehicles immediately to avail themselves of the automobile exception.[44] When police have probable cause to search an entire automobile, they have the right to search "every part of the vehicle and its contents that may conceal the object of the search."[45] However, probable cause to search a container in a vehicle "does not justify a search of the entire" vehicle.[46]

Under current interpretations of the automobile exception, courts are unlikely to curb its use even as automated vehicles become more than mere vehicles that transport us from one place to another. In *California v. Carney*, the Supreme Court, in holding that the automobile exception applied to a motor home situated in an off-the-street lot, stated that its "application of the vehicle exception has never turned on the other uses to which a vehicle might be

---

40. Pennsylvania v. Labron, 518 U.S. 938, 940, 116 S. Ct. 2485, 135 L. Ed. 2d 1031 (1996) ("If a car is readily mobile and probable cause exists to believe it contains contraband, the Fourth Amendment thus permits police to search the vehicle without more." (citation omitted)); California v. Carney, 471 U.S. 386, 390, 105 S. Ct. 2066, 85 L. Ed. 2d 406 (1985); Carroll v. United States, 267 U.S. 132, 45 S. Ct. 280, 69 L. Ed. 543 (1925).

41. S. Dakota v. Opperman, 428 U.S. 364, 367 (1976) ("The reason for [the automobile exception] is twofold. First, the inherent mobility of automobiles creates circumstances of such exigency that, as a practical necessity, rigorous enforcement of the warrant requirement is impossible. But the Court has also upheld warrantless searches where no immediate danger was presented that the car would be removed from the jurisdiction. Besides the element of mobility, less rigorous warrant requirements govern because the expectation of privacy with respect to one's automobile is significantly less than that relating to one's home or office." (internal citations omitted)). The lowered expectations of privacy are based on the "pervasive and continuing government regulation and control" of automobiles and "the obviously public nature of automobile travel." *Id.* at 368.

42. United States v. Jones, 565 U.S. 400 (2012).

43. Michigan v. Thomas, 458 U.S. 259, 261, 102 S. Ct. 3079, 73 L. Ed. 2d 750 (1982).

44. *E.g.*, United States v. Johns, 469 U.S. 478, 487, 105 S. Ct. 881, 83 L. Ed. 2d 890 (1985) (holding that a warrantless search three days after the vehicle was impounded did not violate the Fourth Amendment).

45. United States v. Ross, 456 U.S. 798, 825, 102 S. Ct. 2157, 72 L. Ed. 2d 572 (1982).

46. *Id.* at 824.

put."[47] Thus, under current authority, unique and more homelike automated vehicle configurations will not entitle their users to avoid application of the automobile exception.

However, the availability and the use of the automobile exception may diminish with highly automated vehicles. Fewer traffic stops means fewer opportunities to use the automobile exception. When there are traffic stops, police will have an increasingly hard time finding probable cause to use the automobile exception. Many automobile exception cases arise based on the interactions between the police officer and the driver and other vehicle occupants after a stop for a traffic violation. In unoccupied automated vehicles, police will lose their opportunities to glean clues—such as nervousness, lies, odors, and concealment behavior—to establish probable cause. In occupied automated vehicles, police may not be able to rely upon the actual driving behavior of occupants to help determine probable cause for the stop. Indeed, occupants may be less nervous when they are no longer at risk of receiving a traffic violation.[48] Also, unlike conventional vehicles where the driver is the principal person of inquiry by police and occupants often sink or swim based on that person's conversation (and driving skills), with automated vehicles, there may not be a designated spokesperson and occupants may designate their savviest occupant to interact with police. All of these factors will compound to limit the police's use of the automobile exception.

B. Search Incident to Arrest

Another exception to the warrant requirement is that police officers can search a person incident to arrest. The search-incident-to-arrest concept is based on the need to protect officer safety or to prevent evidence destruction.[49] The latter requirement, the prevention of evidence destruction, is based on the "general interest in gathering evidence relevant to the crime for which the suspect had been arrested."[50]

The Supreme Court has clarified the ability of police officers to search a vehicle incident to arrest. In *Arizona v. Gant*,[51] the defendant was arrested for

47. 471 U.S. 386, 394 (1985).

48. This, of course, assumes that states will change their laws to prohibit the user or occupants of a highly automated vehicle from being responsible for traffic violations. This topic is discussed at length in Chapter 19.

49. Chimel v. California, 395 U.S. 752, 762–63, 89 S. Ct. 2034, 23 L. Ed. 2d 685 (1969) ("When an arrest is made, it is reasonable for the arresting officer to search the person arrested in order to remove any weapons that the latter might seek to use in order to resist arrest or effect his escape. Otherwise, the officer's safety might well be endangered, and the arrest itself frustrated. In addition, it is entirely reasonable for the arresting officer to search for and seize any evidence on the arrestee's person in order to prevent its concealment or destruction.").

50. Thornton v. United States, 541 U.S. 615, 629, 124 S. Ct. 2127, 158 L. Ed. 2d 905 (2004).

51. 556 U.S. 332, 129 S. Ct. 1710, 173 L. Ed. 2d 485 (2009).

driving with a suspended license, handcuffed, and locked in the back of the police vehicle. Police officers then proceeded to search the defendant's vehicle incident to arrest. The Supreme Court examined whether police could search the vehicle incident to arrest. As to the need to prevent the destruction of evidence, the Supreme Court found that there was not a reasonable basis for thinking that the vehicle would contain evidence of the crime of driving with a suspended license. Indeed, the Court stated that "[i]n many cases, as when a recent occupant is arrested for a traffic violation, there will be no reasonable basis to believe the vehicle contains relevant evidence." As to the officer protection justification, the Supreme Court held that this "rationale authorizes police to search a vehicle incident to a recent occupant's arrest only when the arrestee is unsecured and within reaching distance of the passenger compartment at the time of the search." Because the defendant was handcuffed and locked in the back of the police vehicle, he was out of reach of his vehicle, and the search incident to arrest could not be justified as a need to protect officer safety.

Highly automated vehicles may add complications to the use of the search-incident-to-arrest doctrine. For traffic stops that arise from the use of level 4 and level 5 automated vehicles, the occupants of the vehicles may no longer be responsible for these violations and, thus, would not be arrested.[52] Even in situations where a user or occupant of an automated vehicle is the operator or driver, that person may not be subject to arrest for crimes committed by the automated vehicle.[53] If occupants of vehicles are no longer arrested for traffic violations or other more substantive automobile crimes, then the search-incident-to-arrest doctrine would no longer be applicable.[54]

C. Expectation of Privacy and Standing in Automated Ridesharing Vehicles

Standing to challenge the seizure of a vehicle and a search of that vehicle are different inquiries. In essence, a person may be seized but not have a reasonable expectation of privacy in items searched in connection with that seizure.[55] One issue courts will need to grapple with is whether users of automated

52. See Chapter 19 for a discussion of criminal responsibility for traffic violations.

53. An erroneous arrest, in and of itself, under state law is not sufficient to establish that the search-incident-to-arrest doctrine is inapplicable. Virginia v. Moore, 553 U.S. 164, 176, 128 S. Ct. 1598, 170 L. Ed. 2d 559 (2008). The key in these cases would be in establishing that the arrest was not constitutionally permissible.

54. *See* Knowles v. Iowa, 525 U.S. 113 (1998) (refusing to extend the search-incident-to-arrest doctrine to a speeding citation where there was not a custodial arrest).

55. *Compare* Brendlin v. California, 551 U.S. 249 (2007) (holding that vehicle passengers are seized), *with* Rakas v. Illinois, 439 U.S. 128, 99 S. Ct. 421, 58 L. Ed. 2d 387 (1978) (holding that the vehicle passengers lacked a reasonable expectation of privacy).

ridesharing services can challenge searches of these vehicles.[56] In an automated ridesharing vehicle, there may be two types of occupants: the user who hails the ride and that person's passengers. Though, in practice, the line between the requestor and his or her passengers may be blurry given that the occupants could split the fare directly, such as through the ridesharing application, or indirectly, such as through mobile payment applications.

Users of automated ridesharing vehicles will have a legitimate expectation of privacy in their baggage, such as a backpack or a briefcase.[57] This section is about the scope of the other searches, such as searches of the glove compartment, under seats, in the trunk, or in other compartments of the automated vehicle. Standing invites courts to wade into a potential abyss of three lines of authority that may govern whether users of automated ridesharing services have an expectation of privacy in the vehicle: passengers, taxicab users, and lessees and other permissible users of vehicles. In some respects, users resemble passengers of the automated vehicle with the automated driving system as the driver. The occupants who accompany the ride requestor resemble passengers of the requestor. These automated ridesharing vehicles are essentially robo-taxis, inviting a review of the taxicab cases to determine standing for passengers. Like the taxicab cases, the ridesharing users pay a fare and have chosen a more private form of transportation than, for example, a public bus or subway. Additionally, ridesharing users resemble permissible users or lessees of a vehicle because they have permission to use the vehicle and can have complete dominion over the vehicle during their ride (so long as it is not a shared automated ridesharing vehicle).

Rakas v. Illinois[58] is the starting point for determining the privacy rights of passengers in vehicles. In *Rakas*, police stopped a vehicle suspected as the getaway car from a robbery and ordered occupants out of the vehicle so they could search the vehicle. The police discovered a sawed-off rifle under the front passenger seat and a box of rifle shells in the glove compartment. The police arrested the occupants for the bank robbery. The occupants challenged the search. The defendants did not own the vehicle and claimed that they did not own the rifle or the shells. The issue became whether they had standing to challenge the search—the owner of the vehicle was the driver. The Supreme Court ruled that the passengers lacked standing to challenge a search of the vehicle because, by not claiming a property interest or a possessory interest in the vehicle or the items seized, they did not have legitimate expectations of pri-

---

56. *See* Lindsey Barrett, *Herbie Fully Downloaded: Data-Driven Vehicles and the Automobile Exception*, 106 Geo. L.J. 181, 201–02 (2017).

57. *See* Bond v. United States, 529 U.S. 334, 337, 120 S. Ct. 1462, 146 L. Ed. 2d 365 (2000) (stating that "it is undisputed here that petitioner possessed a privacy interest in his bag" on a bus).

58. 439 U.S. 128 (1978).

vacy in the vehicle.[59] The Court stated that "[a] person who is aggrieved by an illegal search and seizure only through the introduction of damaging evidence secured by a search of a third person's premises or property has not had any of his Fourth Amendment rights infringed."[60]

One of the themes in *Rakas* was a belief that the property owners would hold police accountable: "There is no reason to think that a party whose rights have been infringed will not, if evidence is used against him, have ample motivation to move to suppress it. Even if such a person is not a defendant in the action, he may be able to recover damages for the violation of his Fourth Amendment rights or seek redress under state law for invasion of privacy or trespass."[61] However, this observation or presumption may not be true for automated ridesharing vehicles. If all of the occupants are treated as passengers under *Rakas*, then there may not be a party to challenge the search. The automated ridesharing company may not have any reasonable expectations of privacy in a vehicle that it voluntarily rents to strangers.[62] Even if a court were to find that the ridesharing company has standing to challenge an illegal search, a ridesharing company that has billions of rides annually may not find vindicating the rights of a single passenger (who committed a criminal violation) worth challenging the illegal search. Indeed, the ridesharing company may also find it beneficial to cooperate and maintain a friendly relationship with law enforcement and local government. Part of that friendly relationship involves them not filing a section 1983 lawsuit claiming that law enforcement violated their Fourth Amendment rights and asking the local government to compensate them for the harm. Accordingly, there is reason to believe that the property owner would not enforce its rights or have the same incentive to enforce its rights in the manner that the *Rakas* Court envisioned. In any event, the Supreme Court has clarified that "[t]he Court in *Rakas* did not hold that passengers cannot have an expectation of privacy in automobiles."[63]

---

59. *But see* Byrd v. United States, 138 S. Ct. 1518, 1528, 200 L. Ed. 2d 805 (2018) ("The Court in *Rakas* did not hold that passengers cannot have an expectation of privacy in automobiles.").

60. *Rakas*, 439 U.S. at 134.

61. *Id.* (internal citations omitted).

62. *See, e.g.*, United States v. Dunkley, 911 F.2d 522, 526 (11th Cir. 1990) ("By relinquishing possession to another, the owner or lessee of the vehicle evidences an abandonment of his or her privacy interest in the vehicle."); United States v. One 1986 Mercedes Benz, VIN WDBEA30D2GA143459, 846 F.2d 2, 4 (2d Cir. 1988) ("We believe that by lending the Mercedes to Chow, [the vehicle owner] abandoned any legitimate expectation of privacy in the area searched and thus may not now contest the legality of the search."); United States v. Dall, 608 F.2d 910, 914 (1st Cir. 1979) ("Ownership alone is not enough to establish a reasonable and legitimate expectation of privacy. Ownership is relevant to the inquiry, but the total circumstances determine whether the one challenging the search has a reasonable and legitimate expectation of privacy in the locus of the search." (internal citations omitted)); United States v. Nunn, 525 F.2d 958, 959 (5th Cir. 1976).

63. *Byrd*, 138 S. Ct. at 1528.

The *Rakas* Court did not discuss its prior opinions that stated or applied the Fourth Amendment to occupants of taxicabs—people who are seemingly passengers. In *Rios v. United States*,[64] a case from 1960, the Supreme Court applied Fourth Amendment principles to a search of a passenger in a taxicab without explicitly considering whether the passenger had standing to challenge the suit. In its seminal Fourth Amendment case, *Katz v. United States*, the Supreme Court used *Rios* to assume, in dicta, that a passenger in a taxicab has a reasonable expectation of privacy.[65] Courts have acknowledged this potential dichotomy and have dodged addressing the impact of *Rakas* on *Rios* and the statement in *Katz*.[66] Other courts, though not consistently, have found that taxicab passengers have a right of privacy such that they can challenge a search.[67] Courts finding standing have focused on the payment of fares and the right of taxicab passengers to control routes (by selecting the destinations) and to exclude other passengers.[68]

Courts have also considered whether a vehicle renter or vehicle user has a reasonable expectation of privacy in the vehicle. Courts have found that a renter[69] and a permissible user[70] have standing to challenge a search. Courts have held that a user who has control over the vehicle can have an expectation

---

64. 364 U.S. 253 (1960).

65. 389 U.S. 347, 352 (1967) ("No less than an individual in a business office, in a friend's apartment, or in a taxicab, a person in a telephone booth may rely upon the protection of the Fourth Amendment." (citations omitted)).

66. *See, e.g.*, United States v. Harris, 526 F.3d 1334, 1338 (11th Cir. 2008) (stating that "we do not know the effect of *Rakas* on *Katz* and *Rios*" but finding that it need not resolve the issue because the driver consented to the search).

67. For courts that have favorably ruled for standing, see, for example, *Chapa v. State*, 729 S.W.2d 723, 729 (Tex. Crim. App. 1987) ("hold[ing] that appellant qua fare in a taxicab had a legitimate expectation of privacy in, and hence standing to challenge the search of, the area under the front seat of the taxicab"); *United States v. Santiago*, 950 F. Supp. 590, 598 (S.D.N.Y. 1996); *United States v. Harris*, No. CRIM. A.89-0290(RCL), 1989 WL 252648, at *2 (D.D.C. Oct. 12, 1989); *People v. Millan*, 69 N.Y.2d 514, 508 N.E.2d 903, 906 (1987); *Bates v. State*, 64 Md. App. 279, 287, 494 A.2d 976, 980 (1985). For courts that have ruled that the taxicab passenger lacked standing, see, for example, *People v. Millan*, 118 A.D.2d 236, 242, 505 N.Y.S.2d 84, 88 (1986), *aff'd as modified*, 69 N.Y.2d 514, 508 N.E.2d 903 (1987) (The "defendant's claim of a legitimate expectation of privacy in the rear compartment of the taxicab must fail. A passenger in a taxicab cannot rationally expect privacy in an area which, by necessity, he must share with a stranger, the driver."); *United States v. Buckner*, 417 F. Supp. 2d 240, 241 (S.D.N.Y. 2005).

68. *See, e.g.*, *Bates*, 64 Md. App. at 287 (stating that "one who hires a taxicab and contracts to pay a fare well in excess of 15 cents, similarly acquires, for the period of his use and occupancy, a reasonable expectation of privacy"); *Harris*, 1989 WL 252648, at *4 ("The court rejects the government's argument that a taxi passenger has a protected privacy interest only if he has some relationship with the driver. The government fails to recognize the distinction between a taxi—which, like a hotel room, is 'rented' by the passenger, and which, when rented, gives the passenger the right to exercise control and determine the taxi's destination—and a private car, in which the passenger is at all times merely a guest.").

69. *See, e.g.*, Byrd v. United States, 138 S. Ct. 1518 (2018). Indeed, a driver who is not authorized to drive under the rental agreement is not foreclosed from asserting standing. *Id.*

70. *See, e.g.*, United States v. Orrego-Fernandez, 78 F.3d 1497, 1502 (10th Cir. 1996); United States v. Miller, 821 F.2d 546, 548 (11th Cir. 1987); United States v. Williams, 714 F.2d 777, 779 n.1 (8th Cir. 1983);

of privacy even when not driving at the time of the stop.[71] Courts have differed in their justifications for these rulings. Joint control or common authority over property and the right to exclude others have been relevant factors.[72]

Under existing Supreme Court authority, the important factors for determining who has the right to challenge a search depend upon the possessory rights over the vehicle. Passengers have generally lacked such rights because the drivers or the owners of the vehicles effectively control the vehicle. The taxicab cases show that paying a fare can grant persons possessory rights for a certain time period, generally based on their power to control the destination and right to exclude passengers. The rental and other permissible user cases show that ownership of the vehicle is not needed to establish rights and that lessees or users can have standing.

Based on these principles, courts may find standing more readily in nonshared automated ridesharing vehicles than they do with shared automated ridesharing vehicles. A person who uses ridesharing services has the right, when selecting a ride, to exclude additional random passengers. The person selecting a nonshared automated ridesharing vehicle should have a heightened expectation of privacy within the automated vehicle. Nonshared automated ridesharing vehicles also have an advantage over traditional taxicabs in that there will be no person present who is unknown to the occupants—with a traditional taxicab, there is a driver in the vehicle.[73] Thus, perhaps occupants would have a greater expectation of privacy in an automated ridesharing vehicle than in a traditional taxicab.

D. The Scope of Searches of Shared Automated Ridesharing Vehicles

Police need to have probable cause to search shared automated ridesharing vehicles. In many cases, as described previously, police may have difficulty establishing probable cause in unoccupied automated vehicles because of the inability to obtain clues and evidence that lead to probable cause. In occupied

United States v. Portillo, 633 F.2d 1313, 1317 (9th Cir. 1980); United States v. Rose, 731 F.2d 1337, 1343 (8th Cir. 1984); United States v. Posey, 663 F.2d 37, 41 (7th Cir. 1981).

71. *See, e.g., Rose*, 731 F.2d at 1343.

72. *See, e.g.,* United States v. Thomas, 447 F.3d 1191, 1198 (9th Cir. 2006) (stating that "a defendant who lacks an ownership interest may still have standing to challenge a search, upon a showing of 'joint control' or 'common authority' over the property searched"); *Byrd*, 138 S. Ct. at 1527 (noting that the significance of the right to exclude passengers confers property rights).

73. *See* People v. Millan, 118 A.D.2d 236, 242 (1986) (The "defendant's claim of a legitimate expectation of privacy in the rear compartment of the taxicab must fail. A passenger in a taxicab cannot rationally expect privacy in an area which, by necessity, he must share with a stranger, the driver."); United States v. Anderson, 859 F.2d 1171, 1177 (3d Cir. 1988) ("A person cannot be said to retain a unique expectation of privacy in an area which is shared in common with others such as the trunk of the car.").

shared automated ridesharing vehicles, probable cause could become complicated by two distinct issues: (1) whether probable cause exists to search the vehicle as a whole or whether probable cause exists to search the containers of the suspicious occupant, and (2) whether probable cause based on one person's actions in the vehicle gives the officer the right to search all of the occupants, even those who separately hailed the automated ridesharing ride.

Under the automobile exception, police have the right to search "every part of the vehicle and its contents that may conceal the object of the search."[74] The scope of the search will depend upon how the police obtained probable cause to search the vehicle—compare a user admitting that he has drugs in his backpack with police obtaining probable cause that the user has drugs somewhere in the vehicle. Given that police will lack knowledge of the location within the vehicle that the suspicious user accessed, presumably there are instances where the police could gain access to search the entire automated ridesharing vehicle for contraband.

If police obtain the ability to search the entire vehicle and not just a passenger's containers, then the question arises whether police have access to search everyone's containers in the ridesharing vehicle. For example, what if a passenger hails a shared automated ridesharing ride that has another rider with respect to whom the police obtain probable cause to search the entire vehicle? Would the police have the right to search the bag of the passenger who is unassociated with the probable cause passenger? The Supreme Court held in *Wyoming v. Houghton* that the right to search an automobile extends to all containers within that vehicle, even those owned by a passenger.[75] Because occupants of level 4 and level 5 automated vehicles resemble passengers,[76] a police officer may contend that this right to search the contents of a vehicle could extend to all containers within that vehicle, even those owned by passengers who do not know the probable cause passenger. The police in these cases may claim that discerning different users and those who accompany those users would complicate the search.

However, a closer look at the underlying bases of *Houghton* may cut against this argument to search passengers who hailed a separate ride. In *Houghton*, the Supreme Court based its ruling on two facts: (1) the ready mobility of an automobile and (2) the potential common enterprise of the passengers with

74. United States v. Ross, 456 U.S. 798, 825 (1982).

75. Wyoming v. Houghton, 526 U.S. 295, 307, 119 S. Ct. 1297, 143 L. Ed. 2d 408 (1999) ("hold[ing] that police officers with probable cause to search a car may inspect passengers' belongings found in the car that are capable of concealing the object of the search").

76. But see Alyssa Vallar, *Robots on the Road: Fourth Amendment Implications of Stopping and Searching an Autonomous Vehicle*, 26 Geo. Mason L. Rev. 587, 615–21 (2018), for an argument against treating ridesharing users as passengers.

the driver.[77] The Court expounded on the second basis and how passengers and drivers could effectively "game" the system by having their "confederate" passengers "claim everything as their own."[78] The Court emphasized the difficulties police would have in determining "whether the officer should have believed a passenger's claim of ownership, whether he should have inferred ownership from various objective factors, whether he had probable cause to believe that the passenger was a confederate, or to believe that the driver might have introduced the contraband into the package with or without the passenger's knowledge."[79] For shared automated ridesharing vehicles, people who hail different rides are not "confederates" and would have no incentive to "game" the system with people they do not know. A passenger who games the system with someone whom that person has never met plays a dangerous game that any rational person would avoid. Courts will need to determine whether the *Houghton* rule makes sense in these vehicles and whether a factual analysis will be required to determine whether the occupants were in a common enterprise.[80]

E. Mixed Consents in Shared Automated Ridesharing Vehicles

Because nothing prohibits police from asking for and conducting a search based on consent and because the overwhelming majority of people asked do consent, consents to search are an effective tool that police use to search vehicles.[81] Generally, the government can prove consent in one of three ways:

> First, the government can come forward with persuasive evidence of both shared use and joint access to or control over a searched area, which would demonstrate actual authority to consent. Second, it can show that the owner of the property to be searched has expressly authorized a third party to give consent to the search. Finally, it may establish consent by means of the "apparent authority doctrine."[82]

Under the apparent authority doctrine, police have the right to search property so long as the police's reliance on a person's authority is reasonable even if it

77. *Houghton*, 526 U.S. at 304–05.

78. *Id.* at 305.

79. *Id.* at 305–06.

80. *See* Genesis Martinez, *The Constitutional Risks of Ridesharing: Fourth Amendment Protections of Passengers in Uber and Lyft*, 13 FLA. INT'L U. L. REV. 551, 577–80 (2019); Maryland v. Pringle, 540 U.S. 366, 373, 124 S. Ct. 795, 157 L. Ed. 2d 769 (2003) (citing *Houghton* and noting that, based on the facts before the officers, "it was reasonable for the officer to infer a common enterprise among the three men").

81. *See* Schneckloth v. Bustamonte, 412 U.S. 218, 219, 93 S. Ct. 2041, 36 L. Ed. 2d 854 (1973) (stating that it is "well settled that one of the specifically established exceptions to the requirements of both a warrant and probable cause is a search that is conducted pursuant to consent").

82. United States v. Welch, 4 F.3d 761, 764 (9th Cir. 1993) (internal citations omitted).

ends up being erroneous.[83] However, to the extent that someone with apparent or actual authority consents and another person with apparent or actual authority refuses to consent, evidence uncovered in a search cannot be used against the refuser.[84] Although the Supreme Court's apparent authority cases dealt with apparent authority over a residence, courts have applied these rulings to consents to search automobiles.[85]

Consent to search may be complicated with automated ridesharing vehicles. One important issue is who in an automated ridesharing vehicle has an expectation of privacy and the right to control the vehicle. Is the ride requestor the proper party to ask for consent to search the vehicle, or can police ask any party for consent to search the vehicle? Presumably, if a person lacks any possessory interest over the property, they also cannot consent to a search. Shared automated ridesharing vehicles may further complicate the problem. Additionally, occupants may be at odds as to whether to give police consent to search, and the *Randolph* rule—a present, objecting co-occupant can prevent a search from being effective against him or her—may be applicable. But questions arise about the application of the *Randolph* rule to other passengers who did not request the ride. Would they have the right to challenge the search? Can they rely on the objection of one of the co-riders, or do they need to voice a separate objection? Consents to search will be a ripe area for courts to review and defendants to challenge.

---

83. Illinois v. Rodriguez, 497 U.S. 177, 188 (1990).

84. Georgia v. Randolph, 547 U.S. 103, 106, 126 S. Ct. 1515, 164 L. Ed. 2d 208 (2006).

85. *See, e.g.*, United States v. Dunkley, 911 F.2d 522, 526 (11th Cir. 1990) ("We see no reason not to apply *Matlock*'s 'joint access or control' test for valid third-party consent to the search of a vehicle."); United States v. Morales, 861 F.2d 396, 399 (3d Cir. 1988).

Chapter 22

Searches of Automated Vehicle Data

Automated vehicles will collect troves of data on their users, occupants, and surroundings. Modern, nonautomated vehicles generate significant data every hour from their many sensors.[1] One prediction is that automated vehicles will generate approximately four terabytes of data every hour and a half of operation.[2] Police and prosecutors may desire automated vehicle data to prosecute crimes, support case theories, or tell better stories to juries. Data may help reconstruct crashes, the driving pattern of the automated vehicle, or the location of a vehicle (and thus potentially its occupants). Any inward-facing sensors and cameras could help show what the occupants of a vehicle were doing at a specific time. Outward-facing sensors, such as cameras and LiDAR, could establish what bystanders were doing at a specific time.

Chapter 21 explored police seizures and searches of automated vehicles. This chapter focuses on a particular search: the search of data. This chapter initially describes section 2703 of the Stored Communications Act (SCA). This chapter then describes three Fourth Amendment doctrines that will be particularly applicable to these cases: the third-party doctrine, GPS tracking, and

1. Geoffrey A. Fowler, *What Does Your Car Know about You? We Hacked a Chevy to Find Out*, Wash. Post (Dec. 17, 2019), https://www.washingtonpost.com/technology/2019/12/17/what-does-your-car-know-about-you-we-hacked-chevy-find-out/ ("There are multiple, interconnected brains that can generate up to 25 gigabytes of data per hour from sensors all over the car.").

2. Kathy Winter, *For Self-Driving Cars, There's Big Meaning behind One Big Number: 4 Terabytes*, Intel Newsroom (Apr. 14, 2017), https://newsroom.intel.com/editorials/self-driving-cars-big-meaning-behind-one-number-4-terabytes/#gs.pmoas9.

standing. Finally, this chapter explores particular issues that can arise when police search automated vehicle data and obtain warrants to search data.[3]

I. SECTION 2703 OF THE STORED COMMUNICATIONS ACT

Section 2703 of the SCA places limitations on the government's ability to compel communication contents and records. Mere compliance with section 2703 is insufficient to establish a reasonable search under the Fourth Amendment. Conversely, a search may not violate the Fourth Amendment but may be prohibited by section 2703. Thus, in considering a search of contents or records covered by section 2703, courts need to analyze whether the search complies with the Fourth Amendment and section 2703.

Section 2703 applies when a government entity requires the disclosure of a wire or electronic communication from an electronic communication service provider or remote computing service provider. A government entity includes all levels of government.[4] Section 2703 provides an arbitrary distinction between access to communications that have been stored by an electronic communication service provider for shorter than or longer than 180 days.[5] Communications held in storage for fewer than 180 days receive better protection from the government. The government must obtain a warrant supported by probable cause to access a communication stored for fewer than 180 days.[6] If the communication has been stored for longer than 180 days, or if the request is to a remote computing service provider, then the government can obtain a search warrant or gain access by providing prior notice to the subscriber or the customer (except that this notice may be delayed) and obtaining a subpoena or

---

3. Another issue may be the Fifth Amendment privilege against self-incrimination. Because the automated vehicle owner may be treated as the owner of data, police use of that data against a defendant may implicate the Fifth Amendment privilege against self-incrimination. The Fifth Amendment states, in pertinent part, that a person cannot "be compelled in any criminal case to be a witness against himself." The Supreme Court has stated that the Fifth Amendment's prohibition generally applies to testimony or communications but not to real or physical evidence. Schmerber v. California, 384 U.S. 757, 764, 86 S. Ct. 1826, 16 L. Ed. 2d 908 (1966) ("[T]he privilege is a bar against compelling 'communications' or 'testimony,' but that compulsion which makes a suspect or accused the source of 'real or physical evidence' does not violate it."). One court expressly rejected the idea that the prosecution's use of event data recorder data violates the defendant's Fifth Amendment privilege because the evidence is not testimonial. State v. Shabazz, 400 N.J. Super. 203, 218, 946 A.2d 626, 635 (2005).

4. 18 U.S.C. § 2711 ("[T]he term 'governmental entity' means a department or agency of the United States or any State or political subdivision thereof.").

5. *Id.* § 2703(a).

6. *Id.*

a court order.[7] The court order requires less than probable cause and requires "specific and articulable facts showing that there are reasonable grounds to believe that the contents of a wire or electronic communication, or the records or other information sought, are relevant and material to an ongoing criminal investigation."[8]

A wire communication is "any aural transfer made in whole or in part through the use of facilities for the transmission of communications by the aid of wire, cable, or other like connection between the point of origin and the point of reception (including the use of such connection in a switching station) furnished or operated by any person engaged in providing or operating such facilities for the transmission of interstate or foreign communications or communications affecting interstate or foreign commerce."[9] An electronic communication is "any transfer of signs, signals, writing, images, sounds, data, or intelligence of any nature transmitted in whole or in part by a wire, radio, electromagnetic, photoelectronic or photooptical system that affects interstate or foreign commerce."[10] Any transmission of data in an automated vehicle would be an electronic communication. These transmissions may be text messages or e-mails sent through the automated vehicle's infotainment system or other data, such as location data, transferred from the automated vehicle to the manufacturer.

"Electronic communication service" is defined as "any service which provides to users thereof the ability to send or receive wire or electronic communications."[11] A remote computing service is "the provision to the public of computer storage or processing services by means of an electronic communications system."[12] Courts have generally found that traditional sellers of products or services do not provide an "electronic communication service."[13]

---

7. *Id.* § 2703(b).
8. *Id.* § 2703(d).
9. *Id.* § 2510(1).
10. *Id.* § 2510(11).
11. *Id.* § 2510(15).
12. *Id.* § 2711(2).
13. *See, e.g.,* Dyer v. Nw. Airlines Corps., 334 F. Supp. 2d 1196, 1199 (D.N.D. 2004) ("Courts have concluded that 'electronic communication service' encompasses internet service providers as well as telecommunications companies whose lines carry internet traffic, but does not encompass businesses selling traditional products or services online."); *In re* Jetblue Airways Corp. Privacy Litig., 379 F. Supp. 2d 299, 307 (E.D.N.Y. 2005) (stating that "a company such as JetBlue does not become an 'electronic communication service' provider simply because it maintains a website that allows for the transmission of electronic communications between itself and its customers"); Copeland v. Nw. Airlines Corp., No. 04-2156 M1/V, 2005 WL 2365255, at *2 (W.D. Tenn. Feb. 28, 2005) ("Courts have found that the term 'electronic communication service' refers to entities that provide internet access to customers, such as internet service providers ('ISP's'), as opposed to merchants and businesses that sell their products or services over the internet."). A notable outlier was that the Ninth Circuit held that American Airlines, through the use of its computerized travel reservation program called SABRE, "is a provider of wire or electronic communication service." United States v. Mullins, 992 F.2d 1472, 1478 (9th Cir. 1993). Because the Ninth Circuit examined the question within the context of

Thus, the SCA may not apply to government requests to automakers. Instead, these terms—"electronic computing service providers" and "remote computing service providers"—are generally limited to Internet service providers or other entities that to some extent resemble communication-providing companies.[14] Courts so holding may apply different rules to police acquisition of data from e-mail companies than they do for acquisitions from automakers or automated ridesharing companies.

II. FOURTH AMENDMENT

This section has four parts. The first part describes three Fourth Amendment doctrines that may reverberate throughout data searches of automated vehicles. The second part describes the searches of event data recorders (EDRs). An EDR is the so-called black box of a motor vehicle and records performance data typically just before a crash. The limited nature of the information stored on an EDR is the reason this section separately analyzes a search of an EDR and all other data that an automated vehicle may collect. Other data may include location data of the automated vehicle, data about the automated vehicle's surroundings, and data about occupants, including who is in the vehicle and their activities. The third part describes searches of non-EDR data recorded by automated vehicles. The fourth part discusses search warrants for automated vehicle data.

A. Fourth Amendment Doctrines

Many of the Fourth Amendment doctrines applicable to automobiles more generally were described in Chapter 21. This section explores three doctrines and their application to automated vehicle data. The first doctrine is the third-party doctrine. Almost all data recorded by an automated vehicle will be shared with or accessible by a third party, such as a vehicle manufacturer, the driving automation system manufacturer, or an automated ridesharing company. The second doctrine is the *Jones* trespass doctrine. The trespass doctrine has been important in data search cases when police physically enter a vehicle to attach a device to download data. The third doctrine is standing for users of automated ridesharing services and bystanders.

---

the wire or electronic communication interception statute and whether American Airlines employees violated the defendant's Fourth Amendment rights, courts within the Ninth Circuit have not applied *Mullins* to the SCA. Pica v. Delta Air Lines, Inc., No. CV 18-2876-MWF (EX), 2019 WL 1598761, at *7 (C.D. Cal. Feb. 14, 2019).

14. *See supra* note 13; Crispin v. Christian Audigier, Inc., 717 F. Supp. 2d 965, 982 & n.35 (C.D. Cal. 2010) (concluding that Facebook and MySpace are providers of electronic communication services).

1. Third-Party Doctrine

All information collected by an automated vehicle about individuals will be shared with or available to third parties, such as the manufacturer of the vehicle or driving automation system. In many respects, data may be voluntarily shared, as in a vehicle user who voluntarily accepts the sharing of data, or involuntarily shared, as in the activities of bystanders that are captured by the vehicle. The third-party doctrine also has significant application to automated ridesharing services where the user of such services directly shares his or her location data with the service provider. For example, a user shares location data when requesting a ride from one location to another location. The *Katz* test depends upon whether a person has an expectation of privacy that society finds reasonable. Historically, the Supreme Court has, in what is known as the third-party doctrine, found that a person cannot have a reasonable expectation of privacy in information shared with third parties.

The third-party doctrine finds its roots in two United States Supreme Court cases: *United States v. Miller*[15] and *Smith v. Maryland*.[16] In *Miller*, while investigating the defendant for tax evasion, police obtained copies of checks and bank records of the defendant by subpoenaing, pursuant to the Bank Secrecy Act, two banks where the defendant had accounts. The United States Attorney's Office issued the subpoenas. The defendant moved to suppress the checks and bank records, claiming a violation of his Fourth Amendment rights. The Supreme Court held that the defendant did not have a legitimate expectation of privacy in the checks and banknotes for two reasons: first, they are negotiable instruments and not confidential information, and second, they were voluntarily conveyed to the bank. On the latter justification, the court clarified the risks of voluntarily sharing information with the bank:

> The depositor takes the risk, in revealing his affairs to another, that the information will be conveyed by that person to the Government. This Court has held repeatedly that the Fourth Amendment does not prohibit the obtaining of information revealed to a third party and conveyed by him to Government authorities, even if the information is revealed on the assumption that it will be used only for a limited purpose and the confidence placed in the third party will not be betrayed.[17]

Accordingly, the Supreme Court held that the obtainment of the bank records and checks by the subpoena did not violate the defendant's Fourth Amendment rights.

---

15. 425 U.S. 435, 96 S. Ct. 1619, 48 L. Ed. 2d 71 (1976).
16. 442 U.S. 735, 99 S. Ct. 2577, 61 L. Ed. 2d 220 (1979).
17. *Miller*, 425 U.S. at 443 (internal citation omitted).

In *Smith*, the police requested a telephone company to install a pen register to record numbers dialed from the defendant's house. Police used the pen register to determine who was calling a robbery victim; the victim had endured harassment from the robber after she reported the robbery and gave his description to the police. The pen register showed that the defendant called the robbery victim, and this evidence helped police obtain a search warrant and arrest the defendant. The defendant moved to suppress the pen register because police did not obtain a warrant before its installation.

The Supreme Court first noted that the defendant could not claim that his property was invaded and that, therefore, his challenge would fail under the *Katz* test for reasonable expectation of privacy. The Supreme Court emphasized that, due to the limited nature of the pen register's capabilities, the question was whether the defendant had a reasonable expectation of privacy in the numbers he dialed. The Supreme Court held that it doubts someone could have an expectation of privacy in the numbers they dial because they are voluntarily shared with and recorded by the telephone company. The Court also held that even if the defendant had an expectation of privacy in the numbers he dialed, that expectation would not be one that society would find reasonable because the information was voluntarily conveyed to third parties.

The third-party doctrine predated the digital era, but the application of the third-party doctrine in the digital era seems problematic where every person participating in society has to share his or her activities. In 2012, in a concurrence to *United States v. Jones*, Justice Sotomayor stated her belief that the court should revisit the third-party doctrine:

> [I]t may be necessary to reconsider the premise that an individual has no reasonable expectation of privacy in information voluntarily disclosed to third parties. This approach is ill suited to the digital age, in which people reveal a great deal of information about themselves to third parties in the course of carrying out mundane tasks. People disclose the phone numbers that they dial or text to their cellular providers; the URLs that they visit and the e-mail addresses with which they correspond to their Internet service providers; and the books, groceries, and medications they purchase to online retailers. . . . I would not assume that all information voluntarily disclosed to some member of the public for a limited purpose is, for that reason alone, disentitled to Fourth Amendment protection.[18]

In 2018, the United States Supreme Court began to revisit the third-party doctrine in *Carpenter v. United States*.[19]

---

18. United States v. Jones, 565 U.S. 400, 417–18, 132 S. Ct. 945, 957, 181 L. Ed. 2d 911 (2012) (internal citations omitted).

19. 138 S. Ct. 2206, 201 L. Ed. 2d 507 (2018).

In *Carpenter*, police arrested the defendant for robbing four stores. During the prosecution, prosecutors received a subpoena pursuant to the SCA to obtain cell-site location information (CSLI) on the defendant's cell phone. The CSLI is a record created every time a cell phone accesses a cell tower, and the CSLI shows a sector of where the phone accessed the cell tower. The prosecutor sought 152 days of cell-site records from MetroPCS and seven days of cell-site data from Sprint, and the prosecutor received records for 127 days from MetroPCS and two days from Sprint. This information allowed the prosecutor to have 12,898 location points for the defendant. Prosecutors used this evidence to show that the defendant's phone was near all four robberies. The defendant moved to suppress the CSLI evidence because the subpoena, which was issued pursuant to section 2703(d) of the SCA, did not require a showing of probable cause.

The Court expressly rejected the expansion of the third-party doctrine: "We decline to extend *Smith* and *Miller* to cover these novel circumstances. Given the unique nature of cell phone location records, the fact that the information is held by a third party does not by itself overcome the user's claim to Fourth Amendment protection." The Court elaborated and clarified that the third-party doctrine cannot be mechanically applied whenever data is shared with third parties. The Court distinguished *Smith* and *Miller* because of the limited expectation of privacy for call logs and negotiable instruments and the limited scope of those searches.

Because almost all automated vehicles will require or have some data-sharing components to them, the third-party doctrine pre-*Carpenter* might have given police carte blanche access to data. After *Carpenter*, it seems unlikely that police will be able to use warrantless searches of data contained in automated vehicles.

2. *United States v. Jones*

Because of the mobility of vehicles, tracking of vehicles has been a popular police tactic. Police have tracked vehicles by physically following them and have also attempted to use new technologies to obviate the need to physically follow the vehicle.[20] In *United States v. Jones*,[21] the Supreme Court examined whether police conducted a search when they attached a GPS under a vehicle to track its movements. The government's principal argument was that the information was open to the public and that the defendant does not have an expectation of privacy in publicly shared information. The Supreme Court used the trespass

20. *See, e.g., Jones*, 565 U.S. 400 (GPS tracking device on a vehicle). Other tracking cases have involved containers. *See, e.g.*, United States v. Karo, 468 U.S. 705, 104 S. Ct. 3296, 82 L. Ed. 2d 530 (1984) (tracking beeper in a container); United States v. Knotts, 460 U.S. 276, 103 S. Ct. 1081, 75 L. Ed. 2d 55 (1983) (tracking beeper in a container).

21. 565 U.S. 400 (2012).

doctrine to find that police violated the Fourth Amendment by trespassing onto private property by installing a GPS. The majority opinion declined to consider the *Katz* test for reasonable expectation of privacy in ruling that the trespass violated the Fourth Amendment. The Court stated that when "the Government obtains information by physically intruding on a constitutionally protected area, [] a search has undoubtedly occurred."

The *Jones* trespass line of authority will be important when police physically enter an automated vehicle to obtain data from the vehicle. Defendants have used *Jones*'s trespass authority in cases where police physically enter a vehicle and use equipment to obtain data recorded by an EDR.[22] These courts have dodged or found irrelevant the question of whether a defendant has a reasonable expectation of privacy in an EDR. Instead, the courts have found that police commit searches based on the possessory interests of the defendant at the time of the data download, the officer's physical intrusion into the vehicle, the attachment of a computer to the vehicle to obtain data, and the obtainment of data.

3. *Standing*

Two prominent issues arise with standing and data searches for automated vehicles. The first issue is standing for automated ridesharing occupants. A person's relationship with a shared or automated ridesharing vehicle is typically limited to the ride itself. Chapter 21 described standing with respect to a physical search of an automated ridesharing vehicle, concluding that certain passengers could have standing to challenge a search of an automated ridesharing vehicle. The ability to challenge a physical search of an automated ridesharing vehicle could also, but need not, extend to a digital search of the automated ridesharing vehicle. Relevant facts will include whether the automated ridesharing company consented to the search or whether the data search occurred at a time when the occupants had a possessory interest in the automated ridesharing vehicle.

The second issue will be data recorded about bystanders and surroundings. People who lack any relationship to the vehicle, like a bystander whose crime is captured by an automated vehicle's sensors, would need to establish that they have a reasonable expectation of privacy in the area captured. The problem in these cases could be twofold. First, the fact that the vehicle captures the criminal activity on a sensor likely means that the criminal activity was not in private—the person was potentially in public view.[23] Second, except

22. State v. West, 548 S.W.3d 406, 416 (Mo. Ct. App. 2018); Mobley v. State, 307 Ga. 59, 834 S.E.2d 785, 792 (2019).

23. *See Knotts*, 460 U.S. at 281–82 ("A person travelling in an automobile on public thoroughfares has no reasonable expectation of privacy in his movements from one place to another. When [the

for government use of automated vehicles, automated vehicles capturing their surroundings are not conducting Fourth Amendment searches.[24] The vehicles were privately owned at the time of the searches, and absent proof that a government agent caused the searches, criminal activities captured by the sensors would not be Fourth Amendment searches.

Because the recordings of bystanders would not be Fourth Amendment searches, bystanders would need to challenge the actual search by the government—the search of the vehicle's data. Bystanders will have a nearly impossible time challenging these searches. Because police will likely not have possession of the automated vehicle, the search would require either police to obtain a warrant to search the automated vehicle data or vehicle owners to voluntarily share vehicle data with police. In either of these instances, the search would be reasonable under the Fourth Amendment.

B. Event Data Recorder Searches

EDRs are used in almost every modern vehicle. EDRs became common in vehicles in the late 1990s when motor vehicles were required to contain airbags. The National Highway Traffic Safety Administration regulates what data an EDR is required to capture and for how long prior to an event the EDR must record that information.[25] Data required to be captured includes vehicle speed, seat belt use, airbag deployment, ignition cycle, and engine throttle.[26] Courts have considered the admissibility of EDR data[27] and whether police downloading EDR data constitutes a search.[28] Courts are split about whether police can search an EDR without a warrant.

---

driver] travelled over the public streets he voluntarily conveyed to anyone who wanted to look the fact that he was travelling over particular roads in a particular direction, the fact of whatever stops he made, and the fact of his final destination when he exited from public roads onto private property.").

24. *See* United States v. Jacobsen, 466 U.S. 109, 113, 104 S. Ct. 1652, 80 L. Ed. 2d 85 (1984) (The Fourth Amendment "is wholly inapplicable to a search or seizure, even an unreasonable one, effected by a private individual not acting as an agent of the Government or with the participation or knowledge of any governmental official." (internal quotation marks and citation omitted)).

25. 49 C.F.R. § 563.1 *et seq.*

26. *Id.* § 563.7.

27. Admissibility has been considered in both criminal and civil cases. *See, e.g.*, Bachman v. Gen. Motors Corp., 332 Ill. App. 3d 760, 776 N.E.2d 262 (2002); People v. Christmann, 3 Misc. 3d 309, 315–16, 776 N.Y.S.2d 437, 442 (Just. Ct. 2004); Easter v. State, 223 Md. App. 65, 115 A.3d 239, 248 (2015); Commw. v. Zimmermann, 70 Mass. App. Ct. 357, 360, 873 N.E.2d 1215, 1218 (2007); Commw. v. Safka, 2014 Pa. Super. 131, 95 A.3d 304 (2014), *aff'd*, 636 Pa. 169, 141 A.3d 1239 (2016); State v. Shabazz, 400 N.J. Super. 203 (2005); State v. Byard, 2018 WL 2077324 (Del. Super. Ct. May 1, 2018), *as corrected* (May 3, 2018); State v. Diaz, No. 35,563, 2017 WL 1017356, at *3 (N.M. Ct. App. Feb. 27, 2017); *see also* Marjorie A. Shields, *Admissibility of Evidence Taken from Vehicular Event Data Recorders (EDR), Sensing Diagnostic Modules (SDM), or "Black Boxes"*, 40 A.L.R.6TH 595 (2008).

28. *See, e.g.*, State v. Worsham, 227 So. 3d 602, 603 (Fla. Dist. Ct. App. 2017); Mobley v. State, 346 Ga. App. 641, 816 S.E.2d 769 (2018), *cert. granted* (Mar. 4, 2019), *rev'd*, 307 Ga. 59, 834 S.E.2d 785 (2019), *and vacated*, No. A18A0500, 2020 WL 771166 (Ga. Ct. App. Feb. 18, 2020); Mobley v. State, 834 S.E.2d 785

Some courts have found that vehicle owners do not have a reasonable expectation of privacy in data collected by EDRs.[29] This approach is epitomized by *California v. Diaz*.[30] In *Diaz*, the defendant drunkenly drove her vehicle at a high rate of speed and collided with another vehicle, killing its driver. The police arrested the defendant and impounded her vehicle. The police subsequently inspected the vehicle and obtained data from the vehicle's EDR. The court upheld the warrantless search for three reasons. First, the court found that police had probable cause to search the EDR data under the automobile exception to the warrant requirement. Second, the court held that, because the EDR is part of the vehicle and the vehicle was an instrumentality of the crime, the police could search the EDR data under the instrumentality-of-the-crime exception to the warrant requirement. Third, the court held that the defendant lacked a reasonable expectation of privacy in the data "because she was driving on the public roadway, and others could observe her vehicle's movements, braking, and speed, either directly or through the use of technology such as radar guns or automated cameras." Thus, the *Diaz* court upheld the warrantless search of the EDR data.

The other line of authority generally involves newer cases that are reflective of two trends that seem to limit the ability of police to conduct warrantless searches of EDR data.[31] The first is that federal law, the Driver Privacy Protection Act of 2015 (DPPA), established that the vehicle owner is the owner of EDR data and limited how other people, including governments, can obtain access to EDR data.[32] The DPPA prohibits other persons from accessing EDR data unless they (1) receive a court order authorizing the retrieval of EDR data, (2) obtain consent by the vehicle owner or lessee for any reason, (3) perform investigations by federal agencies under sections 1131(a) or 30166 of title 49, (4) need the data to facilitate emergency medical response following a crash, or (5) conduct traffic safety research provided that personally identifiable information is not disclosed. State laws that have been enacted about EDR data also uniformly find that the vehicle owner or lessee is the owner of data.[33]

These laws may also explicitly curtail the ability to use the automobile exception and the inventory exception applying to a vehicle's EDR. Some states explicitly provide that police can have access to an EDR pursuant to a search

---

(2019); People v. Diaz, 213 Cal. App. 4th 743, 153 Cal. Rptr. 3d 90 (2013); *Christmann*, 3 Misc. 3d at 314; People v. Xinos, 121 Cal. Rptr. 3d 496, 515 (Ct. App. 2011).

29. *See, e.g., Mobley*, 816 S.E.2d at 771; *Diaz*, 213 Cal. App. 4th 743; *Christmann*, 3 Misc. 3d 309.

30. 213 Cal. App. 4th 743. In contrast, another California Court of Appeals case ruled that a warrantless search of an EDR violated a defendant's Fourth Amendment rights. *Xinos*, 121 Cal. Rptr. 3d at 500.

31. *See Mobley*, 834 S.E.2d at 792; State v. West, 548 S.W.3d 406 (Mo. Ct. App. 2018); *Worsham*, 227 So. 3d 602.

32. See Chapter 16 for a discussion of the DPPA.

33. See Chapter 17 for a discussion of state EDR laws.

warrant,[34] which further negates the ability of the police to obtain the data under an exception-to-the-warrant requirement. Other states, such as Maine and Virginia, are more ambiguous about the need for a search warrant.[35] States also differ on the effectiveness of consent to retrieve EDR data from one vehicle owner when a vehicle has multiple owners.[36]

An important consideration for automated vehicles is that the DPPA and the state laws would not apply to a user of an automated ridesharing service. These users are not owners of the vehicles. The use of "lessee" in these statutes involves a person who leases the vehicle for an extended time beyond just a single ride.[37] Because users of automated ridesharing services are not considered the owners of the EDR, these statutes would not benefit those users.

The second trend that could call into question the precedential effect of some of the warrantless EDR search cases is the recent United States Supreme Court jurisprudence that limits police ability to have access to a person's life through traditional warrant exceptions as applied to modern technologies. These cases include *Riley v. California*,[38] in which the Supreme Court held that police could not search the digital contents of a cell phone in a warrantless search incident to arrest; *United States v. Jones*,[39] in which the Supreme Court resurrected the trespass doctrine and held that a warrantless attachment of a GPS to a vehicle violated the Fourth Amendment; and *Carpenter v. United States*,[40] in which the Supreme Court held that police could not search cell-site location information

34. MONT. CODE ANN. § 61-12-1004(1)(a) (West) ("Data from a motor vehicle event data recorder may be retrieved or used without the consent of the owner: (a) if a court orders the production of the data pursuant to a valid search warrant."); OR. REV. STAT. § 105.935 ("Data from a motor vehicle event data recorder may be retrieved or used without the consent of the owner after an accident if a court orders the production of the data based on a determination by the court that: (1) A law enforcement officer has probable cause to believe that a crime has occurred and that the data is relevant to the investigation of the crime.").

35. ME. REV. STAT. tit. 29-A, § 1972F ("Data described in section 1971, subsection 1 that are recorded on an event data recorder may not be downloaded or otherwise retrieved by a person other than the owner of the motor vehicle at the time the data are accessed, except under the following circumstances: . . . The data are retrieved by a law enforcement officer acting pursuant to authority recognized under applicable statutory or constitutional law."); VA. CODE ANN. § 46.2-1088.6.B.5 (providing an exception if "[t]he recorded data is accessed by law enforcement in the course of an investigation where constitutionally permissible and in accordance with any applicable law regarding searches and seizures upon probable cause to believe that the recording device contains evidence relating to a violation of the laws of the Commonwealth or the United States").

36. *Compare* OR. REV. STAT. § 105.928 ("If a motor vehicle is owned by more than one person, all owners must consent to the retrieval or use of the data from a motor vehicle event data recorder."), *with* UTAH CODE ANN. § 41-1a-1503(1)(b) (West) ("If a motor vehicle is owned by more than one person, only one owner is required to consent to the retrieval or use of the data from a motor vehicle event data recorder.").

37. *See, e.g.*, OR. REV. STAT. § 105.925 (a lease more than three months).

38. 573 U.S. 373, 134 S. Ct. 2473, 189 L. Ed. 2d 430 (2014).

39. 565 U.S. 400, 132 S. Ct. 945, 181 L. Ed. 2d 911 (2012).

40. 138 S. Ct. 2206, 201 L. Ed. 2d 507 (2018).

on a cell phone without a search warrant. In all of these cases, the Supreme Court examined the use of new technology applied to existing Supreme Court jurisprudence and limited police use of technology to reveal information that they could not have historically obtained without the new technology.

Recent EDR search cases highlight these trends. In *State v. Worsham*,[41] Worsham was the driver of a vehicle that was involved in a high-speed crash. The passenger of the vehicle died as a result of the crash, and police impounded the vehicle. Twelve days later, police downloaded data from the vehicle's EDR without a warrant. Worsham moved to suppress the data, arguing that the warrantless search violated the Fourth Amendment. The trial court granted the motion to suppress, and a Florida court of appeals affirmed. The court of appeals specifically analogized the EDR to other electronic storage devices, like a cell phone, and noted that "the emerging trend is to require a warrant to search these devices." The court of appeals found that a vehicle owner has a reasonable expectation of privacy in EDR data because it is not exposed to the public and it is difficult to extract. The court of appeals ruled that the warrantless search violated the Fourth Amendment.

Another line of authority has relied upon *United States v. Jones* and the trespass doctrine to find that a warrantless search of EDR data violated the Fourth Amendment.[42] Because police need to physically enter a vehicle to attach a computer device to the vehicle to obtain EDR data, two courts have found that such entrance to the vehicle and attachment constitute an impermissible search under the Fourth Amendment, regardless of the expectation of privacy of the defendant. These two courts have found that the automobile exception did not apply for different reasons: one court found that a crash is not a crime for which evidence could be located,[43] and the other court found that the vehicle was completely immobilized at the time of the search.[44] The courts both acknowledged that data could be overwritten, which could justify exigent circumstances, but found that the state failed to prove exigency, in part because, as one court stated, "there was no evidence that mere delay alone—the time it might take, for instance, to procure a warrant—would have caused the loss or corruption of any data."[45]

---

41. 227 So. 3d 602.

42. Mobley v. State, 834 S.E.2d 785 (2019); State v. West, 548 S.W.3d 406 (Mo. Ct. App. 2018).

43. *West*, 548 S.W.3d at 418–20.

44. *Mobley*, 834 S.E.2d at 793 n.10. This holding is questionable in light of Supreme Court cases that find that the exception applies regardless of whether the vehicle has been immobilized. *See* Michigan v. Thomas, 458 U.S. 259, 261, 102 S. Ct. 3079, 73 L. Ed. 2d 750 (1982).

45. *Mobley*, 834 S.E.2d at 793 n.10.

C. Non-EDR Data Searches

Automated vehicles will collect significant data outside of the EDR. They will collect an array of roadway data from sensors, know the vehicle's location, and maintain data about passengers and occupants, such as how the driver monitoring system of GM's Super Cruise monitors the driver's eyes. Automated vehicles may also have computer interfaces that allow users to surf the Internet, watch videos, and store data. Data may also be obtained through connectivity to vehicle users' and owners' other devices, such as cell phone contacts, call history, and text messages.

Non-EDR data has two characteristics that make a search of such data significantly different than the EDR search. First, non-EDR data may reveal substantial information about the occupant's livelihood and predilections. The non-EDR data essentially combines the *Jones v. United States*,[46] *Riley v. California*,[47] and *Carpenter v. United States*[48] searches, potentially allowing police to have full access to a person's private life. Second, a search of non-EDR data resembles a general search, which—unlike the EDR search, which reveals limited, specific data about vehicle performance—would require police to sort through troves of irrelevant data to discover relevant data. These two features of non-EDR data searches should cause courts to treat these searches differently than the EDR searches.

A question arises as to whether police could use the automobile exception or the search-incident-to-arrest exception to search data of the automated vehicle without a search warrant.[49] The automobile exception allows police to search an automobile where they believe they can find evidence of the contraband with respect to which they obtained probable cause to search the vehicle. With the automobile exception, police conceivably could claim that vehicle data could show evidence of the crime. For example, for a drug trafficking crime, vehicle data could show where the defendant obtained the contraband. The search-incident-to-arrest exception allows police to search a person incident to arrest for officer protection and to prevent the destruction or concealment of evidence. If the defendant is arrested for a drug crime, conceivably police may claim that they have the right to search data of the vehicle incident to arrest because it may contain evidence of the crime.

However, this type of extension of the automobile exception and the search-incident-to-arrest exception seems analogous to a similar extension that the Supreme Court rejected in *Riley v. California*.[50] In *Riley*, the issue was whether

46. 565 U.S. 400, 132 S. Ct. 945, 181 L. Ed. 2d 911 (2012).
47. 573 U.S. 373, 134 S. Ct. 2473, 189 L. Ed. 2d 430 (2014).
48. 138 S. Ct. 2206, 201 L. Ed. 2d 507 (2018).
49. These exceptions to the warrant requirement are described in Chapter 21.
50. 573 U.S. 373.

police could conduct a warrantless search of the digital contents of a cell phone under the search-incident-to-arrest exception. The Court acknowledged that "[d]igital data stored on a cell phone cannot itself be used as a weapon to harm an arresting officer or to effectuate the arrestee's escape." Thus, the Court rejected protection of officers as a justification to search the digital contents of a phone. The Court also rejected the need to search the digital contents of a phone incident to arrest to prevent the destruction of evidence. The Court described how cell phones differ from other items that may be found on a person. The Court focused on the breadth of data that may be stored in a cell phone or accessed by a cell phone through the cloud and that a search incident to arrest would "give police officers unbridled discretion to rummage at will among a person's private effects."

Although the mobility justification of the automobile exception would apply equally with respect to automated vehicle data, the reduced expectation of privacy justification may no longer be true for data collected by automated vehicles. Although automated vehicles will be subject to the extensive regulation and inspection that all vehicles are subject to, data itself is not subject to extensive regulation and inspection. The vehicle's data is also not made public. A person cannot monitor the vehicle's data or look into the vehicle's windows to view data. Modern vehicles also access data from areas, such as personal cell phones, where courts have found more heightened justifications of privacy. Accordingly, it would seem that physical and digital searches of automated vehicles should be treated differently. A physical search, which is constrained to the items located within a vehicle, should be subject to the traditional automobile exception and search-incident-to-arrest exception, but searches of data should be treated like searches of computers and cell phones and require a warrant absent exigent circumstances.

D. Warrants to Search Data

A warrant is the default requirement of the Fourth Amendment. A warrant must be supported by probable cause and describe in particularity the place to be searched.[51] The particularity requirement "ensures that the search will be carefully tailored to its justifications, and will not take on the character of the wide-ranging exploratory searches the Framers intended to prohibit."[52] Courts have found that particularity has three components: "First, a warrant must identify the specific offense for which the police have established probable cause. Second, a warrant must describe the place to be searched. Third,

51. Maryland v. Garrison, 480 U.S. 79, 84, 107 S. Ct. 1013, 1016, 94 L. Ed. 2d 72 (1987).
52. *Id.* at 84.

the warrant must specify the 'items to be seized by their relation to designated crimes.'"[53] The question that magistrates may encounter is what does particularity require when the warrant is for a search of automated vehicle data?[54]

Courts have already begun addressing whether a warrant to search an electronic device, such as a computer or a cell phone, satisfies the particularity requirement.[55] As one court aptly described, "[I]n the virtual world, it is not enough to simply permit a search to extend anywhere the targeted electronic objects possibly could be found, as data possibly could be found anywhere within an electronic device."[56] The computer cases have typically allowed police to broadly search computers without requiring a warrant to specify the types of files searched.[57] Courts have been more reluctant to allow broad searches of the digital evidence contained in cell phones.[58] One commentator has suggested that this difference might be due, in part, to the ability to better conceal files on a computer than in a cell phone.[59]

The risk of an unlimited search of automated vehicle data is that it may grant police carte blanche access to find additional crimes in plain view. For example, if data shows that the automated vehicle has traveled in excess of the speed limit, could police write retroactive speeding tickets in light of the new evidence? If inward-facing cameras or sensors show that the occupants of a vehicle engaged in some criminal act, could police retroactively charge the occupants with crimes? Courts will want to carefully delineate the scope of searches of vehicle data to tailor the warrants to the reasons for which the vehicle data needs to be searched.

---

53. United States v. Galpin, 720 F.3d 436, 445–46 (2d Cir. 2013) (internal citations omitted).

54. A related issue that has arisen is ex ante requirements that magistrate judges have imposed when police execute warrants on digital information. *See, e.g.*, United States v. Comprehensive Drug Testing, Inc., 621 F.3d 1162, 1178 (9th Cir. 2010) (Kozinski, J., concurring); *In re* Search Warrant, 2012 VT 102, 193 Vt. 51, 71 A.3d 1158 (2012).

55. For computers, see, for example, *United States v. Burgess*, 576 F.3d 1078 (10th Cir. 2009); *United States v. Otero*, 563 F.3d 1127, 1132 (10th Cir. 2009); *Galpin*, 720 F.3d 436. For cell phone searches, see, for example, *United States v. Russian*, 848 F.3d 1239 (10th Cir. 2017); *State v. Allen*, 288 Or. App. 244, 406 P.3d 89, 90 (2017); *State v. Keodara*, 191 Wash. App. 305, 364 P.3d 777 (2015); *Hedgepath v. Commw.*, 441 S.W.3d 119, 130 (Ky. 2014); *State v. Henderson*, 289 Neb. 271, 854 N.W.2d 616, 632 (2014).

56. Commw. v. Dorelas, 473 Mass. 496, 43 N.E.3d 306, 312 (2016).

57. *See, e.g.*, *Burgess*, 576 F.3d 1078; *but see* State v. Mansor, 279 Or. App. 778, 381 P.3d 930 (2016), *aff'd*, 363 Or. 185, 421 P.3d 323 (2018).

58. *See, e.g.*, *Henderson*, 289 Neb. at 290 (holding that "a catchall phrase stating that they authorized a search of 'any other information that can be gained from the internal components and/or memory Cards'" in a warrant caused the warrant to fail the particularity requirement).

59. Andrew D. Huynh, *What Comes after "Get a Warrant": Balancing Particularity and Practicality in Mobile Device Search Warrants Post-*Riley, 101 Cornell L. Rev. 187, 207–08 (2015).

Table of Authorities

U.S. CONSTITUTION

U.S. Const. amend. IV, passim
U.S. Const. art. I, § 8, cl. 3, 99
U.S. Const. art. VI, 46

FEDERAL STATUTES

5 U.S.C. § 552a, 241
5 U.S.C. § 552a(a)(1), 242
5 U.S.C. § 552a(a)(4), 242
5 U.S.C. § 552a(a)(5), 242
5 U.S.C. § 552a(b), 242
5 U.S.C. § 706(2)(A), 53
6 U.S.C. § 1501 *et seq.*, 237
6 U.S.C. § 1501, 238
6 U.S.C. § 1531, 351
15 U.S.C. § 43, 249
15 U.S.C. § 45(a), 247
15 U.S.C. § 45(a)(1), 85
15 U.S.C. § 45(c), 250, 253
15 U.S.C. § 45(m)(1), 251
15 U.S.C. § 45(n), 248, 253, 254
15 U.S.C. § 46(a), 249
15 U.S.C. § 53(b), 250
15 U.S.C. § 57b, 250
15 U.S.C. § 1221, 96
15 U.S.C. § 1222, 96
15 U.S.C. § 2301 *et seq.*, 86
15 U.S.C. § 2303, 87
15 U.S.C. § 2304(a), 87
15 U.S.C. § 2308(a), (b), 87
18 U.S.C. § 31(a), 342
18 U.S.C. § 33, 342
18 U.S.C. § 39, 343

18 U.S.C. § 1030, 351, 353, 354
18 U.S.C. § 1030(a)(2), 353, 354
18 U.S.C. § 1030(a)(5), 354–357
18 U.S.C. § 1030(a)(7), 357, 358
18 U.S.C. § 1030(c)(2), 353
18 U.S.C. § 1030(c)(4), 357
18 U.S.C. § 1030(e)(1), 352
18 U.S.C. § 1030(e)(2), 352
18 U.S.C. § 1030(e)(6), 351
18 U.S.C. § 1030(e)(8), 355
18 U.S.C. § 1030(e)(11), 357
18 U.S.C. § 1030(g), 351
18 U.S.C. § 1951, 358
18 U.S.C. § 2119, 343
18 U.S.C. § 2510(1), 237, 383
18 U.S.C. § 2510(11), 237, 383
18 U.S.C. § 2510(15), 236, 359, 383
18 U.S.C. § 2701 *et seq.*, 236
18 U.S.C. § 2701, 236, 358, 359
18 U.S.C. § 2702, 236, 237
18 U.S.C. § 2703, 236, 381, 382, 383, 387
18 U.S.C. § 2707, 359
18 U.S.C. § 2711, 236, 382, 383
18 U.S.C. § 2721, 239, 240
18 U.S.C. § 2723, 240
18 U.S.C. § 2724, 240
42 U.S.C. § 300aa-22(b), 204
42 U.S.C. § 2210, 205
42 U.S.C. § 12184, 88
42 U.S.C. § 12184(a), 88
47 U.S.C. § 303(y), 83
47 U.S.C. § 307(e), 84
47 U.S.C. § 308, 84
47 U.S.C. § 309, 84
49 U.S.C. § 14501(c)(2), (3), 82
49 U.S.C. § 30102(a)(1), 86
49 U.S.C. § 30102(a)(3), 71
49 U.S.C. § 30102(a)(7), 50
49 U.S.C. § 30102(a)(8), 50
49 U.S.C. § 30102(a)(9), 53, 72
49 U.S.C. § 30102(a)(11), 86
49 U.S.C. § 30103(b), 93, 152
49 U.S.C. § 30103(e), 93
49 U.S.C. § 30106(a), 210
49 U.S.C. § 30106(a)(1), 214, 215
49 U.S.C. § 30106(a)(2), 212
49 U.S.C. § 30106(d)(2), 215
49 U.S.C. § 30111(a), 52, 245
49 U.S.C. § 30111(b), 52
49 U.S.C. § 30111(d), 54
49 U.S.C. § 30113, 64
49 U.S.C. § 30113(b), 55, 56

49 U.S.C. § 30113(d), 55
49 U.S.C. § 30113(e), 55
49 U.S.C. § 30114, 63
49 U.S.C. § 30114(a), 56
49 U.S.C. § 30118, 71
49 U.S.C. § 30118(c), 67
49 U.S.C. § 30118(d), 67
49 U.S.C. § 30119, 70
49 U.S.C. § 30120(a), 70
49 U.S.C. § 30120(i)(1), 86
49 U.S.C. § 30121(b), 70
49 U.S.C. § 30122, 61
49 U.S.C. § 30122(a), 61
49 U.S.C. § 30161, 54
49 U.S.C. § 30161(b), 55
49 U.S.C. § 30165, 70, 73
49 U.S.C. § 30166, 68
49 U.S.C. § 31141, 80
49 U.S.C. § 31141(c), 80, 81
49 U.S.C. § 32301(1), 145
49 U.S.C. § 140501, 81
49 U.S.C. § 140501(c), 81
American Vision for Safer Transportation through Advancement of Revolutionary Technologies Act (the AV START Act), 101–104
The Autonomous Vehicle Privacy Protection Act of 2015, H.R. 3876 (introduced on November 2, 2015, by Representative Grace Meng), 102
Early Warning Reporting System Improvement Act of 2018, H.R. 7192 (introduced on November 29, 2018, by Representative Matt Cartwright), 101
Moving Ahead for Progress in the 21st Century Act, Pub. L. No. 112-141, § 31402, 126 Stat. 405 (2012), 245
National Defense Authorization Act for Fiscal Year 2001, Pub. L. No. 106-398, § 217 (2000), 4
National Defense Authorization Act for Fiscal Year 2003, Pub. L. No. 107-314, § 2374b (2002), 4
Pub. L. No. 93-579, 88 Stat. 1896 (Dec. 31, 1974), 242
Safely Ensuring Lives Future Deployment and Research in Vehicle Evolution Act (the SELF DRIVE Act), 101–104
Senate Judiciary Committee Report, S. Rep. No. 104-357—The National Information Infrastructure Protection Act of 1995 (1996), 354, 358
Transportation Equity Act for the 21st Century, Pub. L. No. 105–178, § 5206(f), 112 Stat. 107 (1998), 84
The Transportation Workforce Modernization Act, H.R. 6200 (introduced on June 22, 2018, by Representative Eddie Bernice), 101
21st Century Automated Vehicle Safety Act of 2018, H.R. 6058 (introduced on June 8, 2018, by Representative David Loebsack), 101

TREATIES

Geneva Convention, 46–47
Vienna Convention, 46, 48

FEDERAL REGULATIONS AND ADMINISTRATIVE RULINGS

16 C.F.R. § 1.98, 251
16 C.F.R. § 2.31 *et seq.*, 251
16 C.F.R. subchapter G (§ 700 *et seq.*), 87
16 C.F.R. § 700.1, 87
16 C.F.R. § 700.3(a), 87
16 C.F.R. § 700.10, 199
23 C.F.R. § 655.603(a), 74
49 C.F.R. § 37.29(a), 88
49 C.F.R. § 37.29(b), 88
49 C.F.R. § 350.101(a), 75
49 C.F.R. § 355.21, 80
49 C.F.R. § 355.23, 80
49 C.F.R. § 355.25, 81
49 C.F.R. § 379.11, 68
49 C.F.R. § 379.21, 69
49 C.F.R. § 379.5, 68
49 C.F.R. part 380, 77
49 C.F.R. § 380.107, 77
49 C.F.R. § 380.109, 77
49 C.F.R. part 381, 77, 80
49 C.F.R. § 381.200, 80
49 C.F.R. § 381.300, 80
49 C.F.R. § 381.310, 80
49 C.F.R. § 381.315, 80
49 C.F.R. § 390.9, 80
49 C.F.R. § 390.19, 80
49 C.F.R. § 390.21, 80
49 C.F.R. part 391, 77, 80
49 C.F.R. § 391.3, 78
49 C.F.R. § 391.10, 78
49 C.F.R. § 391.11, 78
49 C.F.R. § 391.14, 78
49 C.F.R. § 391.16, 78
49 C.F.R. § 391.2, 78
49 C.F.R. § 391.22, 78
49 C.F.R. § 391.41, 77
49 C.F.R. part 392, 78, 80
49 C.F.R. § 392.5, 78
49 C.F.R. § 392.7, 78
49 C.F.R. § 392.9, 78
49 C.F.R. § 392.81, 78
49 C.F.R. § 392.82, 78
49 C.F.R. part 393, 76, 80
49 C.F.R. § 393.3, 76
49 C.F.R. part 395, 77, 80
49 C.F.R. § 395.3, 77
49 C.F.R. § 395.8, 77
49 C.F.R. § 395.20 *et seq.*, 77
49 C.F.R. part 396, 78, 80

49 C.F.R. § 396.3(a)(1), 76
49 C.F.R. § 396.9, 78
49 C.F.R. § 396.11, 78
49 C.F.R. § 396.18, 78
49 C.F.R. § 396.19, 78
49 C.F.R. part 510, 68
49 C.F.R. § 510.3, 68
49 C.F.R. part 552, 65, 68
49 C.F.R. § 552.6, 65
49 C.F.R. § 552.7, 65
49 C.F.R. § 552.8, 65
49 C.F.R. § 552.9, 65, 68
49 C.F.R. § 552.10, 65
49 C.F.R. part 555, 63, 64
49 C.F.R. § 555.5(b), 63
49 C.F.R. § 555.6, 63
49 C.F.R. § 555.7, 63
49 C.F.R. § 563.1 *et seq.*, 241, 285, 389
49 C.F.R. § 563.5, 240, 241, 286
49 C.F.R. § 563.11, 287
49 C.F.R. part 571, 47
49 C.F.R. § 571, 54
49 C.F.R. § 571.3, 58, 59
49 C.F.R. § 571.101, 54, 57
49 C.F.R. § 571.105, 58
49 C.F.R. § 571.108, 58
49 C.F.R. § 571.109, 58
49 C.F.R. § 571.208, 57, 155
49 C.F.R. § 571.500, 66
49 C.F.R. part 573, 67
49 C.F.R. § 573.5(a), 67
49 C.F.R. § 573.5(b), 67
49 C.F.R. § 573.6, 67
49 C.F.R. § 573.11, 70
49 C.F.R. § 573.12, 70
49 C.F.R. part 577, 70
49 C.F.R. § 577.5, 70
49 C.F.R. § 577.6, 70
49 C.F.R. § 577.7, 70
49 C.F.R. § 577.10, 70
49 C.F.R. § 577.13, 70
49 C.F.R. part 579, 68
Fed. Commc'ns Comm'n, *Dedicated Short Range Communications on Intelligent Transportation Services*, 64 Fed. Reg. 66,405 (Nov. 26, 1999), 84
Fed. Motor Carrier Safety Admin., *Request for Comments Concerning Federal Motor Carrier Safety Regulations (FMCSRs) Which May Be a Barrier to the Safe Testing and Deployment of Automated Driving Systems-Equipped Commercial Motor Vehicles on Public Roads*, 83 Fed. Reg. 12,933 (Mar. 26, 2018), 75
Fed. Motor Carrier Safety Admin., *Safe Integration of Automated Driving Systems-Equipped Commercial Motor Vehicles*, 84 Fed. Reg. 24,449-01 (May 28, 2019), 75, 79
In the Matter of Amendment of the Commission's Rules Regarding Dedicated Short-Range Communication Services in the 5.850-5.925 GHz Band (5.9 GHz Band), Fed. Commc'ns Comm'n, 19 F.C.C. Rcd. 2458 (Feb. 10, 2004), 84

Nat'l Highway Traffic Safety Admin., *Federal Motor Vehicle Safety Standards; Automatic Emergency Braking*, 80 Fed. Reg. 62,487 (Oct. 16, 2015), 66
Nat'l Highway Traffic Safety Admin., *Federal Motor Vehicle Safety Standards; Automatic Emergency Braking*, 82 Fed. Reg. 8391 (Jan. 18, 2017), 66, 94, 157
Nat'l Highway Traffic Safety Admin., *NHTSA Enforcement Guidance Bulletin 2016-02: Safety-Related Defects and Automated Safety Technologies*, 81 Fed. Reg. 65,705-01 (Sept. 23, 2016), 50, 51, 67
Nat'l Highway Traffic Safety Admin., *Nuro, Inc.; Grant of Temporary Exemption for a Low-Speed Vehicle with an Automated Driving System*, 85 Fed. Reg. 7826 (Feb. 11, 2020), 64
Nat'l Highway Traffic Safety Admin., *Occupant Protection for Automated Driving Systems*, 85 Fed. Reg. 17,624 (Mar. 30, 2020), 47, 59, 61
Nat'l Highway Traffic Safety Admin., *Removing Regulatory Barriers for Vehicles with Automated Driving Systems*, 83 Fed. Reg. 2607 (Jan. 18, 2018), 57, 60, 61
Nat'l Highway Traffic Safety Admin., *Removing Regulatory Barriers for Vehicles with Automated Driving Systems*, 84 Fed. Reg. 24,433 (May 28, 2019), 57, 60, 61
Nat'l Telecomms. & Info. Admin., *Developing the Administration's Approach to Consumer Privacy*, 83 Fed. Reg. 48,600-01 (Sept. 26, 2018), 245

FEDERAL RULES OF EVIDENCE, UNIFORM ACT, AND RESTATEMENT

Federal Rule of Evidence 407, 160, 161
Unif. Automated Operation of Vehicles Act, 92, 115–117, 125, 309
RESTATEMENT (SECOND) OF TORTS § 302b, 223
RESTATEMENT (SECOND) OF TORTS § 402a, 134, 147, 148, 149
RESTATEMENT (SECOND) OF TORTS § 491, 221
RESTATEMENT (THIRD) OF TORTS: PRODS. LIAB. § 1, 211
RESTATEMENT (THIRD) OF TORTS: PRODS. LIAB. § 2, passim
RESTATEMENT (THIRD) OF TORTS: PRODS. LIAB. § 3, 134, 138, 139
RESTATEMENT (THIRD) OF TORTS: PRODS. LIAB. § 4(b), 159
RESTATEMENT (THIRD) OF TORTS: PRODS. LIAB. § 5, 141
RESTATEMENT (THIRD) OF TORTS: PRODS. LIAB. § 7, 123
RESTATEMENT (THIRD) OF TORTS: PRODS. LIAB. § 19(a), 131

UNITED STATES SUPREME COURT

Arizona v. Gant, 556 U.S. 332, 129 S. Ct. 1710, 173 L. Ed. 2d 485 (2009), 361, 372
Arizona v. Johnson, 555 U.S. 323, 129 S. Ct. 781, 172 L. Ed. 2d 694 (2009), 369, 370
Ark. Elec. Co-op. Corp. v. Ark. Pub. Serv. Comm'n, 461 U.S. 375, 103 S. Ct. 1905, 76 L. Ed. 2d 1 (1983), 154
Bond v. United States, 529 U.S. 334, 120 S. Ct. 1462, 146 L. Ed. 2d 365 (2000), 374
Brendlin v. California, 551 U.S. 249, 127 S. Ct. 2400, 168 L. Ed. 2d 132 (2007), 363, 364, 373
Brigham City, Utah v. Stuart, 547 U.S. 398, 126 S. Ct. 1943, 164 L. Ed. 2d 650 (2006), 362
Brown-Forman Distillers Corp. v. N.Y. State Liquor Auth., 476 U.S. 573, 106 S. Ct. 2080, 90 L. Ed. 2d 552 (1986), 99
Byrd v. United States, 138 S. Ct. 1518, 200 L. Ed. 2d 805 (2018), 375, 376, 377
California v. Carney, 471 U.S. 386, 105 S. Ct. 2066, 85 L. Ed. 2d 406 (1985), 373, 374
Carpenter v. United States, 138 S. Ct. 2206, 201 L. Ed. 2d 507 (2018), 370, 386, 387, 391, 393

Carroll v. United States, 267 U.S. 132, 45 S. Ct. 280, 69 L. Ed. 543 (1925), 371
Chimel v. California, 395 U.S. 752, 89 S. Ct. 2034, 23 L. Ed. 2d 685 (1969), 372
Cipollone v. Liggett Grp., Inc., 505 U.S. 504, 112 S. Ct. 2608, 120 L. Ed. 2d 407 (1992), 153
City of Columbus v. Ours Garage & Wrecker Serv., Inc., 536 U.S. 424, 122 S. Ct. 2226, 153 L. Ed. 2d 430 (2002), 81
City of Indianapolis v. Edmond, 531 U.S. 32 (2000), 366
Clapper v. Amnesty Int'l USA, 568 U.S. 398, 133 S. Ct. 1138, 185 L. Ed. 2d 264 (2013), 293, 294
Colorado v. Bertine, 479 U.S. 367, 107 S. Ct. 738, 93 L. Ed. 2d 739 (1987), 367
Dan's City Used Cars, Inc. v. Pelkey, 569 U.S. 251, 133 S. Ct. 1769, 185 L. Ed. 2d 909 (2013), 81
Delaware v. Prouse, 440 U.S. 648, 99 S. Ct. 1391, 59 L. Ed. 2d 660 (1979), 364, 367
F.A.A. v. Cooper, 566 U.S. 284, 132 S. Ct. 1441, 182 L. Ed. 2d 497 (2012), 242
F.T.C. v. Henry Broch & Co., 368 U.S. 360, 82 S. Ct. 431, 7 L. Ed. 2d 353 (1962), 250
F.T.C. v. Sperry & Hutchinson Co., 405 U.S. 233, 92 S. Ct. 898, 31 L. Ed. 2d 170 (1972), 247, 254
Florida v. Bostick, 501 U.S. 429, 111 S. Ct. 2382, 115 L. Ed. 2d 389 (1991), 363
Freightliner Corp. v. Myrick, 514 U.S. 280, 115 S. Ct. 1483, 131 L. Ed. 2d 385 (1995), 94, 153, 154
Geier v. Am. Honda Motor Co., 529 U.S. 861, 120 S. Ct. 1913, 146 L. Ed. 2d 914 (2000), 94, 153, 154, 155, 156
Georgia v. Randolph, 547 U.S. 103, 126 S. Ct. 1515, 164 L. Ed. 2d 208 (2006), 380
Holloway v. United States, 526 U.S. 1, 119 S. Ct. 966, 143 L. Ed. 2d 1 (1999), 343
Hudson v. United States, 522 U.S. 93, 118 S. Ct. 488, 139 L. Ed. 2d 450 (1997), 323
Illinois v. Caballes, 543 U.S. 405, 125 S. Ct. 834, 160 L. Ed. 2d 842 (2005), 364, 369
Illinois v. Rodriguez, 497 U.S. 177, 110 S. Ct. 2793, 111 L. Ed. 2d 148 (1990), 366, 380
Indian Towing Co. v. United States, 350 U.S. 61, 76 S. Ct. 122, 100 L. Ed. 48 (1955), 230
Jacob Siegel Co. v. F.T.C., 327 U.S. 608, 66 S. Ct. 758, 90 L. Ed. 888 (1946), 250
Katz v. United States, 389 U.S. 347, 88 S. Ct. 507, 19 L. Ed. 2d 576 (1967), 363, 370, 376, 385, 386, 388
Kennedy v. Mendoza-Martinez, 372 U.S. 144, 83 S. Ct. 554, 9 L. Ed. 2d 644 (1963), 322
Knowles v. Iowa, 525 U.S. 113, 119 S. Ct. 484, 142 L. Ed. 2d 492 (1998), 370, 373
Lujan v. Defs. of Wildlife, 504 U.S. 555, 112 S. Ct. 2130, 119 L. Ed. 2d 351 (1992), 163, 288, 293
Maryland v. Garrison, 480 U.S. 79, 107 S. Ct. 1013, 94 L. Ed. 2d 72 (1987), 394
Maryland v. Pringle, 540 U.S. 366, 124 S. Ct. 795, 157 L. Ed. 2d 769 (2003), 379
Michigan v. Thomas, 458 U.S. 259, 102 S. Ct. 3079, 73 L. Ed. 2d 750 (1982), 371, 392
Morales v. Trans World Airlines, Inc., 504 U.S. 374, 112 S. Ct. 2031, 119 L. Ed. 2d 157 (1992), 81
Motor Vehicle Mfrs. Ass'n of U.S., Inc. v. State Farm Mut. Auto. Ins. Co., 463 U.S. 29, 103 S. Ct. 2856, 77 L. Ed. 2d 443 (1983), 52, 53
New York Cent. & H.R.R. Co. v. United States, 212 U.S. 481, 29 S. Ct. 304, 53 L. Ed. 613 (1909), 312
Pa. Co. v. Roy, 102 U.S. 451, 26 L. Ed. 141 (1880), 220
Pennsylvania v. Labron, 518 U.S. 938, 116 S. Ct. 2485, 135 L. Ed. 2d 1031 (1996), 371
Pike v. Bruce Church, Inc., 397 U.S. 137, 90 S. Ct. 844, 25 L. Ed. 2d 174 (1970), 99, 100
Rakas v. Illinois, 439 U.S. 128, 99 S. Ct. 421, 58 L. Ed. 2d 387 (1978), 373, 374, 375, 376
Reno v. Am. Civil Liberties Union, 521 U.S. 844, 117 S. Ct. 2329, 138 L. Ed. 2d 874 (1997), 352
Rice v. Santa Fe Elevator Corp., 331 U.S. 218, 67 S. Ct. 1146, 91 L. Ed. 1447 (1947), 153
Riley v. California, 573 U.S. 373, 134 S. Ct. 2473, 189 L. Ed. 2d 430 (2014), 362, 391, 393, 395
Rios v. United States, 364 U.S. 253 (1960), 376
Rodriguez v. United States, 575 U.S. 348, 135 S. Ct. 1609, 191 L. Ed. 2d 492 (2015), 361, 364, 366, 380
Rowe v. N.H. Motor Transp. Ass'n, 552 U.S. 364, 128 S. Ct. 989, 169 L. Ed. 2d 933 (2008), 81
S-Cent. Timber Dev., Inc. v. Wunnicke, 467 U.S. 82, 104 S. Ct. 2237, 81 L. Ed. 2d 71 (1984), 99
Schmerber v. California, 384 U.S. 757, 86 S. Ct. 1826, 16 L. Ed. 2d 908 (1966), 382
Schneckloth v. Bustamonte, 412 U.S. 218, 93 S. Ct. 2041, 36 L. Ed. 2d 854 (1973), 379
Scott v. Harris, 550 U.S. 372, 127 S. Ct. 1769, 167 L. Ed. 2d 686 (2007), 337, 366, 368, 369

S. Dakota v. Opperman, 428 U.S. 364, 96 S. Ct. 3092, 49 L. Ed. 2d 1000 (1976), 367, 368, 371
Smith v. Maryland, 442 U.S. 735, 99 S. Ct. 2577, 61 L. Ed. 2d 220 (1979), 385–387
Soldal v. Cook Cnty., Ill., 506 U.S. 56, 113 S. Ct. 538, 121 L. Ed. 2d 450 (1992), 365
Spokeo, Inc. v. Robins, 136 S. Ct. 1540, 194 L. Ed. 2d 635 (2016), 293, 295
Sprietsma v. Mercury Marine, a Div. of Brunswick Corp., 537 U.S. 51, 123 S. Ct. 518, 154 L. Ed. 2d 466 (2002), 94, 153, 154
Susan B. Anthony List v. Driehaus, 573 U.S. 149, 134 S. Ct. 2334, 189 L. Ed. 2d 246 (2014), 293, 294
Terry v. Ohio, 392 U.S. 1, 88 S. Ct. 1868, 20 L. Ed. 2d 889 (1968), 364, 370
Thornton v. United States, 541 U.S. 615, 124 S. Ct. 2127, 158 L. Ed. 2d 905 (2004), 372
United States v. Drayton, 536 U.S. 194, 122 S. Ct. 2105, 153 L. Ed. 2d 242 (2002), 363
United States v. Jacobsen, 466 U.S. 109, 104 S. Ct. 1652, 80 L. Ed. 2d 85 (1984), 363, 365, 389
United States v. Johns, 469 U.S. 478, 105 S. Ct. 881, 83 L. Ed. 2d 890 (1985), 373
United States v. Jones, 565 U.S. 400, 132 S. Ct. 945, 181 L. Ed. 2d 911 (2012), 233, 269, 362, 370, 371, 386, 387, 391, 393
United States v. Karo, 468 U.S. 705, 104 S. Ct. 3296, 82 L. Ed. 2d 530 (1984), 387
United States v. Knotts, 460 U.S. 276, 103 S. Ct. 1081, 75 L. Ed. 2d 55 (1983), 387, 388
United States v. Miller, 425 U.S. 435, 96 S. Ct. 1619, 48 L. Ed. 2d 71 (1976), 385, 387
United States v. Place, 462 U.S. 696, 103 S. Ct. 2637, 77 L. Ed. 2d 110 (1983), 365
United States v. Ross, 456 U.S. 798, 102 S. Ct. 2157, 72 L. Ed. 2d 572 (1982), 371, 378
United States v. Salerno, 481 U.S. 739, 107 S. Ct. 2095, 95 L. Ed. 2d 697 (1987), 323
United States v. S.-E. Underwriters Ass'n, 322 U.S. 533, 64 S. Ct. 1162, 88 L. Ed. 1440 (1944), 92
United States v. Sharpe, 470 U.S. 675, 105 S. Ct. 1568, 84 L. Ed. 2d 605 (1985), 364
Vernonia Sch. Dist. 47J v. Acton, 515 U.S. 646, 115 S. Ct. 2386, 132 L. Ed. 2d 564 (1995), 363
Virginia v. Moore, 553 U.S. 164, 128 S. Ct. 1598, 170 L. Ed. 2d 559 (2008), 373
Warth v. Seldin, 422 U.S. 490, 95 S. Ct. 2197, 45 L. Ed. 2d 343 (1975), 163
Wash. State Grange v. Wash. State Republican Party, 552 U.S. 442, 128 S. Ct. 1184, 170 L. Ed. 2d 151 (2008), 323
Whren v. United States, 517 U.S. 806, 116 S. Ct. 1769, 135 L. Ed. 2d 89 (1996), 361, 364
Williamson v. Mazda Motor of Am., Inc., 562 U.S. 323, 131 S. Ct. 1131, 179 L. Ed. 2d 75 (2011), 94, 153, 155, 156
Wyoming v. Houghton, 526 U.S. 295, 119 S. Ct. 1297, 143 L. Ed. 2d 408 (1999), 378, 379

FIRST CIRCUIT

Cipollone v. Yale Indus. Prod., Inc., 202 F.3d 376 (1st Cir. 2000), 141
Connelly v. Hyundai Motor Co., 351 F.3d 535 (1st Cir. 2003), 130
EF Cultural Travel BV v. Explorica, Inc., 274 F.3d 577 (1st Cir. 2001), 352
Katz v. Pershing, LLC, 672 F.3d 64 (1st Cir. 2012), 298
Mitchell v. Ford Motor Co., 533 F.2d 19 (1st Cir. 1976), 151
Quintana-Ruiz v. Hyundai Motor Corp., 303 F.3d 62 (1st Cir. 2002), 135
United States v. Dall, 608 F.2d 910 (1st Cir. 1979), 375

SECOND CIRCUIT

Berkan v. Penske Truck Leasing Canada, Inc., 535 F. Supp. 2d 341 (W.D.N.Y. 2008), 212
Colon v. Bernabe, No. 07 CIV. 3369 AJP, 2007 WL 2068093 (S.D.N.Y. July 19, 2007), 212
Contini by Contini v. Hyundai Motor Co., 876 F. Supp. 540 (S.D.N.Y. 1995), 54

Crespo v. Chrysler Corp., 75 F. Supp. 2d 225 (S.D.N.Y. 1999), 186
In re Jetblue Airways Corp. Privacy Litig., 379 F. Supp. 2d 299 (E.D.N.Y. 2005), 236, 383
Novovic v. Greyhound Lines, Inc., No. CV-08-3190 (CPS), 2008 WL 5000228 (E.D.N.Y. Nov. 19, 2008), 212
Rein v. CAB E. LLC, No. 08 CIV. 2899PAC, 2009 WL 1748905 (S.D.N.Y. June 22, 2009), 209
In re Sept. 11 Litig., 280 F. Supp. 2d 279 (S.D.N.Y. 2003), 301
Shafran v. Harley-Davidson, Inc., No. 07 CIV. 01365 (GBD), 2008 WL 763177 (S.D.N.Y. Mar. 20, 2008), 297
Stanford v. Kuwait Airways Corp., 89 F.3d 117 (2d Cir. 1996), 301
The T.J. Hooper, 60 F.2d 737 (2d Cir. 1932), 139, 159
United States v. Buckner, 417 F. Supp. 2d 240 (S.D.N.Y. 2005), 376
United States v. Carroll Towing Co., 159 F.2d 169 (2d Cir. 1947), 124
United States v. Galpin, 720 F.3d 436 (2d Cir. 2013), 395
United States v. Morris, 928 F.2d 504 (2d Cir. 1991), 353
United States v. One 1986 Mercedes Benz, VIN WDBEA30D2GA143459, 846 F.2d 2 (2d Cir. 1988), 375
United States v. Santiago, 950 F. Supp. 5908 (S.D.N.Y. 1996), 376
United States v. Valle, 807 F.3d 508 (2d Cir. 2015), 352

THIRD CIRCUIT

F.T.C. v. Wyndham Worldwide Corporation, 799 F.3d 236 (3d Cir. 2015), 252, 254
In re Google Inc. Cookie Placement Consumer Privacy Litig., 806 F.3d 125 (3d Cir. 2015), 359
Granillo v. FCA US LLC, No. CV 16-153 (FLW)(DEA), 2016 WL 9405772 (D.N.J. Aug. 29, 2016), 133
Hughes v. Panasonic Consumer Elecs. Co., No. CIV.A. 10-846 SDW, 2011 WL 2976839 (D.N.J. July 21, 2011), 133
In re Jacoby Airplane Crash Litig., No. 99-cv-6073 (D.N.J. 2007), 170
P.C. Yonkers, Inc. v. Celebrations the Party & Seasonal Superstore, LLC, 428 F.3d 504 (3d Cir. 2005), 351
Reilly v. Ceridian Corp., 664 F.3d 38 (3d Cir. 2011), 298
Sochanski v. Sears, Roebuck & Co., 689 F.2d 45 (3d Cir. 1982), 132
United States v. Anderson, 859 F.2d 1171 (3d Cir. 1988), 377
United States v. Carlson, 209 F. App'x 1814 (3d Cir. 2006), 355
United States v. Morales, 861 F.2d 396 (3d Cir. 1988), 380
United States v. Reader's Digest Ass'n, Inc., 662 F.2d 955 (3d Cir. 1981), 251
Verge v. Ford Motor Co., 581 F.2d 384 (3d Cir. 1978), 141

FOURTH CIRCUIT

Mullins v. Johnson & Johnson, 236 F. Supp. 3d 940 (S.D.W. Va. 2017), 138
Sexton *ex rel.* Sexton v. Bell Helmets, Inc., 926 F.2d 331 (4th Cir. 1991), 193
United States v. Lowe, 65 F.3d 1137 (4th Cir. 1995), 342
United States v. Sullivan, 40 F. App'x 740 (4th Cir. 2002), 355
Vehicle Equip. Safety Comm'n v. Nat'l Highway Traffic Safety Admin., 611 F.2d 53 (4th Cir. 1979), 54
WEC Carolina Energy Sols. LLC v. Miller, 687 F.3d 199 (4th Cir. 2012), 352
Wickersham v. Ford Motor Co., 194 F. Supp. 3d 434 (D.S.C. 2016), 140

FIFTH CIRCUIT

Becker v. Toca, No. CIV. A. 07-7202, 2008 WL 4443050 (E.D. La. Sept. 26, 2008), 355
Bullock v. Tamiami Trail Tours, Inc., 266 F.2d 326 (5th Cir. 1959), 223
Casey v. Toyota Motor Eng'g & Mfg. N. Am., Inc., 770 F.3d 322 (5th Cir. 2014), 133
Coal. for an Airline Passengers' Bill of Rights v. Delta Air Lines, Inc., 693 F. Supp. 2d 667 (S.D. Tex. 2010), 269
Cole v. Gen. Motors Corp., 484 F.3d 717 (5th Cir. 2007), 162
Dykes v. Husqvarna Outdoor Prod., N.A., Inc., 869 F. Supp. 2d 749 (S.D. Miss. 2012), 132
Ford Motor Co. v. Texas Dep't of Transp., 106 F. Supp. 2d 905 (W.D. Tex. 2000), 96
Garcia v. City of Laredo, Tex., 702 F.3d 788 (5th Cir. 2012), 359
Ramos v. Uber Techs., Inc., No. SA-14-CA-502-XR, 2015 WL 758087 (W.D. Tex. Feb. 20, 2015), 88
Trevino v. Yamaha Motor Corp., U.S.A., 882 F.2d 182 (5th Cir. 1989), 141
United States v. John, 597 F.3d 263 (5th Cir. 2010), 352
United States v. Nunn, 525 F.2d 958 (5th Cir. 1976), 375

SIXTH CIRCUIT

Bailey v. Bailey, No. 07-11672, 2008 WL 324156 (E.D. Mich. Feb. 6, 2008), 359
B.F. Goodrich Co. v. Dep't of Transp., 541 F.2d 1178 (6th Cir. 1976), 54
Brouse v. United States, 83 F. Supp. 373 (N.D. Ohio 1949), 170
Chrysler Corp. v. Dep't of Transp., 472 F.2d 659 (6th Cir. 1972), 52, 53, 54
Galaria v. Nationwide Mut. Ins. Co., 663 F. App'x 384 (6th Cir. 2016), 295
Garrison v. Rohm & Haas Co., 492 F.2d 346 (6th Cir. 1974), 222
Hagen v. U-Haul Co. of Tenn., 613 F. Supp. 2d 986 (W.D. Tenn. 2009), 212
Int'l Dairy Foods Ass'n v. Boggs, 622 F.3d 628 (6th Cir. 2010), 100
Morgan v. Preston, No. 3:13-00403, 2013 WL 5963563 (M.D. Tenn. Nov. 7, 2013), 359
Nat'l Truck Equip. Ass'n v. Nat'l Highway Traffic Safety Admin., 711 F.3d 662 (6th Cir. 2013), 53, 54
Nat'l Truck Equip. Ass'n v. Nat'l Highway Traffic Safety Admin., 919 F.2d 1148 (6th Cir. 1990), 53
Pulte Homes, Inc. v. Laborers' Int'l Union of N. Am., 648 F.3d 295 (6th Cir. 2011), 355
Reminga v. United States, 631 F.2d 449 (6th Cir. 1980), 232
Sherman & Co. v. Salton Maxim Housewares, Inc., 94 F. Supp. 2d 817 (E.D. Mich. 2000), 359
United States v. Heightland, 678 F. Supp. 159 (E.D. Ky. 1987), 342

SEVENTH CIRCUIT

Alcala v. Emhart Indus., Inc., 495 F.3d 360 (7th Cir. 2007), 130
Ed Houser Enters., Inc. v. Gen. Motors Corp., 595 F.2d 366 (7th Cir. 1978), 96
Flynn v. FCA US LLC, No. 15-CV-0855-MJR-DGW, 2016 WL 5341749 (S.D. Ill. Sept. 23, 2016), 298, 300
F.T.C. v. Credit Bureau Ctr., LLC, 937 F.3d 764 (7th Cir. 2019), 250
F.T.C. v. World Travel Vacation Brokers, Inc., 861 F.2d 1020 (7th Cir. 1988), 249
H & H Tire Co. v. U.S. Dep't of Transp., 471 F.2d 350 (7th Cir. 1972), 54
Idris v. City of Chicago, Ill., 552 F.3d 564 (7th Cir. 2009), 322
In re Aqua Dots Products Liability Litigation, 654 F.3d 748 (7th Cir. 2011), 164–165
In re Michaels Stores Pin Pad Litig., 830 F. Supp. 2d 518 (N.D. Ill. 2011), 296, 301

Int'l Airport Ctrs., L.L.C. v. Citrin, 440 F.3d 418 (7th Cir. 2006), 352, 355
Janusz v. Symmetry Med. Inc., 256 F. Supp. 3d 995 (E.D. Wis. 2017), 222
Koehring Co. v. Adams, 452 F. Supp. 635 (E.D. Wis. 1978), 50
Martin v. Harrington & Richardson, Inc., 743 F.2d 1200 (7th Cir. 1984), 180
Patrick Patterson Custom Homes, Inc. v. Bach, 586 F. Supp. 2d 1026 (N.D. Ill. 2008), 352
Remijas v. Neiman Marcus Grp., LLC, 794 F.3d 688 (7th Cir. 2015), 295
Spiegel, Inc. v. F.T.C., 540 F.2d 287 (7th Cir. 1976), 247
United States v. JS & A Grp., Inc., 716 F.2d 451 (7th Cir. 1983), 250
United States v. Posey, 663 F.2d 37 (7th Cir. 1981), 377

EIGHTH CIRCUIT

Am. Online, Inc. v. Nat'l Health Care Disc., Inc., 121 F. Supp. 2d 1255 (N.D. Iowa 2000), 352
Dyer v. Nw. Airlines Corp., 334 F. Supp. 2d 1196 (D.N.D. 2004), 236, 383
F.T.C. v. Sec. Rare Coin & Bullion Corp., 931 F.2d 1312 (8th Cir. 1991), 250
In re Target Corp. Customer Data Sec. Breach Litig., 64 F. Supp. 3d 1304 (D. Minn. 2014), 296
Larsen v. Gen. Motors Corp., 391 F.2d 495 (8th Cir. 1968), 72, 145
Linegar v. Armour of America, Inc., 909 F.2d 1150 (8th Cir. 1990), 180
O'Hare v. Merck & Co., 381 F.2d 286 (8th Cir. 1967), 130
Robinson v. Brandtjen & Kluge, Inc., 500 F.3d 691 (8th Cir. 2007), 193
Sperry v. Bauermeister, Inc., 4 F.3d 596 (8th Cir. 1993), 141
United States v. Powers, No. 8:09CR361, 2010 WL 1418172 (D. Neb. Mar. 4, 2010), 353
United States v. Rose, 731 F.2d 1337 (8th Cir. 1984), 377
United States v. Williams, 714 F.2d 777 (8th Cir. 1983), 376

NINTH CIRCUIT

Aetna Cas. & Sur. Co. v. Jeppesen & Co., 642 F.2d 339 (9th Cir. 1981), 232
Beyer v. Symantec Corp., No. 18-CV-02006-EMC, 2019 WL 935135 (N.D. Cal. Feb. 26, 2019), 164, 297, 299
Brocklesby v. United States, 767 F.2d 1288 (9th Cir. 1985), 232
Cahen v. Toyota Motor Corp., 147 F. Supp. 3d 955 (N.D. Cal. 2015), 162, 297, 298
Cahen v. Toyota Motor Corp., 717 F. App'x 720 (9th Cir. 2017), 298
Contreras v. Toyota Motor Sales USA, Inc., No. C 09-06024 JSW, 2010 WL 2528844 (N.D. Cal. June 18, 2010), 162, 163
Crispin v. Christian Audigier, Inc., 717 F. Supp. 2d 965 (C.D. Cal. 2010), 237, 384
Deacon v. Pandora Media, Inc., 901 F. Supp. 2d 1166 (N.D. Cal. 2012), 289
Diva Limousine, Ltd. v. Uber Techs., Inc., 392 F. Supp. 3d 1074 (N.D. Cal. 2019), 220
Doe v. Uber Techs., Inc., 184 F. Supp. 3d 774 (N.D. Cal. 2016), 220
Facebook, Inc. v. Power Ventures, Inc., 844 F.3d 1058 (9th Cir. 2016), 347
F.T.C. v. H.N. Singer, Inc., 668 F.2d 1107 (9th Cir. 1982), 250
In re Apple Processor Litig., No. 5:18-CV-00147-EJD, 2019 WL 3533876 (N.D. Cal. Aug. 2, 2019), 161, 297, 299
In re iPhone Application Litig., 844 F. Supp. 2d 1040 (N.D. Cal. 2012), 359
In re Sony Gaming Networks & Customer Data Sec. Breach Litig., 996 F. Supp. 2d 942 (S.D. Cal. Feb. 10, 2014), 296
In re Toyota Motor Corp., 790 F. Supp. 2d 1152 (C.D. Cal. 2011), 162, 163
In re Toyota Motor Corp. Unintended Acceleration Mktg., Sales Practices, & Prod. Liab. Litig., 754 F. Supp. 2d 1145 (C.D. Cal. 2010), 162, 163, 297

Kalitta Air, L.L.C. v. Cent. Tex. Airborne Sys., Inc., 315 F. App'x 603 (9th Cir. 2008), 296
Mazza v. Am. Honda Motor Co., 666 F.3d 581 (9th Cir. 2012), 162, 297
Moore v. R.G. Indus., Inc., 789 F.2d 1326 (9th Cir. 1986), 180
Nat'l Fed'n of the Blind of Cal. v. Uber Techs., Inc., 103 F. Supp. 3d 1073 (N.D. Cal. 2015), 88
Paccar, Inc. v. Nat'l Highway Traffic Safety Admin., 573 F.2d 632 (9th Cir. 1978), 53, 54
Patel v. Facebook, Inc., 932 F.3d 1264 (9th Cir. 2019), 274
Pica v. Delta Air Lines, Inc., No. CV 18-2876-MWF(EX), 2019 WL 1598671 (C.D. Cal. Feb. 14, 2019), 236
Pub. Citizen Inc. v. Mineta, 343 F.3d 1159 (9th Cir. 2003), 54
Robins v. Spokeo, Inc., 867 F.3d 1108 (9th Cir. 2017), 295
Theofel v. Farey-Jones, 359 F.3d 1066 (9th Cir. 2004), 351
United States v. Comprehensive Drug Testing, Inc., 621 F.3d 1162 (9th Cir. 2010), 395
United States v. Drew, 259 F.R.D. 449 (C.D. Cal. 2009), 352
United States v. Floersheim, No. CV 74-484-RF, 1980 WL 1852 (C.D. Cal. May 1, 1980), 251
United States v. Kurka, 818 F.2d 1427 (9th Cir. 1987), 342
United States v. Mullins, 992 F.2d 1472 (9th Cir. 1993), 236, 383
United States v. Nosal, 676 F.3d 854 (9th Cir. 2012), 353
United States v. Portillo, 633 F.2d 1313 (9th Cir. 1980), 376–377
United States v. Thomas, 447 F.3d 1191 (9th Cir. 2006), 377
United States v. Welch, 4 F.3d 761 (9th Cir. 1993), 379
Van Maanen v. Youth With a Mission-Bishop, 852 F. Supp. 2d 1232 (E.D. Cal. 2012), 219

TENTH CIRCUIT

Bruce v. Martin-Marietta Corp., 544 F.2d 442 (10th Cir. 1976), 151
Butler v. Daimler Trucks N. Am., LLC, No. 2:19-CV-2377-JAR-JPO, 2020 WL 128055 (D. Kan. Jan. 10, 2020), 157, 175
F.T.C. v. Accusearch Inc., 570 F.3d 1187 (10th Cir. 2009), 247
Hanley v. Chrysler Motors Corp., 433 F.2d 708 (10th Cir. 1970), 96
Moe v. Avions Marcel Dassault-Breguet Aviation, 727 F.2d 917 (10th Cir. 1984), 170
Paradigm Alliance, Inc. v. Celeritas Techs., LLC, 248 F.R.D. 598 (D. Kan. 2008), 352
Pryor v. Lee C. Moore Corp., 262 F.2d 673 (10th Cir. 1958), 152
United States v. Burgess, 576 F.3d 1078 (10th Cir. 2009), 395
United States v. Gurule, 461 F.3d 1238 (10th Cir. 2006), 343
United States v. Orrego-Fernandez, 78 F.3d 1497 (10th Cir. 1996), 376
United States v. Otero, 563 F.3d 1127 (10th Cir. 2009), 395
United States v. Russian, 848 F.3d 1239 (10th Cir. 2017), 395
United States v. Willis, 476 F.3d 1121 (10th Cir. 2007), 353
Washington v. Dep't of Transp., 84 F.3d 1222 (10th Cir. 1996), 52
Winzler v. Toyota Motor Sales U.S.A., Inc., 681 F.3d 1208 (10th Cir. 2012), 164

ELEVENTH CIRCUIT

Dubose v. Transp. Enter. Leasing, LLC, No. 608-CV-385-ORL-31DAB, 2009 WL 210724 (M.D. Fla. Jan. 27, 2009), 212
Ferguson v. Bombardier Servs. Corp., 244 F. App'x 944 (11th Cir. 2007), 170
F.T.C. v. U.S. Oil & Gas Corp., 748 F.2d 1431 (11th Cir. 1984), 250
Hack v. SAI Rockville L, LLC, No. CIV. WDQ-14-1985, 2015 WL 795853 (D. Md. Feb. 24, 2015), 212
LabMD, Inc. v. F.T.C., 894 F.3d 1221 (11th Cir. 2018), 252–254
United States v. Dunkley, 911 F.2d 522 (11th Cir. 1990), 375, 380

United States v. Harris, 526 F.3d 1334 (11th Cir. 2008), 376
United States v. Miller, 821 F.2d 546 (11th Cir. 1987), 376
United States v. Rodriguez, 628 F.3d 1258 (11th Cir. 2010), 352
United States v. Steiger, 318 F.3d 1039 (11th Cir. 2003), 359

D.C. CIRCUIT

Attias v. Carefirst, Inc., 865 F.3d 620 (D.C. Cir. 2017), 295
Callas v. Indep. Taxi Owners Assoc., 66 F.2d 192 (D.C. Cir. 1933), 221
Ctr. for Auto Safety, Inc. v. Lewis, 685 F.2d 656 (D.C. Cir. 1982), 69
Electronic Privacy Information Center v. United States Department of Homeland Security, 653 F.3d 1 (D.C. Cir. 2011), 242, 243
Nat'l Tire Dealers & Retreaders Ass'n, Inc. v. Brinegar, 491 F.2d 31 (D.C. Cir. 1974), 53, 54
Pac. Legal Found. v. Dep't of Transp., 593 F.2d 1338 (D.C. Cir. 1979), 52, 53
Parts & Accessories Ass'n v. Boyd, 407 F.2d 330 (D.C. Cir. 1968), 53
United States v. Chrysler Corp., 158 F.3d 1350 (D.C. Cir. 1998), 71
United States v. Ford Motor Co., 453 F. Supp. 1240 (D.D.C. 1978), 71
United States v. Ford Motor Co., 574 F.2d 534 (D.C. Cir. 1978), 71
United States v. Gen. Motors Corp., 518 F.2d 420 (D.C. Cir. 1975), 71, 72
United States v. Gen. Motors Corp., 561 F.2d 923 (D.C. Cir. 1977), 71, 72, 73
United States v. Gen. Motors Corp., 565 F.2d 754 (D.C. Cir. 1977), 71, 73
United States v. Gen. Motors Corp., 574 F. Supp. 1047 (D.D.C. 1983), 69
United States v. Gen. Motors Corp., 656 F. Supp. 1555 (D.D.C. 1987), 71, 73
United States v. Gen. Motors Corp., 841 F.2d 400 (D.C. Cir. 1988), 71, 73
United States v. Harris, No. CRIM. A.89-0290(RCL), 1989 WL 252648 (D.D.C. Oct. 12, 1989), 376

FTC

In re Aspen Way Enters., Inc., No. 112-3151 (F.T.C. Sept. 25, 2012), 256
In re Blu Prods. & Samuel Ohev-Zion, No. C-4657 (F.T.C. Apr. 30, 2018), 263
In re Dave & Buster's, Inc., No. C-4291 (F.T.C. filed Mar. 25, 2010), 262
In re foru Int'l Corp., No. C-4457 (F.T.C. Jan. 7, 2014), 260, 263
In re Franklin's Budget Car Sales, Inc., No. C-4371 (F.T.C. June 7, 2012), 260
In re Lookout Servs., Inc., No. C-4326 (F.T.C. filed May 3, 2011), 261
In re Pfizer Inc., 81 F.T.C. 23 (1972), 248
In re Rite Aid Corp., No. C-4308 (F.T.C. filed July 27, 2010), 264
In re Snapchat, Inc., No. C-4501 (F.T.C. Dec. 23, 2014), 255, 264
In re Sony BMG Music Enm't, No. 062-3019 (F.T.C. Jan. 30, 2007), 256
In re The TJX Cos., Inc., No. C-4227 (F.T.C. Mar. 27, 2008), 264
In re Twitter, Inc., No. C-4316 (F.T.C. June 24, 2010), 260–261
In re Uber Techs., Inc., No. C-4662 (F.T.C. Aug. 15, 2017), 255, 260–261

ALABAMA

ALA. CODE § 8-19-1 *et seq.*, 288
ALA. CODE § 8-19-10(a), 289
ALA. CODE § 8-38-1 *et seq.*, 283
ALA. CODE § 8-38-5(b), 272

Ala. Code §§ 13A-8-112, 13A-8-113, 345
Ala. Code § 32-5A-30, 74
Ala. Code § 32-5A-191(a), 328, 329
2018 Alabama S.B. 318, 282
Gurley *ex rel.* Gurley v. Am. Honda Motor Co., 505 So. 2d 358 (Ala. 1987), 144
Segler v. Ford Motor Co., 438 So. 2d 297 (Ala. 1983), 129
Smith v. State, 470 So. 2d 1365 (Ala. Crim. App. 1985), 329

ALASKA

Alaska Const. art. I, § 22, 268
Alaska Stat. § 11.46.740, 345
Alaska Stat. § 45.48.010 *et seq.*, 283
Alaska Stat. § 45.50.471, 288
Alaska Stat. § 45.50.531(a), 289
Jacobson v. State, 551 P.2d 935 (Alaska 1976), 310
Wainscott v. State, 642 P.2d 1355 (Alaska 1982), 227, 231

ARIZONA

Ariz. Const. art. II, § 8, 268
Ariz. Rev. Stat. Ann. § 12-681(10), 150
Ariz. Rev. Stat. Ann. §§ 13-2316, 13-2316.01, 13-2316.02, 345
Ariz. Rev. Stat. Ann. § 18-551 *et seq.*, 283
Ariz. Rev. Stat. Ann. § 18-552(B), 272
Ariz. Rev. Stat. Ann. § 28-641, 74
Dashi v. Nissan N. Am., Inc., 247 Ariz. 56, 445 P.3d 13 (Ct. App. 2019), 66, 94, 105, 156–158, 175
Kohl v. City of Phoenix, 215 Ariz. 291, 160 P.3d 170 (2007), 230
Varela v. FCA US LLC, No. 1 CA-CV 19-0209, 2020 WL 2123281 (Ariz. Ct. App. May 5, 2020), 158
Wood v. Arizona, No. CV2019-090948 (Ariz. Maricopa Cnty. filed Mar. 20, 2019), 225

ARKANSAS

Ark. Code Ann. § 4-88-113(f), 289
Ark. Code Ann. § 4-110-101 *et seq.*, 283
Ark. Code Ann. § 4-110-103(7), 283
Ark. Code Ann. § 4-110-104(b), 282
Ark. Code Ann. §§ 5-41-101 to -206, 345
Ark. Code Ann. § 5-41-106(a)(1), 345
Ark. Code Ann. § 23-112-107, 285
Ark. Code Ann. § 23-112-107(a)(2), 286
Ark. Code Ann. § 23-112-107(b), 287
Ark. Code Ann. § 23-112-107(e), 287
Ark. Code Ann. § 23-112-107(e)(1), 285
Citizens Coach Co. v. Wright, 228 Ark. 1143, 313 S.W.2d 94 (1958), 223

CALIFORNIA

CAL. CONST. art. I, § 1, 268
CAL. BUS. & PROF. CODE § 17204, 289
CAL. CIV. CODE § 1798.80 *et seq.*, 283
CAL. CIV. CODE § 1798.81.5, 282
CAL. CIV. CODE § 1798.82, 284
CAL. CIV. CODE § 1798.91.04 *et seq.*, 283
CAL. CIV. CODE § 1798.91.04, 283
CAL. CIV. CODE § 1798.91.05, 283
CAL. CIV. CODE § 1798.91.06, 283
CAL. CIV. CODE § 1798.100, 278
CAL. CIV. CODE § 1798.105(a), 278
CAL. CIV. CODE § 1798.105(c), 278
CAL. CIV. CODE § 1798.110(a), 277
CAL. CIV. CODE § 1798.110(b), 277
CAL. CIV. CODE § 1798.115(a), 277
CAL. CIV. CODE § 1798.115(b), 277
CAL. CIV. CODE § 1798.115(d), 278
CAL. CIV. CODE § 1798.120(a), 279
CAL. CIV. CODE § 1798.120(c), 279
CAL. CIV. CODE § 1798.130(b), 279, 280
CAL. CIV. CODE § 1798.135, 279
CAL. CIV. CODE § 1798.140(a), 275
CAL. CIV. CODE § 1798.140(c), 276
CAL. CIV. CODE § 1798.140(e), 277
CAL. CIV. CODE § 1798.140(g), 276
CAL. CIV. CODE § 1798.140(h), 275
CAL. CIV. CODE § 1798.140(o), 275
CAL. CIV. CODE § 1798.140(y), 277
CAL. CIV. CODE § 1798.145, 279
CAL. CIV. CODE § 1798.150(a), 280
CAL. CIV. CODE § 1798.150(b), 280
CAL. CIV. CODE § 1798.150(c), 280
CAL. CIV. CODE § 2100, 220
CAL. CIV. CODE § 2168, 219
CAL. PENAL CODE § 20, 327
CAL. PENAL CODE § 192(c)(1), 334
CAL. PENAL CODE § 192(c)(2), 334
CAL. PENAL CODE § 502, 345, 347–350
CAL. PENAL CODE § 502(a)(5), 347
CAL. PENAL CODE § 502(c), 345
CAL. PENAL CODE § 502(e)(1), 345
CAL. VEH. CODE § 2814, 92
CAL. VEH. CODE § 9951, 285
CAL. VEH. CODE § 9951(a), 287
CAL. VEH. CODE § 10852, 340
CAL. VEH. CODE § 11750-62, 86
CAL. VEH. CODE § 21100, 90
CAL. CODE REGS. tit. 10, § 2632.5, 93

CAL. CODE REGS. tit. 13, § 227.04, 110
CAL. CODE REGS. tit. 13, § 227.16, 110
CAL. CODE REGS. tit. 13, § 227.3, 110, 111
CAL. CODE REGS. tit. 13, § 228.24, 114, 282
Judicial Council of California Criminal Jury Instruction 2241, 311
Barker v. Lull Eng'g Co., 20 Cal. 3d 413, 573 P.2d 443 (1978), 138
Desuza v. Andersack, 63 Cal. App. 3d 694, 133 Cal. Rptr. 920 (Ct. App. 1976), 188
Granite Constr. Co. v. Super. Ct., 149 Cal. App. 3d 465, 197 Cal. Rptr. 3 (Ct. App. 1983), 312
Huang v. Tesla Inc., 1 No. 19-cv-346663 (Cal. Cnty. Santa Clara filed Apr. 26, 2019), 225
In re F.H., 192 Cal. App. 4th 1465, 122 Cal. Rptr. 3d 43 (2011), 312
Kim v. Toyota Motor Corp., 6 Cal. 5th 21, 424 P.3d 290 (2018), 138
Liu v. Uber Techs., Inc., No. CGC-14-536979 (Cal. Super. Ct. filed Jan. 27, 2014), 218
People v. Bussel, 97 Cal. App. 4th Supp. 1, 118 Cal. Rptr. 2d 159 (Cal. App. Dep't Super. Ct. 2002), 334
People v. Diaz, 213 Cal. App. 4th 743, 153 Cal. Rptr. 3d 90 (2013), 390
People v. Xinos, 121 Cal. Rptr. 3d 496 (Ct. App. 2011), 390
Price v. Shell Oil Co., 2 Cal. 3d 245, 466 P.2d 722 (1970), 211
Sea Horse Ranch, Inc. v. Super. Ct., 24 Cal. App. 4th 446, 30 Cal. Rptr. 2d 681 (1994), 312
Shulman v. Grp. W Prods., Inc., 18 Cal. 4th 200, 955 P.2d 469 (July 29, 1998), 269
Soule v. Gen. Motors Corp., 8 Cal. 4th 548, 882 P.2d 298 (1994), 135, 138

COLORADO

COLO. REV. STAT. § 6-1-713.5, 282
COLO. REV. STAT. § 6-1-716, 283
COLO. REV. STAT. § 12-6-401 *et seq.*, 285
COLO. REV. STAT. § 13-21-402, 148, 159
COLO. REV. STAT. § 18-1-501(3), 327
COLO. REV. STAT. § 18-1-501(8), 327
COLO. REV. STAT. § 18-1-606(1), 312
COLO. REV. STAT. § 18-3-106(b)(I), 335
COLO. REV. STAT. §§ 18-5.5-101 to -102, 345
COLO. REV. STAT. § 42-4-104, 74
COLO. REV. STAT. § 42-4-110(2), 111
COLO. REV. STAT. § 42-4-110(6)(b), 112
COLO. REV. STAT. § 42-4-242(1), 110
COLO. REV. STAT. § 42-4-2402, 287
Kern v. Gen. Motors Corp., 724 P.2d 1365 (Colo. App. 1986), 145
People v. Garner, 781 P.2d 87 (Colo. 1989), 335
Roberts v. May, 41 Colo. App. 82, 583 P.2d 305 (1978), 145
Shaw v. Gen. Motors Corp., 727 P.2d 387 (Colo. App. 1986), 141

CONNECTICUT

CONN. GEN. STAT. § 13a-144, 227
CONN. GEN. STAT. § 14-1(28), 311
CONN. GEN. STAT. § 14-164aa, 285
CONN. GEN. STAT. § 14-164aa(d), 285
CONN. GEN. STAT. § 36a-701b, 283

Conn. Gen. Stat. § 38a-999b, 282
Conn. Gen. Stat. § 42-110g(a), 289
Conn. Gen. Stat. § 42-179, 95
Conn. Gen. Stat. § 42-471(a), 282
Conn. Gen. Stat. § 53a-3, 327
Conn. Gen. Stat. § 53a-3(13), 327
Conn. Gen. Stat. §§ 53a-250 to -261, 345
Izzarelli v. R.J. Reynolds Tobacco Co., 321 Conn. 172, 136 A.3d 1232 (2016), 135
Potter v. Chi. Pneumatic Tool Co., 241 Conn. 199, 694 A.2d 1319 (1997), 137–138
State v. Cyr, 291 Conn. 49, 967 A.2d 32 (2009), 311
State v. Ducatt, 22 Conn. App. 88, 575 A.2d 708 (1990), 310
State v. Russo, 38 Conn. Supp. 426, 450 A.2d 857 (Super. Ct. 1982), 334
State v. Swift, 125 Conn. 399, 6 A.2d 359 (Conn. 1939), 310

DELAWARE

Del. Code Ann. tit. 6, § 12B-101 *et seq.*, 283
Del. Code Ann. tit. 6, § 4905, 160
Del. Code Ann. tit. 11, § 281, 312
Del. Code Ann. tit. 11, §§ 931–941, 345
Del. Code Ann. tit. 11, § 933, 346
Del. Code Ann. tit. 11, § 939, 346
Del. Code Ann. tit. 18, § 3918, 285, 287
Del. Code Ann. tit. 18, § 7001, 159
Del. Exec. Order 14, Establishment of the Advisory Council on Connected and Autonomous Vehicles (Sept. 5, 2017), 111
McDuell v. State, 231 A.2d 265 (Del. 1967), 310
State v. Byard, 2018 WL 2077324 (Del. Super. Ct. May 1, 2018), 338, 389

DISTRICT OF COLUMBIA

D.C. Code Ann. § 28-3851 *et seq.*, 283
D.C. Code Ann. § 50-2353, 112, 141
Bodoh v. D.C. Bureau of Motor Vehicle Servs., 377 A.2d 1135 (D.C. 1977), 329

FLORIDA

Fla. Const. art. I, § 23, 268
Fla. Stat. Ann. § 316.85, 97
Fla. Stat. § 316.85(3)(a), 307
Fla. Stat. § 316.192, 331, 334
Fla. Stat. § 501.171, 283
Fla. Stat. § 501.171(1)(g), 283
Fla. Stat. § 501.171(2), 282
Fla. Stat. § 627.748 (2017), 220
Fla. Stat. § 782.071, 333, 334
Fla. Stat. §§ 815.01 to .07, 668.801 to .805, 345

Fla. Ch. 2019-101: Motor Vehicles—Autonomous Vehicles (2019), 111
Aubin v. Union Carbide Corp., 177 So. 3d 489 (Fla. 2015), 136
Commercial Carrier Corp. v. Indian River Cnty., 371 So. 2d 1010 (Fla. 1979), 228
Santisteban v. State, 72 So. 3d 187 (Fla. Dist. Ct. App. 2011), 334
State v. Worsham, 227 So. 3d 602 (Fla. Dist. Ct. App. 2017), 389, 390, 392
Sussman v. Cessna Aircraft Co., No. 2007-20336 CINS (Volusia Cnty. Cir. Ct. Fla. Apr. 10, 2014), 170
Vargas v. Enter. Leasing Co., 60 So. 3d 1037 (Fla. 2011), 214

GEORGIA

Ga. Code Ann. § 10-1-393, 288
Ga. Code Ann. § 10-1-632, 270
Ga. Code Ann. § 10-1-910 *et seq.*, 283
Ga. Code Ann. §§ 16-9-90 to -94, 16-9-150 to -157, 345
Ga. Code Ann. § 16-9-93, 345
Ga. Code Ann. § 16-9-93(g), 345
Ga. Code Ann. § 40-6-25 (West), 341
Georgia Senate Bill 122, Consumer Protection—Motor Vehicles, 2019 Ga. Laws Act 157, 269–270
Am. Ass'n of Cab Cos., Inc. v. Parham, 291 Ga. App. 33, 661 S.E.2d 161 (2008), 221
Banks v. ICI Ams., Inc., 264 Ga. 732, 450 S.E.2d 671 (1994), 136, 137
Dep't of Transp. v. Brown, 267 Ga. 6, 471 S.E.2d 849 (1996), 231
Farmer v. Brannan Auto Parts, Inc., 231 Ga. App. 353, 498 S.E.2d 583 (1998), 143, 213
Flournoy v. Georgia, 128 S.E.2d 528 (Ga. Ct. App. 1962), 310
Giordano v. Ford Motor Co., 165 Ga. App. 644, 299 S.E.2d 897 (1983), 129
Mobley v. State, 307 Ga. 59, 834 S.E.2d 785 (2019), 388, 389, 390, 392
Mobley v. State, 346 Ga. App. 641, 816 S.E.2d 769 (2018), 389, 390
Town of Fort Oglethorpe v. Phillips, 224 Ga. 834, 165 S.E.2d 141 (1968), 231

HAWAII

Haw. Const. art. I, § 6, 269
Haw. Rev. Stat. § 286-2, 311
Haw. Rev. Stat. § 487N-1 *et seq.*, 283
Haw. Rev. Stat. § 626-1, Rule 407, 161
Haw. Rev. Stat. §§ 708-890 to -895.7, 345
Stewart v. Budget Rent-A-Car Corp., 52 Haw. 71, 470 P.2d 240 (1970), 211

IDAHO

Idaho Code §§ 18-2201, -2202, 345
Idaho Code § 28-51-104 *et seq.*, 283
Idaho Code § 48-608(1), 289
Idaho Exec. Order No. 2018-01, Creating the Autonomous and Connected Vehicle and Deployment Committee (Jan. 2, 2018), 111
Aardema v. U.S. Dairy Sys., Inc., 147 Idaho 785, 215 P.3d 505 (2009), 296

ILLINOIS

ILL. CONST. art. I, § 6, 268
625 ILL. COMP. STAT. ANN. 5/11-208 (e-10), 112
625 ILL. COMP. STAT. ANN. 5/11-301, 74
625 ILL. COMP. STAT. ANN. 5/11-503, 331
720 ILL. COMP. STAT. ANN. 5/17-50 to -55, 345
720 ILL. COMP. STAT. ANN. 5/17-51, 346
740 ILL. COMP. STAT. ANN. 14/1, 272
740 ILL. COMP. STAT. ANN. 14/10, 273
740 ILL. COMP. STAT. ANN. 14/15, 273, 274
740 ILL. COMP. STAT. ANN. 14/20, 273
815 ILL. COMP. STAT. ANN. 505/2, 288
815 ILL. COMP. STAT. ANN. 530/5, 283
815 ILL. COMP. STAT. ANN. 530/45, 282
Ill. Exec. Order 2018-13: Executive Order Establishing the Autonomous Illinois Initiative (Oct. 25, 2018), 110
Bachman v. Gen. Motors Corp., 332 Ill. App. 3d 760, 776 N.E.2d 262 (2002), 338, 389
Calles v. Scripto-Tokai Corp., 224 Ill. 2d 247, 864 N.E.2d 249 (2007), 137
Doe v. Rockdale Sch. Dist., No. 84, 287 Ill. App. 3d 791, 679 N.E.2d 771 (1997), 219
Fischetti v. Vill. of Schaumburg, 2012 IL App (1st) 111008, 967 N.E.2d 950 (2012), 322
Long v. Muncie Aviation Co., 2006L006992 (Cook Cnty. Cir. Ct. of Ill. Nov. 7, 2016), 170
Moorman Mfg. Co. v. Nat'l Tank Co., 91 Ill. 2d 69, 435 N.E.2d 443 (1982), 296
Rosenbach v. Six Flags Entm't Corp., 2019 IL 123186, 129 N.E.3d 1197 (2019), 274
Rosenstein v. Standard & Poor's Corp., 264 Ill. App. 3d 818, 636 N.E.2d 665 (1993), 232
Woods v. Graham Eng'g Corp., 183 Ill. App. 3d 337, 539 N.E.2d 316 (1989), 141

INDIANA

IND. CODE § 9-21-2-2, 74
IND. CODE § 9-21-22-1 *et seq.*, 112
IND. CODE § 24-4.9-1-1 *et seq.*, 283
IND. CODE § 24-4.9-2-10, 283
IND. CODE § 24-4.9-3-3.5, 282
IND. CODE § 24-4.9-3-3.5(c), 282
IND. CODE § 34-20-3-1(a), 199
IND. CODE § 34-20-3-1(b), 199
IND. CODE § 34-20-6-4, 148
IND. CODE §§ 35-43-1-8, 35-43-2-3, 345
IND. CODE § 35-43-2-3(b), 345
Corl v. State, 544 N.E.2d 211 (Ind. Ct. App. 1989), 310
Ford Motor Co. v. Rushford, 868 N.E.2d 806 (Ind. 2007), 143, 213
Jackson v. State, 576 N.E.2d 607 (Ind. Ct. App. 1991), 332

IOWA

IOWA CODE § 321.514 *et seq.*, 110
IOWA CODE § 321.515(3), 111

Iowa Code § 321.518, 110
Iowa Code § 613.18, 159
Iowa Code § 714H.5, 289
Iowa Code § 715C.1 *et seq.*, 283
Iowa Code § 716.6B, 345, 346
Iowa R. Civ. P. 5.407, 161
Emp'rs Mut. Cas. Co. v. Chicago & N.W. Transp. Co., 521 N.W.2d 692 (Iowa 1994), 219
Iowa City v. Nolan, 239 N.W.2d 102 (Iowa 1976), 321

KANSAS

Kan. Stat. Ann. § 8-2003, 74
Kan. Stat. Ann. § 21-5839, 345
Kan. Stat. Ann. § 50-7a01, 283
Kan. Stat. Ann. § 50-6139b, 282
Kan. Stat. Ann. § 60-3306, 159
Delaney v. Deere & Co., 268 Kan. 769, 999 P.2d 930 (2000), 149
Jones v. Yellow Cab & Baggage Co., 176 Kan. 558, 271 P.2d 249 (1954), 223
State v. Baker, 1 Kan. App. 2d 568, 571 P.2d 65 (1977), 318

KENTUCKY

Ky. Rev. Stat. Ann. § 365.732, 283
Ky. Rev. Stat. Ann. § 367.220(1), 289
Ky. Rev. Stat. Ann. § 411.340, 159
Ky. Rev. Stat. Ann. §§ 434.840, .845, .850, .851, .853, .855, .860, 345
Ky. Rev. Stat. Ann. § 434.845(1), 346
Cox v. General Motors Corp., 514 S.W.2d 197 (Ky. 1974), 185
Hedgepath v. Commw., 441 S.W.3d 119 (Ky. 2014), 395

LOUISIANA

La. Const. Ann. art. I, § 5, 269
La. Stat. Ann. §§ 14:73.1 to .8, 345
La. Stat. Ann. § 14:73.7, 346
La. Stat. Ann. § 51:1409(A), 289
La. Stat. Ann. § 51:3071 *et seq.*, 283
La. Stat. Ann. § 51:3073(4), 283
La. Stat. Ann. § 51:3074(B), 282
La. Stat. Ann. § 400.1 *et seq.*, 112
La. Stat. Ann. § 3074, 282
Lipscomb v. News Star World Pub. Corp., 5 So. 2d 41 (La. Ct. App. 1941), 326
Morales v. Parish of Jefferson, 13-486 (La. App. 5 Cir. 4/30/14), 140 So. 3d 375, 322

MAINE

ME. STAT. tit. 5, § 213(1), 289
ME. STAT. tit. 10, § 1175, 160
ME. STAT. tit. 10, § 1346 *et seq.*, 283
ME. STAT. tit. 17-A, § 60(1)(A), 312
ME. STAT. tit. 17-A, §§ 431–435, 345
ME. STAT. tit. 29-A, § 1971 *et seq.*, 285
ME. STAT. tit. 29-A, § 1972, 288
ME. STAT. tit. 29-A, § 1972F, 291
ME. STAT. tit. 29-A, § 2075, 74
Maine H.P. 1204, 111

MARYLAND

MD. CODE ANN., COM. LAW § 14-3501 to -3503, 282, 283
MD. CODE ANN., COM. LAW § 14-3503(a), 282
MD. CODE ANN., COM. LAW § 14-3504(b)(3), 272
MD. CODE ANN., CTS. & JUD. PROC. § 5-405, 159
MD. CODE ANN., CRIM. LAW § 7-302, 345
MD. CODE ANN., CRIM. LAW § 7-302(c)(1), 345
Bates v. State, 64 Md. App. 279, 494 A.2d 976 (1985), 376
Beatty v. Trailmaster Prod., Inc., 330 Md. 726, 625 A.2d 1005 (1993), 159
Easter v. State, 223 Md. App. 65, 115 A.3d 239 (2015), 338, 389
Nolan v. Dillon, 261 Md. 516, 276 A.2d 36 (1971), 144
Thomas v. State, 277 Md. 314, 353 A.2d 256 (1976), 310

MASSACHUSETTS

MASS. GEN. LAWS ch. 90, § 24G(c), 333
MASS. GEN. LAWS ch. 93A, § 2, 288
MASS. GEN. LAWS ch. 93H, § 1 *et seq.*, 283
MASS. GEN. LAWS ch. 93H, § 1, 283
MASS. GEN. LAWS ch. 93H, § 2(a), 282
MASS. GEN. LAWS ch. 266, § 33A, 345
MASS. GEN. LAWS ch. 266, § 127, 341
Altman v. Aronson, 231 Mass. 588, 121 N.E. 505 (1919), 334
Commw. v. Dorelas, 473 Mass. 496, 43 N.E.3d 306 (2016), 395
Commw. v. Ober, 286 Mass. 25, 189 N.E. 601 (1934), 321
Commw. v. Plowman, 28 Mass. App. Ct. 230, 548 N.E.2d 1278 (1990), 310
Commw. v. Welansky, 316 Mass. 383, 55 N.E.2d 902 (1944), 326
Commw. v. Zimmermann, 70 Mass. App. Ct. 357, 873 N.E.2d 1215 (2007), 338, 389
Massachusetts v. Uski, 160 N.E. 305 (Mass. 1928), 310
Minicost Car Rental, Inc., 354 Mass. 746, 242 N.E.2d 411 (1968), 321
Mitchell v. Sky Climber, Inc., 396 Mass. 629, 487 N.E.2d 1374 (1986), 141
Quigley v. Wilson Line of Mass., Inc., 338 Mass. 125, 154 N.E.2d 77 (1958), 223
Uloth v. City Tank Corp., 376 Mass. 874, 384 N.E.2d 1188 (1978), 149

MICHIGAN

Mich. Comp. Laws § 257.6a, 325
Mich. Comp. Laws § 257.13, 306
Mich. Comp. Laws § 257.36, 307
Mich. Comp. Laws § 257.40, 306
Mich. Comp. Laws § 257.414, 350
Mich. Comp. Laws § 257.665, 110, 111
Mich. Comp. Laws § 257.665a, 112, 141
Mich. Comp. Laws § 257.665b, 110, 114
Mich. Comp. Laws §§ 445.63, .72, 383
Mich. Comp. Laws § 500.3135, 203
Mich. Comp. Laws § 600.2947, 148, 185
Mich. Comp. Laws § 691.1402, 227
Mich. Comp. Laws § 691.1403, 230
Mich. Comp. Laws §§ 752.791, .792, .793, .794, .795, .796, .797, 345
Mich. Pub. Act No. 332 (2016), 111
Hendricks v. DSW Shoe Warehouse, Inc., 444 F. Supp. 2d 775 (W.D. Mich. 2006), 297
People v. Gen. Dynamics Land Sys., Inc., 175 Mich. App. 701, 438 N.W.2d 359 (1989), 312
People v. Jones, 132 Mich. App. 368, 347 N.W.2d 235 (1984), 318
People v. Yamat, 475 Mich. 49, 714 N.W.2d 335 (2006), 312

MINNESOTA

Minn. Stat. § 325E.61, 283
Minn. Stat. § 325M.05, 282
Minn. Stat. § 544.41, 159
Minn. Stat. Ann. § 604.03, 199
Minn. Stat. §§ 609.87 to .893, 345
Minn. Stat. § 609.2112, 334
Minn. Exec. Order 18-04, Establishing the Governor's Council on Connected and Automated Vehicles (Apr. 1, 2019), 111
Condux Int'l, Inc. v. Haugum, No. CIV 08-4824 ADM/JSM, 2008 WL 5244818 (D. Minn. Dec. 15, 2008), 355
Glorvigen v. Cirrus Design Corp., 816 N.W.2d 572 (Minn. 2012), 144
In re Target Corp. Data Security Breach Litigation, 66 F. Supp. 3d 1154 (D. Minn. 2014), 296
State v. Al-Naseer, 690 N.W.2d 744 (Minn. 2005), 334
State v. Kremer, 262 Minn. 190, 114 N.W.2d 88 (1962), 318, 319

MISSISSIPPI

Miss. Code Ann. § 63-3-103(k), 112
Miss. Code Ann. § 75-24-15(1), 289
Miss. Code Ann. § 75-24-29, 283
Miss. Code Ann. §§ 97-45-1 to -33, 345
Miss. Cent. v. Roberts, 173 Miss. 487, 160 So. 604 (1935), 188
Nunnally v. R.J. Reynolds Tobacco Co., 869 So. 2d 373 (Miss. 2004), 137
Warren *ex rel.* Warren v. Glascoe, 852 So. 2d 634 (Miss. Ct. App. 2003), 188

MISSOURI

Mo. Rev. Stat. § 407.025(1), 289
Mo. Rev. Stat. § 407.1500, 283
Mo. Rev. Stat. §§ 537.525, 569.095, 569.097, 569.099, 345
Mo. Rev. Stat. § 537.762, 159
Mo. Rev. Stat. § 537.764, 150
Mo. Rev. Stat. § 569.080, 340
Mo. Rev. Stat. § 577.010, 328
Bloemer v. Art Welding Co., 884 S.W.2d 55 (Mo. Ct. App. 1994), 222
City of Kansas City v. Hertz Corp., 499 S.W.2d 449 (Mo. 1973), 321
Johnson v. Pac. Intermountain Exp. Co., 662 S.W.2d 237 (Mo. 1983), 221
Jones v. Dir. of Revenue, 189 S.W.3d 187 (Mo. Ct. App. 2006), 330
Tupper v. City of St. Louis, 468 S.W.3d 360 (Mo. 2015), 321
State v. West, 548 S.W.3d 406 (Mo. Ct. App. 2018), 388, 390, 392

MONTANA

Mont. Const. art. II, § 10, 268
Mont. Code Ann. § 30-14-133(1), 289
Mont. Code Ann. § 30-14-1701 to -1702, -1704, 284
Mont. Code Ann. § 45-6-311, 345
Mont. Code Ann. § 45-6-311(2), 346
Mont. Code Ann. § 61-8-202, 74
Mont. Code Ann. § 61-12-1001 *et seq.*, 285
Mont. Code Ann. § 61-12-1001(1), 286
Mont. Code Ann. § 61-12-1003, 287
Mont. Code Ann. § 61-12-1003(1), 285
Mont. Code Ann. § 61-12-1004(1)(a), 391

NEBRASKA

Neb. Rev. Stat. Ann. §§ 28-1341 to -1348, 345
Neb. Rev. Stat. Ann. § 60-3302, 110
Neb. Rev. Stat. Ann. § 60-3305, 110
Neb. Rev. Stat. Ann. § 60-3306, 111
Neb. Rev. Stat. Ann. § 60-3308, 112
Neb. Rev. Stat. Ann. § 87-801 to -807, 282, 284
Neb. Rev. Stat. Ann. § 87-808, 282
State v. Henderson, 289 Neb. 271, 854 N.W.2d 616 (2014), 395

NEVADA

Nev. Rev. Stat. §§ 205.473 to .513, 345
Nev. Rev. Stat. § 482A.070, 110
Nev. Rev. Stat. § 484A.080, 307, 308, 310

Nev. Rev. Stat. § 482A.090, 112, 141
Nev. Rev. Stat. § 482A.200, 111
Nev. Rev. Stat. § 484D.485, 285
Nev. Rev. Stat. § 603A.010 *et seq.*, 284
Nev. Rev. Stat. § 603A.210, 603A.215(2), 282
Stackiewicz v. Nissan Motor Corp. in U.S.A., 100 Nev. 443, 686 P.2d 925 (1984), 132

NEW HAMPSHIRE

N.H. Rev. Stat. Ann. § 357-G:1, 285
N.H. Rev. Stat. Ann. § 359-C:19-C:21, 284
N.H. Rev. Stat. Ann. § 507:8-g, 150
N.H. Rev. Stat. Ann. § 626:1, 317
N.H. Rev. Stat. Ann. § 626:2, 327
N.H. Rev. Stat. Ann. §§ 638:16, :17, :18, :19, 345
N.H. Rev. Stat. Ann. § 638:17, 345, 346
State v. Goding, 126 N.H. 50, 489 A.2d 579 (1985), 329

NEW JERSEY

N.J. Stat. Ann. §§ 2A:38A-1 to -3, 2C:20-2, 2C:20-23 to -34, 345
N.J. Stat. Ann. § 2A:58C-9, 159
N.J. Stat. Ann. § 2C:11-5, 333
N.J. Stat. Ann. § 39:4-50(a), 330
N.J. Stat. Ann. § 39:10B-7 to -9, 285
N.J. Stat. Ann. § 39:10B-9, 285
N.J. Stat. Ann. § 56:8-19, 289
N.J. Stat. Ann. § 56:8-161, -163, -165 to -166, 284
Cintrone v. Hertz Truck Leasing & Rental Serv., 45 N.J. 434, 212 A.2d 769 (1965), 211
Lewis v. Am. Cyanamid Co., 155 N.J. 544, 715 A.2d 967 (1998), 149
State v. Hammond, 118 N.J. 306, 571 A.2d 942 (1990), 329
State v. Packin, 107 N.J. Super. 93, 257 A.2d 120 (App. Div. 1969), 318, 319
State v. Shabazz, 400 N.J. Super. 203, 946 A.2d 626 (2005), 338, 382, 389
State v. Stas, 212 N.J. 37, 50 A.3d 632 (2012), 330
State v. Zanger, 370 N.J. Super. 360, 851 A.2d 134 (Law. Div. 2004), 330
Zaza v. Marquess & Nell, Inc., 144 N.J. 34, 675 A.2d 620 (1996), 141

NEW MEXICO

N.M. Stat. §§ 30-45-1 to -7, 345
N.M. Stat. § 57-12C-1 to 57-12C-12, 284
N.M. Stat. § 57-12C-4, 57-12C-5, 282
N.M. Stat. § 57-12C-6(A), 272
N.M. Stat. § 66-8-102, 328
Brooks v. Beech Aircraft Corp., 1995-NMSC-043, 120 N.M. 372, 902 P.2d 54 (1995), 159
Stang v. Hertz Corp., 83 N.M. 730, 497 P.2d 732 (1972), 211
State v. Diaz, No. 35,563, 2017 WL 1017356 (N.M. Ct. App. Feb. 27, 2017), 338, 389

NEW YORK

N.Y. Gen. Bus. Law § 899-aa, 284
N.Y. Penal Law § 15.05, 327
N.Y. Penal Law § 15.10, 317
N.Y. Penal Law § 20.20, 313
N.Y. Penal Law § 20.20(2)(a), 312
N.Y. Penal Law § 145.00, 342
N.Y. Penal Law §§ 156.00 to .50, 345
N.Y. Veh. & Traf. Law § 113, 306
N.Y. Veh. & Traf. Law § 131, 306, 310
N.Y. Veh. & Traf. Law § 148-a, 99
N.Y. Veh. & Traf. Law § 416-b, 285
N.Y. Veh. & Traf. Law § 1101, 326
N.Y. Veh. & Traf. Law § 1212, 331
N.Y. Veh. & Traf. Law § 1691, 98
N.Y. Veh. & Traf. Law § 1692, 220
17 N.Y. S.B. 2005 (Jan. 23, 2017), 110
Abramowitz v. Maghsoudlou, 295 A.D.2d 548, 745 N.Y.S.2d 36 (2002), 143, 213
Auld v. Sears, Roebuck & Co., 261 A.D. 918, 25 N.Y.S.2d 491 (App. Div. 1941), 151
Betzold v. Town of Babylon, 18 A.D.3d 787, 796 N.Y.S.2d 680 (2005), 230
Cioffi v. S.M. Foods, Inc., 129 A.D.3d 888, 10 N.Y.S.3d 620 (N.Y. App. Div. 2015), 209
Cukoviq v. Iftikhar, 169 A.D.3d 766, 93 N.Y.S.3d 710 (N.Y. App. Div. 2019), 212
Daly v. Metro. Life Ins. Co., 4 Misc. 3d 887, 782 N.Y.S.2d 530 (Sup. Ct. 2004), 296
Kinney Car Corp. v. City of New York, 28 N.Y.2d 741, 269 N.E.2d 829 (1971), 321
Landahl v. Chrysler Corp., 144 A.D.2d 926, 534 N.Y.S.2d 245 (1988), 132
MacPherson v. Buick Motor Company, 217 N.Y. 382, 111 N.E. 1050 (1916), 126
Moreau v. Josaphat, 42 Misc. 3d 345, 975 N.Y.S.2d 851 (Sup. Ct. 2013), 214, 215
Murphy v. Pontillo, 12 Misc. 3d 1146, 820 N.Y.S.2d 743 (Sup. Ct. 2006), 215
New York v. Shaughnessy, 66 Misc. 2d 19, 319 N.Y.S.2d 626 (Dist. Ct. 1971), 317
People v. Alamo, 34 N.Y.2d 453, 315 N.E.2d 446 (1974), 310
People v. Christmann, 3 Misc. 3d 309, 776 N.Y.S.2d 437 (Just. Ct. 2004), 338, 389, 390
People v. Garo, 208 Misc. 496, 144 N.Y.S.2d 107 (Cnty. Ct. 1955), 331
People v. Marzulli, 76 Misc. 2d 971, 351 N.Y.S.2d 775 (App. Term 1973), 330
People v. Millan, 118 A.D.2d 2362, 505 N.Y.S.2d 848 (1986), 376, 377
People v. Millan, 69 N.Y.2d 514, 508 N.E.2d 9036 (1987), 376
People v. Prescott, 95 N.Y.2d 655, 745 N.E.2d 1000 (2001), 310
Voss v. Black & Decker Mfg. Co., 59 N.Y.2d 102, 450 N.E.2d 204 (1983), 137

NORTH CAROLINA

N.C. Gen. Stat. §§ 14-453 to -458, 345
N.C. Gen. Stat. § 20-107, 340
N.C. Gen. Stat. § 20-140, 331
N.C. Gen. Stat. § 20-140(a), 331
N.C. Gen. Stat. § 20-152(c), 112
N.C. Gen. Stat. § 75-1.1, 288
N.C. Gen. Stat. § 75-61, 75-65, 284
N.C. Gen. Stat. § 99B-4, 148
N.C. Gen. Stat. § 99B-2, 159

N.C. Gen. Stat. § 136-30, 74
Jackson v. Stancil, 253 N.C. 291, 116 S.E.2d 817 (1960), 219, 220

NORTH DAKOTA

N.D. Cent. Code § 8-02-02, 220
N.D. Cent. Code § 12.1-06.1-08, 345
N.D. Cent. Code § 28-01.3-05, 159
N.D. Cent. Code § 39-13-06, 74
N.D. Cent. Code § 51-07-28, 285, 287
N.D. Cent. Code § 51-30-01 to -07, 284
N.D. Cent. Code § 51-30-01, 285
N.D. Cent. Code § 51-30-01.2, 308
N.D. Cent. Code § 51-30-01.18, 308

OHIO

Ohio Rev. Code Ann. §§ 1347.12(B)(2), 1349.19(B)(2), 272
Ohio Rev. Code Ann. § 1349.19 to .192, 284
Ohio Rev. Code Ann. § 1354.01 to 1354.05, 282
Ohio Rev. Code Ann. § 2307.78, 159
Ohio Rev. Code Ann. § 2744.02, 227, 229
Ohio Rev. Code Ann. § 2744.02(B)(1), 229
Ohio Rev. Code Ann. §§ 2913.01 to .04, 345
Ohio Rev. Code Ann. § 2913.03, 350
Ohio Rev. Code Ann. § 4511.18(A), 341
Ohio Rev. Code Ann. § 4511.99, 325
City of Kettering v. Greene, 9 Ohio St. 2d 26, 222 N.E.2d 638 (1966), 318
Kobza v. General Motors Corp., 63 Ohio App. 3d 742, 580 N.E.2d 47 (1989), 185

OKLAHOMA

Okla. Stat. tit. 13, § 32, 220
Okla. Stat. tit. 21, § 1951 *et seq.*, 345
Okla. Stat. tit. 24, §§ 161-166, 284
Okla. Stat. tit. 47, § 1-144, 306
Okla. Stat. tit. 74, § 2285, 74
Coleman v. Hertz Corp., 1975 OK CIV APP 5, 534 P.2d 940 (Okla. Ct. App. 1975), 211

OREGON

Or. Rev. Stat. § 105.925 *et seq.*, 285, 394
Or. Rev. Stat. § 105.925, 391
Or. Rev. Stat. § 105.928, 391
Or. Rev. Stat. § 105.932, 285, 287

Or. Rev. Stat. § 105.935, 391
Or. Rev. Stat. § 164.377, 345
Or. Rev. Stat. § 646A.600 to .628, 284
Or. Rev. Stat. § 646A.602(12)(a), 283
Or. Rev. Stat. § 646A.604(3)(a), 272
Or. Rev. Stat. § 646A.622, 282
Or. Rev. Stat. § 646.638(1), 289
Or. Rev. Stat. § 811.140, 331
Or. Rev. Stat. § 813.010(1), 328
Or. House Bill 4063, 111
Cole v. Ford Motor Co., 136 Or. App. 45, 900 P.2d 1059 (1995), 169
State v. Allen, 288 Or. App. 244, 406 P.3d 89 (2017), 395
State v. Mansor, 279 Or. App. 778, 381 P.3d 930 (2016), 395
Util. Equip., Inc. v. Morbark Indus., Inc., 308 Or. 209, 779 P.2d 139 (1989), 50

PENNSYLVANIA

18 Pa. Cons. Stat. Ann. § 5741, 345
66 Pa. Cons. Stat. Ann. § 102, 220
73 Pa. Cons. Stat. Ann. § 201-9.2(a), 289
73 Pa. Cons. Stat. Ann. § 2301 *et seq.*, 284
75 Pa. Cons. Stat. Ann. § 850, 111
75 Pa. Cons. Stat. Ann. § 3802(a)(1), 328
75 Pa. Cons. Stat. Ann. § 6121, 74
Commw. v. Kallus, 212 Pa. Super. 504, 243 A.2d 483 (1968), 311
Commw. v. Safka, 2014 Pa. Super. 131, 95 A.3d 304 (2014), 338, 389
Dep't of Transp., Bureau of Driver Licensing v. Hoover, 161 Pa. Commw. 517, 637 A.2d 721 (1994), 312
Gaudio v. Ford Motor Co., 2009 Pa. Super. 102, 976 A.2d 524 (2009), 159
Phatak v. United Chair Co., 756 A.2d 690 (Pa. Super. Ct. 2000), 186

RHODE ISLAND

6 R.I. Gen. Laws § 6-13.1-5.2(a), 289
11 R.I. Gen. Laws § 11-49.3-1 to 11-49.3-6, 284
11 R.I. Gen. Laws § 11-49.3-2, 282
11 R.I. Gen. Laws § 11-49.3-4(a)(2), 272
11 R.I. Gen. Laws § 11-52-1 to -8, 345
31 R.I. Gen. Laws § 31-9-3, 340
31 R.I. Gen. Laws § 31-27-2, 328
State v. Benoit, 650 A.2d 1230 (R.I. 1994), 336

SOUTH CAROLINA

S.C. Const. art. I, § 10, 268
S.C. Code Ann. § 16-3-910, 350
S.C. Code Ann. § 16-16-10 *et seq.*, 345

S.C. Code Ann. § 16-21-60(A), 350
S.C. Code Ann. § 38-99-10 to -100, 282
S.C. Code Ann. § 39-1-90, 284
S.C. Code Ann. § 39-5-140(a), 289
S.C. Code Ann. § 56-5-730, 326
S.C. Code Ann. § 56-5-1390, 112
S.C. Code Ann. § 56-5-2910, 333
S.C. Code Ann. § 56-5-2945, 336
S.C. Code Ann. § 56-5-2933(A), 328
S.C. Code Ann. § 56-5-6190, 325
Branham v. Ford Motor Co., 390 S.C. 203, 701 S.E.2d 5 (2010), 136
Mickle v. Blackmon, 252 S.C. 202, 166 S.E.2d 173 (1969), 152
State v. Grampus, 288 S.C. 395, 343 S.E.2d 26 (1986), 336
State v. Watson, 349 S.C. 372, 563 S.E.2d 336 (2002), 334

SOUTH DAKOTA

S.D. Codified Laws § 20-9-9, 159
S.D. Codified Laws § 22-40-19 to -26, 284
S.D. Codified Laws § 32-6B-80, 96
S.D. Codified Laws § 43-43B-1 *et seq.*, 345
Dahl v. Sittner, 474 N.W.2d 897 (S.D. 1991), 326

TENNESSEE

Tenn. Code Ann. § 29-28-106, 159
Tenn. Code Ann. § 29-28-108, 151, 185
Tenn. Code Ann. § 39-14-106, 350
Tenn. Code Ann. § 39-14-601 *et seq.*, 345
Tenn. Code Ann. § 39-14-602, 346
Tenn. Code Ann. § 47-18-109(a)(1), 289
Tenn. Code Ann. § 47-18-2105 to -2107, 284
Tenn. Code Ann. § 47-18-2107(b), 272
Tenn. Code Ann. § 55-8-101, 306
Tenn. Code Ann. § 55-10-205, 331
Tenn. Code Ann. § 55-10-401, 328
Tenn. Code Ann. § 55-30-103, 110
Tenn. Code Ann. § 55-30-105, 112
Tenn. Code Ann. § 55-30-106(b)(1), 113
Tenn. Code Ann. § 55-30-106(b)(2), 307
City of Knoxville v. Brown, 284 S.W.3d 330 (Tenn. Ct. App. 2008), 322
Copeland v. Nw. Airlines Corp., No. 04-2156 M1/V, 2005 WL 2365255 (W.D. Tenn. Feb. 28, 2005), 236, 383
Shoemake v. Omniquip Int'l, Inc., 152 S.W.3d 567 (Tenn. Ct. App. 2003), 129

TEXAS

Tex. Bus. & Com. Code Ann. § 17.46, 288
Tex. Bus. & Com. Code Ann. § 503.001, 272
Tex. Bus. & Com. Code Ann. § 503.001(a), 273
Tex. Bus. & Com. Code Ann. § 503.001(c)(2), 274
Tex. Bus. & Com. Code Ann. § 503.001(d), 273
Tex. Bus. & Com. Code Ann. § 521.002, 521.053, 521.151 to .152, 284
Tex. Bus. & Com. Code Ann. § 521.002(a)(2), 283
Tex. Bus. & Com. Code Ann. § 521.052, 282
Tex. Civ. Prac. & Rem. Code Ann. § 101.021, 229
Tex. Civ. Prac. & Rem. Code Ann. § 101.060, 230
Tex. Occ. Code Ann. § 2402.001(5), 98
Tex. Occ. Code Ann. § 2402.002, 220
Tex. Penal Code Ann. § 6.01, 323
Tex. Penal Code Ann. § 33.02, 345
Tex. Penal Code Ann. § 33.02(a), 345
Tex. Transp. Code Ann. § 544.006, 341
Tex. Transp. Code Ann. § 545.401, 331
Tex. Transp. Code Ann. § 545.452(b), 112
Tex. Transp. Code Ann. § 545.453, 110, 111, 307, 320
Tex. Transp. Code Ann. § 547.615, 285
Barton v. State, 882 S.W.2d 456 (Tex. App. 1994), 310
Bren-Tex Tractor Co. v. Massey-Ferguson, Inc., 97 S.W.3d 155 (Tex. App. 2002), 143
Chamberlain v. State, 163 Tex. Crim. 529, 294 S.W.2d 719 (1956), 311
Chapa v. State, 729 S.W.2d 723 (Tex. Crim. App. 1987), 376
City of Midland v. Sullivan, 33 S.W.3d 1 (Tex. App. 2000), 231
Ford Motor Co. v. Ledesma, 242 S.W.3d 32 (Tex. 2007), 132
Uniroyal Goodrich Tire Co. v. Martinez, 977 S.W.2d 328 (Tex. 1998), 136, 149

UTAH

Utah Code Ann. § 13-37-101 *et seq.*, 268, 271, 282, 284
Utah Code Ann. § 13-44-201(1), 282
Utah Code Ann. § 41-1a-1501 *et seq.*, 285
Utah Code Ann. § 41-1a-1502(2), 286
Utah Code Ann. § 41-1a-1503(1)(b), 391
Utah Code Ann. § 41-1a-1504(1), 285
Utah Code Ann. § 41-1a-1504(2), 287
Utah Code Ann. § 41-6a-301, 74
Utah Code Ann. § 41-26-104(1)(a), 307
Utah Code Ann. § 41-26-106, 97
Utah Code Ann. § 41-26-108, 112
Utah Code Ann. § 76-2-204(1), 312
Utah Code Ann. § 76-6-702 to -705, 345
Sanns v. Butterfield Ford, 2004 UT App. 203, 94 P.3d 301 (2004), 143

VERMONT

VT. STAT. ANN. tit. 9, §§ 2430, 2435, 284
VT. STAT. ANN. tit. 9, § 2430, 270–271
VT. STAT. ANN. tit. 9, § 2433, 270
VT. STAT. ANN. tit. 9, § 2435(b)(1), 272
VT. STAT. ANN. tit. 9, §§ 2446–2447, 283
VT. STAT. ANN. tit. 9, § 2446, 271
VT. STAT. ANN. tit. 9, § 2447, 271
VT. STAT. ANN. tit. 13, § 4101 *et seq.*, 345
VT. STAT. ANN. tit. 23, § 1025, 74
In re Search Warrant, 2012 VT 102, 193 Vt. 51, 71 A.3d 1158 (2012), 395

VIRGINIA

VA. CODE ANN. § 18.2-152.1 *et seq.*, 345
VA. CODE ANN. § 18.2-186.6, § 32.1-127.1:05, § 58.1-341.2, 284
VA. CODE ANN. § 38.2-2212.C, 287, 288
VA. CODE ANN. § 38.2-2213.1, 288
VA. CODE ANN. § 46.2-852, 331
VA. CODE ANN. § 46.2-1088.6, 285, 287, 288, 391
VA. CODE ANN. § 46.2-1572, 96
Dugger v. Commw., 40 Va. App. 586, 580 S.E.2d 477 (2003), 312
Hines v. Garrett, 131 Va. 125, 108 S.E. 690 (1921), 223
Holiday Motor Corp. v. Walters, 292 Va. 461, 790 S.E.2d 447 (2016), 145
Williams v. City of Petersburg, 216 Va. 297, 217 S.E.2d 893 (1975), 310

WASHINGTON

WASH. CONST. art. I, § 7, 268
WASH. REV. CODE § 7.72.040, 159
WASH. REV. CODE § 9A.08.030(2)(a), 312
WASH. REV. CODE § 9A.48.090, 341
WASH. REV. CODE § 9A.90.010 *et seq.*, 345
WASH. REV. CODE § 19.86.090, 289
WASH. REV. CODE § 19.255.010, 273
WASH. REV. CODE § 19.255.010(16), 272
WASH. REV. CODE § 19.375.010 *et seq.*, 272
WASH. REV. CODE § 19.375.010(4)(a), 274
WASH. REV. CODE § 19.375.030, 273
WASH. REV. CODE § 42.56.590(15), 272
WASH. REV. CODE § 46.35.010 *et seq.*, 285
WASH. REV. CODE § 46.35.010, 286, 287
Wash. Exec. Order 17-02, Autonomous Vehicle Testing & Technology in Washington State and Autonomous Vehicle Work Group (June 7, 2017), 110, 111
Baughn v. Honda Motor Co., 107 Wash. 2d 127, 727 P.2d 655 (1986), 180
Evangelical United Brethren Church of Adna v. State, 67 Wash. 2d 246, 407 P.2d 440 (1965), 227, 228

State v. Keodara, 191 Wash. App. 305, 364 P.3d 777 (2015), 395
State v. Votava, 149 Wash. 2d 178, 66 P.3d 1050 (2003), 330

WEST VIRGINIA

W. Va. Code § 17A-8-4, 350
W. Va. Code § 46A-2A-101 *et seq.*, 284
W. Va. Code § 46A-6-106, 289
W. Va. Code § 61-3C-3 *et seq.*, 345
Anderson v. Chrysler Corp., 184 W. Va. 641, 403 S.E.2d 189 (1991), 132

WISCONSIN

Wis. Stat. § 100.18(11)(b)(2), 289
Wis. Stat. § 134.98, 284
Wis. Stat. § 134.98(3)(a), 272
Wis. Stat. § 218.0121(2m), 96
Wis. Stat. § 346.14, 112
Wis. Stat. § 943.70, 345
Wis. Stat. § 995.50, 269
Allstate Insurance Co. v. Volkswagen of America, 2004 WI App 109, 273 Wis. 2d 786, 680 N.W.2d 833 (2004), 185

WYOMING

Wyo. Stat. Ann. § 6-3-501 *et seq.*, 345
Wyo. Stat. Ann. § 31-5-112, 74
Wyo. Stat. Ann. § 40-12-501, -502, 284
Noonan v. Texaco, Inc., 713 P.2d 160 (Wyo. 1986), 141

About the Author

Jeff Gurney practices law with Nelson Mullins Riley & Scarborough, LLP. He has advised on different aspects of advanced automation features, serving a variety of clients from small businesses to businesses as large as an automaker. He is a frequent writer and speaker on different aspects of automated vehicle law. His publications have included *Imputing Driverhood: Applying a Reasonable Driver Standard to Accidents Caused by Autonomous Vehicles*, *in* ROBOT ETHICS 2.0: THE ETHICAL AND SOCIAL IMPLICATIONS OF ROBOTICS (Patrick Lin, Keith Abney & George A. Bekey eds., 2017); *Crashing into the Unknown: An Examination of Crash-Optimization Algorithms through the Two Lanes of Ethics and Law*, 79 ALBANY L. REV. 183 (2016); *Driving into the Unknown: Examining the Crossroads of Criminal Law and Autonomous Vehicles*, 5 WAKE FOREST J.L. & POL'Y 393 (2015); and *Sue My Car Not Me: Products Liability and Accidents Involving Autonomous Vehicles*, 2013 U. ILL. J.L. TECH. & POL'Y 247 (2013). He has spoken throughout the United States and internationally about legal aspects of automated vehicles.